Selected Titles in This Series

13 **Michael Baake and Robert V. Moody, Editors,** Directions in mathematical quasicrystals, 2000
12 **Masayoshi Miyanishi,** Open algebraic surfaces, 2001
11 **Spencer J. Bloch,** Higher regulators, algebraic K-theory, and zeta functions of elliptic curves, 2000
10 **James D. Lewis,** A survey of the Hodge conjecture, Second Edition, 1999
9 **Yves Meyer,** Wavelets, vibrations and scaling, 1998
8 **Ioannis Karatzas,** Lectures on the mathematics of finance, 1996
7 **John Milton,** Dynamics of small neural populations, 1996
6 **Eugene B. Dynkin,** An introduction to branching measure-valued processes, 1994
5 **Andrew Bruckner,** Differentiation of real functions, 1994
4 **David Ruelle,** Dynamical zeta functions for piecewise monotone maps of the interval, 1994
3 **V. Kumar Murty,** Introduction to Abelian varieties, 1993
2 **M. Ya. Antimirov, A. A. Kolyshkin, and Rémi Vaillancourt,** Applied integral transforms, 1993
1 **D. V. Voiculescu, K. J. Dykema, and A. Nica,** Free random variables, 1992

Volume 13

CRM MONOGRAPH SERIES

Centre de Recherches Mathématiques
Université de Montréal

Directions in Mathematical Quasicrystals

Michael Baake
Robert V. Moody
Editors

The Centre de Recherches Mathématiques (CRM) of the Université de Montréal was created in 1968 to promote research in pure and applied mathematics and related disciplines. Among its activities are special theme years, summer schools, workshops, postdoctoral programs, and publishing. The CRM is supported by the Université de Montréal, the Province of Québec (FCAR), and the Natural Sciences and Engineering Research Council of Canada. It is affiliated with the Institut des Sciences Mathématiques (ISM) of Montréal, whose constituent members are Concordia University, McGill University, the Université de Montréal, the Université du Québec à Montréal, and the Ecole Polytechnique. The CRM may be reached on the Web at www.crm.umontreal.ca.

American Mathematical Society
Providence, Rhode Island USA

The production of this volume was supported in part by the Fonds pour la Formation de Chercheurs et l'Aide à la Recherche (Fonds FCAR) and the Natural Sciences and Engineering Research Council of Canada (NSERC).

2000 *Mathematics Subject Classification*. Primary 52C23, 43A25, 28A80, 46L80, 47B25, 11R52, 52C17, 82B20.

Library of Congress Cataloging-in-Publication Data

Directions in mathematical quasicrystals / Michael Baake, Robert V. Moody, editors.
  p. cm. — (CRM monograph series, ISSN 1065-8599 ; v. 13)
Includes bibliographical references and index.
ISBN 0-8218-2629-8 (alk. paper)
  1. Quasicrystals. 2. Crystallography, Mathematical. I. Baake, Michael. II. Moody, R. V., 1941–. III. Series.
QD926.D57 2000
548′.7—dc21   00-051097

Copying and reprinting. Material in this book may be reproduced by any means for educational and scientific purposes without fee or permission with the exception of reproduction by services that collect fees for delivery of documents and provided that the customary acknowledgment of the source is given. This consent does not extend to other kinds of copying for general distribution, for advertising or promotional purposes, or for resale. Requests for permission for commercial use of material should be addressed to the Assistant to the Publisher, American Mathematical Society, P. O. Box 6248, Providence, Rhode Island 02940-6248. Requests can also be made by e-mail to reprint-permission@ams.org.

Excluded from these provisions is material in articles for which the author holds copyright. In such cases, requests for permission to use or reprint should be addressed directly to the author(s). (Copyright ownership is indicated in the notice in the lower right-hand corner of the first page of each article.)

© 2000 by the American Mathematical Society. All rights reserved.
The American Mathematical Society retains all rights
except those granted to the United States Government.
Printed in the United States of America.

∞ The paper used in this book is acid-free and falls within the guidelines
established to ensure permanence and durability.
Visit the AMS home page at URL: http://www.ams.org/

10 9 8 7 6 5 4 3 2 1   05 04 03 02 01 00

Contents

Preface	vii
Self-Similar Measures for Quasicrystals *Michael Baake and Robert V. Moody*	1
Fourier Analysis of Deformed Model Sets *Guillaume Bernuau and Michel Duneau*	43
Mathematical Quasicrystals and the Problem of Diffraction *Jeffrey C. Lagarias*	61
Designer Quasicrystals: Cut-and-Project Sets with Pre-Assigned Properties *Peter A. B. Pleasants*	95
Generalized Model Sets and Dynamical Systems *Martin Schlottmann*	143
On Shelling Icosahedral Quasicrystals *Alfred Weiss*	161
Tilings, C^*-algebras, and K-theory *Johannes Kellendonk and Ian F. Putnam*	177
Hulls of Aperiodic Solids and Gap Labeling Theorems *J. Bellissard, D. J. L. Herrmann, and M. Zarrouati*	207
Quasicrystals, Parametric Density, and Wulff-Shape *Károly Böröczky, Jr., Uwe Schnell, and Jörg M. Wills*	259
Gordon-Type Arguments in the Spectral Theory of One-Dimensional Quasicrystals *David Damanik*	277
The Planar Dimer Model with Boundary: A Survey *Richard Kenyon*	307
Digit Tiling of Euclidean Space *Andrew Vince*	329
A Guide to Quasicrystal Literature *Michael Baake and Uwe Grimm*	371
Index	375

Preface

The experimental discovery of real-world quasicrystalline materials by several groups in the early 1980's brought into question a number of long-standing assumptions in crystallography regarding the relationship between long-range order and periodic order. By now it is clear that there is indeed a very real world of aperiodic long-range order, and there are many fascinating questions to ask and to answer about the vast territory that lies between crystallographic order and random (or amorphous) (dis)order.

Once the dust had settled a bit, after the initial outburst of activities in the world of physics, it turned out that, on the one hand, there were a number of important mathematical predecessors (some of which were painfully reestablished), and that, on the other hand, many new questions emerged that needed (and still need) professional mathematical attention.

The latter started with a NATO ASI workshop at the Fields Institute in fall 1995. Two books with research articles grew out of that: *The mathematics of long-range aperiodic order* (R. V. Moody, ed.), NATO ASI Ser. C: Math. Phys. Sci., vol. 489, 1997, and *Quasicrystals and discrete geometry* (J. Patera, ed.), Fields Inst. Monogr., vol. 10, 1998.

In spring 1998, a meeting on aperiodic order took place in Oberwolfach, and it was the stimulating atmosphere of this meeting together with the large number of new results that led to the decision to edit this volume. It is not meant as a proceedings of this meeting, and in fact it isn't. The reader will find a number of contributions that emerged from talks given at that meeting, as well as a number of additional ones that we have solicited.

All articles were especially written for this volume, with the aim of giving an account of our present knowledge and of the open questions of the field (or at least a substantial part of it). Taken together with the other two volumes, we hope that a rather coherent picture will emerge and that this will serve as a guide and inspiration to those interested in learning more about the mathematics of aperiodic order.

We would like to express our gratitude to our contributors for taking the time and care to produce these accounts of their work and that of their fellow coworkers. Particular thanks also go to Uwe Grimm and Moritz Höffe for providing several figures and for helping us in editing some of the contributions. We also thank our

©2000 Michael Baake and Robert V. Moody

referees for their competent and timely responses, as well as André Montpetit from CRM and the publisher for excellent cooperation.

Tübingen and Edmonton,
March 2000

Michael Baake and Robert V. Moody

Self-Similar Measures for Quasicrystals

Michael Baake and Robert V. Moody

Dedicated to Peter A. B. Pleasants on the occasion of his 60th birthday.

ABSTRACT. We study self-similar measures of Hutchinson type, defined by compact families of contractions, both in a single and multi-component setting. The results are applied in the context of general model sets to infer, via a generalized version of Weyl's Theorem on uniform distribution, the existence of invariant measures for families of self-similarities of regular model sets.

Introduction

There have been two very successful approaches to generate aperiodic sets with features of long-range internal order. The first is by creating tilings by the method of inflation followed by decomposition using a finite set of proto-tiles. The second is by creating point sets through the method of cut and project sets, or model sets as we call them here. Neither theory subsumes the other and they both have their own particular virtues. However, they have a considerable overlap. It is easy to replace a tiling by an equivalent set of points (by selecting suitable points from each type of tile) and in many cases the result is a model set or, more generally, a union of several model sets, one for each type of point. Conversely, there are many ways to obtain a tiling from a point set (for instance by using the Voronoï cells, or the dual Delone cells determined by them), and the equivalence concept of mutual local derivability, see [**B, P**] and references given there, is an adequate tool to make this connection precise. So it is natural to study the most notable feature of inflation tilings, namely their self-similarity, in the context of model sets, and indeed, even when no simple tiling is in sight, many interesting model sets have striking self-similarity.

The objective of this paper is to set up some of the machinery that makes such a study possible and to show how naturally it can be associated with families of self-similar measures on locally compact Abelian groups.

There is one initial hurdle which is not usually considered in the study of model sets. In order to have any sort of reasonable correspondence with the tiling world,

---

2000 *Mathematics Subject Classification.* Primary 52C23, 28A80; Secondary 37A05, 43A25.

This work was supported by the Natural Sciences and Engineering Research Council of Canada (NSERC) and by the German Research Council (DFG).

©2000 Michael Baake and Robert V. Moody

and to create a useful theory, we need to have a *multi-component* model set context in which there are a finite number of model sets, all based on the same cut and project scheme, that are mutually coupled by the self-similarities. After all, almost all inflation tiling systems have several types of tiles, and the decomposition of inflated tiles typically involves all of these various types simultaneously.

Thus the situation that we envision consists of a *family* of model sets $\Lambda_1, \ldots, \Lambda_n$ in some real space $\mathbb{R}^m$, all based on the same cut and project scheme, and a set of families of inflationary mappings F_{ij} on $\mathbb{R}^m$ with the property that

$$(0.1) \qquad \Lambda_i = \bigcup_{j=1}^n \bigcup_{f \in F_{ij}} f(\Lambda_j), \quad 1 \leq i \leq n.$$

There are two new features that are different from the tiling situation. In inflation tilings, the usual idea is that the tiles only overlap on their boundaries which are of measure 0. That would be equivalent to some form of disjointness in (0.1) which we do *not* wish to assume. Secondly, in tilings, the sets F_{ij} of mappings are assumed finite. Again, we do *not* wish to make this assumption. In fact, it is quite useful to make F_{ij} consist of *all* possible mappings of a certain type (for example all affine mappings) that are consistent with (0.1), and this is in general an infinite set.

As soon as overlapping is allowed, there naturally arises the question of whether there is implicit in (0.1) a corresponding set of relative weights which measure the "frequency" of occurrence of the points of Λ_i coming from the substitution process (0.1). If we were counting, this could be loosely construed as noting the occurrences of points of Λ_i with their multiplicities as they appear in the right-hand side of (0.1). This links up to a recent approach of Lagarias and Wang [**LW**]. However, in our situation, there is no reason to assume that our weights can be normalized to be integral nor that there are only a finite number of contributions to Λ_i in the right-hand side of (0.1).

This then is the primary goal of this contribution: to discuss the existence and nature of self-similar densities on systems of model sets which are coupled together by a substitution system of type (0.1). Our definition of model sets is based on arbitrary locally compact Abelian groups as internal spaces, and so is more flexible and handles more situations (see for instance [**BMS, LM**]) than the usual method of projection from Euclidean spaces. This extra generality requires a little more care than usual but it is remarkable how much of the theory is natural in this context.

The method by which we attack the problem is to use the formalism of the underlying cut and project scheme to pass everything over to the internal side of the picture, i.e., onto the locally compact Abelian group that is controlling the projection. The advantage of doing this is that the system of inflationary mappings turns into a family of contractions, and what has started off as a problem in the domain of discrete mathematics turns into one of analysis. A primary virtue of systems of contractions is the ready-built Hutchinson theory of iterated function systems with their attractors and self-similar measures. In our multi-component setting with infinite families of mappings, we need a slight variation on this theme, which occupies the first five Sections of the paper. These parts of the paper have nothing in particular to do with model sets, but rather are a development of Hutchinson's theory in the multi-component situation where the coupling is by *compact* families of contractions each of which has its own, essentially arbitrarily pre-given positive Borel measure. An important part of this is determining some conditions under which the self-similar measures are in fact absolutely continuous and, more importantly, when the representing L^1-functions, the Radon-Nikodym densities, are actually continuous (continuously representable self-similar measures), for it is only then that we can bring back information to the discrete side again.

Section 6 of this article brings in the model sets and develops the mathematics that allows us to pass information back from the internal side to the model set side in $\mathbb{R}^m$. The primary tool here is Weyl's theory of uniform projection, but we have to redevelop this in the context of locally compact Abelian groups and model sets.

In Section 7 we are, at last, set to tackle the problem of determining self-similar distribution of weights (also called self-similar densities from now on) on the model sets themselves. The Weyl theory applies only to continuous functions, so we can only refer to it when our self-similar measures have continuously representable Radon-Nikodym densities. Fortunately, this is the case in a number of interesting situations. We provide a general description of the situations in which such self-similar densities exist, and then, in Section 8, we offer a number of examples which illustrate what we have achieved.

The results obtained here extend those of two previous papers [**BM1, BM2**]. There, we considered only model sets based on Euclidean internal spaces, and the context was not primarily measure-theoretical as it is here. The method of dealing with multi-component model sets also differed from the product approach that we have adopted in this article. There are nonetheless several examples of self-similar densities (called invariant densities there) in those papers which the reader may find of interest.

1. Compact Families of Contractions and Attractors

Let us first review some basic facts from the theory of iterated function systems, both finite and compact, in a way that is adequate for our needs.

1.1. Hutchinson's Contraction Principle. Let X be a complete metric space with metric d. We denote by $\mathcal{K}X$ the set of all nonempty compact subsets of X. Let $d(x, U) := \inf\{d(x, u) \mid u \in U\}$ be the distance of x from U. Note that $d(x, U) = 0$ implies $x \in \overline{U}$. For $\varepsilon > 0$, the ε-*fringe* of a subset $U \in \mathcal{K}X$ is

$$[U]_\varepsilon := \{x \in X \mid d(x, U) < \varepsilon\}. \tag{1.1}$$

In view of the set $\mathcal{K}X$, we introduce the *Hausdorff metric*: for $U, V \in \mathcal{K}X$, it is defined by

$$d_H(U, V) := \inf\{\varepsilon > 0 \mid U \subset [V]_\varepsilon \text{ and } V \subset [U]_\varepsilon\}. \tag{1.2}$$

Note that for singletons $U = \{u\}$ and $V = \{v\}$, one has $d_H(\{u\}, \{v\}) = d(u, v)$. An alternative way to determine $d_H(U, V)$ is

$$d_H(U, V) = \sup\{d(u, V), d(v, U) \mid u \in U, v \in V\}.$$

Relative to the Hausdorff metric, $\mathcal{K}X$ is again a complete metric space. If now $U_j, V_j \in \mathcal{K}X$, $j \in J$, are two sets of compact subsets of X then it is easy to see [**Wi**, Note 2.1.6] that

$$d_H\left(\bigcup_{j \in J} U_j, \bigcup_{j \in J} V_j\right) \leq \sup_{j \in J} d_H(U_j, V_j). \tag{1.3}$$

Given complete metric spaces X, Y, we consider the space $C(X,Y)$ of all continuous mappings of X into Y, endowed with the compact-open topology. This topology has the property of making the evaluation maps

$$\text{eval}_x \colon C(X,Y) \to Y; \quad f \mapsto f(x) \tag{1.4}$$

continuous [**Kel**, Thm. 7.4]. There is a natural extension of mappings that leads to

$$\mathcal{K}(\cdot) \colon C(X,Y) \to C(\mathcal{K}X, \mathcal{K}Y) \tag{1.5}$$

with $(\mathcal{K}(f))(U) := f(U)$. This mapping is continuous [**Wi**, Prop. 2.5.1][1].

A mapping $f \colon X \to Y$ of metric spaces is *Lipschitz* if there is an $r > 0$ with $d_Y(f(x_1), f(x_2)) \leq r\, d_X(x_1, x_2)$ for all $x_1, x_2 \in X$. If f is Lipschitz then the infimum of all such r is the *Lipschitz constant* r_f of f. If $r_f < 1$ then f is called a *contraction*. The set of all Lipschitz maps from X to Y with Lipschitz constant equal to r (resp. at most r) is denoted by $\text{Lip}(r, X, Y)$ (resp. $\text{Lip}(\leq r, X, Y)$). Lipschitz functions are clearly uniformly continuous mappings. We observe:

$$f \in \text{Lip}(r, X, Y) \implies \mathcal{K}(f) \in \text{Lip}(r, \mathcal{K}X, \mathcal{K}Y). \tag{1.6}$$

Note in particular that $r_f = r_{\mathcal{K}(f)}$, as is seen by looking at singleton sets.

We are interested in *union maps*. Let X, Y be complete metric spaces and let $F = \{f_1, f_2, \ldots, f_N\}$ be a set of continuous maps from X to Y. We define two mappings, both given the same name:

$$\begin{aligned} {}^\cup F \colon X \to \mathcal{K}Y; &\quad {}^\cup F(x) := \bigcup_{i=1}^N \{f_i(x)\} \\ {}^\cup F \colon \mathcal{K}X \to \mathcal{K}Y; &\quad {}^\cup F(U) := \bigcup_{i=1}^N f_i(U). \end{aligned} \tag{1.7}$$

In view of (1.3), the union map ${}^\cup F$ is Lipschitz if all mappings in F are Lipschitz, and we have $r_{\cup F} \leq \sup\{r_f \mid f \in F\}$.

When $X = Y$ and F consists of contractions, then F is called an *iterated function system* (IFS). The principal result, see [**Hut**, §3.1] or [**Wi**, Prop. 3.1.1], is based on the general Banach contraction principle and reads as follows.

THEOREM 1.1 (Hutchinson's contraction principle). *Let F be an IFS on a complete metric space X. Then there is a unique $W \in \mathcal{K}X$ which is a fixed point of ${}^\cup F$, i.e., $W = \bigcup_{f \in F} f(W)$. Furthermore, for any $Z \in \mathcal{K}X$, $({}^\cup F)^\ell(Z)$ converges to $W \in \mathcal{K}X$ in the Hausdorff metric, as $\ell \to \infty$.* □

W is the *attractor* of the IFS. Our aim, in this and in the following Section, is to generalize this result in two directions:

(1) to compact sets of contractions (which is also well known);
(2) to products of metric spaces (in what we call the multi-component situation).

---

[1] Wicks' book [**Wi**] is a great place to look for information on spaces of compact sets, and it was an important source for the part of this paper on compact spaces of maps. Unfortunately, or fortunately, according to one's taste, Wicks uses nonstandard analysis to streamline his presentation, so 'standard' readers need to look elsewhere for the proofs, or to adapt them.

1.2. Compact Sets of Contractions.

Let X, Y be complete metric spaces. We form $\mathcal{K}X$ as above. We remark that the union of any compact subset of $\mathcal{K}X$ is a compact subset of X, i.e.,

$$(1.8) \qquad C \in \mathcal{K}\mathcal{K}X \implies \bigcup C := \bigcup_{U \in C} U \in \mathcal{K}X,$$

and that the union map $\bigcup \colon \mathcal{K}\mathcal{K}X \to \mathcal{K}X$ is continuous, cf. [**Wi**, Chapter 1.5].

Consider the space $C(X, \mathcal{K}Y)$ equipped with the compact-open topology. Let $F \in \mathcal{K}C(X, \mathcal{K}Y)$, i.e., F is a compact subset of continuous mappings from X into the space of compact subsets of Y. In view of (1.4), we have, for all $x \in X$,

$$(1.9) \qquad F(x) := \{f(x) \mid f \in F\} \in \mathcal{K}\mathcal{K}Y,$$

i.e., $F(x)$ is compact. From (1.8), we deduce that $\bigcup_{f \in F} f(x)$ is also compact. Thus, as in (1.7), we have a new mapping

$$(1.10) \qquad \begin{aligned} {}^\cup F \colon X &\to \mathcal{K}Y \\ x &\mapsto \bigcup_{f \in F} f(x). \end{aligned}$$

With these preliminary definitions out of the way, we define G to be a *compact admissible family of Lipschitz mappings* [**Wi**, §3.1] from X to Y if

C1. G is a compact set of Lipschitz mappings from X to Y;
C2. there exists $r > 0$ such that $r_g \leq r$ for all $g \in G$.

If $r < 1$ then we call G a *compact admissible family of contractions*.

Let G be a compact admissible family of Lipschitz mappings from X to Y. From (1.5), $\mathcal{K}(G) \subset C(\mathcal{K}X, \mathcal{K}Y)$ is compact, and since the $\mathcal{K}$ operator preserves Lipschitz constants, $\mathcal{K}(G)$ is itself a compact admissible family of Lipschitz mappings. Thus, from (1.10) we obtain

$$(1.11) \qquad {}^\cup\mathcal{K}G := {}^\cup(\mathcal{K}(G)) \in C(\mathcal{K}X, \mathcal{K}Y) \colon \quad U \mapsto \bigcup_{g \in G} g(U).$$

PROPOSITION 1.2. ${}^\cup\mathcal{K}G \in C(\mathcal{K}X, \mathcal{K}Y)$ *is Lipschitz with Lipschitz constant*

$$r_{{}^\cup\mathcal{K}G} \leq \sup\{r_{\mathcal{K}(g)} \mid g \in G\} = \sup\{r_g \mid g \in G\} \leq r. \qquad \square$$

If G consists of contractions and $X = Y$ then we call G a *compact iterated function system* (IFS). This generalizes the previous definition which will now be referred to as a *finite* IFS. Hutchinson's theorem evidently generalizes:

PROPOSITION 1.3. *Let G be a compact admissible family of contractions from X to Y with uniform Lipschitz bound $r < 1$. Then ${}^\cup\mathcal{K}G \colon \mathcal{K}X \to \mathcal{K}Y$ is a contraction with Lipschitz constant at most r. If, in addition, $X = Y$ (so G is a compact IFS), then there is a unique $W \in \mathcal{K}X$ (the attractor of the IFS) which is invariant under G: $W = \bigcup_{g \in G} g(W)$. For arbitrary $Z \in \mathcal{K}X$, the iterates $({}^\cup\mathcal{K}G)^\ell(Z)$ converge to $W \in \mathcal{K}X$ as $\ell \to \infty$.*

PROOF. For $Z \in \mathcal{K}X$, $\bigl(({}^\cup\mathcal{K}G)^\ell(Z)\bigr)_{\ell \in \mathbb{N}}$ is a Cauchy sequence in $\mathcal{K}X$ since ${}^\cup\mathcal{K}G$ is a contraction. Completeness shows the existence of a limit, which is a fixed point, and uniqueness follows immediately from contractivity. $\square$

2. Self-Similar Measures for Compact Sets of Contractions

One of the important contributions of Hutchinson [**Hut**] was the realization that the attractor of an IFS carries a *measure* that is likewise an invariant of the IFS, and indeed a far finer one than the attractor itself. In this Section, we re-establish this result in the context of compact iterated function systems. The basic assumption that we need is that our compact space of contractions carries a measure of its own. We do not need to specify in advance what this measure is. In the original case studied by Hutchinson, where the IFS was finite, this supplementary measure was (effectively) counting measure.

Let X be a compact metric space. We denote by $\mathcal{P}(X)$ the space of all probability measures on X—that is, positive regular Borel measures μ with total measure $\mu(X) = 1$, see [**RS**, Chapter IV.4] for background material. Note that

$$\mathcal{P}(X) \subset \mathcal{M}_+(X) \subset \mathcal{M}(X) \subset \mathcal{M}_\mathbb{C}(X),$$

where $\mathcal{M}_+(X)$, $\mathcal{M}(X)$, and $\mathcal{M}_\mathbb{C}(X)$ denote the spaces of positive, signed (or real), and complex regular Borel measures, respectively. For later use, we also define

$$\mathcal{M}_+^m(X) := \{\mu \in \mathcal{M}_+(X) \mid \mu(X) = m\}$$

so that $\mathcal{P}(X) = \mathcal{M}_+^1(X)$. Since the Riesz-Markov theorem [**RS**, Thms. IV.14 and IV.18] states that regular Borel measures are in one-to-one correspondence with linear functionals on the space $C(X, \mathbb{R})$ (resp. $C(X, \mathbb{C})$), equipped with the compact-open topology, we shall usually identify these pictures. So, we shall write $\mu(E)$ for the measure of a Borel set E, but often also $\mu(g)$ instead of $\int_X g \, d\mu$ for the measure (= integral) of a function.

In view of this, it is natural to equip $\mathcal{M}(X)$ (resp. $\mathcal{M}_\mathbb{C}(X)$) with the weak-$*$ topology, called the *vague topology* in this context [**RS**, p. 114], the weakest topology that makes all the mappings $\mu \mapsto \mu(g)$ of $\mathcal{M}(X) \to \mathbb{R}$ (resp. of $\mathcal{M}_\mathbb{C}(X) \to \mathbb{C}$) continuous, where $g \in C(X, \mathbb{R})$ (resp. $g \in C(X, \mathbb{C})$). Let us mention that, since X is compact, $C(X, \mathbb{C})$ is actually a Banach space with the sup-norm $\|\cdot\|_\infty$ (which induces the compact-open topology), hence $\mathcal{M}_\mathbb{C}(X) = C(X, \mathbb{C})^*$ can also be viewed as a Banach space, with induced norm

$$\|\mu\| := \sup\{|\mu(g)| \mid g \in C(X, \mathbb{C}), \|g\|_\infty = 1\}$$

(see [**RS**, Thm. III.2]). With Hahn's decomposition theorem [**RS**, Thm. IV.16], one gets $\|\mu\| = |\mu|(X)$, where $|\mu| \in \mathcal{M}_+(X)$ denotes the total variation measure of μ. The analogous statement also holds for $\mathcal{M}(X) = C(X, \mathbb{R})^*$. We shall need both ways to look at $\mathcal{M}(X)$. Another result in this context, using the Banach-Alaoglu Theorem [**RS**, Thm. IV.21], is that the unit balls in $\mathcal{M}_\mathbb{C}(X)$ and $\mathcal{M}(X)$ are compact in the vague topology. The closed subspace $\mathcal{P}(X) = \{\mu \in \mathcal{M}_+(X) \mid \|\mu\| = 1\}$ is then also compact. For an alternative derivation of the last statement, without reference to the Banach-Alaoglu Theorem, see [**A**, Prop. 8.1].

Following Hutchinson, we now define a *metric* on $\mathcal{P}(X)$:

(2.1) $$L(\mu, \nu) := \sup\{|\mu(\phi) - \nu(\phi)| \mid \phi \in \text{Lip}(1, X, \mathbb{R})\}.$$

In fact, it is hardly clear that this *is* a metric[2], but we shall show that below. It is useful to observe that, in this definition, we can also replace $\text{Lip}(1, X, \mathbb{R})$ by

---

[2]This is well known among experts, see [**Hut**, §4.3], but we could not find an explicit proof in the literature. Since it is an important part of our argument and the proof is not entirely trivial, we include it here. Note that the restriction to $\mathcal{P}(X)$ (or to $\mathcal{M}_+^m(X)$ for some $m > 0$) is vital.

Lip($\leq 1, X, \mathbb{R}$) without altering the resulting function L. We often make use of this in the sequel. Since X is compact, diam(X) := sup$\{d(x,y) \mid x, y \in X\}$, the *diameter* of X, is finite, and we can state another property explicitly.

LEMMA 2.1. *Let L be defined, on $\mathcal{P}(X)$, by (2.1). Then we have*

$$L(\mu, \nu) = \sup\{|\mu(\psi) - \nu(\psi)| \mid \psi \in \mathcal{L}\}$$

where $\mathcal{L} := \{\psi \in \mathrm{Lip}(1, X, \mathbb{R}) \mid \|\psi\|_\infty \leq \mathrm{diam}(X)\}$. Furthermore, $\mathcal{L} \subset C(X, \mathbb{R})$ is compact (in the compact-open topology).

PROOF. Let $\mu, \nu \in \mathcal{P}(X)$, $\phi \in \mathrm{Lip}(1, X, \mathbb{R})$. Then, for any $c \in \mathbb{R}$, we have $|\mu(\phi - c) - \nu(\phi - c)| = |\mu(\phi) - \nu(\phi)|$ since $\mu(c) = \nu(c)$ ($= c$). Let $a \in X$. Choosing $c = \phi(a)$ we obtain $|\phi(x) - \phi(a)| \leq r_\phi d(x, a) \leq \mathrm{diam}(X)$. So, $\psi(x) := \phi(x) - \phi(a)$ is a function in $\mathcal{L}$, and the restriction to $\mathcal{L}$ does not change the supremum value of $|\mu(\phi) - \nu(\phi)|$. This establishes the first assertion.

Note that $\mathcal{L}$ is clearly closed in $C(X, \mathbb{R})$. Also, since the $\psi \in \mathcal{L}$ are uniformly bounded, we have $\overline{\mathcal{L}(x)} = \overline{\{\psi(x) \mid \psi \in \mathcal{L}\}} \subset [-\mathrm{diam}(X), \mathrm{diam}(X)]$ for every $x \in X$, so each $\overline{\mathcal{L}(x)}$ is compact. Finally, $\mathcal{L}$ is equi-continuous since it consists of Lipschitz functions with uniformly bounded Lipschitz constants. By Ascoli's theorem, see [**Kel**, Thms. 7.21 and 7.22], $\mathcal{L}$ itself is then compact in $C(X, \mathbb{R})$ (in the compact-open topology). $\square$

PROPOSITION 2.2. *L is a metric on $\mathcal{P}(X)$ and induces the vague topology on $\mathcal{P}(X)$. In particular, $\mathcal{P}(X)$ is a complete metric space.*

PROOF. That $\mathcal{P}(X)$ is a complete metric space follows from its compactness (see above) as soon as we have shown L to be a metric.

We use Lemma 2.1. If $\mu, \nu \in \mathcal{P}(X)$, then $\mu - \nu : \mathcal{L} \to \mathbb{R}$ is continuous and so has compact image. This shows that $L(\mu, \nu)$ is finite. Nonnegativity and symmetry are obvious, as is the triangle inequality. Thus L is certainly a pseudo-metric. It remains to be shown that $L(\mu, \nu) = 0$ implies $\mu = \nu$. Assume the converse and set $\omega = \mu - \nu$. Then $c := \|\omega\| > 0$ and there is a $\delta > 0$ and a function $g \in C(X, \mathbb{R})$ such that $|\omega(g)| \geq \delta > 0$.

Since Lipschitz functions are dense in $C(X, \mathbb{R})$, see Lemma A.2 in the Appendix, we can choose ϕ Lipschitz with $\|g - \phi\|_\infty < \delta/2c$. Then,

$$|\omega(g - \phi)| \leq \|\omega\| \cdot \|g - \phi\|_\infty < \frac{\delta}{2}$$

and thus $|\omega(\phi)| \geq \delta/2 > 0$. Now, we don't know the Lipschitz constant r_ϕ of ϕ, but $\phi' := \phi/r_\phi$ is in $\mathrm{Lip}(1, X, \mathbb{R})$ and still $|\omega(\phi')| > 0$, so $L(\omega, 0) = L(\mu, \nu) > 0$, which contradicts the assumption. This shows that L is a metric.

Finally, we compare the topologies. Let $\mu_n \to \mu$ vaguely as $n \to \infty$. Since $\mu_n \in \mathcal{P}(X)$, we get $|\mu_n(\phi_1) - \mu_n(\phi_2)| \leq \|\phi_1 - \phi_2\|_\infty$ independently of n, so the μ_n constitute a family of equi-continuous mappings in $C(C(X, \mathbb{R}), \mathbb{R})$. Consequently, by Lemma A.3 of the Appendix, $\mu_n(\phi) \to \mu(\phi)$ uniformly on $\mathcal{L}$ because $\mathcal{L}$ is compact (Lemma 2.1), and hence $L(\mu_n, \mu) \to 0$. Conversely, observe that $\mathcal{L}$ is compact in the vague topology and Hausdorff under the metric L. But the identity is a one-to-one mapping, and (due to the previous argument) also continuous, when viewed as a mapping from $\mathcal{L}$ with the vague topology to $\mathcal{L}$ with the metric topology. Therefore, it is a homeomorphism [**Kel**, Thm. 5.8], and the topologies coincide. $\square$

Let F be a compact IFS on the compact metric space X and let $W \in \mathcal{K}X$ be its attractor. Let $r_F := \sup\{r_f \mid f \in F\} < 1$.

For each $f \in F$, $f(W) \subset W$, and we obtain a bounded linear operator $f.(\cdot)$ on the space of all signed Borel measures $\mathcal{M}(W)$ of W by

(2.2) $$\mu \mapsto f.\mu; \quad f.\mu(\phi) := \mu(\phi \circ f)$$

for all $\phi \in C(W, \mathbb{R})$. Evidently, $f.(\cdot) \colon \mathcal{P}(W) \to \mathcal{P}(W)$, i.e., if μ is a probability measure, so is $f.\mu$. Note that the matching definition for Borel sets E reads $f.\mu(E) = \mu(f^{-1}(E))$ where $f^{-1}(E)$ is the preimage of E under f.

PROPOSITION 2.3. *The mapping $F \times \mathcal{P}(W) \to \mathcal{P}(W)$ defined by $(f, \mu) \mapsto f.\mu$ is continuous.*

PROOF. $\mathcal{P}(W)$ is a compact metric space, so certainly Hausdorff. Fix $f \in F$ and consider the mapping $\mu \mapsto f.\mu$. Then, for $\phi \in \text{Lip}(1, W, \mathbb{R})$, we clearly have $|f.\mu(\phi) - f.\nu(\phi)| = |(\mu - \nu)(\phi \circ f)| \leq L(\mu, \nu)$ because $\phi \circ f$ has Lipschitz constant ≤ 1 due to the definition of F. So $\mu \mapsto f.\mu$ is Lipschitz and thus (uniformly) continuous on $\mathcal{P}(W)$. It follows that $F \times \mathcal{P}(W) \to \mathcal{P}(W)$ is jointly continuous [**Kel**, Thm. 7.5]. $\square$

PROPOSITION 2.4. *Let F be a compact IFS on the compact metric space X, with attractor $W \in \mathcal{K}X$. Let $\nu \in \mathcal{P}(F)$. Then the ν-averaging mapping*

$$\mathcal{A}_\nu \colon \mathcal{P}(W) \to \mathcal{P}(W)$$
(2.3)
$$\mu \mapsto \int_F (f.\mu) \, d\nu(f)$$

is a contraction relative to the L-metric on $\mathcal{P}(W)$, with contraction constant at most $r := r_F = \sup\{r_f \mid f \in F\}$. In particular, there is a unique measure $\rho^{(\nu)} \in \mathcal{P}(W)$ which satisfies

(2.4) $$\rho^{(\nu)} = \int_F \bigl(f.\rho^{(\nu)}\bigr) \, d\nu(f).$$

PROOF. Let $\omega_1, \omega_2 \in \mathcal{P}(W)$ and let $\phi \in \text{Lip}(1, W, \mathbb{R})$. Then

$$\bigl|\bigl(\mathcal{A}_\nu(\omega_1)\bigr)(\phi) - \bigl(\mathcal{A}_\nu(\omega_2)\bigr)(\phi)\bigr| = \left|\int_F \omega_1(\phi \circ f) \, d\nu(f) - \int_F \omega_2(\phi \circ f) \, d\nu(f)\right|$$

$$= \left|\int_F \bigl(\omega_1(\phi \circ f) - \omega_2(\phi \circ f)\bigr) \, d\nu(f)\right|$$

$$\leq r \int_F |\omega_1(r^{-1}\phi \circ f) - \omega_2(r^{-1}\phi \circ f)| \, d\nu(f)$$

$$\leq r \int_F L(\omega_1, \omega_2) \, d\nu(f)$$

$$= r \, L(\omega_1, \omega_2) \nu(F) = r \, L(\omega_1, \omega_2),$$

since $r^{-1}\phi \circ f \in \text{Lip}(\leq 1, W, \mathbb{R})$. This being true for all $\phi \in \text{Lip}(1, W, \mathbb{R})$, we have

(2.5) $$L\bigl(\mathcal{A}_\nu(\omega_1), \mathcal{A}_\nu(\omega_2)\bigr) \leq r L(\omega_1, \omega_2)$$

which is what we wanted. The existence of the unique measure $\rho^{(\nu)}$ follows now, once again, from the general contraction principle. $\square$

REMARK. More generally, we may replace W in Proposition 2.4 by any $W^+ \in \mathcal{K}X$ that satisfies $f(W^+) \subset W^+$ for all $f \in F$. Of course, $W \subset W^+$ and the invariant measure for W^+ is supported on W, hence is effectively the same as $\rho^{(\nu)}$.

3. Affine Mappings in Locally Compact Abelian Groups

In this Section, we treat the foregoing material in the setting of a locally compact Abelian group (LCAG). A convenient source for the results on LCAGs that we need is [**Ru**, Chapter 1].

3.1. Affine Mappings.
Let H be an additive LCAG that is equipped with a translation invariant metric d with respect to which H is complete. For more information on metrizability, see [**HR**, Chapter 2, § 8]. We also assume that an automorphism A is given (in particular, $A(0) = 0$) and that a Haar measure θ on H has been fixed. It is unique up to normalization.

Since $A.\theta := \theta \circ A^{-1}$ is also H-invariant, it is another Haar measure, and we thus have $A.\theta = \alpha \theta$ for some $\alpha > 0$, the *modulus* of A. If $H = \mathbb{R}^n$, θ is Lebesgue measure, A is simply an invertible linear map, and $\alpha = |\det(A^{-1})| = 1/|\det(A)|$.

LEMMA 3.1. *Let H be as described with metric d. If A is a contraction on H relative to d, then A has modulus $\alpha > 1$.*

PROOF. Since H is locally compact, it contains a compact neighborhood U of 0. By assumption, the topology given on H agrees with the metric topology. So, we can choose U such that inside it we can find balls $B_r(0)$ and $B_s(0)$ with $r > s > 0$ and the property that $B_r(0) \setminus B_s(0)$ contains a nonempty open set which must then have positive measure.

On the other hand, A is a contraction, $A(0) = 0$, and $d(0, A^n(U)) \to 0$ as $n \to \infty$ for any compact $U \subset H$. So, there must be some $m \in \mathbb{N}$ such that $A^m(B_r(0)) \subset B_s(0)$. Combined with the previous argument, this says $A^m(B_r(0))$ has smaller measure than $B_r(0)$. Consequently, the modulus of A^m is $\alpha^m > 1$, and then also $\alpha > 1$. $\square$

Clearly, the converse of Lemma 3.1 is not true. In view of the general context of this paper, we assume from now on that A is a contraction on H relative to d. A mapping $f: H \to H$ of the form

(3.1) $$f: x \mapsto A(x) + v$$

with $v \in H$ is called an *affine map* with *automorphism* A and *translation* v. We will sometimes denote this mapping by A_v.

Let $\mathcal{M}_{\mathbb{C}}(H)$ be the space of all bounded regular complex Borel measures λ on H, i.e., measures with $\|\lambda\| = |\lambda|(H) < \infty$. We recall that the convolution of two measures $\lambda_1, \lambda_2 \in \mathcal{M}_{\mathbb{C}}(H)$ is defined by

(3.2) $$(\lambda_1 * \lambda_2)(\phi) = \int_{H \times H} \phi(x+y) \, d\lambda_1(x) \, d\lambda_2(y),$$

for $\phi \in C(H, \mathbb{C})$. If formulated for Borel sets E, the matching equation is

(3.3) $$(\lambda_1 * \lambda_2)(E) = (\lambda_1 \otimes \lambda_2)(E^{(2)})$$

where $E^{(2)} := \{(x, y) \in H \times H \mid x + y \in E\}$.

The Fourier-Stieltjes transform of $\lambda \in \mathcal{M}_{\mathbb{C}}(H)$ is the function $\widehat{\lambda}$ defined on the dual group $\widehat{H}$ of H by

$$\widehat{\lambda}(k) = \int_H \overline{\langle k, x\rangle}\, \mathrm{d}\lambda(x) \tag{3.4}$$

where $x \mapsto \langle k, x \rangle$ is the continuous *character* on H defined by $k \in \widehat{H}$.

The automorphism A on H induces an automorphism A^T on $\widehat{H}$: $k \mapsto A^T k$ where $A^T k$, in turn, defines the character $x \mapsto \langle k, Ax\rangle$ on H.

We collect now some basic facts that we need. These are all elementary consequences of the definitions, whence we omit proofs. We write $A.h$ for the function defined by $x \mapsto h(A^{-1}(x))$ in analogy to $A.\mu = \mu \circ A^{-1}$ for measures, and $h\mu$, with $h \in L^1(H)$, for the measure defined by $(h\mu)(\phi) = \mu(h\phi)$. Thus, $h\mu$ is absolutely continuous with respect to μ, and h is the corresponding Radon-Nikodym density (also called Radon-Nikodym derivative).

PROPOSITION 3.2. *Let H, θ, A, α be as defined above. Let $\lambda_1, \lambda_2 \in \mathcal{M}_{\mathbb{C}}(H)$ and $h \in L^1(H)$. Then we have*

(1) $\mathrm{d}\theta(A^{-1}x) = \alpha\, \mathrm{d}\theta(x)$
(2) $A.(h\theta) = \alpha(A.h)\theta$
(3) $A.(\lambda_1 * \lambda_2) = A.\lambda_1 * A.\lambda_2$
(4) $\widehat{A.\lambda} = \widehat{\lambda} \circ A^T = (A^T)^{-1}.\widehat{\lambda}$ □

3.2. Compact Families of Affine Mappings. Assume now that F is a compact family of contractions on the LCAG H, each $f \in F$ being of the form

$$A_v \colon x \mapsto Ax + v$$

for some $v \in H$ (but all having the same A, namely our contractive automorphism fixed above). Evidently, F is a compact admissible family of mappings from H to H. Define

$$F_H = \{v \mid A_v \in F\} = \{f(0) \mid f \in F\} = F(0) \subset H. \tag{3.5}$$

In view of (1.4) and (1.5), the mapping $F \to F_H$ induced by $f \mapsto f(0)$ is continuous and hence F_H is compact and homeomorphic to F. In particular, there is a natural isomorphism between $\mathcal{M}_{\mathbb{C}}(F)$ and the space of regular measures on H that are supported on F_H.

Let $\nu_F \in \mathcal{P}(F)$ and $\nu \in \mathcal{P}(H)$ be such a corresponding pair of (probability) measures. We then have an averaging operator $\mathcal{A}_\nu \colon \mathcal{P}(W^+) \to \mathcal{P}(W^+)$ whenever $W^+ \subset H$ is any compact subset of H for which $FW^+ \subset W^+$.

Let $\lambda \in \mathcal{P}(W^+)$. Then, for all Borel sets $E \subset H$,

$$\begin{aligned}
\mathcal{A}_\nu \lambda(E) &= \int_F f.\lambda(E)\, \mathrm{d}\nu_F(f) = \int_F \lambda(f^{-1}(E))\, \mathrm{d}\nu_F(f) \\
&= \int_H \lambda(A^{-1}(E - v))\, \mathrm{d}\nu(v) = \int_H (A.\lambda)(E - v)\, \mathrm{d}\nu(v) \\
&= \int_{H \times H} \mathbf{1}_E(u + v)\, \mathrm{d}(A.\lambda)(u)\, \mathrm{d}\nu(v) = (\nu * A.\lambda)(E)
\end{aligned} \tag{3.6}$$

where $\mathbf{1}_E$ is the characteristic function of E. Since $\mathrm{supp}(\lambda) \subset W^+$, we have $f(\mathrm{supp}(\lambda)) \subset W^+$ for all $f \in F$. This implies $\mathrm{supp}(\mathcal{A}_\nu \lambda) \subset W^+$ and, more

generally, $\operatorname{supp}(\mathcal{A}_\nu^\ell \lambda) \subset W^+$ for all $\ell \geq 0$. In particular, we can also infer
$$\operatorname{supp}(\mathcal{A}_\nu \lambda) = \operatorname{supp}(\nu * A.\lambda) \subset \operatorname{supp}(\nu) + A\operatorname{supp}(\lambda) \subset W^+.$$
It is clear that we can now iterate (3.6) to get
$$\mathcal{A}_\nu^\ell \lambda = \nu * A.\nu * \cdots * A^{\ell-1}.\nu * A^\ell.\lambda$$
together with the inclusion relation
$$\operatorname{supp}(\mathcal{A}_\nu^\ell \lambda) \subset \operatorname{supp}(\nu) + A\operatorname{supp}(\nu) + \cdots + A^{\ell-1}\operatorname{supp}(\nu) + A^\ell \operatorname{supp}(\lambda) \subset W^+.$$
Since A is a contraction, $A^\ell \operatorname{supp}(\lambda) \to \{0\}$ (in $\mathcal{K}H$) as $\ell \to \infty$ and we have
$$\sum_{\ell=0}^\infty A^\ell \operatorname{supp}(\nu) \subset W^+.$$
In particular, since W is F-invariant, we must also have
$$\sum_{\ell=0}^\infty A^\ell \operatorname{supp}(\nu) \subset W.$$
Define W^+ to be the smallest compact subset of H which is F-invariant and contains $\sum_{j=0}^\ell A^j \operatorname{supp}(\nu)$ for all $\ell \geq 0$.

Define $\omega^{(\ell)} \in \mathcal{P}(W^+)$ by $\omega^{(0)} = \nu$, and (for $\ell \geq 0$)
$$\omega^{(\ell+1)} = \mathcal{A}_\nu \omega^{(\ell)} = \nu * A.\omega^{(\ell)}.$$

Next, let ω be the unique $\mathcal{A}_\nu$-invariant measure of $\mathcal{P}(W^+)$, see Proposition 2.4 and the Remark following it. We know, again by Proposition 2.4, that $\mathcal{A}_\nu$ is a contraction on $\mathcal{P}(W^+)$. Moreover, by (3.6), $\{\omega^{(\ell)}\}$ contracts to ω as $\ell \to \infty$. Thus
$$\lim_{\ell \to \infty} \omega^{(\ell)} = \omega$$
in $\mathcal{P}(W^+)$, with convergence in the vague topology.

PROPOSITION 3.3. *Under the above assumptions, we have*
  (1) $\omega = \bigstar_{\ell=0}^\infty (A^\ell.\nu) \in \mathcal{P}(W)$, *which converges in the vague topology, is the unique self-similar probability measure for the compact admissible family of contractions* $F = \{A_v \mid v \in F_H\}$ *with respect to the measure* ν_F *on* F.
  (2) $\widehat{\omega} = \prod_{\ell=0}^\infty (A^T)^{-\ell}.\widehat{\nu}$, *convergence being uniform convergence on compact sets (compact convergence)*.
  (3) *If the convolution product for* ω *converges also in the* $\|\cdot\|$-*topology on* $\mathcal{P}(W)$, *the convergence of* $\widehat{\omega}$ *is (globally) uniform*.

PROOF. Part (1) is clear from the discussion above. The support of ω is inside W by Proposition 2.4. Part (2) follows from Proposition 3.2 and the continuity of the Fourier transform, sending measures μ to bounded and uniformly continuous functions $\widehat{\mu}$. The convergence statement is a direct consequence of Lévy's continuity theorem, see Theorem A.5 of the Appendix. Finally, the third assertion follows directly from $\|\widehat{\mu}\|_\infty \leq \|\mu\|$, see (3.4), without reference to Part (2). □

REMARK. It is only a matter of convenience to start the above iteration with $\omega^{(0)} = \nu$. Any other choice $\lambda \in \mathcal{P}(W^+)$ is equally admissible and will lead to the same result, because $A^\ell.\lambda \to \delta_0$, as $\ell \to \infty$, and δ_0, the unit point measure at 0, is the neutral element of convolution, i.e., $\mu * \delta_0 = \mu$ for all measures μ.

3.3. Self-Similar Functions.
If we assume that the measure ν on our compact family of affine contractions is absolutely continuous with respect to Haar measure, then Proposition 3.3 gets re-interpreted in terms of functions rather than measures.

We suppose the same notation as in Section 3.2 and assume in addition that the measure ν derived from ν_F on F is absolutely continuous w.r.t. θ, so ν is of the form $\nu = h\theta$, where $h \in L^1(H)$ and $\operatorname{supp}(h) \subset F_H$, with F_H compact. For such measures, convolution matches the usual convolution of functions. Thus, using Proposition 3.2 and Proposition 3.3(1), we obtain

$$\omega^{(\ell)} = \nu * A.\nu * \cdots * A^\ell.\nu = h\theta * \alpha(A.h)\theta * \cdots * \alpha^\ell(A^\ell.h)\theta = \left(\underset{j=0}{\overset{\ell}{\text{\Large *}}} \alpha^j(A^j.h) \right)\theta$$

and $\omega = (\text{\Large *}_{j=0}^\infty \alpha^j(A^j.h))\theta$, with convergence so far only in the vague topology. However, as the brackets already imply, convergence in the $\|\cdot\|$-topology would be preferable. The situation is as follows. If we identify $L^1(H)$ with a subspace of $\mathcal{M}_{\mathbb{C}}(H)$ via $f \mapsto f\theta$, this is a *closed subspace* of $\mathcal{M}_{\mathbb{C}}(H)$ in the $\|\cdot\|$-topology, see [**Ru**, §1.3.5]. Consequently, the $\|\cdot\|$-convergence of absolutely continuous measures is equivalent to the L^1-convergence of their Radon-Nikodym densities in $L^1(H)$.

To establish also the $\|\cdot\|$-convergence in our case, recall first the following result [**Ru**, Thm. 1.1.8] on approximate units in the commutative convolution Banach algebra $L^1(H)$ (with norm $\|\cdot\|_1$).

LEMMA 3.4. *Given $f \in L^1(H)$ and $\varepsilon > 0$, there exists a neighborhood V of 0 in H with the following property: if u is a nonnegative Borel function which vanishes outside V, and if $\|u\|_1 = \int_H u(x)\,\mathrm{d}\theta(x) = 1$, then*

$$\|f - f * u\|_1 < \varepsilon. \qquad \square$$

We are now in the following situation. Our starting function is $h \in L^1(H)$, with $\operatorname{supp}(h)$ compact, $h \geq 0$ and $\int_H h\,\mathrm{d}\theta = \|h\|_1 = 1$. Let $f_\ell = \alpha^\ell(A^\ell.h)$ for $\ell \geq 0$, so that $f_\ell \geq 0$ and $\|f_\ell\|_1 = 1$. Also, $\operatorname{supp}(f_\ell) = A^\ell \operatorname{supp}(h)$, and we have the relation $\|f_\ell * f_{\ell+1} * \cdots * f_{\ell+k}\|_1 = 1$ for all $k \geq 0$.

PROPOSITION 3.5. *Let F be a compact family of affine mappings, with contractive automorphism A, modulus α and attractor $W \subset H$. Let F_H be the corresponding set of translations and let $\nu = h\theta$ be an absolutely continuous probability measure on F_H, where $h \in L^1(H)$. Then, the infinite convolution product $\text{\Large *}_{\ell=0}^\infty f_\ell$ converges to an L^1-function, hence $\text{\Large *}_{\ell=0}^\infty f_\ell \theta$ converges also in the $\|\cdot\|$-topology.*

PROOF. Since $L^1(H)$ is complete, it suffices to show that $(\text{\Large *}_{\ell=0}^n f_\ell)_{n\geq 0}$ is a Cauchy sequence. Fix $\varepsilon > 0$ and let V be the neighborhood for the L^1-function $f = f_0 = h$ according to Lemma 3.4. Since A is a contraction, there exists an integer N so that $\sum_{\ell \geq N} \operatorname{supp}(f_\ell) \subset V$. In particular, any finite convolution of the form $\text{\Large *}_{\ell=N}^{N+k} f_\ell$, $k \geq 0$, is then an approximate unit for h with bound ε.

Let now $n, m \geq N$ and define $u = \text{\Large *}_{\ell=N}^n f_\ell$ and $v = \text{\Large *}_{\ell=N}^m f_\ell$. Then

$$\left\| \underset{\ell=0}{\overset{n}{\text{\Large *}}} f_\ell - \underset{\ell=0}{\overset{m}{\text{\Large *}}} f_\ell \right\|_1 = \left\| \left(\underset{\ell=0}{\overset{N-1}{\text{\Large *}}} f_\ell \right) * (u-v) \right\|_1 \leq \left\| \underset{\ell=1}{\overset{N-1}{\text{\Large *}}} f_\ell \right\|_1 \cdot \|h*u - h*v\|_1$$

$$= \|(h*u - h) + (h - h*v)\|_1 \leq \|h - h*u\|_1 + \|h - h*v\|_1 < 2\varepsilon$$

by application of Lemma 3.4. This gives part one of the claim, while the rest follows, once again, from the Radon-Nikodym theorem. □

PROPOSITION 3.6. *Under the general assumptions of Proposition 3.5, we have:*

(1) *There is a unique nonnegative function $g \in L^1(H)$ which satisfies*[3]

$$g = \alpha \int_H g(A^{-1}(x-v))h(v)\,d\theta(v)$$

with normalization $\int_H g\,d\theta = 1$.

(2) $g = *_{\ell=0}^{\infty} \alpha^\ell (A^\ell.h)$, *with convergence in the L^1-norm, and* $\mathrm{supp}(g) \subset W$.

(3) *The Fourier transform of g is the continuous function*

$$\hat{g} = \prod_{\ell=0}^{\infty} \alpha^\ell \bigl(\hat{h}.(A^T)^\ell\bigr),$$

with uniform convergence of the product.

(4) *If $h \in L^1(H) \cap L^\infty(H)$, then g is continuous on H.*

PROOF. The convergence claimed in Part (2) follows from Proposition 3.5, so $g \in L^1(H)$ and $\hat{g}$ is then continuous.

From $\omega = \nu * A.\omega$, we then have, by Proposition 3.2, $g = \alpha h * A.g$, which gives Part (1) by applying Proposition 3.3(1), and also the statement that $\mathrm{supp}(g) \subset W$.

The situation for Fourier transforms is even easier since $\widehat{g\theta} = \hat{g}$ and we get the product formula in Part (3) from Proposition 3.3(2) with uniform convergence by means of Proposition 3.3(3).

Finally, suppose that $h \in L^1(H) \cap L^\infty(H)$. Since $h \in L^\infty(H)$ and $A.g \in L^1(H)$, we obtain ([**Ru**, Thm. 1.1.6]) the continuity of $h * A.g$, hence of g itself. □

REMARK. If $H = \mathbb{R}^n$, we can actually iterate the last argument and arrive at the stronger statement that $h \in L^1(\mathbb{R}^n) \cap L^\infty(\mathbb{R}^n)$ implies that g is a C^∞-function with compact support contained in W. Furthermore, if ν is Lebesgue measure, then the self-similar function g enjoys remarkable properties with respect to the averaging operator $\mathcal{A}_\nu$, namely its partial derivatives are eigenfunctions for the refinement operator with eigenvalues directly related to the spectrum of A. This is the situation in our previously studied examples [**BM1, BM2**] and these results may be found there.

4. Multi-Component Families of Contractions

In this section, we consider the generalization of the previous material to the case in which we have several compact metric spaces and sets of contractions between these spaces. This is the multi-component situation. The approach here is to consider the product of the various spaces in question. The basic theorem on the existence of attractors then reduces at once to the single-component situation already dealt with. The question of self-similar measures also fits naturally into the product formalism, though the situation now acquires some new features that did not appear before.

---

[3]In [**JLS**], a mapping on functions of this form is called a continuous refinement operator.

4.1. Contractions and Attractors.
Let $(X_1, d_1), \ldots, (X_n, d_n)$ be n complete metric spaces and define $N := \{1, \ldots, n\}$. Also, let $d_{i,H}$ denote the corresponding Hausdorff metric for $\mathcal{K}X_i$, $i \in N$. We set

$$(4.1) \qquad X_N := X_1 \times \cdots \times X_n$$

and write $x = (x_1, \ldots, x_n)$ for the elements of X_N. We endow X_N with the metric

$$(4.2) \qquad d(x, y) := \sup\{d_i(x_i, y_i) \mid i \in N\}$$

relative to which it is also complete, and denote by d_H the attached Hausdorff metric on $\mathcal{K}X_N$.

For each pair $(i, j) \in N \times N$, let F_{ij} be a compact admissible family of contractions $f = f_{ij} \colon X_j \to X_i$. We extend this to allow the possibility that F_{ij} is empty, though we require that for each i there is at least one j for which $F_{ij} \neq \emptyset$. We let $0 < r < 1$ be a uniform upper bound on the contractivity factors of all these mappings. For each pair (i, j), we have from (1.11) the mapping $^{\cup}\mathcal{K}F_{ij} \colon \mathcal{K}X_j \to \mathcal{K}X_i$. We define

$$(4.3) \qquad {}^{\cup}\mathcal{K}F \colon \mathcal{K}X_1 \times \cdots \times \mathcal{K}X_n \to \mathcal{K}X_1 \times \cdots \times \mathcal{K}X_n$$

where

$$(4.4) \qquad {}^{\cup}\mathcal{K}F(U_1, \ldots, U_n) := \left(\bigcup_j {}^{\cup}(\mathcal{K}F_{1j})(U_j), \ldots, \bigcup_j {}^{\cup}(\mathcal{K}F_{nj})(U_j) \right)$$
$$= \left(\bigcup_j \bigcup_{f \in F_{1j}} f(U_j), \ldots, \bigcup_j \bigcup_{f \in F_{nj}} f(U_j) \right).$$

Note that we write $(U_1, \ldots, U_n)$ rather than $U_1 \times \cdots \times U_n$ and that $\mathcal{K}X_1 \times \cdots \times \mathcal{K}X_n$ is a strict subset of $\mathcal{K}X_N$.

PROPOSITION 4.1. $^{\cup}\mathcal{K}F \colon \mathcal{K}X_1 \times \cdots \times \mathcal{K}X_n \to \mathcal{K}X_1 \times \cdots \times \mathcal{K}X_n$ is a contraction with Lipschitz constant at most r.

PROOF. For $U_i, V_i \in \mathcal{K}X_i$, $i \in N$, we find

$d_H\left({}^{\cup}\mathcal{K}F(U_1, \ldots, U_n), {}^{\cup}\mathcal{K}F(V_1, \ldots, V_n)\right)$

$= d_H\left(\left(\ldots, \bigcup_j {}^{\cup}(\mathcal{K}F_{ij})(U_j), \ldots\right), \left(\ldots, \bigcup_j {}^{\cup}(\mathcal{K}F_{ij})(V_j), \ldots\right)\right)$

$= \sup_i \left\{ d_{i,H}\left(\bigcup_j {}^{\cup}(\mathcal{K}F_{ij})(U_j), \bigcup_j {}^{\cup}(\mathcal{K}F_{ij})(V_j)\right) \right\}$ (by definition)

$\leq \sup_i \sup_j \left\{ d_{i,H}\left({}^{\cup}(\mathcal{K}F_{ij})(U_j), {}^{\cup}(\mathcal{K}F_{ij})(V_j)\right) \right\}$ (by (1.3))

$\leq \sup_i \sup_j \{ r_{\mathcal{K}F_{ij}} d_{j,H}(U_j, V_j) \} \leq r \sup_j \{ d_{j,H}(U_j, V_j) \}$ (by Prop. 1.2)

$= r \, d_H\left((U_1, \ldots, U_n), (V_1, \ldots, V_n)\right)$

which establishes our assertion. □

We conclude, using the usual contraction principle, that there is a unique attractor for $^\cup\mathcal{K}F$, in $\mathcal{K}X_1 \times \cdots \times \mathcal{K}X_n$, say $W_1 \times \cdots \times W_n$. The W_i thus form the unique solution (in compact sets) to the system of equations:

$$(4.5) \qquad W_i = \bigcup_{j=1}^{n} \bigcup_{f \in F_{ij}} f(W_j), \quad i \in N.$$

4.2. Multi-Component Invariant Measures.

The idea behind the invariant measures in the multi-component setting is straightforward in its conception but looks complicated in its details. We start with n compact metric spaces X_i, $i \in N$, that are coupled by families F_{ij} of contractions $f \colon X_j \to X_i$. For the moment we can take each set of mappings F_{ij} to be finite.

Each $f \in F_{ij}$ determines a transformation $\mu_j \mapsto f.\mu_j$ (see (2.2) for notation) of measure spaces $\mathcal{M}(X_j) \to \mathcal{M}(X_i)$. Basically, we want to find a family of measures $\{\mu_1, \ldots, \mu_n\}$ that is invariant under the average of these transformations:

$$(4.6) \qquad \mu_i = \sum_{j=1}^{n} \frac{1}{\mathrm{card}(F_{ij})} \sum_{f \in F_{ij}} f.\mu_j.$$

There are some extensions and modifications that make this picture both more useful and easier to cope with mathematically:

(1) We are at liberty to give each set of mappings F_{ij} its own weighting.
(2) We need not restrict ourselves to *finite* sets F_{ij}, nor need we assume that our averaging is uniform within each of these sets. In what follows, we only assume that the sets F_{ij} are compact spaces of mappings. We then deal with these points simultaneously by assigning positive[4] measures σ_{ij} to each of these spaces of mappings.
(3) It is mathematically easiest to deal with all of the measures $\{\mu_1, \ldots, \mu_n\}$ as a single entity. Thus we prefer to deal with product measures $\mu_1 \otimes \cdots \otimes \mu_n$ on the space $X_1 \times \cdots \times X_n$. This means that we will be deriving a product form of the invariance equation (4.6).

After these considerations, the mathematics unfolds in much the same way as before, with one exception. Invariant measures $\mu_1 \otimes \cdots \otimes \mu_n$ can exist only if a certain eigenvector condition involving the total measures of the μ_i and the σ_{ij} is met (see (4.8) below).

Let N and (X_i, d_i), $i \in N$, be as above. For each $J = (j_1, \ldots, j_n) \in N^n$, we define $X_J := X_{j_1} \times \cdots \times X_{j_n}$ and adopt standard multi-index notation, e.g., $x_J = (x_{j_1}, \ldots, x_{j_n})$. In particular, $X_N = X_1 \times \cdots \times X_n$ in agreement with our previous definition. We then define the metric d_J on X_J by $d_J(x_J, y_J) = \sup_{k=1}^{n} d_{j_k}(x_{j_k}, y_{j_k})$. For measures $\mu_i \in \mathcal{M}(X_i)$, $i \in N$, we write $\mu_J = \mu_{j_1} \otimes \cdots \otimes \mu_{j_n} \in \mathcal{M}(X_J)$ and $\mathrm{d}\mu_J = \mathrm{d}\mu_{j_1} \ldots \mathrm{d}\mu_{j_n}$.

For each $(i, j) \in N \times N$, let F_{ij} be a compact admissible family of contractions $f \colon X_j \to X_i$ (allowing, as above, the possibility that F_{ij} is empty). We let $0 < r < 1$ be a uniform upper bound on the contractivity factors of all these mappings.

We let $F = \bigtimes_{i,j} F_{ij}$ be the product of all these spaces of maps, a typical element being a matrix of maps $\boldsymbol{f} = (f_{ij})$. For each such $\boldsymbol{f}$, and for all $J, K \in N^n$,

---

[4]Strictly speaking, we should say nonnegative measures, but we will always explicitly mention when the 0-measure occurs.

let $f_{KJ}\colon X_J \to X_K$ be given by

$$(4.7) \qquad f_{KJ}(x_{j_1},\ldots,x_{j_n}) = \bigl(f_{k_1 j_1}(x_{j_1}),\ldots,f_{k_n j_n}(x_{j_n})\bigr).$$

We write f_J for the special case $f_{NJ}\colon X_J \to X_N$ and $F_J := \{f_J \mid \boldsymbol{f} \in F\}$. Note that now $f_{KJ}.\mu_J = (f_{k_1 j_1}.\mu_{j_1}) \otimes \cdots \otimes (f_{k_n j_n}.\mu_{j_n})$. Consequently, $f_{KJ}.\mu_J \in \mathcal{M}(X_K)$ and $f_J.\mu_J \in \mathcal{M}(X_N)$.

We assume that each space F_{ij} is equipped with a positive Borel measure σ_{ij} and define $s_{ij} := \sigma_{ij}(F_{ij})$, or $s_{ij} = 0$ if F_{ij} is empty. For each $J, K \in N^n$, we define the measure $\sigma_J := \sigma_{NJ} = \sigma_{1 j_1} \otimes \cdots \otimes \sigma_{n j_n}$ and $s_{KJ} := s_{k_1 j_1} \cdot \ldots \cdot s_{k_n j_n}$.

The matrix $\boldsymbol{s} := (s_{ij})$ is a nonnegative matrix. We now make the following compatibility assumption:

CA. The total measures $m_i = \mu_i(X_i)$ of the μ_i are all (strictly) positive, and $\boldsymbol{m} := (m_1,\ldots,m_n)^T$ is an eigenvector of $\boldsymbol{s}$ with eigenvalue 1:

$$(4.8) \qquad \boldsymbol{sm} = \boldsymbol{m}.$$

REMARK. If $\boldsymbol{s}$ is nonnegative, but $\boldsymbol{sm} = \boldsymbol{m}$ for a vector $\boldsymbol{m}$ with all $m_i > 0$ as we assume in CA, the eigenvalue 1 is also the spectral radius of $\boldsymbol{s}$ (see Appendix 2 of [**KT**], and Corollary 2.2 of it in particular) and thus its Perron-Frobenius (PF) eigenvalue. Under the additional assumption of irreducibility of $\boldsymbol{s}$ (which we do not make!), $\boldsymbol{m}$ would be the unique PF eigenvector, and primitivity of $\boldsymbol{s}$ would further imply that all other eigenvalues of $\boldsymbol{s}$ were less than 1 in absolute value.

Let us also mention that there is no need to choose any particular normalization here, but a convenient one would be $m_N := m_1 \cdot \ldots \cdot m_n = 1$.

Define $\mathcal{P}^m(X_N)$ to be the space of all *product* measures $\mu = \mu_1 \otimes \cdots \otimes \mu_n$ where $\mu_i \in \mathcal{M}_+^{m_i}(X_i)$, i.e., μ_i is a positive measure of total variation $\|\mu_i\| = \mu_i(X_i) = m_i$. For each $\boldsymbol{f} = (f_{ij}) \in F$, we define the operator

$$(4.9) \qquad \begin{aligned} \mathcal{A}_{\boldsymbol{f}}\colon \mathcal{P}^m(X_N) &\to \mathcal{M}(X_N) \\ \mu &\mapsto \mathcal{A}_{\boldsymbol{f}}(\mu) := \sum_{J \in N^n} (f_J.\mu_J). \end{aligned}$$

For any $\phi \in C(X_N, \mathbb{R})$, we have

$$(4.10) \qquad \mathcal{A}_{\boldsymbol{f}}(\mu)(\phi) = \sum_{J \in N^n} \mu_J(\phi \circ f_J).$$

In particular, if $\phi(x_1,\ldots,x_n) = \phi_1(x_1) \cdot \ldots \cdot \phi_n(x_n)$ for some $\phi_i \in C(X_i, \mathbb{R})$, then this can be rewritten as

$$\begin{aligned} \mathcal{A}_{\boldsymbol{f}}(\mu)(\phi) &= \sum_{j_1,\ldots,j_n} \mu_{j_1}(\phi_1 \circ f_{1 j_1}) \cdot \ldots \cdot \mu_{j_n}(\phi_n \circ f_{n j_n}) \\ &= \left(\sum_j (f_{1j}.\mu_j)(\phi_1)\right) \cdot \ldots \cdot \left(\sum_j (f_{nj}.\mu_j)(\phi_n)\right) \\ &= \sum_{J \in N^n} (f_{1 j_1}.\mu_{j_1} \otimes \cdots \otimes f_{n j_n}.\mu_{j_n})(\phi), \end{aligned}$$

which, since the linear span of the product functions $\phi = (\phi_1,\ldots,\phi_n)$ is dense in $C(X_N, \mathbb{R})$, shows that

$$(4.11) \qquad \mathcal{A}_{\boldsymbol{f}}(\mu) = \sum_{J \in N^n} f_{1 j_1}.\mu_{j_1} \otimes \cdots \otimes f_{n j_n}.\mu_{j_n} = \sum_{J \in N^n} f_J.\mu_J.$$

We define the *averaging* operator $\mathcal{A}$ on $\mathcal{P}^m(X_N)$ by

$$(4.12) \quad \mathcal{A}(\mu) := \int_F \mathcal{A}_{\boldsymbol{f}}(\mu) \, d\boldsymbol{\sigma}(\boldsymbol{f})$$

$$= \int_F \sum_{J \in N^n} (f_{1j_1}.\mu_{j_1} \otimes \cdots \otimes f_{nj_n}.\mu_{j_n}) \, d\sigma_{1j_1}(f_{1j_1}) \cdot \cdots \cdot d\sigma_{nj_n}(f_{nj_n})$$

$$= \left(\sum_j \int_{F_{1j}} (f_{1j}.\mu_j) \, d\sigma_{1j}(f_{1j}) \right) \cdot \cdots \cdot \left(\sum_j \int_{F_{nj}} (f_{nj}.\mu_j) \, d\sigma_{nj}(f_{nj}) \right)$$

$$= \sum_{J \in N^n} \int_{F_J} (f_J.\mu_J) \, d\sigma_J(f_J),$$

for $\mu = \mu_1 \otimes \cdots \otimes \mu_n$. For any $\phi \in C(X_N, \mathbb{R})$, this reads:

$$(4.13) \quad \int_F \mathcal{A}_{\boldsymbol{f}}(\mu)(\phi) \, d\boldsymbol{\sigma}(\boldsymbol{f}) = \sum_{J \in N^n} \int_{F_J} \mu_J(\phi \circ f_J) \, d\sigma_J(f_J) \in \mathcal{M}(X_N),$$

and, if $\phi(x_1, \ldots, x_n) = \phi_1(x_1) \cdot \cdots \cdot \phi_n(x_n)$ for some $\phi_i \in C(X_i, \mathbb{R})$, this becomes

$$\int_F \mathcal{A}_{\boldsymbol{f}}(\mu)(\phi) \, d\boldsymbol{\sigma}(\boldsymbol{f})$$
$$= \left(\sum_j \int_{F_{1j}} (f.\mu_j) \, d\sigma_{1j}(f)(\phi_1) \right) \cdot \cdots \cdot \left(\sum_j \int_{F_{nj}} (f.\mu_j) \, d\sigma_{nj}(f)(\phi_n) \right).$$

This shows that the averaging process is of the sort envisaged in (4.6) and

$$(4.14) \quad \mathcal{A}(\mu) = \left(\sum_j \int_{F_{1j}} (f.\mu_j) \, d\sigma_{1j}(f) \right) \otimes \cdots \otimes \left(\sum_j \int_{F_{nj}} (f.\mu_j) \, d\sigma_{nj}(f) \right).$$

Furthermore, consider

$$(4.15) \quad \sum_j \int_{F_{ij}} (f.\mu_j) \, d\sigma_{ij}(f) \in \mathcal{M}(X_i).$$

Since $f \in F_{ij}$ implies $\mathbf{1}_{X_i} \circ f = \mathbf{1}_{X_j}$, (4.15) satisfies

$$\sum_j \int_{F_{ij}} (f.\mu_j) \, d\sigma_{ij}(f)(\mathbf{1}_{X_i}) = \sum_j \int_{F_{ij}} \mu_j(\mathbf{1}_{X_i} \circ f) \, d\sigma_{ij}(f)$$
$$= \sum_j m_j \int_{F_{ij}} d\sigma_{ij}(f) = \sum_j s_{ij} m_j = m_i.$$

This shows that our averaging operator stabilizes the space of product measures that we are considering:

PROPOSITION 4.2. *The averaging operator* $\mathcal{A} := \int_F \mathcal{A}_{\boldsymbol{f}} \, d\boldsymbol{\sigma}(\boldsymbol{f})$ *of* (4.12) *maps the space* $\mathcal{P}^{\boldsymbol{m}}(X_N)$ *of product measures with mass vector* $\boldsymbol{m}$ *into itself.* □

If $\phi \in C(X_N, \mathbb{R})$ is a contraction, then so is $\phi \circ f_J : X_J \to \mathbb{R}$ for every $\boldsymbol{f} \in F$:

$$|\phi \circ f_J(x) - \phi \circ f_J(x')| \leq r_\phi d_N(f_J(x), f_J(x')) = r_\phi \sup_{i \in N} d_i(f_{ij_i}(x_{j_i}), f_{ij_i}(x'_{j_i}))$$
$$\leq r_\phi r \sup_{i \in N} d_{j_i}(x_{j_i}, x'_{j_i}) = r_\phi r \, d_J(x, x')$$

for all $x, x' \in X_J$.

Define a metric L_J on $\mathcal{M}_+^{m_J}(X_J)$, the space of positive measures of total measure $m_J := \prod_{i=1}^n m_{j_i} > 0$, by

$$(4.16) \qquad L_J(\mu, \nu) = \frac{1}{m_J} \sup\{|\mu(\psi) - \nu(\psi)| \mid \psi \in \text{Lip}(\leq 1, X_J, \mathbb{R})\}.$$

This makes $\mathcal{M}_+^{m_J}(X_J)$ into a complete metric space by Proposition 2.2.

Define L on $\mathcal{P}^m(X_N)$ by

$$(4.17) \qquad L(\mu, \nu) = \sup\{L_J(\mu_J, \nu_J) \mid J \in N^n\}.$$

PROPOSITION 4.3. *The operator* $\mathcal{A}: \mathcal{P}^m(X_N) \to \mathcal{P}^m(X_N)$ *is a contraction with respect to the metric L, with contractivity factor at most r.*

PROOF. Let $\mu, \nu \in \mathcal{P}^m(X_N)$. In order to determine the $L_K(\mathcal{A}(\mu), \mathcal{A}(\nu))$, we have to determine $\mathcal{A}(\mu)_K$ for any $K \in N^n$. Now, $\mathcal{A}\mu = \mathcal{A}(\mu)$ is a product measure, and from (4.14) we find

$$(\mathcal{A}\mu)_K = \left(\sum_{j_1} \int_{F_{k_1 j_1}} (f.\mu_{j_1}) \, d\sigma_{k_1 j_1}(f)\right) \otimes \cdots \otimes \left(\sum_{j_n} \int_{F_{k_n j_n}} (f.\mu_{j_n}) \, d\sigma_{k_n j_n}(f)\right)$$

$$= \sum_J \int_{F_{KJ}} (f.\mu_J) \, d\sigma_{KJ}(f).$$

Thus

$$L_K\big((\mathcal{A}\mu)_K, (\mathcal{A}\nu)_K\big) = L_K\left(\sum_J \int_{F_{KJ}} (f.\mu_J) \, d\sigma_{KJ}(f), \sum_J \int_{F_{KJ}} (f.\nu_J) \, d\sigma_{KJ}(f)\right)$$

$$= \frac{1}{m_K} \sup_\psi \left|\sum_J \int_{F_{KJ}} ((f.\mu_J) - (f.\nu_J))(\psi) \, d\sigma_{KJ}(f)\right|$$

$$\leq \frac{1}{m_K} \sum_J \sup_\psi \int_{F_{KJ}} |(\mu_J - \nu_J)(\psi \circ f)| \, d\sigma_{KJ}(f)$$

$$\leq \frac{r}{m_K} \sum_J \int_{F_{KJ}} \sup_\psi |(\mu_J - \nu_J)(r^{-1}\psi \circ f)| \, d\sigma_{KJ}(f)$$

$$\leq \frac{r}{m_K} \sum_J \int_{F_{KJ}} m_J L_J(\mu_J, \nu_J) \, d\sigma_{KJ}(f)$$

$$\leq \frac{rL(\mu, \nu)}{m_K} \sum_J m_J \int_{F_{KJ}} d\sigma_{KJ}(f) = \frac{rL(\mu, \nu)}{m_K} \sum_J s_{KJ} m_J,$$

where ψ runs through $\text{Lip}(\leq 1, X_K, \mathbb{R})$. Finally,

$$L(\mathcal{A}\mu, \mathcal{A}\nu) = \sup_K L_K\big((\mathcal{A}\mu)_K, (\mathcal{A}\nu)_K\big) \leq rL(\mu, \nu) \sup_K \frac{1}{m_K} \sum_J s_{KJ} m_J$$

$$= rL(\mu, \nu) \sup_K \frac{1}{m_K} \prod_{i=1}^n \sum_{j_i=1}^n s_{k_i j_i} m_{j_i}$$

$$= rL(\mu, \nu) \sup_K \frac{1}{m_K} \prod_{i=1}^n m_{k_i} = rL(\mu, \nu),$$

where we have used (4.8) in the last line. $\square$

We have thus established the following result.

THEOREM 4.4. *Let $X_1, \ldots, X_n$ be compact metric spaces and, for each pair $(i,j) \in N \times N$, let F_{ij} be a compact admissible family of contractions (possibly empty) from X_j to X_i. Assume that each F_{ij} is equipped with a positive Borel measure σ_{ij} and define $s_{ij} = \sigma_{ij}(F_{ij})$, with $s_{ij} := 0$ if F_{ij} is empty. Assume that $\boldsymbol{s} := (s_{ij})$ has a positive 1-eigenvector $\boldsymbol{m} = (m_1, \ldots, m_n)^T$.*

Then there exists a unique family of measures $\omega_i \in \mathcal{M}(X_i)_+^{m_i}$ which satisfy

$$(4.18) \qquad \omega_i = \sum_{j=1}^n \int_{F_{ij}} (f.\omega_j) \, d\sigma_{ij}(f), \quad i \in N.$$

The support of ω is contained in W, the attractor of (4.5). □

We call $\omega = \omega_1 \otimes \cdots \otimes \omega_n$ the $(F, \boldsymbol{\sigma}, \boldsymbol{m})$-invariant measure on X_N, or simply $(F, \boldsymbol{\sigma})$-invariant measure, if $\boldsymbol{m}$ is understood from the context.

5. Multi-Component Families of Affine Mappings

In this section, H is an LCAG and A is an automorphism of H with modulus $\alpha > 1$. The Haar measure on H is denoted by θ. H is assumed to be complete with respect to a metric d, relative to which A is a contraction.

We assume that we are given n copies of H, which we call $H_1, \ldots, H_n$, and compact families F_{ij}, $1 \leq i,j \leq n$, of affine mappings $f_{ij} \colon H_j \to H_i$, all of the form $f_{ij}(x) = Ax + u_{ij}$ with $u_{ij} \in H_i$. Our objective is to understand the multi-component system formed by

$$H^n = H_1 \times \cdots \times H_n = H \times \cdots \times H$$

and the admissible family of contractions $F = \times F_{ij}$ in a way that parallels our previous analysis in Section 3.

By Proposition 4.1, F has a unique attractor $W = W_1 \times \cdots \times W_n \in (\mathcal{K}H^n)$. We wish to describe the unique F-self-similar measure $\omega = \omega_1 \otimes \cdots \otimes \omega_n$ on H^n with respect to a system $\boldsymbol{\sigma} = (\sigma_{ij})$ of measures on F.

The mapping $F_{ij} \to H_i$ defined by $f_{ij} \mapsto f_{ij}(0) = u_{ij}$ is continuous and produces a homeomorphism between F_{ij} and a compact subset $F'_{ij} := F_{ij}(0)$ of H_i.

We assume that each compact space F_{ij} is supplied with a positive regular Borel measure σ_{ij}, supported on F_{ij} with $s_{ij} := \sigma_{ij}(F_{ij})$. We may identify σ_{ij} with a regular Borel measure on H_i supported on F'_{ij}. It is understood that F_{ij} may be empty, in which case $s_{ij} := 0$. Furthermore, we assume the existence of a mass vector $\boldsymbol{m} = (m_1, \ldots, m_n)^T > 0$ satisfying the compatibility assumption CA, i.e., $\boldsymbol{sm} = \boldsymbol{m}$.

We define $X_i \in \mathcal{K}H_i$, $i = 1, \ldots, n$, to be compact subspaces with the following properties:

(1) $X_1 \times \cdots \times X_n$ is invariant under the family of mappings F;
(2) $\sum_{k=0}^\ell A^k \operatorname{supp}(\boldsymbol{\sigma}) \subset X_1 \times \cdots \times X_n$, for all $\ell \geq 0$.

It is easy to see that such sets exist because the mappings of F and the automorphism A are all contractive. The F-invariance already forces $W_i \subset X_i$.

Let notation be as in Section 4, so $X_N = X_1 \times \cdots \times X_n$ and $\mathcal{P}^m(X_N)$ is the space of all product measures $\mu_1 \otimes \cdots \otimes \mu_n$ on X_N for which $\mu_i \in \mathcal{M}_+^{m_i}(X_i)$. We know that the averaging operator $\mathcal{A}$ of (4.12) is a contraction on $\mathcal{P}^m(X_N)$.

Let $E_i \subset X_i$, $i \in N$, be measurable sets. Then, from (4.14), we obtain

$$
(5.1) \quad \mathcal{A}\mu(E_1 \times \cdots \times E_n) = \prod_{i=1}^n \sum_j \int_{F_{ij}} (f.\mu_j)(E_i) \, \mathrm{d}\sigma_{ij}(f)
$$

$$
= \prod_{i=1}^n \sum_j \int_{F_{ij}} \mu_j(f^{-1}(E_i)) \, \mathrm{d}\sigma_{ij}(f)
$$

$$
= \prod_{i=1}^n \sum_j \int_{F'_{ij}} (A.\mu_j)(E_i - u) \, \mathrm{d}\sigma_{ij}(u)
$$

$$
= \prod_{i=1}^n \sum_j (\sigma_{ij} * A.\mu_j)(E_i).
$$

Thus

$$
\mathcal{A}\mu = \left(\sum_j \sigma_{1j} * A.\mu_j\right) \otimes \cdots \otimes \left(\sum_j \sigma_{nj} * A.\mu_j\right).
$$

Adopting matrix notation, with $\mu = \mu_1 \otimes \cdots \otimes \mu_n$ written as $(\mu_1, \ldots, \mu_n)^T$ and $\boldsymbol{\sigma} = (\sigma_{ij})$, this reads

$$
(5.2) \quad \mathcal{A}\mu = \boldsymbol{\sigma} * A.\mu
$$

where $A.\mu := (A.\mu_1, \ldots, A.\mu_m)^T$. If we now define $A.\boldsymbol{\sigma} := (A.\sigma_{ij})$, we can iterate (5.2). Observing Proposition 3.2(3), we obtain

$$
(5.3) \quad \mathcal{A}^\ell \mu = \boldsymbol{\sigma} * A.\boldsymbol{\sigma} * \cdots * A^{\ell-1}.\boldsymbol{\sigma} * A^\ell.\mu.
$$

We now proceed as in Section 3 to define a suitable sequence of measures $\left(\omega^{(\ell)} \in \mathcal{P}^m(X)\right)_{\ell \geq 0}$. First, let

$$
(5.4) \quad \omega^{(0)} := \delta^m = (m_1 \delta_0, \ldots, m_n \delta_0)^T
$$

where δ_0 is the unit point measure supported at $\{0\}$. Clearly, $\delta^m \in \mathcal{P}^m(X)$, but since A is a contraction, we also have $A.\delta^m = \delta^m$, and $A^\ell.\mu \to \delta^m$ as $\ell \to \infty$, for any $\mu \in \mathcal{P}^m(X)$. Define iteratively, as before, $\omega^{(\ell+1)} = \mathcal{A}\omega^{(\ell)}$ for $\ell \geq 0$. Then

$$
\omega^{(\ell+1)} = \boldsymbol{\sigma} * A.\boldsymbol{\sigma} * \cdots * A^\ell.\boldsymbol{\sigma} * \delta^m.
$$

We have $\mathrm{supp}(\omega^{(\ell)}) \subset \sum_{k=0}^\ell A^k \mathrm{supp}(\boldsymbol{\sigma} * \delta^m)$ and $\mathcal{AP}^m(X) \subset \mathcal{P}^m(X)$, so we know, since $\delta^m \in \mathcal{P}^m(X)$, that $\omega^{(\ell)} \in \mathcal{P}^m(X)$. Consequently, $\omega^{(\ell)}$ vaguely converges, as $\ell \to \infty$, to the unique $(F, \boldsymbol{\sigma}, \boldsymbol{m})$-self-similar measure $\omega = \omega_1 \otimes \cdots \otimes \omega_n \in \mathcal{P}^m(X)$. Since $\mathrm{supp}(\omega) \subset W$ by Theorem 4.4, we find that $\omega \in \mathcal{P}^m(W)$. To summarize:

PROPOSITION 5.1. *Let H be an LCAG which is a complete metric space. Let A be a contractive automorphism on H and let F_{ij}, $1 \leq i, j \leq n$, with attractor W, be compact admissible families of affine maps on H, all of the form $x \mapsto Ax + v$ with $v \in H$. Let σ_{ij} be a positive regular Borel measure on F_{ij}, identified with a regular Borel measure on H supported on $F_{ij}(0)$ (with $\sigma_{ij} := 0$ if $F_{ij} = \varnothing$). Let $\boldsymbol{s} = (s_{ij}) = \bigl(\sigma_{ij}(H)\bigr)$ and, finally, let $\boldsymbol{m} = (m_1, \ldots, m_n)^T > 0$ satisfy $\boldsymbol{sm} = \boldsymbol{m}$. Then*

(1) *$\omega = (\bigstar_{\ell=0}^\infty A^\ell.\boldsymbol{\sigma}) * \delta^m$ is the unique $(F, \boldsymbol{\sigma}, \boldsymbol{m})$-self-similar measure of (4.18), with $\omega \in \mathcal{P}^m(H^n)$, $\mathrm{supp}(\omega) \subset W$, and δ^m as in (5.4).*
(2) *$\widehat{\omega} = (\prod_{\ell=0}^\infty (A^T)^{-\ell}.\widehat{\boldsymbol{\sigma}}) \mathbf{1}^m$, where $\mathbf{1}^m = (m_1 \mathbf{1}_H, \ldots, m_n \mathbf{1}_H)^T$, the convergence of the product being uniform on compact sets.* □

If we assume that the measures σ_{ij} are absolutely continuous with respect to Haar measure θ on H, then $\sigma_{ij} = h_{ij}\theta$ where $h_{ij} \in L^1(H)$ due to the Radon-Nikodym theorem. In particular, we have $\operatorname{supp}(h_{ij}) \subset F_{ij}(0) \subset H$, $h_{ij} \geq 0$ and $\|h_{ij}\|_1 = \int_H h_{ij}\,\mathrm{d}\theta = s_{ij}$, for all $1 \leq i, j \leq n$. Then

$$\begin{aligned}
\omega^{(\ell+1)} &= \boldsymbol{\sigma} * A.\boldsymbol{\sigma} * \cdots * A^\ell.\boldsymbol{\sigma} * \delta^{\boldsymbol{m}} \\
&= \boldsymbol{h}\Theta * A.(\boldsymbol{h}\Theta) * \cdots * A^\ell.(\boldsymbol{h}\Theta) * \delta^{\boldsymbol{m}} \\
&= \boldsymbol{h}\Theta * \alpha(A.\boldsymbol{h})\Theta * \cdots * \alpha^\ell(A^\ell.\boldsymbol{h})\Theta * \delta^{\boldsymbol{m}} \\
&= \boldsymbol{h} * \alpha(A.\boldsymbol{h}) * \cdots * \alpha^\ell(A^\ell.\boldsymbol{h})(\Theta * \delta^{\boldsymbol{m}})
\end{aligned}$$

where $\boldsymbol{h} = (h_{ij})$ and $\Theta = \operatorname{diag}(\theta,\ldots,\theta)$ is a diagonal matrix. Thus we have $\Theta * \delta^{\boldsymbol{m}} = (m_1\theta,\ldots,m_n\theta)^T$ and

$$(5.5) \qquad \omega^{(\ell+1)} = \left(\underset{k=0}{\overset{\ell}{\LARGE *}} \alpha^k(A^k.\boldsymbol{h}) \right)(m_1\theta,\ldots,m_n\theta)^T \in \mathcal{P}^{\boldsymbol{m}}(H^n).$$

Vague convergence of this sequence is clear, but the results of Section 4 suggest that we can expect more. However, $\|\cdot\|$-convergence is technically more involved here. Let us thus first postpone this question and state first the result on the self-similar functions.

PROPOSITION 5.2. *Let notation and assumptions be as in Proposition 5.1, and suppose that the measures $\sigma_{ij} = h_{ij}\theta$ are absolutely continuous with respect to Haar measure θ. Assume that the convolution in (5.5), as $\ell \to \infty$, converges also in the $\|\cdot\|$-topology. Then there is a unique vector $\boldsymbol{g} = (g_1,\ldots,g_n)^T$ of nonnegative functions in $L^1(H)$ that satisfies*

(1) $\boldsymbol{g} = (\underset{\ell=0}{\overset{\infty}{\LARGE *}} \alpha^\ell(A^\ell.\boldsymbol{h}))\boldsymbol{m}$
(2) $g_i(x) = \sum_{j=1}^n \int_H h_{ij}(x-v)g_j(A^{-1}v)\,\mathrm{d}\theta(v)$, $i=1,\ldots,n$, *with normalization $\int_H g_i\,\mathrm{d}\theta = m_i$ and $\operatorname{supp}(g_i) \subset W_i$.*

Furthermore, if the h_{ij} are functions in $L^1(H) \cap L^\infty(H)$, then the functions g_i are continuous on H.

PROOF. Part (1) is a direct reformulation of (5.5), and Part (2) is a component-wise recounting of Part (1). The continuity follows from the properties of the convolution product, as in Proposition 3.6(4). $\square$

Infinite convolution products like that of Proposition 5.2(1) also appear in the context of matrix continuous refinement operators. These are introduced in [**JL**] (with H being $\mathbb{R}^n$). In our paper [**BM2**], we relied on the results of [**JL**] for the existence of our self-similar densities. However, the methods of [**JL**] are from functional analysis and do not lend themselves to the general measure theoretic situation that we are trying to address here.

Let us come back to the convergence issue in (5.5). Unlike the situation in Section 3.3, with Lemma 3.4 and Proposition 3.5, the $\|\cdot\|$-convergence in (5.5) is not entirely automatic. Note that the vector notation for the measures is handy for the formulation of the iteration, but it still represents a product measure. We are interested in the $\|\cdot\|$-convergence of the sequence of product measures (5.5). For this, it is sufficient, but not necessary, that the sequence $K_\ell = \underset{k=0}{\overset{\ell}{\LARGE *}} \alpha^k(A^k.\boldsymbol{h})$, seen as a sequence of linear operators, converges in the operator norm. However, for fixed i, j, $\|(K_\ell)_{ij}\|_1 = (\boldsymbol{s}^{\ell+1})_{ij}$, and convergence of this, for $\ell \to \infty$, does not follow from our general assumptions on the matrix $\boldsymbol{s}$, see the remark after (4.8), because we did not assume primitivity of $\boldsymbol{s}$.

Nevertheless, there is an analogue of Proposition 3.5 which we will now derive. To this end, define $\boldsymbol{h}^{(k)} = \alpha^k(A^k.\boldsymbol{h})$ for $k \geq 0$. In particular, $\boldsymbol{h}^{(0)} = \boldsymbol{h} = (h_{ij})$, which is a matrix of functions in $L^1(H)$, and also each $(\boldsymbol{h}^{(k)})_{ij}$ is a nonnegative L^1-function of norm s_{ij}. Recall that (5.5) means $\omega^{(\ell)} = f_1^{(\ell)}\theta \otimes \cdots \otimes f_n^{(\ell)}\theta$, with L^1-functions $f_i^{(\ell)} = \sum_j (\boldsymbol{*}_{k=0}^{\ell-1} \boldsymbol{h}^{(k)})_{ij} m_j$ of norm $\|f_i^{(\ell)}\|_1 = \sum_j (\boldsymbol{s}^\ell)_{ij} m_j = m_i$. Consequently, showing that (5.5) converges also in the $\|\cdot\|$-topology means showing that $f_i^{(\ell)}$ converges in $L^1(H)$ for each i as $\ell \to \infty$.

Fix $\varepsilon > 0$, and let V_{ij} be the corresponding neighborhood for the function h_{ij} according to Lemma 3.4. Let $V = \bigcap_{i,j} V_{i,j}$ and choose an integer M such that, for all $k \geq 0$ and all i,j, the nonnegative L^1-function $(\boldsymbol{*}_{\ell=M}^{M+k} \boldsymbol{h}^{(\ell)})_{ij}$, of norm $(\boldsymbol{s}^{k+1})_{ij}$, has support inside V. Such an M clearly exists. With Lemma 3.4, we then find, for all i', j' simultaneously, the approximation formula

$$\left\| \left(\underset{\ell=M}{\overset{M+k}{\boldsymbol{*}}} \boldsymbol{h}^{(\ell)} \right)_{ij} * h_{i'j'} - (\boldsymbol{s}^{k+1})_{ij} h_{i'j'} \right\|_1 \leq (\boldsymbol{s}^{k+1})_{ij}\, \varepsilon.$$

Note that this formulation remains valid even in the limiting case that $(\boldsymbol{s}^{k+1})_{ij}$ happens to vanish.

Let now $n, m \geq M$ and define $u_{ij} = (\boldsymbol{*}_{\ell=M}^n \boldsymbol{h}^{(\ell)})_{ij}$ and $v_{ij} = (\boldsymbol{*}_{\ell=M}^m \boldsymbol{h}^{(\ell)})_{ij}$. Then, we can calculate as follows

$$\|f_i^{(n)} - f_i^{(m)}\|_1 = \left\| \sum_{k,\ell,j} (\boldsymbol{h}^{(1)} * \cdots * \boldsymbol{h}^{(M-1)})_{k\ell} * h_{ik} * (u_{\ell j} - v_{\ell j}) m_j \right\|_1$$

$$\leq \sum_{k,\ell} \left(\|(\boldsymbol{h}^{(1)} * \cdots * \boldsymbol{h}^{(M-1)})_{k\ell}\|_1 \cdot \left\| \sum_j h_{ik} * (u_{\ell j} - v_{\ell j}) m_j \right\|_1 \right)$$

$$\leq \sum_{k,\ell} (\boldsymbol{s}^{M-1})_{k\ell} \left\| \sum_j h_{ik} * (u_{\ell j} - v_{\ell j}) m_j \right\|_1$$

where we have used that convolution on the level of functions is commutative. Observe next that

$$\left\| \sum_j h_{ik} * (u_{\ell j} - v_{\ell j}) m_j \right\|_1$$

$$\leq \left\| h_{ik} \sum_j ((\boldsymbol{s}^{n-M+1})_{\ell j} - (\boldsymbol{s}^{m-M+1})_{\ell j}) m_j \right\|_1$$

$$+ \sum_j (\|h_{ik} * u_{\ell j} - h_{ik}(\boldsymbol{s}^{n-M+1})_{\ell j}\|_1 + \|h_{ik} * v_{\ell j} - h_{ik}(\boldsymbol{s}^{m-M+1})_{\ell j}\|_1) m_j$$

$$\leq \varepsilon \left(\sum_j (\boldsymbol{s}^{n-M+1})_{\ell j} m_j + \sum_j (\boldsymbol{s}^{m-M+1})_{\ell j} m_j \right) = 2 m_\ell\, \varepsilon$$

where we have used the above approximation formula and the equation $\boldsymbol{sm} = \boldsymbol{m}$. This finally gives

$$\|f_i^{(n)} - f_i^{(m)}\|_1 \leq 2\varepsilon \sum_{k,\ell} (\boldsymbol{s}^{M-1})_{k\ell} m_\ell = 2(m_1 + \cdots + m_n)\varepsilon,$$

independently of i. This shows that all sequences $\left(f_i^{(n)}\right)_{n\geq 0}$ are Cauchy, and we have thus established the expected analogue of Proposition 3.5 and strengthening of Proposition 5.2:

PROPOSITION 5.3. *Let notation and assumptions be as in Proposition 5.1, and suppose that the measures $\sigma_{ij} = h_{ij}\theta$ are absolutely continuous with respect to Haar measure θ. Then, the sequence of product measures in (5.5), as $\ell \to \infty$, converges not only vaguely, but also in the $\|\cdot\|$-topology of $\mathcal{P}^m(H \times \cdots \times H)$.* □

6. Model Sets and Weyl's Theorem

To link our previous analysis to quasicrystals, let us now summarize some of the key ingredients to their mathematical description. A *cut and project scheme* consists of the following set of data:

- a real space $\mathbb{R}^m$
- a locally compact Abelian group H
- a lattice $\tilde{L} \subset \mathbb{R}^m \times H$

which satisfies the following properties. If π and π_H are the natural projections of $\mathbb{R}^m \times H$ onto $\mathbb{R}^m$ and H, respectively, then

- $\pi|_{\tilde{L}}$ is one-to-one.
- $\pi_H(\tilde{L})$ is dense in H.

This is summarized in the following diagram.

(6.1)
$$\mathbb{R}^m \xleftarrow{\pi} \mathbb{R}^m \times H \xrightarrow{\pi_H} H$$
$$\underset{1-1}{\nwarrow} \quad \underset{\cup}{\big|} \quad \underset{\text{dense}}{\nearrow}$$
$$\tilde{L}$$

To say that $\tilde{L}$ is a *lattice* in $\mathbb{R}^m \times H$ means that $\tilde{L}$ is a discrete subgroup of $\mathbb{R}^m \times H$ such that $(\mathbb{R}^m \times H)/\tilde{L}$ is compact.

We set $L = \pi(\tilde{L})$, a subgroup of $\mathbb{R}^m$, and define the *star map* $(.)^*: L \to H$ by $x^* = \pi_H \circ \left(\pi|_{\tilde{L}}\right)^{-1}(x)$. Although $(.)^*$ is a group homomorphism, it has, in general, no natural extension to $\mathbb{R}^m$ and, indeed, it is typically totally discontinuous in the topology on L induced by $\mathbb{R}^m$. In fact, it is this property that makes it useful!

Given any subset $U \subset H$, we define

(6.2) $\quad \Lambda(U) := \{x \in L \mid x^* \in U\} = \{\pi(\tilde{x}) \mid \tilde{x} \in \tilde{L}, \pi_H(\tilde{x}) \in U\} \subset L \subset \mathbb{R}^m.$

A set $\Lambda \subset \mathbb{R}^m$ is a *model set* relative to (6.1), if $\Lambda = \Lambda(W)$ for some $W \subset H$ that is compact and equals the closure of its nonempty interior.[5]

Model sets have remarkable properties that make them important objects of study in the theory of mathematical quasicrystals. We refer the reader to [**B, M, P**] and references therein for more details, but we mention here a few of the key points:

(1) If $\Lambda \subset \mathbb{R}^m$ is a model set, then it is a *Delone set* in $\mathbb{R}^m$, i.e., Λ is both uniformly discrete and relatively dense.

---

[5]There are variations on the exact conditions imposed on W depending on the delicateness of the results required. Our assumptions imply the Delone property of Λ, and are rather convenient for many other purposes. There is still something unsatisfying about our present understanding of model sets. To say that $\Lambda \subset \mathbb{R}^m$ is a model set is to say that it arises from some cut and project scheme. But we still lack a useful direct characterization of such sets, compare [**Sch1**].

(2) Generically, model sets have no translational symmetries, although they certainly still have a high degree of long-range order.
(3) If W is *Riemann measurable*, i.e., if ∂W has vanishing Haar measure in H, then Λ has a well-defined density (see Proposition 6.1 below).
(4) If W is *Riemann measurable*, then Λ is pure point diffractive [**Hof1, Sch2**].

Model sets appeared very early in the theory of quasicrystals (under the name of cut and project sets, see [**B**] and references given there), but originally only with the internal group H being another real space. However, model sets had been defined much earlier and in full generality in a totally different context by Y. Meyer [**Mey**]. Recent papers [**M, BMS, Sch2**] show that the more general setting is completely relevant in the mathematical theory of quasicrystals, aperiodic tilings and substitution systems, both geometric and algebraic.

A key feature of a model set is that it puts together a discrete geometric object, $\Lambda(W)$, with a relatively compact set $W \subset H$ on which we can use the powerful array of tools from analysis on locally compact Abelian groups. The essential mathematical link is Weyl's Theorem on uniform distribution [**We**] which connects densities on $\Lambda(W)$ to measures on W. We refer to this theory in the more general setting of LCAG's, see [**KN**, Chapter 4.4] for background material.

In its usual form, Weyl's Theorem is stated for real spaces, but it works at the level of locally compact Abelian groups, too. Here we establish the theorem in this more general setting. The basis of the theorem in the context of model sets is the phenomenon of 'uniformity of projection', which we state here in the generality that we will need. In fact, our proof demonstrates that Weyl's Theorem, in this context, is actually equivalent to the uniformity of projection. For an approach to uniformity of projection via ergodic theory, see [**Hof2, Sch2**].

In the following two Propositions, it is understood that a cut and project scheme according to (6.1) has been given. In addition, let $\tilde{\theta}$ denote the product measure on $\mathbb{R}^m \times H$, formed from Lebesgue measure on $\mathbb{R}^m$ and our fixed Haar measure θ on H. Let now T be any measurable fundamental domain of $\mathbb{R}^m \times H$ with respect to the action of its discrete subgroup $\tilde{L}$, and define $|\tilde{L}| := \tilde{\theta}(T)$ as its volume. Note that the value of $|\tilde{L}|$ does not depend on the actual choice of T. Its meaning really is the averaged number of lattice points per unit volume (in Haar measure).

PROPOSITION 6.1 (Schlottmann [**Sch1**]). *Let a cut and project setup according to diagram* (6.1) *be given, with* $|\tilde{L}|$ *as described above. Let* $U \subset H$ *be totally bounded and Riemann measurable* (*i.e.,* U *measurable with* $\theta(\partial U) = 0$). *Then*

$$\lim_{r \to \infty} \frac{1}{\operatorname{vol} B_r(0)} \left(\sum_{x \in \Lambda(U) \cap B_r(a)} 1 \right) = \frac{\theta(U)}{|\tilde{L}|}.$$

Furthermore, the limit is uniform in a. □

This limit is called the *density*, $\operatorname{den}(\Lambda(U))$, of $\Lambda(U)$. Note that the totally bounded set U in this Proposition need not be closed. It is only demanded that its boundary has vanishing Haar measure. If U itself is of measure 0, the density of $\Lambda(U)$ vanishes.

REMARK. There is another way to explain the meaning of $|\tilde{L}|$ which is perhaps more natural from the group theoretic point of view. Consider the factor group $T' := (\mathbb{R}^m \times H)/\tilde{L}$ (which is compact) and let μ be its normalized Haar measure. If now f is a continuous function on $\mathbb{R}^m \times H$ with compact support, define a new function by $F(x) = \sum_{u \in \tilde{L}} f(x+u)$. So, F results from f by averaging over the canonical Haar measure of the lattice $\tilde{L}$, which is counting measure. The function F can then also be viewed as a function on T', and we can determine its integral, $\mu(F)$. If we then define a new measure on $\mathbb{R}^m \times H$ by $\tilde{\theta}'(f) := \mu(F)$, it is another Haar measure on $\mathbb{R}^m \times H$, and we must have $\tilde{\theta}' = c\tilde{\theta}$. The constant c is nothing but $|\tilde{L}|$, see [**D**, Chapter XIV.4] for background material.

THEOREM 6.2 (Weyl's Theorem for general model sets). *Let $\Lambda = \Lambda(W)$ be a model set in the above sense, with compact, Riemann measurable $W \subset H$. Let $f \colon H \to \mathbb{R}$ be continuous with $\mathrm{supp}(f) \subset W$. Let $p \colon L \to \mathbb{R}$ be defined by $p(x) = f(x^*)$. Then*

$$\lim_{r \to \infty} \frac{1}{\mathrm{vol}\, B_r(0)} \sum_{x \in \Lambda \cap B_r(a)} p(x) = \frac{1}{|\tilde{L}|} \int_H f(y)\, \mathrm{d}\theta(y)$$

uniformly in a.

PROOF. The strategy will be to derive this more general result from Proposition 6.1. To this end, we approximate f by a step function ψ on a (Riemann) admissible partition $\{U_1, \ldots, U_n\}$ of W, i.e., $W = \bigcup_{i=1}^n U_i$ with pairwise disjoint sets $U_i \subset W$ that are all Riemann measurable.

Fix $\varepsilon > 0$. By Lemma A.6, there is a step function ψ on such an admissible partition (with suitable $n = n(\varepsilon)$) of W, $\psi = \sum_{i=1}^n c_i \mathbf{1}_{U_i}$, with $\|f - \psi\|_\infty < \varepsilon$. Choose a radius R big enough so that we have, for all $r > R$,

$$\left| \frac{1}{\mathrm{vol}\, B_r(0)} \left(\sum_{x \in \Lambda \cap B_r(a)} 1 \right) - \frac{\theta(W)}{|\tilde{L}|} \right| < \varepsilon$$

and also

$$\left| \frac{1}{\mathrm{vol}\, B_r(0)} \left(\sum_{x \in \Lambda(U_i) \cap B_r(a)} 1 \right) - \frac{\theta(U_i)}{|\tilde{L}|} \right| < \frac{\varepsilon}{n}$$

for all $1 \le i \le n$, uniformly in a. Such a radius clearly exists. Then we have, since $p(x) = f(x^*)$, for all $r > R$ the following 3ε-type argument,

$$\left| \frac{1}{\mathrm{vol}\, B_r(0)} \left(\sum_{x \in \Lambda \cap B_r(a)} p(x) \right) - \frac{1}{|\tilde{L}|} \int_W f(y)\, \mathrm{d}\theta(y) \right|$$

$$\le \frac{1}{\mathrm{vol}\, B_r(0)} \left(\sum_{x \in \Lambda \cap B_r(a)} |f(x^*) - \psi(x^*)| \right) + \frac{1}{|\tilde{L}|} \left| \int_W (f(y) - \psi(y))\, \mathrm{d}\theta(y) \right|$$

$$+ \sum_{i=1}^n \left| \frac{c_i}{\mathrm{vol}\, B_r(0)} \left(\sum_{x \in \Lambda(U_i) \cap B_r(a)} 1 \right) - \frac{1}{|\tilde{L}|} \int_{U_i} \psi(y)\, \mathrm{d}\theta(y) \right|$$

$$< \frac{\varepsilon}{\mathrm{vol}\, B_r(0)} \left(\sum_{x \in \Lambda \cap B_r(a)} 1 \right) + \frac{\theta(W)}{|\tilde{L}|}\varepsilon + \sum_{i=1}^n |c_i| \left| \frac{1}{\mathrm{vol}\, B_r(0)} \left(\sum_{x \in \Lambda(U_i) \cap B_r(a)} 1 \right) - \frac{\theta(U_i)}{|\tilde{L}|} \right|$$

$$< \left(2\frac{\theta(W)}{|\tilde{L}|} + 1 \right)\varepsilon + \sum_{i=1}^n \|\psi\|_\infty \frac{\varepsilon}{n} < \left(2\frac{\theta(W)}{|\tilde{L}|} + \|f\|_\infty + 2 \right)\varepsilon,$$

from which the theorem follows. □

REMARK. Weyl's Theorem also extends to all functions that are only continuous on the compact set W, see the remark following the proof of Lemma A.6.

7. Self-Similar Densities on Model Sets

Finally, we have collected all results that we need to construct self-similar measures for model sets on their "internal" side, and then, under certain circumstances, also invariant densities on the model sets themselves.

7.1. Self-Similar Systems. An affine self-similar system of model sets consists of the the following data (SS1–SS4):

- **SS1.** a cut and project scheme (6.1) whose internal space H, in addition to being an LCAG, is a complete metric space with translation invariant metric d.
- **SS2.** a family of regular model sets $\Lambda_i = \Lambda(W_i)$, $i = 1, \ldots, n$, for this cut and project scheme, with each W_i compact.
- **SS3.** an invertible linear mapping $Q \colon \mathbb{R}^m \to \mathbb{R}^m$ which satisfies $Q(L) \subset L$, where L is the projection $\pi(\tilde{L})$ of the lattice $\tilde{L}$ in (6.1).
- **SS4.** sets F_{ij}, $1 \leq i, j \leq n$, some of which may be empty, of affine mappings

$$C = C_a \colon x \mapsto Qx + a \quad (a \in L)$$

which map Λ_j to Λ_i and satisfy

$$(7.1) \qquad \Lambda_i = \bigcup_{j=1}^{n} \bigcup_{C \in F_{ij}} C(\Lambda_j), \quad 1 \leq i \leq n.$$

The sets F_{ij} may (and usually will) be infinite. Because all the affine mappings involved have the same linear part, Q, each F_{ij} is parameterized by the translational parts $a \in L$. In the sequel, we will thus mostly view F_{ij} as a subset of L. In this case, we will, from now on, use the notation F'_{ij}, i.e., $F'_{ij} := F_{ij}(0)$.

Such systems of model sets can arise quite naturally in the study of self-similar tilings. Each proto-tile is marked in some suitable way with a finite set of points, call them proto-points or *control points*. This provides a marking of the tiling by points, and the sets Λ_i are then taken to be the set of points that correspond to each class of control points. In this case, the union in (7.1) would typically be disjoint, but in our study we definitely wish to include nondisjoint unions as well.

The idea of this section is to pass the self-similar system to the internal side H of the cut and project scheme, to apply our theory of self-similar measures there, and finally to pull back the results to the physical side, namely to the model sets Λ_i themselves. We will see that pulling back is not automatically possible, and we need to make various types of assumptions to guarantee it. Still, these assumptions are not unnatural, and they are actually met in many interesting cases.

The situation for simple model sets $\Lambda = \Lambda(W)$ is just the special case of the general situation here, where $n = 1$. In this case, the matrix s that appears below is simply the unit matrix (1).

We can directly lift $Q \colon L \to L$ to a group homomorphism $\tilde{Q} \colon \tilde{L} \to \tilde{L}$, and then to a group homomorphism $Q^* \colon L^* \to L^*$. We assume

- **SS5.** Q^* is contractive with respect to the metric d.

In this case, since L^* is dense in H, Q^* extends to a continuous contractive automorphism

$$A \colon H \to H$$

with $A|_{L^*} = Q^*$. Due to Lemma 3.1, the modulus α of A with respect to the Haar measure θ on H satisfies $\alpha > 1$, see Section 3 for details.

For each affine map $C_a : x \mapsto Qx + a$, $a \in L$, we define the affine mapping C_a^* on H by $y \mapsto Ay + a^*$. In this way, we arrive at admissible families of contractions F_{ij}^* on H (see Section 1). We let $\mathcal{F}_{ij}$ be the closure of F_{ij}^* in the space $C(H,H)$ of continuous mappings on H. If we identify the mappings in F_{ij} with their translational parts in L, then $F_{ij}^{\prime *}$ is viewed as a subset of L^* and $\mathcal{F}_{ij}'$ is the closure of this in H (see Section 3.2 where we did a similar thing). Let us summarize our notation in the following diagram, where $*$ stands for the $*$-map and $'$ for the mapping that links affine transformations with their translational parts.

$$\begin{array}{ccccccc} F_{ij} & \xleftrightarrow{*} & F_{ij}^* & \subset & \mathcal{F}_{ij} & \subset & C(H,H) \\ {\scriptstyle '}\updownarrow & & {\scriptstyle '}\updownarrow & & {\scriptstyle '}\updownarrow & & \\ F_{ij}' & \xleftrightarrow{*} & F_{ij}'^* & \subset & \mathcal{F}_{ij}' & \subset & H. \end{array}$$

From (7.1), we have, for all $1 \leq i \leq n$,

(7.2) $$\Lambda_i^* = \bigcup_{j=1}^{n} \bigcup_{C \in F_{ij}} C^*(\Lambda_j^*)$$

and taking closures gives us

(7.3) $$W_i \supset \bigcup_{j=1}^{n} \bigcup_{C^* \in F_{ij}^*} C^*(W_j).$$

Since the W_i are compact and $C^*(W_j) = AW_j + a^* \subset W_i$ for $C = C_a$, we see that the translational parts F_{ij}' of the affine mappings are bounded (with respect to d). Thus we have that $\mathcal{F}_{ij}'$ is compact in H and $\mathcal{F}_{ij}$ is compact in $C(H,H)$.

PROPOSITION 7.1. *Under the above conditions, we have*

(7.4) $$W_i = \bigcup_{j=1}^{n} \bigcup_{D \in \mathcal{F}_{ij}} D(W_j), \quad i = 1, \ldots, n.$$

and $W_1 \times \cdots \times W_n$ is the attractor of $\mathcal{F}$.

To prove this result, we first establish

LEMMA 7.2. *Let F be a relatively compact set of continuous mappings from H to H. Suppose that U, V are compact subsets of H such that $C(U) \subset V$ for all $C \in F$. Then, $D(U) \subset V$ for all $D \in \overline{F}$.*

PROOF. Let $D \in \overline{F}$. Fix any $\varepsilon > 0$ and let $K = K(U, B_\varepsilon(0))$ be the set of all continuous mappings of H to itself that map U inside $B_\varepsilon(0)$. This is an open neighborhood of 0 in $C(H,H)$, so $D - C \in K$ for some $C \in F$. Thus

$$D(U) \subset C(U) + B_\varepsilon(0) \subset V + B_\varepsilon(0) \subset [V]_\varepsilon.$$

This being true for all $\varepsilon > 0$, we have $D(U) \subset V$. □

PROOF OF PROPOSITION 7.1. Consider (7.3). Using Lemma 7.2, we get

$$W_i \supset \bigcup_{j=1}^{n} \bigcup_{D \in \mathcal{F}_{ij}} D(W_j).$$

The right hand side is compact by (1.10) and contains Λ_i^* by (7.2), hence also $\overline{\Lambda_i^*} = W_i$. □

REMARK. A solution to (7.4) is guaranteed by the general theory of contractions of Section 4. However, in the present situation, we know more: the W_i have nonempty interiors, since the Λ_i are model sets. In general, it is hard to know when such a self-similar system of mappings leads to an attractor with nonempty interior.

7.2. Self-Similar Measures. We now assume that each $\mathcal{F}_{ij}$ is equipped with a positive regular Borel measure σ_{ij}, with $\sigma_{ij} := 0$ if $\mathcal{F}_{ij} = \varnothing$ as before. As in Proposition 5.1, we usually identify σ_{ij} with a positive Borel measure on H that is supported on $\mathcal{F}'_{ij}$. We set $s_{ij} = \sigma_{ij}(\mathcal{F}'_{ij}) = \sigma_{ij}(H)$ and restate the compatibility assumption **CA** for the matrix $\boldsymbol{s} = (s_{ij})$, namely that it has a positive 1-eigenvector:

SS6. There is a positive vector $\boldsymbol{m} = (m_1, \ldots, m_n)^T$ which satisfies $\boldsymbol{sm} = \boldsymbol{m}$.

By Proposition 5.1, we have the existence of a positive measure $\omega_1 \otimes \cdots \otimes \omega_n$, supported on $W_1 \times \cdots \times W_n$, which satisfies

$$(7.5) \qquad \omega_i = \sum_{j=1}^{n} \int_{\mathcal{F}_{ij}} (D.\omega_j) \, d\sigma_{ij}(D),$$

with $\omega_i(H) = m_i$, for all $i = 1, \ldots, n$, and which is explicitly given by the infinite product formula in Proposition 5.1. The next task is to convert (7.5) into a statement about densities on the model sets $\Lambda_1, \ldots, \Lambda_n$. Our basic assumption is

SS7. Each ω_i is *continuously representable* in the sense that, for $i = 1, \ldots, n$,

$$(7.6) \qquad \omega_i = g_i \theta$$

where $g_i \geq 0$ is a function on H which is supported on W_i and has the property that its restriction to W_i is continuous on W_i.

Given (7.6), we define the corresponding weights or *densities*

$$p_i \colon L \to \mathbb{R}, \quad p_i(x) = g_i(x^*), \quad i = 1, \ldots, n.$$

Since g_i is supported on W_i, p_i is supported on $\Lambda_i = \{x \in L \mid x^* \in W_i\}$. Our assumption that each Λ_i is regular, i.e., that each W_i is Riemann measurable, allows us to apply Weyl's Theorem (Theorem 6.2) to prove the existence of the average density for each p_i:

$$(7.7) \qquad \lim_{r \to \infty} \frac{1}{\mathrm{vol}(B_r(a))} \sum_{x \in L \cap B_r(a)} p_i(x) = \frac{1}{|\tilde{L}|} \int_H g_i \, d\theta = \frac{m_i}{|\tilde{L}|}$$

where the convergence of the limit is uniform in a.

For any affine mapping $D \colon x \mapsto Ax + a$ and any $h \in L^1(H)$, we have

$$D.(h\theta) = \alpha(D.h)\theta$$

as easily follows by applying both sides to a test function and then using the formula $d\theta(A^{-1}y) = \alpha \, d\theta(y)$, see Proposition 3.2(1). Plugging this into (7.5) gives

$$g_i = \alpha \sum_{j=1}^{n} \int_{\mathcal{F}_{ij}} (D.g_j) \, d\sigma_{ij}(D)$$

and then, for $i = 1, \ldots, n$,

$$(7.8) \qquad g_i(x) = \alpha \sum_{j=1}^{n} \int_{\mathcal{F}'_{ij}} g_j\bigl(A^{-1}(x-u)\bigr) \, d\sigma_{ij}(u).$$

To pass this to the physical side, we need to be able to deal with the integral. There are two situations in which we know how to do this, namely when the F_{ij} are finite and we basically use counting measures on the $\mathcal{F}'_{ij}$, and when the sets $\mathcal{F}'_{ij}$ are Riemann measurable subsets of H and the measures σ_{ij} are basically restrictions of the Haar measure on H. Let us now discuss these cases.

7.2.1. *F_{ij} Finite, σ_{ij} Counting Measure.* If each set F_{ij} is finite, then so is F^*_{ij} and $\mathcal{F}_{ij} = \overline{F^*_{ij}} = F^*_{ij}$. The model sets Λ_i are linked by a finite collection of finite unions as follows,

$$(7.9) \qquad \Lambda_i = \bigcup_{j=1}^{n} \bigcup_{k=1}^{N_{ij}} (Q\Lambda_j + a_{ijk}).$$

We suppose that σ_{ij} is counting measure normalized to total measure s_{ij} satisfying SS2. Then

$$(7.10) \qquad g_i = \alpha \sum_{j=1}^{n} \frac{s_{ij}}{\operatorname{card}(F_{ij})} \sum_{a \in F'_{ij}} C^*_a \cdot g_j$$

$$(7.11) \qquad p_i(x) = \alpha \sum_{j=1}^{n} \frac{s_{ij}}{\operatorname{card}(F_{ij})} \sum_{a \in F'_{ij}} p_j\bigl(Q^{-1}(x-a)\bigr).$$

We do not know many conditions that guarantee the existence of the representing functions g_i.

However, suppose that the unions in (7.2) are *nonoverlapping*; more precisely, we assume that

NO1. F_{ij} is finite
NO2. $m_i = \theta(W_i)$
NO3. σ_{ij} is counting measure scaled by α^{-1}, i.e., $s_{ij} = \sigma_{ij}(F_{ij}) = \alpha^{-1} \operatorname{card}(F_{ij})$.
NO4. for each i, the sets $D(W_j)$ entering into the union in (7.4) intersect at most on sets of measure 0, i.e., they are *just touching*.

Taking measures in (7.4), we see how the compatibility condition SS3 fits in:

$$m_i = \sum_{j=1}^{n} \sum_{D \in \mathcal{F}_{ij}} \alpha^{-1} m_j = \alpha^{-1} \sum_{j=1}^{n} \operatorname{card}(F_{ij}) m_j = \sum_{j=1}^{n} s_{ij} m_j.$$

The self-similar measures ω_i are easy to find, $\omega_i = \mathbf{1}_{W_i} \theta$. In fact, the equation

$$\mathbf{1}_{W_i} = \sum_{j=1}^{n} \sum_{D \in \mathcal{F}_{ij}} \mathbf{1}_{D(W_j)} \quad \text{(a.e.)}$$

(which is what nonoverlapping means in (7.4)) is equivalent to

$$\mathbf{1}_{W_i} \theta = \alpha^{-1} \sum_{j=1}^{n} \sum_{D \in \mathcal{F}_{ij}} D.(\mathbf{1}_{W_j} \theta) = \sum_{j=1}^{n} \int_{\mathcal{F}_{ij}} D.(\mathbf{1}_{W_j} \theta) \, d\sigma_{ij}(D).$$

Consequently, the self-similar densities p_i are simply

$$p_i(x) = \begin{cases} 1 & \text{if } x \in \Lambda_i \\ 0 & \text{otherwise} \end{cases}$$

and (7.11), for $i = 1, \ldots, n$, reads as

(7.12) $$p_i(x) = \sum_{j=1}^{n} \sum_{a \in F'_{ij}} p_j(Q^{-1}(x-a)) \quad \text{(a.e.)}$$

where (a.e.) means that equality holds for $x \in L$, possibly up to a set of density 0.

Equation (7.12) is the point set analogue of the situation in an inflation tiling where the measure of the inflated tile is the sum of the measures of the tiles into which it decomposes. Here, density replaces measure and the "tiling" condition is effectively put on the attractor in our assumptions in equations (7.4).

Let us briefly mention that Lagarias and Wang [**LW**] have recently begun an investigation of multi-component point sets with self-similarities. Their paper is taken from the discrete and combinatorial point of view, but does address the issue of counting multiplicities due to overlapping in the substitution process, and hence implicitly the question of self-similar measures.

7.2.2. *$\mathcal{F}'_{ij}$ Riemann Measurable, σ_{ij} Haar Measure.* By definition,

$$F'_{ij} \subset \{a \in L \mid Q\Lambda_j + a \subset \Lambda_i\},$$

and so F'^{*}_{ij} is a subset of

(7.13) $$\mathcal{G}'_{ij} := \{b \in H \mid AW_j + b \subset W_i\}.$$

Since $\mathcal{G}'_{ij} = \bigcap_{w \in W_j}(W_i - Aw)$ is closed (and compact), we have $\mathcal{F}'_{ij} \subset \mathcal{G}'_{ij}$ for all i, j. Thus the $\mathcal{G}'_{ij}$ give us an upper bound on the $\mathcal{F}'_{ij}$ and, in any case,

$$F'_{ij} \subset \{a \in L \mid a^* \in \mathcal{G}'_{ij}\}.$$

The right hand side has the interesting property that, provided that $\mathcal{G}'_{ij}$ is equal to the closure of its interior, it constitutes a model set of the cut and project scheme (6.1). If, furthermore, the $\mathcal{G}'_{ij}$ are Riemann measurable, then we have access to Weyl's Theorem again. This suggests that we may use *all possible* self-similarities of a given collection of model sets, replacing the F'_{ij} by the sets

(7.14) $$G'_{ij} = \{a \in L \mid a^* \in \mathcal{G}'_{ij}\}.$$

Thus given regular model sets $\Lambda_i = \Lambda(W_i)$, $i = 1, \ldots, n$, and an inflation Q satisfying SS1–SS4 above, we can define $\mathcal{G}'_{ij}$ by (7.13) and replace the given system $\mathcal{F}$ of affine mappings by the new set $\mathcal{G}$ with

(7.15) $$G_{ij} := \{C : x \mapsto Qx + a \mid a \in L, a^* \in \mathcal{G}'_{ij}\}.$$

With this motivation in mind, we go back to our original setup with SS1–SS5. We now assume in addition that

SS8. each $\mathcal{F}'_{ij}$ is Riemann measurable and, if $\theta(\mathcal{F}'_{ij}) > 0$, we have

$$\sigma_{ij} = \frac{s_{ij}}{\theta(\mathcal{F}'_{ij})} \mathbf{1}_{\mathcal{F}'_{ij}} \theta,$$

i.e., σ_{ij} is Haar measure restricted to $\mathcal{F}'_{ij}$ and normalized to total measure s_{ij} where $\mathbf{s} = (s_{ij})$ is an arbitrary nonnegative matrix satisfying SS2. If $\theta(\mathcal{F}'_{ij}) = 0$, σ_{ij} is defined as the 0-measure.

We now apply the results of Section 5. The σ_{ij} are absolutely continuous with respect to Haar measure θ since $\sigma_{ij} = h_{ij}\theta$ where $h_{ij} = \frac{s_{ij}}{\theta(\mathcal{F}'_{ij})}\mathbf{1}_{\mathcal{F}'_{ij}}$ and the density is $h_{ij} \in L^1(H) \cap L^\infty(H)$.

Assuming that the convolution (5.5) converges and using Proposition 5.2, we obtain a family of *continuous* nonnegative functions $g_1, \ldots, g_n$ which satisfy the self-similarity equations

$$(7.16) \qquad g_i(x) = \alpha \sum_{j=1}^{n} w_{ij} \int_{\mathcal{F}_{ij}} g_j\big(A^{-1}(x-u)\big) \, d\theta(u)$$

where $w_{ij} := s_{ij}/\theta(\mathcal{F}'_{ij})$. Applying Theorem 6.2 we obtain

THEOREM 7.3. *Let $\Lambda_1, \ldots, \Lambda_n$ be a self-similar system of model sets which satisfy the assumptions SS1–SS6 and SS8. Then there exist nonnegative functions $p_i \colon L \to \mathbb{R}$, supported on Λ_i, $i = 1, \ldots, n$, with the following properties*

$$(7.17) \qquad m_i = \lim_{r\to\infty} \frac{|\tilde{L}|}{\mathrm{vol}(B_r(a))} \sum_{x \in L \cap B_r(a)} p_i(x),$$

$$(7.18) \qquad p_i(x) = \lim_{r\to\infty} \frac{\alpha\,|\tilde{L}|}{\mathrm{vol}(B_r(a))} \sum_{j=1}^{n} w_{ij} \sum_{x \in L \cap B_r(a)} p_j\big(Q^{-1}(x-u)\big),$$

for all $x \in L$ and for all $i = 1, \ldots, n$, where the limits are uniform in $a \in \mathbb{R}^m$. □

This may be compared with the similar formula that we derived in [**BM2**]. There, $\tilde{Q}$ was assumed to be an automorphism of $\tilde{L}$, and the set of Q-affine mappings was the set of all possible mappings. There, however, the scaling constants ν^{ij} (which are our s_{ij} here) had no general interpretation.

REMARK. For simplicity, the assumptions made in SS8 are actually a little stronger than necessary—there is no general need to exclude the case that some $\mathcal{F}'_{ij}$ are singleton sets, but still carry a positive (point) measure. Though this is then not absolutely continuous, it is 'harmless' in the convolution process because the convolution of a function with a unit point mass only results in a shift of the function. We will meet this case in Section 8.1.4 below.

8. Concrete Examples

A number of explicit examples have been presented earlier, in [**BM1**] and [**BM2**]. The case of the planar Penrose pattern in its rhombic version is an example of a multi-component model set, because the vertex points fall into 4 classes. This was described in detail in [**BM2**] and will not be repeated here. Instead, let us look into a few other examples. First, we describe the silver mean chain in one dimension and look at it in different ways, both as a single and as a multi-component model set. Next, we briefly describe the Ammann-Beenker pattern [**AGS**] in the plane, an eightfold symmetric relative of the Penrose pattern. This example appears also in other contributions to this volume. Finally, we look at a more unusual example that involves the 3-adic numbers.

8.1. The Silver Mean Chain.
The silver mean chain is a 2-sided sequence on the alphabet $\{a, b\}$. It can generated by iterating the substitution

$$a \mapsto aba, \quad b \mapsto a$$

which, when starting from $a|a$, leads to the palindromic fixed point

(8.1) $\qquad \ldots abaaabaabaabaaaba|abaaabaabaabaaaba\ldots$

where $|$ simply marks the center. With a and b interpreted as intervals ($=$ tiles) of length $\alpha := 1 + \sqrt{2}$ and 1 respectively, this gives rise to a tiling of the line, see [**HRB**] for details of its structure. The number α is called the *silver mean*.

Let Λ_1 (resp. Λ_2) denote the coordinates of the left end points of the tiles of type a (resp. b), assuming that the initial block $a|a$ was centered at 0, i.e., its left end points are located at $-\alpha$ and 0. Then Λ_1, Λ_2, and $\Lambda := \Lambda_1 \cup \Lambda_2$ are model sets for the following cut and project scheme:

(8.2)
$$\begin{array}{ccccc} \mathbb{R} & \xleftarrow{\pi} & \mathbb{R} \times \mathbb{R} & \xrightarrow{\pi_\mathbb{R}} & \mathbb{R} \\ \cup & & \cup & & \cup \\ \mathbb{Z}[\sqrt{2}] & \longleftarrow & \tilde{L} & \longrightarrow & \mathbb{Z}[\sqrt{2}] \end{array}$$

where $\mathbb{Z}[\sqrt{2}] = \mathbb{Z} \oplus \mathbb{Z}\sqrt{2}$ is the ring of integers in the quadratic field $\mathbb{Q}(\sqrt{2})$, the $*$-map from $\mathbb{Z}[\sqrt{2}]$ to $\mathbb{Z}[\sqrt{2}]$ is the algebraic conjugation defined by $\sqrt{2} \mapsto -\sqrt{2}$, and the lattice is $\tilde{L} := \{(x, x^*) \mid x \in \mathbb{Z}[\sqrt{2}]\}$. Explicitly, we obtain

(8.3)
$$\begin{aligned} \Lambda_1 &= \{x \in \mathbb{Z}[\sqrt{2}] \mid x^* \in W_1\} \\ \Lambda_2 &= \{x \in \mathbb{Z}[\sqrt{2}] \mid x^* \in W_2\} \\ \Lambda &= \{x \in \mathbb{Z}[\sqrt{2}] \mid x^* \in W\} \end{aligned}$$

where the corresponding windows are intervals, namely

(8.4) $\quad W_1 := \left[\frac{1}{\sqrt{2}} - 1, \frac{1}{\sqrt{2}}\right], \quad W_2 := \left[-\frac{1}{\sqrt{2}}, \frac{1}{\sqrt{2}} - 1\right], \quad W := \left[-\frac{1}{\sqrt{2}}, \frac{1}{\sqrt{2}}\right],$

where $W = W_1 \cup W_2$ and $W_1 \cap W_2 = \{\alpha^*/\sqrt{2}\}$.

This can be verified in the following way. Let Q be the linear mapping which is scalar multiplication by $\alpha = 1 + \sqrt{2}$ and let A be its conjugate map, which is scalar multiplication by $\alpha^* = -\alpha^{-1} = 1 - \sqrt{2}$. With the explicit coordinatization given above, the substitution rules say that

(8.5)
$$\begin{aligned} \Lambda_1 &= Q\Lambda_1 \cup Q\Lambda_2 \cup (Q\Lambda_1 + \alpha + 1) \\ \Lambda_2 &= Q\Lambda_1 + \alpha \end{aligned}$$

so that we now have a system of Q-inflations which, in the notation of SS4, is defined by

(8.6)
$$F' = (F'_{ij}) = \begin{pmatrix} \{0, \alpha+1\} & \{0\} \\ \{\alpha\} & \varnothing \end{pmatrix}.$$

The corresponding contractions satisfy:

(8.7)
$$\begin{aligned} W_1 &= AW_1 \cup AW_2 \cup (AW_1 + \alpha^* + 1) \\ W_2 &= AW_1 + \alpha^* \end{aligned}$$

which shows that $W_1 \times W_2$ is the attractor for the system of A-affine contractions given by F'^*.

Since the generators of Λ (which correspond to the a-tiles with coordinates 0 and $-\alpha$) are mapped into W_1 by $*$, all subsequent points generated from them are $*$-mapped into W. Thus the model set $\Lambda(W)$ assuredly contains our set Λ. On the other hand, it is easy to see that the minimum separations between the points of $\Lambda(W)$ are 1 and $1+\sqrt{2}$, and no point can be added to Λ without violating this. In short, $\Lambda = \Lambda(W)$, $\Lambda_1 = \Lambda(W_1)$, $\Lambda_2 = \Lambda(W_2)$.

We now examine this situation in four different ways, namely as single component and multi-component case, and each then with minimal and maximal families of affine contractions. By this we mean either the case when the window system is minimally generated by affine contractions or when we use the entire set of affine contractions available.

We content ourselves with a few remarks about each case and one figure illustrating the continuous case. We indicate the appropriate Sections of the paper as we go along and freely use the notation from these Sections. Note that $\alpha = 1 + \sqrt{2}$ as we have defined it in this Section is the modulus of the contraction A, in keeping with previous notation.

8.1.1. *The Single Model Set $\Lambda = \Lambda(W)$ with F Minimal* (see Section 3.2). The contractivity factor is α^*. Since $|\alpha^*| < \frac{1}{2}$, we need at least three affine mappings to get the full window as the attractor—we would end up with a Cantor subset of W if we would start with only two. One possible choice is

$$W = (AW + \alpha^*) \cup AW \cup (AW - \alpha^*)$$

and our family of mappings is then $F_H = F_\mathbb{R} = \{\alpha^*, 0, -\alpha^*\}$. We take the simplest of all probability measures on $F_\mathbb{R}$, i.e., counting measure,

$$\nu = \tfrac{1}{3}(\delta_{\alpha^*} + \delta_0 + \delta_{-\alpha^*}),$$

so that this is an example of a finite IFS. The corresponding invariant measure on W and its Fourier transform are given by

(8.8)
$$\omega = \underset{\ell=1}{\overset{\infty}{\Huge *}} \tfrac{1}{3}\big(\delta_{(\alpha^*)^\ell} + \delta_0 + \delta_{-(\alpha^*)^\ell}\big)$$
$$\widehat{\omega} = \prod_{\ell=1}^{\infty} \tfrac{1}{3}\big(1 + 2\cos(2\pi i(\alpha^*)^\ell k)\big).$$

This measure is similar to those studied in the context of the binary addressing problem, see [**So**] and references therein. Although most of them are absolutely continuous if the IFS covers the full interval (which is does here), exceptions emerge, see [**BoGi**, Thm. 4], if the scaling factor of the IFS is the inverse of a Pisot-Vijayaraghavan number (which is the case here, too). These exceptional self-similar measures will then be purely singular continuous. Note that this might be very difficult to detect if one is not aware of it—the fractal dimension of such a measure can be tantalizingly close to 1, see [**Lal**, §8]. In any case, we cannot pull back such a measure to the physical side, and thus do not gain much insight into the structure of our model set from it.

8.1.2. *The Single Model Set $\Lambda = \Lambda(W)$ with F Maximal* (see Section 3.3). Here, we start with the observation

$$W = \bigcup_{u \in [\alpha^*, -\alpha^*]} AW + u.$$

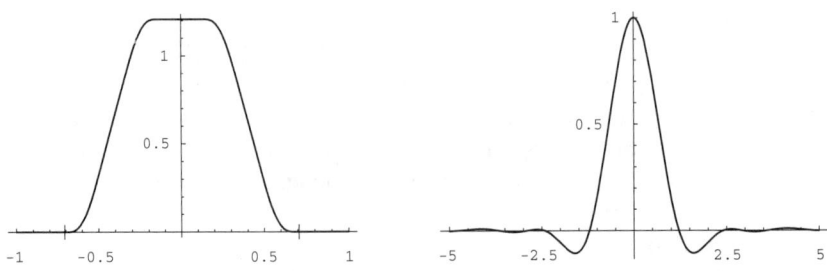

FIGURE 1. Invariant density (left) and its Fourier transform (right) for the silver mean chain. The support of the C^∞-density g of (8.9) is the window W of (8.4), indicated by extra markers.

We now use the complete set of available self-similarities $\mathcal{F}' = [\alpha^*, -\alpha^*]$. For our pre-assigned probability measure ν, we choose

$$\nu = \frac{\mathbf{1}_{[\alpha^*, -\alpha^*]}}{2|\alpha^*|}\theta = \frac{\mathbf{1}_{\mathcal{F}'}}{2|\alpha^*|}\theta$$

where θ is Lebesgue measure on $\mathbb{R}$. Then we are the situation of Proposition 3.6:

(8.9)
$$g = \mathop{\mbox{\Large$*$}}_{\ell=0}^{\infty}\left(\frac{\mathbf{1}_{|\alpha^*|^\ell \mathcal{F}'}}{2|\alpha^*|^{\ell+1}}\right)$$
$$\hat{g}(k) = \prod_{\ell=1}^{\infty}\left(\frac{\sin(2\pi(\alpha^*)^\ell k)}{2\pi(\alpha^*)^\ell k}\right).$$

This is illustrated in Figure 1.

8.1.3. *The System of Model Sets $\{\Lambda_1, \Lambda_2\}$ with Windows $\{W_1, W_2\}$, with F Minimal* (see Section 7.2, nonoverlapping case). We consider $\Lambda = \Lambda_1 \cup \Lambda_2$ as a multi-component model set through (8.5) and (8.6) with the windows W_1 and W_2 given by (8.4) and related by (8.7). The nonoverlapping conditions NO1 and NO4 hold. To obtain NO2 and NO3, we define $\boldsymbol{m} = (1, |\alpha^*|)^T$ and (since $|\alpha^*| = \alpha - 2 = 1/\alpha$)

(8.10)
$$\boldsymbol{s} = \begin{pmatrix} 2/\alpha & 1/\alpha \\ 1/\alpha & 0 \end{pmatrix},$$

which is clearly primitive. The self-similar measures are $\omega_1 = \mathbf{1}_{W_1}\theta$ and $\omega_2 = \mathbf{1}_{W_2}\theta$, and the corresponding densities on the physical side are p_1 and p_2, which are the characteristic functions (defined on $\mathbb{Z}[\sqrt{2}]$) of Λ_1 and Λ_2, respectively.

8.1.4. *The System of Model Sets $\{\Lambda_1, \Lambda_2\}$ with Windows $\{W_1, W_2\}$, with F Maximal* (see Section 7.2, Riemann measurable case). We continue with the multi-component picture of Section 8.1.3, but now we use all available α-affine self-similarities. These are easily computed from (8.4)) and (8.7), resulting in

$$\mathcal{F}' = \begin{pmatrix} [0, 1+\alpha^*] & [\alpha^*, -\alpha^*] \\ \{\alpha^*\} & \varnothing \end{pmatrix}.$$

The sets appearing here are all measurable. Let us fix the same vector $\boldsymbol{m}$ and matrix $\boldsymbol{s}$ as in Section 8.1.3. If we use Lebesgue measure θ on $\mathbb{R}$, we obtain SS7 in

the form
$$\sigma = \begin{pmatrix} (1-\alpha^*)\mathbf{1}_{[0,1+\alpha^*]}\theta & \frac{1}{2}\mathbf{1}_{[\alpha^*,-\alpha^*]}\theta \\ -\alpha^*\delta_{\alpha^*} & 0 \end{pmatrix}.$$

Using Proposition 5.1, we have the self-similar measure

(8.11) $$\omega = \underset{\ell=0}{\overset{\infty}{\LARGE *}}\, \alpha^\ell \begin{pmatrix} (1-\alpha^*)\mathbf{1}_{(\alpha^*)^\ell[0,1+\alpha^*]}\theta & \frac{1}{2}\mathbf{1}_{(\alpha^*)^\ell[\alpha^*,-\alpha^*]}\theta \\ -\alpha^*\delta_{(\alpha^*)^{\ell+1}} & 0 \end{pmatrix} * \begin{pmatrix} \delta_0 \\ |\alpha^*|\delta_0 \end{pmatrix}.$$

The solution here is mildly different from the one appearing in SS8 because of the appearance of the point measure, compare the Remark following Theorem 7.3. However, the convolution of a delta and a function simply translates the function. Furthermore, the convolution of two functions f and g is point-wise bounded by $\|f\|_\infty \|g\|_1$. Thus the partial convolution products in $\omega^{(n)}$ of ω in (8.11), for all $n \geq 1$, are *functions*
$$\begin{pmatrix} f_{11}^{(n)} & f_{12}^{(n)} \\ f_{21}^{(n)} & f_{22}^{(n)} \end{pmatrix}$$
where the $f_{ij}^{(n)}$ are supported on W_j, are uniformly bounded, and are increasingly differentiable as $n \to \infty$. Alternatively, one could also work with the square of the averaging operator here, which would match SS8 from the beginning.

The resulting self-similar measure is represented by two C^∞ functions, g_1 and g_2. They are of the kind shown on the left of Figure 1, but now supported on W_1 and W_2, respectively, and with total mass 1 and $|\alpha^*|$, in accordance with $\boldsymbol{m}$.

8.2. The Ammann-Beenker Model Set. This relative of the rhombic Penrose tiling is usually described as a model set obtained from the primitive cubic lattice $\mathbb{Z}^4$ in 4-space, see [**BJ**] and references therein for details. Here, we prefer the number theoretic approach given in [**HRB**], which is more compatible with the above description of the silver mean chain. With $\xi := e^{2\pi i/8}$, we use the following cut and project scheme

(8.12)
$$\begin{array}{ccccc} \mathbb{R}^2 & \xleftarrow{\pi} & \mathbb{R}^2 \times \mathbb{R}^2 & \xrightarrow{\pi_{\mathbb{R}^2}} & \mathbb{R}^2 \\ \cup & & \cup & & \cup \\ \mathbb{Z}[\xi] & \longleftarrow & \tilde{L} & \longrightarrow & \mathbb{Z}[\xi] \end{array}$$

where $\mathbb{Z}[\xi]$ is the ring of cyclotomic integers generated by the primitive 8th roots of unity. It is the maximal order in the cyclotomic field $\mathbb{Q}(\xi)$. Then, the lattice is
$$\tilde{L} = \{(x, x^*) \mid x \in \mathbb{Z}[\xi]\}$$
where the $*$-map is given by algebraic conjugation $\xi \mapsto \xi^3$.

In this setting, the window W of the Ammann-Beenker model set $\Lambda = \Lambda(W)$ is simply a regular octagon of edge length 1, centered at the origin. Its area is 2α, with $\alpha = 1 + \sqrt{2}$ as above. With this choice, the model set Λ is both *regular* (W is a polytope, hence Riemann measurable) and *generic* ($L = \mathbb{Z}[\xi]$ does not intersect ∂W). A symmetric patch and its lift to internal space is shown in Figure 2.

Let us now consider $\Lambda = \Lambda(W)$ as a single model set and let us determine the invariant density on W that results from the set of *all* self-similarities of the form $\Lambda \mapsto \alpha\Lambda + v \subset \Lambda$ (with $v \in \mathbb{Z}[\xi]$). So, Q is again multiplication by α, and A multiplication by α^*, a contraction. It then follows that
$$\mathcal{F}' = \{u \mid \alpha^* W + u \subset W\} = (2 - \sqrt{2})W,$$

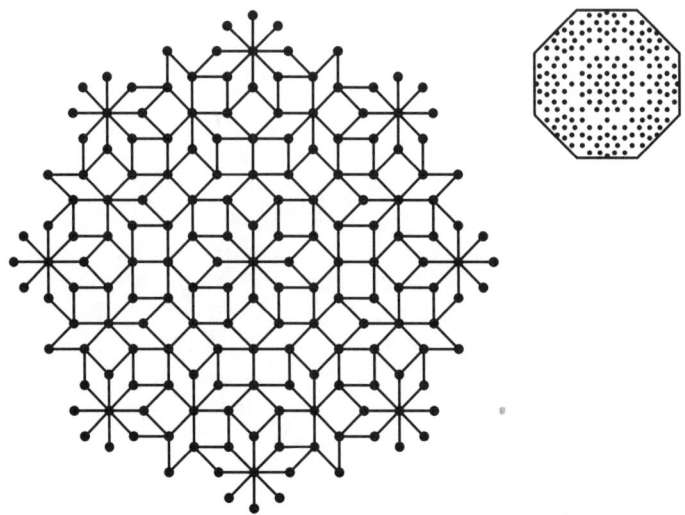

FIGURE 2. The Ammann-Beenker model set, seen as the vertex set of a tiling with squares and rhombi. The window in internal space is a regular octagon. It is shown to the right with the $*$-images of the vertex points from the patch to the left.

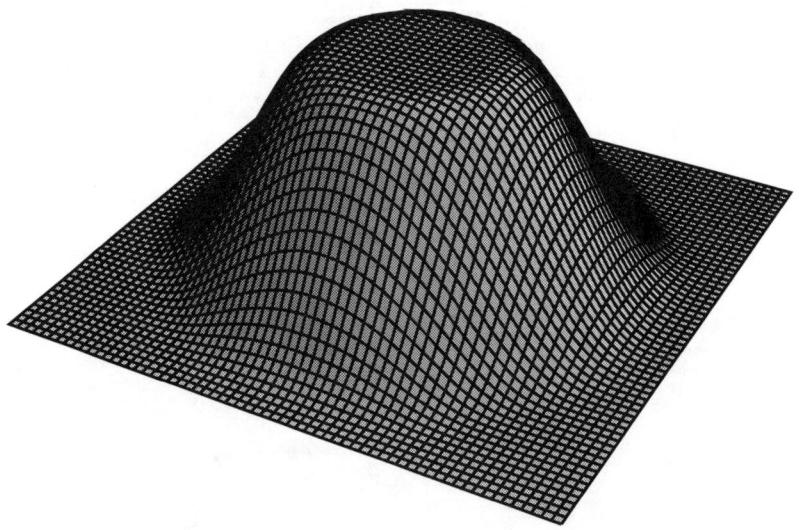

FIGURE 3. The self-similar density g for the Ammann-Beenker model set, according to (8.13). The support of the C^∞-function g is a regular octagon, the window W in the upper right corner of Figure 2.

i.e., $\mathcal{F}' \subset \mathbb{R}^2$ is another octagon centered at the origin, but with reduced edge length $|\alpha^*|\sqrt{2} = 2 - \sqrt{2}$. Consequently, $\mathcal{F}'$ has area $4|\alpha^*|$. For the a priori probability measure on $\mathcal{F}'$, we choose

$$\nu = \frac{\mathbf{1}_{\mathcal{F}'}}{4|\alpha^*|}\theta$$

where θ is now Lebesgue measure on $\mathbb{R}^2$. Proposition 3.6 then gives

(8.13) $$g = \overset{\infty}{\underset{\ell=0}{*}} \left(\frac{\mathbf{1}_{|\alpha^*|^\ell \mathcal{F}'}}{4|\alpha^*|^{2\ell+1}} \right).$$

The invariant density is a C^∞-function supported on the window W, see Figure 3. The deviations from circular symmetry are rather faint. However, in contrast to the Penrose case investigated in [**BM2**], it has a central plateau and then rolls off smoothly towards the boundary of W. The same phenomenon is actually visible for the invariant density of the silver mean chain in Figure 1, in contrast to the one for the Fibonacci chain in [**BM1**]. It is due to the larger absolute value of α in comparison to the golden ratio $\tau = (1 + \sqrt{5})/2$ which appears there. This is interesting in relation to the rather widespread experimental finding that "real world" quasicrystals are to be described by window functions with a smooth roll-off. Although this is usually explained as a random tiling effect, our above examples show that other mechanisms are possible as well, and the attractive feature is then that they result from some residual (or statistical) inflation symmetry.

8.3. A 3-adic Example. In this final example, which will be fully developed along with some other p-adic examples in [**HM**], we indicate what invariant measures can look like in a very different situation. The result is quite surprising.

This time, we begin with the 2-sided chain on the ternary alphabet $\{a, b, c\}$, generated by the substitution rule

$$a \mapsto ab \quad , \quad b \mapsto abc \quad , \quad c \mapsto abcc.$$

Starting from $c|a$, iteration leads to the 2-sided fixed point

(8.14) $\quad \ldots babcabccababcabccabcc|ababcababcabccababcababc \ldots$

With a, b, c assigned intervals of length 1, 2, 3 respectively, which is the natural geometric realization here, this again gives rise to a tiling of the line. This system was studied in [**BMS**] where it was shown that, when it is coordinatized, the resulting sets of points $\Lambda_1, \Lambda_2, \Lambda_3$ are model sets based on the 3-adic integers $\widehat{\mathbb{Z}}_3$. The coordinatization starting at 0 and using *right* end points is

(8.15) $$\Lambda_1 = \bigcup_{k=2}^{\infty} \left(\left(\sum_{i=0}^{k-2} 3^i \right) + 3^k \mathbb{Z} \right); \quad \Lambda_2 = \bigcup_{k=2}^{\infty} \left(\left(2 + \sum_{i=0}^{k-2} 3^i \right) + 3^k \mathbb{Z} \right);$$

$$\Lambda_3 = 3^2 \mathbb{Z} \cup \left(\bigcup_{k=3}^{\infty} \left(\left(-\sum_{i=1}^{k-2} 3^i \right) + 3^k \mathbb{Z} \right) \right).$$

The corresponding model sets are

$$\Lambda_i = \{x \in \mathbb{Z} \mid x \in W_i\}$$

where the windows are given by

$$W_1 = \bigcup_{k=2}^{\infty}\left(\left(\sum_{i=0}^{k-2} 3^i\right) + 3^k\widehat{\mathbb{Z}}_3\right); \quad W_2 = \bigcup_{k=2}^{\infty}\left(\left(2 + \sum_{i=0}^{k-2} 3^i\right) + 3^k\widehat{\mathbb{Z}}_3\right);$$
(8.16)
$$W_3 = 3^2\widehat{\mathbb{Z}}_3 \cup \left(\bigcup_{k=3}^{\infty}\left(\left(-\sum_{i=1}^{k-2} 3^i\right) + 3^k\widehat{\mathbb{Z}}_3\right)\right).$$

We consider this as a multi-component model set and the basic inflationary maps as affine mappings of the form $x \mapsto 3x + u$, $u \in \mathbb{Z}$. These maps transfer over to the internal side without any symbolic change, but there they are contractions with respect to the standard 3-adic metric topology. The most interesting case seems to be when we allow all possible mappings of this type between the various windows (this corresponds to the case of the silver mean example in Section 8.1.4). These are quite straightforward to work out and give rise to a 3×3 matrix $\mathcal{F}$. The invariant measures on $\widehat{\mathbb{Z}}_3$ are then given by the matrix equation

(8.17)
$$\omega = \left(\underset{\ell=0}{\overset{\infty}{*}} Y^{(\ell)}\right) * \delta^m$$

where

(8.18)
$$Y^{(\ell)} := \left(\frac{s_{ij}}{\theta(3^\ell \mathcal{F}_{ij})} \mathbf{1}_{3^\ell \mathcal{F}_{ij}}\right)_{1 \leq i,j \leq 3},$$

θ is the Haar measure on $\widehat{\mathbb{Z}}_3$ normalized to total measure 1, and $\boldsymbol{sm} = \boldsymbol{m}$.

Remarkably, this convolution can be explicitly computed and yields

(8.19)
$$\omega = 3^2 \begin{pmatrix} \mathbf{1}_{3^2\widehat{\mathbb{Z}}_3+1} m_1 \theta \\ \mathbf{1}_{3^2\widehat{\mathbb{Z}}_3+3} m_2 \theta \\ \mathbf{1}_{3^2\widehat{\mathbb{Z}}_3} m_3 \theta \end{pmatrix}.$$

In particular, the self-similar measures are absolutely continuous and give rise to the following self-similar densities on our three model sets $\Lambda_{\{1,2,3\}}$ in $\mathbb{Z}$:

$$p_{\{1,2,3\}}(\ell) = \begin{cases} 9\, m_{\{1,2,3\}} & \text{if } \ell \equiv \{1,3,0\} \bmod 9, \\ 0 & \text{otherwise.} \end{cases}$$

Thus, the self-similar densities are *periodic* although the point sets Λ_i are aperiodic. In fact, the support of the densities consists of three different cosets of $9\mathbb{Z}$, each of which lies entirely inside one of the three point sets Λ_i, $i = 1, 2, 3$. We do not know many other mechanisms that result in periodic states, although it is not clear whether this inflation induced periodicity could be interpreted physically.

Appendix A

For the sake of completeness, we collect a number of results in this appendix. Let us first show that real (complex) Lipschitz functions are dense in $C(X, \mathbb{R})$ (in $C(X, \mathbb{C})$) if X a compact metric space. We first need the following result which is essentially stated in [**Bou**, Prop. IX.2.2.3].

LEMMA A.1. *Let X be an arbitrary metric space. For every nonempty $A \subset X$, the function $\phi \colon x \mapsto d(x, A)$ is in $\mathrm{Lip}(\leq 1, X, \mathbb{R})$. The Lipschitz constant is $r_\phi = 0$ iff $\phi \equiv 0$ on X, and $r_\phi = 1$ otherwise.*

PROOF. Recall that $d(x, A) = \inf\{d(x, z) \mid z \in A\}$. Let $x, y \in X$. Fix $\varepsilon > 0$ and choose $z \in A$ such that $d(y, z) \leq d(y, A) + \varepsilon$. Then

$$d(x, A) - d(y, A) \leq d(x, A) - d(y, z) + \varepsilon \leq d(x, z) - d(y, z) + \varepsilon \leq d(x, y) + \varepsilon.$$

Now, since $\varepsilon > 0$ was arbitrary and the argument is essentially symmetric in x and y, we can conclude

$$|\phi(x) - \phi(y)| = |d(x, A) - d(y, A)| \leq d(x, y).$$

This means that $\phi : x \mapsto d(x, A)$ is Lipschitz with constant $r_\phi \leq 1$.

If $\phi \equiv 0$, we trivially have $r_\phi = 0$. Otherwise, there exists an $x \in X \setminus A$ with $\phi(x) = d(x, A) > 0$. We know from above that $r_\phi \leq 1$. Fix $\varepsilon > 0$ and $y \in A$ with $d(x, y) \leq d(x, A) + \varepsilon$. Then, we have $d(y, A) = 0$ and get

$$0 < d(x, y) \leq d(x, A) + \varepsilon = d(x, A) - d(y, A) + \varepsilon \leq r_\phi d(x, y) + \varepsilon.$$

Since $\varepsilon > 0$ was arbitrary, this is only possible with $r_\phi \geq 1$, hence $r_\phi = 1$. □

LEMMA A.2. *If X is a compact metric space, the real (complex) Lipschitz functions are dense in $C(X, \mathbb{R})$ (in $C(X, \mathbb{C})$).*

PROOF. This is a straight-forward application of the Stone-Weierstrass theorems. The real Lipschitz functions form a subalgebra of $C(X, \mathbb{R})$ under pointwise addition and multiplication because $r_{\phi+\psi} \leq r_\phi + r_\psi$ and $r_{\phi\psi} \leq \|\phi\|_\infty r_\psi + \|\psi\|_\infty r_\phi$. Furthermore, the constant functions are Lipschitz, and the separation property follows from Lemma A.1. So, the real variant, see [**RS**, Thm. IV.9], gives the one claim, while the complex variant, [**RS**, Thm. IV.10], gives the other. In the latter case, the additional requirement that the algebra of complex Lipschitz functions is closed under complex conjugation is obvious. □

We also need a number of convergence results, which we collect here. The following standard result is a special case of [**Q**, Thm. 14.22 and Cor. 14.23] or [**Kel**, Thms. 7.14 and 7.15], see also [**Q**, A 14.8].

LEMMA A.3. *Let X be a locally compact space and $\{g_n\}$ an equi-continuous sequence of functions in $C(X, \mathbb{C})$ which converges pointwise. Then the limit is a continuous function, $g \in C(X, \mathbb{C})$, and the convergence is uniform on compact subsets of X (compact convergence).* □

LEMMA A.4. *Let W be a compact subset of an LCAG H, with dual group $\widehat{H}$. Let $(\mu_n)_{n \in \mathbb{N}}$ be a vaguely convergent sequence of probability measures in $\mathcal{P}(W)$, with limit μ. If $K \subset \widehat{H}$ is compact, the family $\{\widehat{\mu}_n\}$ of functions from $C(H, \mathbb{C})$ is equi-continuous (and even equi-uniformly continuous) on K.*

PROOF. Let $\varepsilon > 0$ and $V_\varepsilon := \{k \in \widehat{H} \mid \sup_{x \in W} |\langle k, x \rangle - 1| < \varepsilon\}$. Note that $0 \in V_\varepsilon$ (since the trivial character is $\langle 0, x \rangle \equiv 1$) and that V_ε is a neighborhood of 0 in $\widehat{H}$ because it is a typical open set in the compact-open topology of $\widehat{H} \subset C(H, \mathbb{C})$. Fix some $k_1 \in K$ and choose $k_2 \in K$ so that $k_1 - k_2 \in V_\varepsilon$. Then we have, for all $x \in W$,

$$|\overline{\langle k_1, x \rangle} - \overline{\langle k_2, x \rangle}| = |\langle k_1 - k_2, x \rangle - 1| < \varepsilon,$$

where we used $|\langle k_2, x \rangle| = 1$ and the multiplication rule for characters.

Let $\hat\mu_n$ be an arbitrary element of our sequence. We then get

$$\left|\hat\mu_n(k_1) - \hat\mu_n(k_2)\right| = \left|\int_H \left(\overline{\langle k_1, x\rangle} - \overline{\langle k_2, x\rangle}\right) \mathrm{d}\mu_n(x)\right|$$
$$\leq \int_W \left|\overline{\langle k_1, x\rangle} - \overline{\langle k_2, x\rangle}\right| \mathrm{d}\mu_n(x) < \varepsilon \|\mu_n\| = \varepsilon.$$

Since this is independent both of n and of $k_1 \in K$, equi-uniform continuity of $\{\hat\mu_n\}$ on K follows. $\square$

REMARK. With little extra complication, the result can be extended to vaguely convergent sequences of measures $\mu_n \in \mathcal{M}_+^m(H)$, see [**Bau**, Lemma 23.7] for a proof that can easily be adapted to this case.

THEOREM A.5 (Continuity theorem of P. Lévy). *Let W, H, $\widehat{H}$ be defined as in Lemma A.4. If $(\mu_n)_{n\in\mathbb{N}}$ is a sequence of measures in $\mathcal{P}(W)$ that vaguely converges to $\mu \in \mathcal{P}(W)$, then the corresponding sequence of Fourier-Stieltjes transforms, $(\hat\mu_n)_{n\in\mathbb{N}}$, converges compactly to $\hat\mu$.*

PROOF. Each of the functions $x \mapsto \langle k, x\rangle$, $k \in \widehat{H}$, lies in $C(H, \mathbb{C})$. Therefore, $\lim_{n\to\infty} \hat\mu_n(k) = \hat\mu(k)$ by the very definition of vague convergence. So, we have a sequence of pointwise converging functions in $C(H, \mathbb{C})$ that are equi-continuous on compact subsets $K \subset \widehat{H}$ due to Lemma A.4, hence the convergence of $(\hat\mu_n)_{n\in\mathbb{N}}$ is uniform on K by Lemma A.3 which proves the assertion. $\square$

REMARK. Once again, the restriction to $\mathcal{P}(W)$ is not essential, and the result holds for sequences in $\mathcal{M}_+^m(H)$ as well, compare [**Bau**, Thm. 23.8], but we do not need the stronger result here.

Finally, we formulate the approximation property that we need in Section 6 to prove Theorem 6.2.

LEMMA A.6. *Let H be an LCAG with Haar measure μ, and let W be a Riemann measurable compact subset of H. Then each $f \in C(H, \mathbb{C})$ with $\mathrm{supp}(f) \subset W$ can be uniformly approximated by a sequence of step functions $\psi^{(\ell)}$ of the form*

$$\psi^{(\ell)} = \sum_{i=1}^{n(\ell)} c_i^{(\ell)} \mathbf{1}_{U_i}$$

where $c_i^{(\ell)} \in \mathbb{C}$ and $W = \bigcup_{i=1}^{n(\ell)} U_i$ is a partition into pairwise disjoint sets $U_i \subset W$ that are all Riemann measurable.

PROOF. Although this is a standard type of result, we give an explicit proof because the additional condition of Riemann measurability of the partition requires some attention. It is sufficient to prove the statement for real functions because any f may be split into its real and imaginary parts, both of which are continuous and supported on W.

Assume f is real. For all $s \in \mathbb{R}$, $f^{-1}(s)$ is a closed subset of H, hence measurable. For $0 \neq s \neq t \neq 0$, $f^{-1}(s)$ and $f^{-1}(t)$ are disjoint subsets of W, and we thus have $\sum_{s\in\mathbb{R}\setminus\{0\}} \mu(f^{-1}(s)) \leq \mu(W) < \infty$. But this means, as usual, that

$$P := \{s \in \mathbb{R} \mid \mu(f^{-1}(s)) > 0\}$$

is at most a *countable* set, called the set of "bad" points. Since f is continuous, $f(W) \subset \mathbb{R}$ is compact, hence $f(W) \subset (a, b)$ for some $a, b \in \mathbb{R}$.

Fix some $\varepsilon > 0$ and choose an integer $n > (b-a)/\varepsilon$. We can then subdivide $[a,b)$ into nonempty intervals $I_1, \ldots, I_n$, each of the form $[\alpha_i, \beta_i)$ with $\alpha_i, \beta_i \notin P$ and $0 < \text{length}(I_i) = \beta_i - \alpha_i < \varepsilon$. Define $U_i = f^{-1}(I_i) \cap W$. Then $\{U_1, \ldots, U_n\}$ clearly is a partition of W. In addition, we have, for all $1 \leq i \leq n$,

$$\left.\begin{array}{l} f^{-1}((\alpha_i, \beta_i)) \text{ is open} \\ f^{-1}([\alpha_i, \beta_i]) \text{ is closed} \\ \mu\{f^{-1}(\alpha_i) \cup f^{-1}(\beta_i)\} = 0 \end{array}\right\} \implies \text{each } U_i \text{ is Riemann measurable}.$$

Now, $\overline{U}_i \subset W$ is closed and hence compact. Define $c_i := \inf\{f(x) \mid x \in U_i\}$ and $\psi := \sum_{i=1}^n c_i \mathbf{1}_{U_i}$. Then ψ is supported on W. If $x \in W$, then x is in precisely one of the sets of the partition, U_i say, and $|f(x) - \psi(x)| = |f(x) - c_i| \leq \text{length}(I_i) < \varepsilon$, independently of x. Consequently, $\|f - \psi\|_\infty < \varepsilon$. Finally, we can construct a sequence $\psi^{(\ell)}$ this way (e.g., via $\varepsilon^{(\ell)} = 1/\ell$) which establishes our claim. □

REMARK. The result of this Lemma extends to all continuous functions on W, even if they do not define continuous functions on H. This follows from Tietze's extension theorem [**RS**, Thm. IV.11].

References

[A] E. Akin, *The general topology of dynamical systems*, Grad. Stud. Math., vol. 1, Amer. Math. Soc., Providence, RI, 1993; corrected reprint, 1996.

[AGS] R. Ammann, B. Grünbaum, and G. C. Shephard, *Aperiodic tiles*, Discrete Comput. Geom. **8** (1992), 1–25.

[B] M. Baake, *A guide to mathematical quasicrystals*, Quasicrystals (J.-B. Suck, M. Schreiber, and P. Häussler, eds.), Springer, Berlin, 2001 (to appear); math-ph/9904014.

[BJ] M. Baake and D. Joseph, *Ideal and defective vertex configurations in the planar octagonal quasilattice*, Phys. Rev. B **42** (1990), 8091–8102.

[BM1] M. Baake and R. V. Moody, *Self-similarities and invariant densities for model sets*, Algebraic Methods and Theoretical Physics (Y. Saint-Aubin, ed.), Springer, New York, 2000 (in press); math-ph/9809006.

[BM2] _____, *Multi-component model sets and invariant densities*, Aperiodic '97 (M. de Boissieu, J.-L.Verger-Gaugry, and R. Currat, eds.) World Scientific, Singapore, 1998, pp. 9–20; math-ph/9809005.

[BMS] M. Baake, R. V. Moody, and M. Schlottmann, *Limit-(quasi)periodic point sets as quasicrystals with p-adic internal spaces*, J. Phys. A **31** (1998), 5755–5765; math-ph/9901008.

[Bau] H. Bauer, *Probability theory*, de Gruyter Stud. Math., vol. 23, de Gruyter, Berlin, 1996.

[BoGi] J. M. Borwein and R. Girgensohn, *Functional equations and distribution functions*, Results in Math. **26** (1994), 229–237.

[Bou] N. Bourbaki, *Elements of mathematics. General topology*, Addison-Wesley, Reading, MA, 1966; reprint, Springer, Berlin, 1989.

[D] J. Dieudonné, *Treatise on analysis*, vol. II, Pure Appl. Math., vol. 10-II, Academic Press, New York, 1970.

[HRB] J. Hermisson, C. Richard, and M. Baake, *A guide to the symmetry structure of quasiperiodic tiling classes*, J. Physique I **7** (1997), 1003–1018.

[HR] E. Hewitt and K. A. Ross, *Abstract harmonic analysis*, vol. I, 2nd ed., Grundlehren Math. Wiss., vol. 115, Springer, New York, 1979; 1997 (corrected 3rd printing).

[HM] M. Höffe and R. V. Moody, *Self-similar densities for p-adic model sets* (in preparation).

[Hof1] A. Hof, *On diffraction by aperiodic structures*, Commun. Math. Phys. **169** (1995), 25–43.

[Hof2] _____, *Uniform distribution and the projection method*, Quasicrystals and Discrete Geometry (Toronto, ON, 1995) (J. Patera, ed.), Fields Inst. Monogr., vol. 10, Amer. Math. Soc., Providence, RI, 1998, pp. 201–206.

[Hut] J. E. Hutchinson, *Fractals and self-similarity*, Indiana Univ. Math. J. **30** (1981), 713–743.

[JLS] R. Q. Jia, S. L. Lee, and A. Sharma, *Spectral properties of continuous refinement operators*, Proc. Amer. Math. Soc. **126** (1998), 729–737.

[JL] Q. Jiang and S. L. Lee, *Spectral properties of matrix continuous refinement operators*, Adv. Comput. Math. **7** (1997), 361–382.

[KT] S. Karlin and H. M. Taylor, *A first course in stochastic processes*, 2nd ed., Academic Press, Boston, MA, 1975.

[Kel] J. L. Kelley, *General topology*, Van Nostrand, Princeton, NJ, 1955; reprint, Grad. Texts Math., vol. 27, Springer, New York, 1975.

[KN] L. Kuipers and H. Niederreiter, *Uniform distribution of sequences*, Wiley, New York, 1974.

[LW] J. C. Lagarias and Y. Wang, *Substitution Delone sets*, 2000, (preprint).

[Lal] S. P. Lalley, *Random series in powers of algebraic integers: Hausdorff dimension of the limit distribution*, J. London Math. Soc. (2) **57** (1998), 629–654.

[LM] J.-Y. Lee and R. V. Moody, *Lattice substitution systems and model sets*, Discr. Comput. Geom. (to appear); math.MG/0002019.

[Mey] Y. Meyer, *Algebraic numbers and harmonic analysis*, North-Holland Math. Library, vol. 2, North-Holland, Amsterdam, 1972.

[M] R. V. Moody, *Meyer sets and their duals*, in: The Mathematics of Long-Range Aperiodic Order (Waterloo, ON, 1995) (R. V. Moody, ed.), NATO ASI Ser. C: Math. Phys. Sci., vol. 489, Kluwer, Dordrecht, 1997, pp. 403–441.

[P] P. A. B. Pleasants, *Designer quasicrystals: cut-and-project sets with pre-assigned properties*, Directions in Mathematical Quasicrystals, CRM Monogr. Series, vol. 13, Amer. Math. Soc., Providence, RI, 2000, pp. 95–141, this volume.

[Q] B. von Querenburg, *Mengentheoretische Topologie*, 2nd ed., Hochschultext, Springer, Berlin, 1979.

[RS] M. Reed and B. Simon, *Methods of modern mathematical physics. I. Functional analysis*, 2nd ed., Academic Press, San Diego, CA, 1980.

[Ru] W. Rudin, *Fourier analysis on groups*, Interscience Tracts in Pure and Applied Mathematics, vol. 12, Wiley, New York, 1962; reprint, Wiley Classics Lib., New York, 1990.

[Sch1] M. Schlottmann, *Cut-and-project sets in locally compact Abelian groups*, Quasicrystals and Discrete Geometry (Toronto, ON, 1995) (J. Patera, ed.), Fields Inst. Monogr., vol. 10, Amer. Math. Soc., Providence, RI, 1998, pp. 247–264.

[Sch2] _____, *Generalized model sets and dynamical systems*, Directions in Mathematical Quasicrystals, CRM Monogr. Series, vol. 13, Amer. Math. Soc., Providence, RI, 2000, pp. 143–159, this volume.

[So] B. Solomyak, *On the random series $\sum \pm \lambda^n$ (an Erdős problem)*, Ann. of Math. (2) **142** (1995), 611–625.

[We] H. Weyl, *Über die Gleichverteilung von Zahlen mod. Eins*, Math. Ann. **77** (1916), 313–352.

[Wi] K. R. Wicks, *Fractals and hyperspaces*, Lecture Notes in Math., vol. 1492, Springer, Berlin, 1991.

INSTITUT FÜR THEORETISCHE PHYSIK, UNIVERSITÄT TÜBINGEN, AUF DER MORGENSTELLE 14, 72076 TÜBINGEN, GERMANY

E-mail address: `michael.baake@uni-tuebingen.de, baake@miles.math.ualberta.ca`

DEPARTMENT OF MATHEMATICAL SCIENCES, UNIVERSITY OF ALBERTA, EDMONTON, ALBERTA, CANADA T6G 2G1

E-mail address: `rvm@miles.math.ualberta.ca`

Centre de Recherches Mathématiques
CRM Monograph Series
Volume **13**, 2000

Fourier Analysis of Deformed Model Sets

Guillaume Bernuau and Michel Duneau

1. Introduction

Model Sets, Quasicrystals, and Harmonic Analysis. Model sets are mathematical idealization of physical quasicrystals [**M3, B**] (see for instance [**J**] for a physical survey). These sets enjoy the Delaunay (or Delone) property [**M3**] and can be defined using the cut and projection method ([**M3, M2**]). Among all Delaunay sets, model sets X share the additional property that the set $X - X$ is uniformly discrete [**L**]. Another theorem, [**M2, Sch**], states that they are characterized by the fact that X and $X - X$ do not differ too much: There exists a finite set F satisfying
$$X - X \subset X + F.$$
Lattices enjoy this property with $F = \{0\}$ and therefore model sets are also called quasi-lattices. From the harmonic analysis point of view, these strong additive properties are illustrated by four different concepts as soon as one considers regular model sets. The first one is based on a generalized Poisson summation formula for regular model sets. More precisely, a countable dense subgroup S of $\mathbb{R}^n$, also called the spectrum, can be associated to a regular model set X. It allows one to write the Fourier transform (in the sense of tempered distributions) of the measure $\mu_X = \sum_{x \in X} \delta_x$ as a sum of Dirac masses

(1.1) $$\widehat{\mu_X} = \sum_{k \in S} c_k \delta_k,$$

where the r.h.s. of the last equality can be given a nontrivial meaning in the sense of tempered distributions for some suitable complex coefficients c_k, $k \in S$ (see [**M2, M1**]). The second concept is the Fourier-Bohr analysis and it involves the same set S and the same coefficients c_k as above. It allows one to define some discrete component for $\widehat{\mu_X}$ in the sense that for any k in S

(1.2) $$\lim_{|B| \to \infty} |B|^{-1} \sum_{x \in X \cap B} e^{-2\pi i k.x} = c_k,$$

whereas if q is not in S

(1.3) $$\lim_{|B| \to \infty} |B|^{-1} \sum_{x \in X \cap B} e^{-2\pi i q k.x} = 0,$$

2000 *Mathematics Subject Classification.* Primary 42B05, 52C23; Secondary 43A25, 82D25.

©2000 American Mathematical Society

where the limits are reached uniformly with respect to the position of the ball B in $\mathbb{R}^n$ (see [**Sch, M1**]).

The third concept is the autocorrelation γ_X defined as any vague limit of the positive measures, [**H1**]:

$$|B|^{-1}\left(\sum_{x \in X \cap B} \delta_x\right) * \left(\sum_{x' \in X \cap B} \delta_{x'}\right)$$

as the volume $|B|$ goes to infinity. Regular model sets admit a unique autocorrelation which is a discrete positive measure given by

(1.4) $$\gamma_X = \sum_{a \in X - X} n(a)\delta_a,$$

where the coefficients $n(a)$, $a \in X - X$, are explicit [**B**].

The last spectral concept is the diffraction pattern $\widehat{\gamma}_X$ defined as the Fourier transform (in the sense of tempered distributions) of the autocorrelation γ_X. It can be shown, [**H1**], that $\widehat{\gamma}_X$ is always a positive measure. Once again the spectrum S and the coefficients c_k happen to play a key role for regular model sets since one has [**B**]:

(1.5) $$\widehat{\gamma}_X = \sum_{k \in S} |c_k|^2 \delta_k.$$

Deformed Model Sets and General Statements of the Results. In this paper we study more general sets than model sets: deformed model sets. Each model set X_0, in association with a function θ on X_0, gives rise to a deformed model set by setting

$$X = \{x + \theta(x); x \in X_0\}.$$

If the variation of θ is small enough, X remains a Delaunay set. Hence deformed model sets can be considered as model sets with local perturbations. The aim of this article is to show that, for a wide class of suitable functions θ, three of the explicit spectral results described above remain valid for deformed model sets derived from a regular model set X_0: Fourier transform (1.1), Fourier-Bohr analysis (1.2), (1.3) and diffraction (1.5). This means that deforming regular model sets with a suitable deformation does not alter long range order. As it will be proved in the sequel, the only major difference appears with the computation of the autocorrelation γ_X: For the class of deformed model sets introduced in the following, γ_X will still be explicit but no longer a discrete measure. These results will be proved with the assumption that the internal space of the cut and project scheme is $\mathbb{R}^m$, for some integer $m > 0$, a special case of a locally compact Abelian group.

As a consequence these results give explicit examples of Delaunay sets X with purely discrete diffraction pattern and such that the group generated by X

$$\left\{\sum_{j=1}^{N} m_j x_j; N \in \mathbb{N}, m_j \in \mathbb{Z}, x_j \in X\right\}$$

is not finitely generated.

Related Works. The general idea on which this article relies is the Poisson summation formula. This formula can appear under various forms all of which are direct consequences of the following one:

$$\sum_{k \in \mathbb{Z}^n} \hat{f}(k) = \sum_{x \in \mathbb{Z}^n} f(x)$$

where f is in the Schwartz class $\mathcal{S}(\mathbb{R}^n)$ and $\hat{f}$ is the Fourier transform of f defined by $\hat{f}(q) = \int_{\mathbb{R}^n} f(x) e^{-2\pi i q \cdot x} \, dx$.

In this paper, we adapt and improve harmonic analysis techniques that have been introduced in previous works. Our main results are stated in Theorems 2.6, 3.1 and 4.4.

In Section 2, we adapt a harmonic analysis technique introduced in [**M2, M1, Sch**] to compute the Fourier-Bohr analysis. One can mention related works concerning the Fourier-Bohr analysis of deformed model sets. In [**H2**] and references quoted therein, the same conclusions as in Theorem 2.6 are proved but with some restricting hypothesis. It is also mentioned, in [**H2**], that Theorem 3.3 can be lifted to a statement including our Theorem 2.6.

In Section 3, we prove a fundamental lemma concerning regular model sets and we compute the autocorrelation of deformed model sets by a direct method. The same approach was used in [**B**] to compute the autocorrelation of any regular model set. Other explicit results on the autocorrelation have been proved for tilings obtained by the projection method, [**H2**], and for tilings obtained by primitive substitution, [**LP, GH**].

In Section 4, we compute the diffraction pattern by a direct method where the Poisson summation formula is closely involved. This technique is a generalization of the harmonic analysis arguments introduced in [**B**] to compute the diffraction pattern of regular model sets. Other techniques were introduced to study the diffraction pattern of different types of model sets. In [**So, D**], the diffraction pattern of self-similar tilings is determined using properties of an associated dynamical system. In [**H1, GK**], self-similar tilings are studied and only the discrete part of the diffraction pattern is determined.

Concerning deformed model sets, let us mention an important result, due to Hof [**H2**], stating that for any tiling X_0 and for a very wide class of deformations which depend measurably on the set X_0 (see [**H2**, p. 258] for further details), the associated deformed set X has purely discrete spectrum if X_0 has. But the spectrum of the deformed structure is not made explicit. The arguments invoked are linked to ergodic theory and suppose the study of an associated dynamical system.

General Definitions and Notations. Let us first recall the general definition of model sets in $\mathbb{R}^n$ (see [**Sch, M2**] for further details). We start by setting a cut and project scheme for $\mathbb{R}^n$ i.e., a locally compact Abelian group G and a subgroup L of $\mathbb{R}^n \times G$ enjoying the following properties:

(i) L is discrete and the quotient group $(\mathbb{R}^n \times G)/L$ is compact
(ii) the canonical projection p from L into $\mathbb{R}^n$ is injective
(iii) the canonical projection $\tilde{p}$ from L into G is dense.

Then a model set of $\mathbb{R}^n$ is defined as any subset X satisfying the following two conditions:

(i) there exists an open ball B such that $X + B = \mathbb{R}^n$

(ii) there exists a cut and project scheme (G, L) for $\mathbb{R}^n$ and a bounded set K in G such that $X = \{p(\xi); \xi \in L, \tilde{p}(\xi) \in K\}$.

The property (i), which means that X is uniformly dense, is satisfied if K has nonempty interior. The notation is $X = \mathfrak{M}(L, K)$. If, in addition, K is a Riemann integrable[1] compact subset of G with nonempty interior, X is called *regular model set*. We shall restrict our attention to regular model sets where $G = \mathbb{R}^m$ for some integer $m > 0$. Therefore the subgroup L becomes a lattice in the usual $(n + m)$-dimensional real space and X is a Delaunay set ([**M2**]).

We now introduce notations for deformed model sets. Let $X_0 = \mathfrak{M}(L, K)$ be a regular model set in $\mathbb{R}^n$ where K is a compact subset of $\mathbb{R}^m$ and L is a lattice of rank $m + n$ in $\mathbb{R}^{m+n} \approx \mathbb{R}^n \times \mathbb{R}^m$. The canonical projections from $\mathbb{R}^{n+m}$ onto $\mathbb{R}^n \times \{0\} \approx \mathbb{R}^n$ and onto $\{0\} \times \mathbb{R}^m \approx \mathbb{R}^m$ will be noted p and $\tilde{p}$ respectively. If $\xi \in L$ we put $\xi = (x, \tilde{x})$ where x is in $p(L)$ and $\tilde{x}$ is in $\tilde{p}(L)$. The Lebesgue measure of a fundamental domain of L is denoted $\operatorname{vol} L$. L^* is the dual lattice of L i.e the set of all κ in $\mathbb{R}^{m+n}$ such that each inner product $\kappa.\xi$, $\xi \in L$, belongs to $\mathbb{Z}$. We also put $\kappa = (k, \tilde{k})$ for $\kappa \in L^*$, where $k = p(\kappa)$ and $\tilde{k} = \tilde{p}(\kappa)$.

Let φ be a continuous function with compact support from $\mathbb{R}^m$ into $\mathbb{R}^n$. We assume that $\sup_{y \in \mathbb{R}^m} \|\varphi(y)\|$ is small enough to make sure the set

(1.6) $$X_\varphi = \{x + \varphi(\tilde{x}); x \in X_0\}$$

is a Delaunay set. This restriction will be used in order to bound the number of points close to the boundary of a box $B \subset \mathbb{R}^n$ by $C|\partial B|$. Such a set X_φ will be called a *deformed model set*. One may observe that model sets can be seen as deformed model sets where the associated function φ is identically 0. Also notice that the support of φ is not supposed to be included in the bounded set K. A canonical measure μ_φ is associated to the set X_φ by

(1.7) $$\mu_\varphi = \sum_{\substack{\xi \in L \\ \tilde{x} \in K}} \delta_{x+\varphi(\tilde{x})}.$$

2. Fourier-Bohr Analysis of Deformed Model Sets

Roughly speaking the Fourier-Bohr analysis of a Delaunay set X consists in looking for frequencies $q \in \mathbb{R}^n$ for which all the phases $e^{-2\pi i q.x}$, $x \in X$, are coherent. This means that the averages

$$|B_R|^{-1} \sum_{x \in X \cap B_R} e^{-2\pi i q.x}$$

converge to a limit $c_q \neq 0$ when R goes to infinity. B_R denotes the closed ball of radius R centered at the origin, of volume $|B_R|$.

Let us give a rigorous definition of this concept:

DEFINITION 2.1. Let X be a Delaunay set in $\mathbb{R}^n$, let $(\alpha_x)_{x \in X}$ be a complex sequence and let q be in $\mathbb{R}^n$. For any $R > 0$ and t in $\mathbb{R}^n$, we define the translated ball $B_{R,t} = B_R + t$. One says that the measure $\mu = \sum_{x \in X} \alpha_x \delta_x$ admits a Fourier-Bohr coefficient at the point q, equal to complex number C, if the averages

$$|B_R|^{-1} \sum_{x \in X \cap B_{R,t}} \alpha_x e^{-2\pi i q.x}$$

---

[1] K Riemann integrable means that the topological boundary ∂K has Lebesgue measure 0.

converge to C, uniformly with respect to t in $\mathbb{R}^n$, as R goes to infinity, i.e.:

$$\lim_{R \to \infty} \sup_{t \in \mathbb{R}^n} \left| C - |B_R|^{-1} \sum_{x \in X \cap B_{R,t}} \alpha_x e^{-2\pi i q \cdot x} \right| = 0.$$

The notation is $C = c_q(\mu)$.

When all coefficients α_x equal 1, Fourier-Bohr coefficients will be noted $c_q(X)$. We are now going to show that the Fourier-Bohr analysis of regular deformed model sets can be made explicit. The main idea used in the proof was initially introduced in [**M2, Sch**]. It states that when X_0 is a regular model set, the measure $\sum_{x \in X_0} \delta_x$ is a limit in the Schwartz class of tempered distributions $\mathcal{S}'(\mathbb{R}^n)$ of a sequence $\{\nu_i\}$ of atomic measures such that $\widehat{\nu_i}$ is also an atomic measure. We are going to check that this approach remains valid for deformed model sets defined by (1.6) and (1.7).

The four following preliminary lemmas implement regularizations of the characteristic function of K and assume that φ is a C^∞ function. This last restriction will be dropped in the proof of Theorem 2.6.

LEMMA 2.2. *Let K be a compact subset of $\mathbb{R}^m$ and let $\varepsilon > 0$. Then $K_\varepsilon = K + B_\varepsilon$ is compact and the topological boundary ∂K_ε has Lebesgue measure 0.*

PROOF. The distance $d_K(x) = \inf\{\|x - x'\|; x' \in K\}$ with respect to K is a continuous function. Then $K_\varepsilon = d_K^{-1}([0, \varepsilon])$ is closed and obviously compact. $d_K^{-1}([0, \varepsilon))$ is open, hence contained in the interior $\overset{\circ}{K_\varepsilon}$ of K_ε. It follows that the boundary ∂K_ε is contained in the compact set $S_K(\varepsilon) = d_K^{-1}(\{\varepsilon\})$. The characteristic function $\chi_{S_K(\varepsilon)}$ is integrable and [**R**, Theorem 7.8] states that for almost every x for the Lebesgue measure we have

$$(2.1) \qquad \chi_{S_K(\varepsilon)}(x) = \lim_{r \to 0} \frac{|S_K(\varepsilon) \cap B_{r,x}|}{|B_{r,x}|}$$

where $B_{r,x}$ is the closed ball of radius r centered at x in $\mathbb{R}^m$. Now, for any $x \in S_K(\varepsilon)$ there exists at least one point $x' \in K$ such that $\|x - x'\| = \varepsilon$. The open ball $\overset{\circ}{B}_{\varepsilon,x'}$ does not intersect $S_K(\varepsilon)$ so that for r small enough $S_K(\varepsilon) \cap B_{r,x} \subset B_{r,x} \setminus \overset{\circ}{B}_{\varepsilon,x'}$. The r.h.s. of equation (2.1) is therefore bounded by $\frac{1}{2}$ while $\chi_{S_K(\varepsilon)}(x) = 1$. This implies that the Lebesgue measure of $S_K(\varepsilon)$ is 0. Notice that it is not required here that ∂K has zero Lebesgue measure. $\square$

For any $\varepsilon > 0$, $\beta_\varepsilon(y)$ will denote a C^∞ function on $\mathbb{R}^m$, with values in $[0, 1]$, taking value 1 on K with support in $K_\varepsilon = K + B_\varepsilon$. Let $\mu_{\varphi,\varepsilon}$ be the measure defined by

$$(2.2) \qquad \mu_{\varphi,\varepsilon} = \sum_{\tilde{\xi} \in L} \beta_\varepsilon(\tilde{x}) \, \delta_{x + \varphi(\tilde{x})}.$$

The measure $\mu_{\varphi,\varepsilon}$ is carried by the following deformed model set (K_ε is Riemann integrable by Lemma 2.2)

$$(2.3) \qquad X_{\varphi,\varepsilon} = \{x + \varphi(\tilde{x}); (x, \tilde{x}) \in L, \tilde{x} \in K_\varepsilon\}$$

which still satisfies the Delaunay property for ε small enough (depending on the variation of φ). We mean to show that the measures $\mu_{\varphi,\varepsilon}$ enjoy an explicit Fourier-Bohr analysis which, in a way to be defined, converges to the Fourier-Bohr analysis of X_φ as ε tends to 0. To achieve this goal, the first step is to compute the Fourier transform $\widehat{\mu_{\varphi,\varepsilon}}$ (in the sense of tempered distributions):

LEMMA 2.3. *The tempered distribution $\widehat{\mu_{\varphi,\varepsilon}}$ is an atomic measure given by*

$$\widehat{\mu_{\varphi,\varepsilon}} = \sum_{\kappa \in L^*} r_{\varphi,\varepsilon}(\kappa) \delta_k$$

where

$$r_{\varphi,\varepsilon}(u,v) = (\text{vol } L)^{-1} \int_{\mathbb{R}^m} \beta_\varepsilon(y) e^{-2\pi i (u.\varphi(y) - v.y)} \, dy.$$

Moreover, for any function f in the Schwartz class $\mathcal{S}(\mathbb{R}^n)$, $f\widehat{\mu_{\varphi,\varepsilon}}$ is a bounded measure.

PROOF. Let f be a function in the Schwartz class $\mathcal{S}(\mathbb{R}^n)$ and let F be the function in $\mathcal{S}(\mathbb{R}^{n+m})$ whose Fourier transform is

$$\widehat{F}(x,y) = \hat{f}(x + \varphi(y))\beta_\varepsilon(y).$$

Then, by the definition (2.2)) of $\mu_{\varphi,\varepsilon}$,

$$\int_{\mathbb{R}^n} \hat{f} \, d\mu_{\varphi,\varepsilon} = \sum_{\xi \in L} \widehat{F}(x, \tilde{x})$$

and applying the Poisson summation formula adapted to the lattice L, one gets

$$\int_{\mathbb{R}^n} \hat{f} \, d\mu_{\varphi,\varepsilon} = (\text{vol } L)^{-1} \sum_{\kappa \in L^*} F(k, \tilde{k}).$$

Let us compute F:

$$F(u,v) = \iint_{\mathbb{R}^{n+m}} \hat{f}(x + \varphi(y))\beta_\varepsilon(y) e^{2\pi i (u.x + v.y)} \, dx \, dy.$$

One makes the following change of variables $(x, y) \to (x', y') = (x + \varphi(y), y)$ which leads to

$$F(u,v) = \iint_{\mathbb{R}^{n+m}} \hat{f}(x)\beta_\varepsilon(y) e^{2\pi i (u.(x-\varphi(y)) + v.y)} \, dxdy$$

$$= f(u) \int_{\mathbb{R}^m} \beta_\varepsilon(y) e^{-2\pi i (u.\varphi(y) - v.y)} \, dy$$

$$= (\text{vol } L) f(u) r_{\varphi,\varepsilon}(u,v).$$

Hence

$$\int_{\mathbb{R}^n} f \, d\widehat{\mu_{\varphi,\varepsilon}} = \sum_{\kappa \in L^*} f(k) \, r_{\varphi,\varepsilon}(\kappa).$$

The r.h.s. of the above equation is an absolutely convergent series. However, to prove that $\widehat{\mu_{\varphi,\varepsilon}}$ is a measure, it is necessary to show that for any ball B in $\mathbb{R}^n$ the series $\{r_{\varphi,\varepsilon}(\kappa), \kappa \in L^*, k \in B\}$, is summable. This can be checked by choosing a function f such that $|f(k)| \geq 1$ for $k \in B$ since we get

$$\sum_{\substack{\kappa \in L^* \\ k \in B}} |r_{\varphi,\varepsilon}(\kappa)| \leq \sum_{\kappa \in L^*} |f(k)||r_{\varphi,\varepsilon}(\kappa)| < +\infty.$$

Finally, we see that for any function f belonging to $\mathcal{S}(\mathbb{R}^n)$, $f\widehat{\mu_{\varphi,\varepsilon}}$ is a bounded measure since the series $\sum_{\kappa \in L^*} f(k) r_{\varphi,\varepsilon}(\kappa)$ is absolutely convergent. $\square$

A first consequence of this lemma is the existence of a generalized Poisson summation formula for deformed model sets when φ is a C^∞ function. Indeed, for any g in the Schwartz class $\mathcal{S}(\mathbb{R}^n)$, it is clear that $\int_{\mathbb{R}^n} g \, d\mu_{\varphi,\varepsilon}$ tends to $\int_{\mathbb{R}^n} g \, d\mu_\varphi$ as ε tends to 0. Therefore $\mu_{\varphi,\varepsilon}$ converges to μ_φ in the sense of tempered distributions and we get

$$\sum_{x \in X_\varphi} \hat{f}(x) = \lim_{\varepsilon \to 0} \int_{\mathbb{R}^n} \hat{f} \, d\mu_{\varphi,\varepsilon} = \lim_{\varepsilon \to 0} \sum_{\kappa \in L^*} f(k) \, r_{\varphi,\varepsilon}(\kappa).$$

Hence, the preceding limit process is the meaning that we can give to the identity

$$\sum_{x \in X_\varphi} \hat{f}(x) = \sum_{\kappa \in L^*} f(k) r_{\varphi,0}(\kappa).$$

One has to notice that the r.h.s of this identity is not an absolutely convergent series.

The following lemma states that the Fourier-Bohr coefficients of measures such as $\mu_{\varphi,\varepsilon}$ are well defined and are related to the discrete component of their Fourier transforms. This is achieved by using a second regularization in $\mathbb{R}^n$.

LEMMA 2.4. *Let μ be a sum of uniformly bounded Dirac masses located on a Delaunay set X in $\mathbb{R}^n$. Assume that $\hat{\mu}$ is a measure and that, for any function f in $\mathcal{S}(\mathbb{R}^n)$, $f \hat{\mu}$ is a bounded measure. Then, for any q in $\mathbb{R}^n$, the Fourier-Bohr coefficient $c_q(\mu)$ exists and is given by*

$$c_q(\mu) = \hat{\mu}(\{q\}).$$

PROOF. Put $\mu = \sum_{x \in X} b_x \delta_x$. Let $q \in \mathbb{R}^n$ and let α be a function in the space $C_0^\infty(\mathbb{R}^n)$ of C^∞ functions with compact support such that $\hat{\alpha}(q) = 1$. We define the measure $\nu = \alpha * \mu$ and therefore the Fourier transform $\hat{\nu} = \hat{\alpha}\hat{\mu}$ is a bounded measure. Let B_r be a ball containing the support of α. Then for any $R > r$ and $t \in \mathbb{R}^n$:

$$(2.4) \quad \int_{B_{R,t}} e^{-2\pi i q.u} \, d\nu(u)$$
$$= \sum_{x \in X} b_x \int_{B_{R,t}} \alpha(u-x) e^{-2\pi i q.u} \, du$$
$$= \sum_{x \in X \cap B_{R-r,t}} b_x e^{-2\pi i q.x} + \sum_{x \in X \cap (B_{R+r,t} \setminus B_{R-r,t})} b_x \int_{B_{R,t}} \alpha(u-x) e^{-2\pi i q.u} \, du.$$

The shell $B_{R+r,t} \setminus B_{R-r,t}$ can be covered by unit balls whose number is $\mathcal{O}(R^{n-1})$. Since X is a Delaunay set the cardinal of $X \cap (B_{R+r,t} \setminus B_{R-r,t})$ can be bounded by AR^{n-1} where A only depends on n, r and X. The second term in the r.h.s. can be therefore bounded by:

$$A\|\alpha\|_{L^1} R^{n-1} \sup_{x \in X} |b_x|.$$

Consequently $c_q(\mu)$ exists if and only if $c_q(\nu)$ exists, with equal values in case of existence.

We show now that the Fourier-Bohr coefficients of ν exist and are given by $c_q(\nu) = \hat{\nu}(\{q\})$. We first observe that the l.h.s. of equation 2.4 can be rewritten as

$$(2.5) \quad \int_{\mathbb{R}^n} \chi_{B_R}(u-t) e^{-2\pi i q.u} \, d\nu(u) = \int_{\mathbb{R}^n} \widehat{\chi_{B_R}}(q-k) e^{2\pi i t.(k-q)} \, d\hat{\nu}(k)$$

where χ_B denotes the characteristic function of B. This can be checked by using regular approximations of the characteristic function χ_{B_R} and Lebesgue's dominated convergence theorem. Thus, we get:

$$(2.6) \quad |B_R|^{-1} \int_{B_{R,t}} e^{-2\pi i q \cdot u} \, d\nu(u) = \int_{\mathbb{R}^n} |B_R|^{-1} \widehat{\chi_{B_R}}(q-k) e^{2\pi i t \cdot (k-q)} \, d\widehat{\nu}(k).$$

Define $L(k)$ as the function taking value 1 at 0 and vanishing for $k \neq 0$. Then we have

$$\left| |B_R|^{-1} \int_{B_{R,t}} e^{-2\pi i q \cdot u} \, d\nu(u) - \widehat{\nu}(\{q\}) \right|$$
$$= \left| \int_{\mathbb{R}^n} |B_R|^{-1} \widehat{\chi_{B_R}}(q-k) e^{2\pi i t \cdot (k-q)} \, d\widehat{\nu}(k) - \int_{\mathbb{R}^n} L(k-q) e^{2\pi i t \cdot (k-q)} \, d\widehat{\nu}(k) \right|$$
$$\leq \int_{\mathbb{R}^n} \left| |B_R|^{-1} \widehat{\chi_{B_R}}(q-k) - L(k-q) \right| d|\widehat{\nu}|(k).$$

Notice that the last integral does not depend on t.

For each k, $\left| |B_R|^{-1} \widehat{\chi_{B_R}}(q-k) - L(k-q) \right|$ converges to 0 as R tends to $+\infty$ and is bounded by the $|\widehat{\nu}|$-integrable function g: $g(k) \equiv 1$. Hence, by Lebesgue's dominated convergence theorem

$$\lim_{R \to \infty} \int_{\mathbb{R}^n} \left| |B_R|^{-1} \widehat{\chi_{B_R}}(q-k) - L(k-q) \right| d|\widehat{\nu}|(k) = 0.$$

Thus we have proved that $c_q(\nu)$ exists and is equal to $\widehat{\nu}(\{q\})$. Finally, since $\widehat{\alpha(q)} = 1$ we have

$$c_q(\mu) = \widehat{\mu}(\{q\})$$

and Lemma 2.4 is proved. □

The third lemma pertains to the upper density of deformed model sets. Recall that the upper density of a Delaunay set X is defined by

$$\overline{\mathrm{dens}}\, X = \limsup_{R \to \infty} |B_R|^{-1} \sup_{t \in \mathbb{R}^n} \mathrm{card}(X \cap B_{R,t}).$$

LEMMA 2.5. *Let X_φ be a deformed model set defined with the regular model set $X_0 = \mathfrak{M}(L, K)$ and the C^∞ function on $\mathbb{R}^m$ with compact support φ. Then*

$$\overline{\mathrm{dens}}(X_\varphi) \leq (\mathrm{vol}\, L)^{-1} |K|.$$

PROOF. Starting from

$$|B_R|^{-1} \mathrm{card}(X_\varphi \cap B_{R,t}) \leq |B_R|^{-1} \int_{B_{R,t}} d\mu_{\varphi,\varepsilon}(x)$$

we have

$$|B_R|^{-1} \sup_{t \in \mathbb{R}^n} \mathrm{card}(X_\varphi \cap B_{R,t}) \leq |B_R|^{-1} \sup_{t \in \mathbb{R}^n} \int_{B_{R,t}} d\mu_{\varphi,\varepsilon}(x).$$

Using the previous lemmas, we know that the Fourier-Bohr analysis of measures $\mu_{\varphi,\varepsilon}$ is explicit. In particular, one has, by Lemmas 2.3 and 2.4,

$$c_0(\mu_{\varphi,\varepsilon}) := \lim_{R \to \infty} |B_R|^{-1} \int_{B_{R,t}} d\mu_{\varphi,\varepsilon} = \widehat{\mu_{\varphi,\varepsilon}}(\{0\})$$

where the limit is uniform with respect to t in $\mathbb{R}^n$. Using now Lemma 2.3 where $r_{\varphi,\varepsilon}(0)$ is made explicit we obtain

$$c_0(\mu_{\varphi,\varepsilon}) = (\operatorname{vol} L)^{-1}\widehat{\beta_\varepsilon}(0).$$

It follows that

$$\overline{\operatorname{dens}}(X_\varphi) \leq (\operatorname{vol} L)^{-1}|K_\varepsilon|$$

and then let $\varepsilon \to 0$

$$\overline{\operatorname{dens}}(X_\varphi) \leq (\operatorname{vol})^{-1}|K|$$

which ends the proof of Lemma 2.5. $\square$

THEOREM 2.6. *Let X_φ be a deformed model set constructed with a regular model set $X_0 = \mathfrak{M}(L, K)$ and a continuous function φ with compact support. Then, for any q in $\mathbb{R}^n$, the Fourier-Bohr coefficient $c_q(X_\varphi)$ exists and its value is given by the following rule. For any $\kappa = (k, \tilde{k}) \in L^*$, one has*

$$c_k(X_\varphi) = (\operatorname{vol})^{-1} \int_K e^{-2\pi i(k.\varphi(y) - \tilde{k}.y)}\, dy$$

and $c_q(X_\varphi) = 0$ if q does not belong to the $\mathbb{Z}$-module $p(L^*)$.

PROOF. The proof is now straightforward but some notation is needed. Let

$$m_B(q) := |B|^{-1} \int_B e^{-2\pi i q.x}\, d\mu_\varphi(x)$$

$$m_{B,\varepsilon}(q) := |B|^{-1} \int_B e^{-2\pi i q.x}\, d\mu_{\varphi,\varepsilon}(x)$$

$$c(q) := \begin{cases} (\operatorname{vol})^{-1} \int_K e^{-2\pi(ik.\varphi(y) - i\tilde{k}.y)}\, dy & \text{if } q = k \text{ with } \kappa = (k, \tilde{k}) \in L^* \\ 0 & \text{otherwise.} \end{cases}$$

1ST CASE. φ is a C^∞ function. Let us show that $m_B(q)$ converges to $c(q)$ as $|B| \to \infty$, uniformly with respect to the position of B:

$$|m_B(q) - c(q)| \leq |m_B(q) - m_{B,\varepsilon}(q)| + |m_{B,\varepsilon}(q) - c_q(\mu_{\varphi,\varepsilon})| + |c_q(\mu_{\varphi,\varepsilon}) - c(q)|.$$

Notice first that

$$|m_B(q) - m_{B,\varepsilon}(q)| \leq |B|^{-1} \operatorname{card}\{(x, \tilde{x}) \in L; x + \varphi(\tilde{x}) \in B, \tilde{x} \in K_\varepsilon \setminus K\}.$$

It follows from Lemma 2.5 applied to the deformation of the model set $\mathfrak{M}(L, K_\varepsilon \setminus K)$ that

$$\varlimsup_{R \to \infty} \sup_{t \in \mathbb{R}^n} |m_{B_{R,t}}(q) - m_{B_{R,t},\varepsilon}(q)| \leq (\operatorname{vol})^{-1}|K_\varepsilon \setminus K|.$$

By Lemma 2.4, we know that $m_{B,\varepsilon}(q)$ converges uniformly with respect to the position of B to the limit $c_q(\mu_{\varphi,\varepsilon}) = \widehat{\mu_{\varphi,\varepsilon}}(\{q\})$.

Using Lemma 2.3 we get $|c_q(\mu_{\varphi,\varepsilon}) - c(q)| \leq (\operatorname{vol})^{-1}|K_\varepsilon \setminus K|$. Therefore, taking the superior limit as $R \to \infty$, we have

$$\varlimsup_{R \to \infty} \sup_{t \in \mathbb{R}^n} |m_{B_{R,t}}(q) - c(q)| \leq 2(\operatorname{vol})^{-1}|K_\varepsilon \setminus K|.$$

Now, since K is Riemann-integrable we can take the limit $\varepsilon \to 0$ and we obtain

$$\varlimsup_{R \to \infty} \sup_{t \in \mathbb{R}^n} |m_{B_{R,t}}(q) - c(q)| = 0.$$

Hence $c_q(X_\varphi) = c(q)$ for all $q \in \mathbb{R}^n$.

2ND CASE. φ is a continuous function. We consider a sequence of $C_0^\infty(\mathbb{R}^n)$ functions $(\varphi_j)_{j\in\mathbb{N}}$ such that
$$\|\varphi - \varphi_j\|_\infty = \sup_{y\in K}|\varphi(y) - \varphi_j(y)| \to 0, \quad \text{as } j \to \infty.$$
The corresponding terms are denoted $m_{B,j}(q)$ and $c_j(q)$. Then
$$|m_B(q) - c(q)| \le |m_B(q) - m_{B,j}(q)| + |m_{B,j}(q) - c_j(q)| + |c_j(q) - c(q)|.$$
First we have
$$|m_B(q) - m_{B,j}(q)| \le |B|^{-1} \sum_{\xi\in L, \tilde{x}\in K} \chi_B(x + \varphi(\tilde{x}))|e^{2\pi i q\cdot\varphi(\tilde{x})} - e^{2\pi i q\cdot\varphi_j(\tilde{x})}|$$
$$+ |B|^{-1} \sum_{\xi\in L, \tilde{x}\in K} |\chi_B(x + \varphi(\tilde{x})) - \chi_B(x + \varphi_j(\tilde{x}))|.$$

The first summation is bounded by $|B|^{-1}\operatorname{card}(X_\varphi \cap B)2\pi|q|\|\varphi - \varphi_j\|_\infty$. The second summation involves points of X_φ which lie within a distance $\|\varphi - \varphi_j\|_\infty$ from the boundary ∂B of B. This shell can be covered by balls of fixed radius whose number is proportional to $|\partial B|$. Using the Delaunay property of X_φ, we see that this summation can be bounded by $C|\partial B||B|^{-1}$. Therefore, by Lemma 2.5,
$$\varlimsup_{R\to\infty} \sup_{t\in\mathbb{R}^n} |m_{B_{R,t}}(q) - m_{B_{R,t},j}(q)| \le (\mathrm{vol})^{-1}2\pi|q|\|K\|\|\varphi - \varphi_j\|_\infty.$$
For any fixed j the previous proof applies to φ_j and gives
$$\varlimsup_{R\to\infty} \sup_{t\in\mathbb{R}^n} |m_{B_{R,t},j}(q) - c_j(q)| = 0.$$
Then we have
$$|c_j(q) - c(q)| \le (\mathrm{vol})^{-1}2\pi|q|\|K\|\|\varphi - \varphi_j\|_\infty.$$
Hence, for any $j \in \mathbb{N}$
$$\varlimsup_{R\to\infty} \sup_{t\in\mathbb{R}^n} |m_{B_{R,t}}(q) - c(q)| \le 4\pi(\mathrm{vol})^{-1}|q|\|K\|\|\varphi - \varphi_j\|_\infty.$$
Then taking the limit as $j \to \infty$ ends the proof of Theorem 2.6. □

The existence of Fourier-Bohr coefficients has a direct consequence on the ergodicity properties of deformed model sets. If f is a function defined on X_φ such that $g(\tilde{x}) = f(x + \varphi(\tilde{x}))$ for $x \in X_0$ extends to a continuous function g on K then the averages
$$|B|^{-1} \sum_{x\in X_\varphi \cap B} f(x)$$
converge to the integral of g on K. This is proved in the following lemmas. The second one, an extension of Weyl's theorem ([**KN**]), is connected to uniform projection (see [**H3**, **S1**, **BM**]).

LEMMA 2.7. *Let U be a compact subset of $\mathbb{R}^m$. Then any continuous function on U is a uniform limit on U of a sequence of trigonometric sums whose spectra lie in the $\mathbb{Z}$-module $\tilde{p}(L^*)$.*

PROOF. Let g be a continuous function on the compact set U and put $\|g\|_\infty = \sup_{y\in U}|g(y)|$. It is well known that g is a uniform limit on U of a sequence of trigonometric sums $P_j(y) = \sum_{q\in S_j} c_{q,j} e^{2\pi i q\cdot y}$. Let $\varepsilon > 0$ and choose an integer j such that
$$\|g - P_j\|_\infty \le \varepsilon.$$

Since $\tilde{p}(L^*)$ is dense in $\mathbb{R}^m$, we can find a finite family $(\tilde{k}_q)_{q \in S_j}$ included in $\tilde{p}(L^*)$ such that
$$\sup_{q \in S_j} |\tilde{k}_q - q| \leq \frac{\varepsilon}{2\pi \sup_{y \in U} |y|(1 + \sum_{q \in S_j} |c_{q,j}|)}.$$
Then choosing $Q_j(y) = \sum_{q \in S_j} c_{q,j} e^{2\pi i \tilde{k}_q \cdot y}$, we get
$$\|g - Q_j\|_\infty \leq \|g - P_j\|_\infty + \|P_j - Q_j\|_\infty$$
$$\leq \varepsilon + 2\pi \sup_{y \in U} |y| \sum_{q \in S_j} |c_{q,j}| \sup_{q \in S_j} |\tilde{k}_q - q| \leq 2\varepsilon. \qquad \square$$

LEMMA 2.8. *Let $X_U = \mathfrak{M}(L, U)$ be a regular model set in $\mathbb{R}^n$ where U is a Riemann integrable compact subset of $\mathbb{R}^m$. Then for any continuous function g on U, one has*
$$\lim_{|B| \to \infty} |B|^{-1} \sum_{x \in X_U \cap B} g(\tilde{x}) = (\mathrm{vol})^{-1} \int_U g(y)\, dy.$$

PROOF. The result is clear when $g(y) = e^{2\pi i \tilde{k} \cdot y}$ with $\kappa = (k, \tilde{k}) \in L^*$, since we have $e^{2\pi i \tilde{k} \cdot \tilde{x}} = e^{-2\pi i k \cdot x}$ for any $\xi = (x, \tilde{x}) \in L$, and we can apply Theorem 2.6. By linearity it is then clear for any trigonometric sum whose spectrum lies is $\tilde{p}(L^*)$. For the general case, we put $m_B(g) = |B|^{-1} \sum_{x \in X_U \cap B} g(\tilde{x})$ and $c = (\mathrm{vol}\, L)^{-1}$. Then for any trigonometric sum P, we have

$$\left| m_B(g) - c \int_U g(y)\, dy \right|$$
$$\leq |m_B(g) - m_B(P)| + \left| m_B(P) - c \int_U P(y)\, dy \right| + c \left| \int_U P(y)\, dy - \int_U g(y)\, dy \right|$$
$$\leq |B|^{-1} \mathrm{card}(X_U \cap B) \|g - P\|_\infty + \left| m_B(P) - c \int_U P(y)\, dy \right| + c|U| \|g - P\|_\infty.$$

Then let $\varepsilon > 0$ and choose P to be a trigonometric sum with spectrum in $\tilde{p}(L^*)$ such that $\|g - P\|_\infty \leq \varepsilon/(2c|U|)$. Taking the superior limit, as $|B| \to \infty$, in the last inequality we obtain
$$\overline{\lim_{|B| \to \infty}} \left| m_B(g) - c \int_U g(y)\, dy \right| \leq 2c|U| \|g - P\|_\infty \leq \varepsilon.$$
Since ε is arbitrary, Lemma 2.8 is proved. $\qquad \square$

3. Autocorrelation of Deformed Model Sets

Recall that an autocorrelation of a Delaunay set X is any vague limit γ of a sequence of measures γ_j given by
$$\gamma_j = |B_j|^{-1} \left(\sum_{x \in X \cap B_j} \delta_x \right) * \left(\sum_{x' \in X \cap B_j} \delta_{-x'} \right)$$
where $(B_j)_{j \in \mathbb{N}}$ is an increasing sequence of balls in $\mathbb{R}^n$, whose union is $\mathbb{R}^n$ (see [**H1**] for more details). Usually a Delaunay set enjoys several autocorrelations. The case where there is a unique autocorrelation is rather remarkable and it may reflect the fact that X has long range order. We are going to show that deformed model sets share this property. The approach is to operate a direct study of γ_j as it was done in [**B**] for regular model sets.

THEOREM 3.1. *Let X_φ be a deformed model set in $\mathbb{R}^n$ constructed with a regular model set $X_0 = \mathfrak{M}(L, K)$ and a continuous function φ with compact support:*

$$X_\varphi = \{x + \varphi(\tilde{x}); x \in X_0\}.$$

Then X_φ admits a unique autocorrelation γ given by:

$$\int_{\mathbb{R}^n} f \, d\gamma = (\text{vol})^{-1} \sum_{a \in p(L)} \int_{K \cap (K - \tilde{a})} f\bigl(a + \varphi(\tilde{a} + y) - \varphi(y)\bigr) \, dy$$

for any function f in the Schwartz class $\mathcal{S}(\mathbb{R}^n)$.

PROOF. Let B be a ball in $\mathbb{R}^n$ and let γ_B be the autocorrelation of the finite set $X_\varphi \cap B$:

$$\gamma_B = |B|^{-1} \left(\sum_{x \in X_\varphi \cap B} \delta_x \right) * \left(\sum_{x' \in X_\varphi \cap B} \delta_{-x'} \right).$$

We first observe that a measure γ is an autocorrelation of X_φ if and only if γ is a vague limit of a sequence of the form

$$\nu_B = |B|^{-1} \left(\sum_{x \in X_0 \cap B} \delta_{x + \varphi(\tilde{x})} \right) * \left(\sum_{x' \in X_0 \cap B} \delta_{-x' - \varphi(\tilde{x}')} \right)$$

as $|B| \to \infty$. In fact, the difference between γ_B and ν_B involves pairs of points such that at least one of them lies within a distance $\|\varphi\|_\infty$ from the boundary of B. Then, if f is a continuous function with compact support we can get the following bound using the Delaunay property of X_0

$$\left| \int_{\mathbb{R}^n} f \, d\gamma_B - \int_{\mathbb{R}^n} f \, d\nu_B \right| \le C |B|^{-1} |\partial B| \|f\|_\infty.$$

Now let us study $\int_{\mathbb{R}^n} f \, d\nu_B$. One has

$$\int_{\mathbb{R}^n} f \, d\nu_B = |B|^{-1} \sum_{x' \in X_0 \cap B} \sum_{x \in X_0 \cap B} f\bigl(x' - x + \varphi(\tilde{x}') - \varphi(\tilde{x})\bigr).$$

Put $a = x' - x$ and write

$$\int_{\mathbb{R}^n} f \, d\nu_B = |B|^{-1} \sum_{a \in X_0 - X_0} \sum_{\substack{x \in X_0 \cap B \\ a + x \in X_0 \cap B}} f\bigl(a + \varphi(\tilde{a} + \tilde{x}) - \varphi(\tilde{x})\bigr).$$

Since $X_0 - X_0$ is a Delaunay set and f has compact support, there are only finitely many points a in $X_0 - X_0$ involved in the first summation. Therefore it suffices to study separately each term

$$R_a(B) = |B|^{-1} \sum_{\substack{x \in X_0 \cap B \\ a + x \in X_0 \cap B}} f\bigl(a + \varphi(\tilde{a} + \tilde{x}) - \varphi(\tilde{x})\bigr).$$

We first observe that conditions $x \in X_0 \cap B$ and $a + x \in X_0 \cap B$ in the summation are equivalent to the single condition $x \in X_0 \cap (X_0 - a) \cap B \cap (B - a)$. Moreover $X_0 \cap (X_0 - a)$ is the regular model set $\mathfrak{M}(L, K \cap (K - \tilde{a}))$. Since $|B \cap (B - a)| \sim |B|$ when $|B| \to \infty$, the existence of a limit for $R_a(B)$ and the possible limit itself remain the same if we consider the expression

$$S_a(B) = |B|^{-1} \sum_{x \in X_0 \cap (X_0 - a) \cap B} f\bigl(a + \varphi(\tilde{a} + \tilde{x}) - \varphi(\tilde{x})\bigr).$$

Applying Lemma 2.8 to the regular model set $\mathfrak{M}(L, K \cap (K - \tilde{a}))$ and the continuous function $h(y) = f\big(a + \varphi(\tilde{a} + y) - \varphi(y)\big)$, we get

$$\lim_{|B| \to \infty} R_a(B) = \lim_{|B| \to \infty} S_a(B) = (\text{vol})^{-1} \int_{K \cap (K-\tilde{a})} f\big(a + \varphi(\tilde{a}+y) - \varphi(y)\big)\, dy.$$

We end the proof of Theorem 3.1 by observing that the preceding limit is 0 as soon as $a \in p(L)$ does not belong to $X_0 - X_0$. $\square$

4. Diffraction Pattern of Deformed Model Sets

Diffraction is a fundamental concept involved in the study of aperiodic Delaunay sets. Let X be a Delaunay set which admits a unique autocorrelation γ. We recall that the *diffraction pattern* of X is the positive measure $\widehat{\gamma}$ defined as the Fourier transform (in the sense of tempered distributions) of the autocorrelation γ [**H1**]. Of great importance is the study of the discrete and continuous parts of $\widehat{\gamma}$. The existence of a discrete part is a sign of long range order for X. Hof ([**H1**]) has proved that if the Fourier-Bohr coefficient $c_k(X)$ exists then $\widehat{\gamma}(\{k\}) = |c_k(X)|^2$. The continuous part of $\widehat{\gamma}$ is much more difficult to determine and no general result seems to be available for the moment. Concerning regular model sets $X_0 = \mathfrak{M}(L, K)$, it has been proved in [**B**] by a direct computation that the diffraction pattern is purely discrete

$$\widehat{\gamma}_{X_0} = \sum_{k \in p(L^*)} |c_k(X_0)|^2\, \delta_k.$$

In this section, we are going to compute the diffraction pattern of deformed model sets by the same direct approach. The proof requires several preliminary lemmas which follow.

LEMMA 4.1. *Let f be any function in the Schwartz class $\mathcal{S}(\mathbb{R}^n)$ and let g be the function defined for $\xi = (x, y) \in \mathbb{R}^{n+m}$ by*

$$g(\xi) = \int_{K \cap (K-y)} f\big(x + \varphi(y+t) - \varphi(t)\big)\, dt.$$

Then g is continuous on $\mathbb{R}^{n+m}$.

PROOF. Let $\xi = (x,y) \in \mathbb{R}^{n+m}$ and $(\xi_j)_{j \in \mathbb{N}}$ be a sequence with limit ξ. We assume that $|\xi_j - \xi| \leq 1$ for all j. Define

$$h(\xi, t) = \chi_K(t) \chi_K(y+t) f\big(x + \varphi(y+t) - \varphi(t)\big).$$

Then $g(\xi) = \int_{\mathbb{R}^m} h(\xi, t)\, dt$ and we are going to check that the Lebesgue dominated convergence theorem applies to the sequence of functions $\big(h(\xi_j, \cdot)\big)_{j \in \mathbb{N}}$. We first observe that if $y + t$ does not belong to the boundary ∂K of K then

$$\lim_{j \to \infty} h(\xi_j, t) = h(\xi, t).$$

Since the compact set K is Riemann integrable, $|\partial K| = 0$ and therefore

$$\lim_{j \to \infty} h(\xi_j, t) = h(\xi, t), \quad \text{for almost all } t \in \mathbb{R}^m.$$

Then, the functions $h(\xi_j, \cdot)$ are dominated by

$$|h(\xi_j, t)| \leq \|f\|_\infty \chi_K(t) \in L^1(\mathbb{R}^m)$$

and Lemma 4.1 is proved. $\square$

LEMMA 4.2. *For any positive integer N, there exist two positive constants c_N and r_N, only depending on N, f and φ, such that for any $\xi = (x,y)$ in $\mathbb{R}^{n+m}$*

$$|g(\xi)| \leq c_N |K|(|x| + r_N)^{-N} \chi_{(K-K)}(y).$$

PROOF. If $y \notin K - K$, $g(\xi)$ is clearly 0. If $y \in K - K$, we use the fast decay of f

$$|f(x)| \leq c'_N (1 + |x|)^{-N}$$

which gives two kinds of estimate, uniformly with respect to t:

$$|f(x + \varphi(y+t) - \varphi(t))| \leq c'_N (1 + |x| - 2\|\varphi\|_\infty)^{-N} \quad \text{if } |x| \geq 2\|\varphi\|_\infty$$
$$|f(x + \varphi(y+t) - \varphi(t))| \leq c'_N \quad \text{if } |x| \leq 2\|\varphi\|_\infty.$$

Then for any (ξ, t) in $\mathbb{R}^{n+m} \times \mathbb{R}^m$

$$|f(x + \varphi(y+t) - \varphi(t))| \leq c_N (|x| + r_N)^{-N}$$

as soon as c_N and r_N fulfill the following conditions:

$$\begin{cases} (c_N/c'_N)^{1/N} = A_N > \max(1, 2\|\varphi\|_\infty) \\ r_N \in \,]0, A_N - 2\|\varphi\|_\infty[. \end{cases}$$

Finally we get

$$|g(\xi)| \leq c_N |K| (|x| + r_N)^{-N} \chi_{(K-K)}(y). \qquad \square$$

We deduce from Lemmas 4.1 and 4.2 that the following function is well defined and continuous on $\mathbb{R}^{n+m}$

$$F(\zeta) = \sum_{\xi \in L} g(\zeta + \xi).$$

In particular, it follows from Theorem 3.1 that for $\zeta = 0$

$$F(0) = \sum_{\xi \in L} g(\xi) = \sum_{\xi \in L} \int_{K \cap (K - \tilde{x})} f(x + \varphi(\tilde{x} + t) - \varphi(t))\, dt = (\text{vol}\, L) \int_{\mathbb{R}^n} f\, d\gamma$$

where γ is the autocorrelation of X_φ. Now F is clearly L-periodic and the key argument to prove our theorem is

LEMMA 4.3. *The Fourier coefficients of F are given by*

$$a_\kappa = (\text{vol}\, L) \hat{f}(k) |c_k(X_\varphi)|^2, \quad \kappa = (k, \tilde{k}) \in L^*$$

where $c_k(X_\varphi)$ is the Fourier-Bohr coefficient of X_φ. Moreover, one has

(4.1) $$\sum_{\kappa \in L^*} |a_\kappa| < +\infty.$$

PROOF. Let Ω be a fundamental domain of L and let a_κ, $\kappa \in L^*$, be the Fourier coefficients of F defined by

$$a_\kappa = \frac{1}{\text{vol}\, L} \int_\Omega F(\zeta) e^{-i2\pi \kappa \cdot \zeta}\, d\zeta.$$

Thanks to Lemma 4.2, there exist two positive constants c''_N and r''_N such that for any $\xi = (x, \tilde{x}) \in L$ and for any $\zeta \in \Omega$

$$|g(\xi + \zeta)| \leq c''_N \chi_{K - K - \tilde{\Omega}}(\tilde{x})(|x| + r''_N)^{-N}$$

where $\tilde{\Omega} = \tilde{p}(\Omega)$. We deduce that

$$\int_\Omega \left(\sum_{\xi \in L} g(\xi + \zeta)\right) e^{-2\pi i \kappa \cdot \zeta}\, d\zeta = \sum_{\xi \in L} \int_\Omega g(\xi + \zeta) e^{-2\pi i \kappa \cdot \zeta}\, d\zeta = \int_{\mathbb{R}^{n+m}} g(\zeta) e^{-2\pi i \kappa \cdot \zeta}\, d\zeta$$

and consequently, the Fourier coefficients of F are given by

$$a_\kappa = (\mathrm{vol}\, L)^{-1} \hat{g}(\kappa).$$

Then we write

$$\hat{g}(\kappa) = \int_{\mathbb{R}^{n+m}} \left(\int_{\mathbb{R}^m} \chi_K(t) \chi_K(y+t) f\big(x + \varphi(y+t) - \varphi(t)\big)\, dt\right) e^{-2\pi i \kappa \cdot \xi}\, d\xi.$$

At this stage, two cases can occur:

1ST CASE. φ is a C^1 function on $\mathbb{R}^m$. Since

$$\int_{\mathbb{R}^{n+m}} \left|\int_{\mathbb{R}^m} \chi_K(t) \chi_K(y+t) f\big(x + \varphi(y+t) - \varphi(t)\big)\, dt\right|\, d\xi < +\infty$$

we can operate the following change of variables in the integral defining $\hat{g}(\kappa)$:

$$\begin{cases} x' = x + \varphi(y+t) - \varphi(t) \\ y' = y. \end{cases}$$

Then

$$(4.2) \qquad \hat{g}(\kappa) = \iiint \chi_K(t) \chi_K(y+t) f(x) e^{-2\pi(ik \cdot (x - \varphi(y+t) + \varphi(t)) + i\tilde{k} \cdot y)}\, dx\, dy\, dt.$$

Applying Fubini's theorem one gets

$$\hat{g}(\kappa) = \int_{\mathbb{R}^n} f(x) e^{-2\pi i k \cdot x}\, dx \iint_{\mathbb{R}^m \times \mathbb{R}^m} \chi_K(t) e^{-2\pi i k \cdot \varphi(t)} \chi_K(s) e^{2\pi i k \cdot \varphi(s)} e^{2\pi i \tilde{k} \cdot (t-s)}\, ds\, dt$$

$$= (\mathrm{vol}\, L)^2 \hat{f}(k)\, c_k(X_\varphi) \overline{c_k(X_\varphi)}$$

and finally,

$$a_\kappa = (\mathrm{vol}\, L) \hat{f}(k) |c_k(X_\varphi)|^2.$$

2ND CASE. φ is continuous but not C^1 on $\mathbb{R}^m$. Take a sequence of C^1 functions with compact support, $(\varphi_j)_{j \in \mathbb{N}}$, converging uniformly on $\mathbb{R}^m$ to φ. Then one can easily check that Lebesgue's dominated convergence theorem applies in each side of equation (4.2) written for φ_j.

Let us prove the second part of Lemma 4.3. We first suppose that $\hat{f}$ is a positive function. Let ρ be a function in the Schwartz class $\mathcal{S}(\mathbb{R}^{n+m})$ such that

(i) $\hat{\rho}$ is a positive function
(ii) $\hat{\rho}(0) = 1$
(iii) for any $\xi \in \mathbb{R}^{n+m}$, $t \in [0, +\infty[\, \to \hat{\rho}(t\xi)$ is decreasing.

Then it is well known that, since F is a continuous function, the approximation sums

$$\sum_{\kappa \in L^*} \hat{\rho}(\varepsilon \kappa) a_\kappa e^{2\pi i \kappa \cdot \zeta}$$

converge uniformly on $\mathbb{R}^{n+m}$ to $F(\zeta)$ as ε decays to 0 (see [**S**, p. 203] for a proof of this Fejer-Poisson-Dirichlet type theorem). In particular, we get for each $\varepsilon > 0$

$$\sum_{\kappa \in L^*} \hat{\rho}(\varepsilon \kappa) a_\kappa \leq F(0)$$

and therefore using the theorem of Beppo-Levi
$$\sum_{\kappa \in L^*} a_\kappa \leq F(0).$$

The case where $\hat{f}$ is not a positive function is straightforward since a function in $\mathcal{S}(\mathbb{R}^n)$ can always be written as the difference of two positive functions in the Schwartz class. Hence (4.1) holds true for any f in $\mathcal{S}(\mathbb{R}^n)$. □

The final result can be stated as follows

THEOREM 4.4. *Let X_φ be a deformed model set in $\mathbb{R}^n$ constructed with a regular model set $X_0 = \mathfrak{M}(L, K)$ and a continuous function with compact support φ:*
$$X_\varphi = \{x + \varphi(\tilde{x}), x \in X_0\}.$$
Then the diffraction pattern of X_φ is a purely discrete measure given by
$$\hat{\gamma} = \sum_{k \in p(L^*)} |c_k(X_\varphi)|^2 \delta_k$$
where $c_k(X_\varphi)$ is the Fourier-Bohr coefficient of X_φ at k.

PROOF. The proof is now straightforward: F is a continuous L-periodic function whose Fourier series $\sum_{\kappa \in L^*} a_\kappa e^{2\pi i \kappa . \zeta}$ is absolutely convergent. Then it is clear that the following identity holds for each ζ in $\mathbb{R}^{n+m}$
$$F(\zeta) = \sum_{\kappa \in L^*} a_\kappa e^{2\pi i \kappa . \zeta}.$$

In particular we can take $\zeta = 0$ which leads to
$$\int_{\mathbb{R}^n} f \, d\gamma = \sum_{k \in p(L^*)} |c_k(X_\varphi)|^2 \hat{f}(k).$$
Since this identity holds for any f in $\mathcal{S}(\mathbb{R}^n)$, we finally obtain
$$\hat{\gamma} = \sum_{k \in p(L^*)} |c_k(X_\varphi)|^2 \delta_k$$
which ends the proof of Theorem 4.4. □

5. Example: Application to the Deformations of the Fibonacci Chain

We end this work with a simple application to the deformations of the Fibonacci chain. Here $n = m = 1$ and the two-dimensional lattice L is generated by $\varepsilon_1 = (1, -\tau^{-1})$ and $\varepsilon_2 = (\tau^{-1}, 1)$ where $\tau = \frac{1}{2}(1 + \sqrt{5})$ is the golden mean. We put $\varepsilon_i = (e_i, \tilde{e}_i)$ for $i = 1, 2$. The bounded set K is the interval $[\alpha, \alpha + \tau)$ where α is a constant. The corresponding regular model set X_0 can be viewed as a tiling of the real line with two tiles T_1 and T_2 of length $l_1 = 1$ and $l_2 = \tau^{-1}$ which coincide with the projections of ε_1 and ε_2 respectively. We shall consider a deformation given by a continuous function φ with compact support in $\mathbb{R}$ and we will suppose that φ does not alter the order of the points:
$$x \in X_0, x' \in X_0, x < x' \implies x + \varphi(\tilde{x}) < x' + \varphi(\tilde{x}').$$
A tile of type T_i ($i = 1, 2$) occurs between two points x and x' of X_0 whenever $x = p(\xi)$ and $x' = p(\xi + \varepsilon_i)$ with $\xi \in L$ and $\tilde{x}, \tilde{x}' \in K$. Consequently, the deformed tile is defined by the two points $x + \varphi(\tilde{x})$ and $x' + \varphi(\tilde{x}')$ and its length is

$l_i + \varphi(\tilde{x} + \tilde{e}_i) - \varphi(\tilde{x})$. Therefore, unless φ is an affine function, we see that these lengths will be dense in a nontrivial interval. As a consequence, the autocorrelation function has a continuous component.

A special case of deformation occurs when φ is linear (or affine) on K since the length of the deformed tile is given by $l_{\varphi,i} = l_i + \varphi(\tilde{e}_i)$, which is independent of x. This deformed Fibonacci chain F_φ remains a tiling with at the most two tiles and the analysis presented in this work shows that the diffraction pattern is a discrete measure carried by the spectrum of X_0, namely

$$p(L^*) = \frac{1}{1+\tau^{-2}}\mathbb{Z} + \frac{\tau^{-1}}{1+\tau^{-2}}\mathbb{Z}.$$

A very special case occurs when $l_{\varphi,1} = l_{\varphi,2}$ which is obtained when $\varphi(y) = \tau^{-3}y$ for $y \in K$. In this case the deformed set F_φ is periodic and provides an average lattice of the Fibonacci chain. Precisely $F_\varphi = h_\alpha + (3\tau^{-1} - 1)\mathbb{Z}$ for some real number h_α in $[0, 3\tau^{-1} - 1[$. Applying Theorem 4.4, we know that the diffraction pattern of this deformed Fibonacci chain is a pure Bragg measure of the form

(5.1) $$\widehat{\gamma} = \sum_{k \in p(L^*)} |c_k(F_\varphi)|^2 \delta_k.$$

But since F_φ is a lattice, $\widehat{\gamma}$ should be supported by its dual lattice and this does not appear yet in (5.1). This apparent contradiction disappears as soon as we use Theorem 2.6 which gives

$$c_k(F_\varphi) = (\text{vol } L)^{-1} \int_\alpha^{\alpha+\tau} e^{-2\pi i(\tau^{-3}k - \tilde{k})y}\, dy$$

for each k in $p(L^*)$. It is an easy check to see that $c_k(F_\varphi)$ vanishes as soon as there exist two distinct integers m_1 and m_2 such that

$$k = \frac{1}{1+\tau^{-2}}m_1 + \frac{\tau^{-1}}{1+\tau^{-2}}m_2.$$

Moreover when m_1 equals m_2, $c_k(F_\varphi) = \tau/(\text{vol } L)$. Hence we get

$$\widehat{\gamma} = \left(\frac{\tau}{\text{vol } L}\right)^2 \sum_{m \in \mathbb{Z}} \delta_{m(1+\tau^{-1})/(1+\tau^{-2})}$$

and $\widehat{\gamma}$ now appears to be carried by a "dual" lattice which is contained in $p(L^*)$. Furthermore, one can observe that $h_\alpha = 0$ since there always exists $(x, \tilde{x})$ in L such that $x + \varphi(\tilde{x}) = 0$ and $\tilde{x} \in [\alpha, \alpha + \tau]$. Therefore

$$F_\varphi = (3\tau^{-1} - 1)\mathbb{Z}.$$

Notice that the deformed set does not depend on α: for any window $K = [\alpha, \alpha+\tau]$, the deformed Fibonacci chain is always centered around 0. This actually follows from the linearity of the deformation.

References

[BM] M. Baake and R. V. Moody, *Self-similar measures for quasicrystals*, Directions in Mathematical Quasicrystals, CRM Monogr. Series, vol. 13, Amer. Math. Soc., Providence, RI, 2000, pp. 1–42, this volume.

[B] G. Bernuau, *Propriétés spectrales et géométriques des quasicristaux. Ondelettes adaptées aux quasicristaux*, Ph.D. Thesis, Ceremade, Université Paris IX Dauphine, France, 1998.

[D] S. Dworkin, *Spectral theory and X-ray diffraction*, J. Math. Phys. **34** (1993), 2965–2967.

[GK] F. Gähler and R. Klitzing, *The diffraction pattern of self-similar tilings*, The Mathematics of Long-Range Aperiodic Order (Waterloo, ON, 1995) (R. V. Moody, ed.), NATO ASI Ser. C: Math. Phys. Sci., vol. 489, Kluwer, Dordrecht, 1997, pp. 141–174.

[GH] C. P. M. Geerse and A. Hof, *Lattice gas models on self-similar aperiodic tilings*, Rev. Math. Phys. **3** (1991), 163–221.

[H1] A. Hof, *On diffraction by aperiodic structures*, Commun. Math. Phys. **169** (1995), 25–43.

[H2] _____, *Diffraction by aperiodic structures*, The Mathematics of Long-Range Aperiodic Order (Waterloo, ON, 1995) (R. V. Moody, ed.), NATO ASI Ser. C: Math. Phys. Sci., vol. 489, Kluwer, Dordrecht, 1997, pp. 239–268.

[H3] A. Hof, *Uniform distribution and the projection method*, Quasicrystals and Discrete Geometry (Toronto, ON, 1995) (J. Patera, ed.), Fields Inst. Monogr., vol. 10, Amer. Math. Soc., Providence, RI, 1998, pp. 201–206.

[J] C. Janot, *Quasicrystals. A primer*, 2nd ed., Monographs on the Physics and Chemistry of Materials, Clarendon Press, Oxford, 1994.

[KN] L. Kuipers and H. Niederreiter, *Uniform distribution of sequences*, Pure Appl. Math., Wiley, New York, 1974.

[L] J. C. Lagarias, *Meyer's concept of quasicrystal and quasiregular sets*, Commun. Math. Phys. **179** (1996), 365–376.

[LP] W. F. Lunnon and P. A. B. Pleasants, *Quasicrystallographic tilings*, J. Math. Pures Appl. (9) **66** (1987), 217–263.

[M1] Y. Meyer, *Nombres de Pisot, nombres de Salem et analyse harmonique*, Lecture Notes in Math., vol. 117, Springer, Berlin, 1970.

[M2] _____, *Algebraic numbers and harmonic analysis*, North-Holland Math. Library, vol. 2, North-Holland, Amsterdam, 1972.

[M3] _____, *Quasicrystals, Diophantine approximations and algebraic numbers*, Beyond Quasicrystals (Les Houches, 1994) (F. Axel and D. Gratias, eds.), Springer, Berlin, 1995, pp. 3–16.

[R] W. Rudin, *Real and complex analysis*, 3rd ed., McGraw-Hill, New York, 1987.

[S1] M. Schlottmann, *Cut-and-project sets in locally compact Abelian groups*, Quasicrystals and Discrete Geometry (Toronto, ON, 1995) (J. Patera, ed.), Fields Inst. Monogr., vol. 10, Amer. Math. Soc., Providence, RI, 1998, pp. 247–264.

[S2] _____, *Generalized model sets and dynamical systems*, Directions in Mathematical Quasicrystals, CRM Monogr. Series, vol. 13, Amer. Math. Soc., Providence, RI, 2000, pp. 143–159, this volume.

[Sch] J. P. Schreiber, *Approximations diophantiennes et problèmes additifs dans les groupes abéliens localement compacts*, Ph.D. Thesis, Faculté des Sciences d'Orsay, France, 1972.

[S] L. Schwartz, *Analyse*, Part 4. Applications à la théorie de la mesure, Hermann, Paris, 1993.

[So] B. Solomyak, *Dynamics of self-similar tilings*, Ergodic Theory & Dynam. Systems **17** (1997), 695–738.

Centre de physique théorique, École Polytechnique, 91128 Palaiseau Cedex, France
E-mail address: `duneau@cpht.polytechnique.fr`

Mathematical Quasicrystals and the Problem of Diffraction

Jeffrey C. Lagarias

ABSTRACT. This paper studies three mathematical idealizations of quasicrystals which embody a notion of perfectly sharp diffraction spectrum. These idealizations consist of Delone sets that satisfy additional conditions. The first concept (Patterson set) is based on having a pure point diffraction spectrum. To each Patterson set which is a Delone set of finite type there corresponds a summation formula, which can be viewed as generalizing the Poisson summation formula. The second and third concepts (Bohr almost periodic Delone set and Besicovitch almost periodic Delone set) are based on almost-periodicity conditions imposed on their Fourier transform. The latter two concepts are proposed to extract "phase information" for quasicrystals. The paper concludes with a list of open problems.

1. Introduction

The discovery of quasicrystalline materials in 1982 by Shechtman et al. (published two years later, [**SBGC**]) led to extensive theoretical and empirical efforts to understand their structure, see [**Jan, Sen**]. The intuitive notion of a quasicrystal is a (very large) discrete set of atoms in space whose X-ray diffraction pattern exhibits sharp spots. This condition requires that the interatomic distance vectors exhibit long-range order under translations, in a statistical sense. This paper considers mathematical idealizations of such structures which are infinite discrete sets which have perfect diffraction patterns (pure delta functions), rather than the slightly diffuse spots of actual X-ray diffraction patterns. These structures could model pure point diffractive quasicrystalline materials. In this framework we view ideal crystals as a special kind of pure point diffractive quasicrystal.

The basic mathematical object is a set in $\mathbb{R}^n$ which models an infinite limit of a physical structure consisting of a discrete set of atoms.

DEFINITION 1.1. A *Delone set* Λ in $\mathbb{R}^n$ is a set with the properties:

(1) *Uniform Discreteness.* There is $r > 0$ such that each ball of radius r contains at most one element of Λ.

2000 *Mathematics Subject Classification.* Primary 52C23, 42A75; Secondary 11M26, 46F10, 52C07.

I am indebted to M. Baake, A. Hof, R. V. Moody, P. A. B. Pleasants, A. Robinson, Jr., and B. Solomyak for helpful comments. I am grateful to the referees for many constructive suggestions.

©2000 American Mathematical Society

(2) *Relative Denseness.* There is $R > 0$ such that each ball of radius R contains at least one element of Λ.

Such sets are sometimes called *(r,R)-sets*. These sets are named after the Russian crystallographer and number theorist B. N. Delone [**D**].

We consider Delone sets that usually satisfy additional conditions. Recently Lagarias [**La99a**] formulated the notion of Delone set of finite type as a model for Delone sets having weak translational order.

DEFINITION 1.2.   (i) A *Delone set of finite type* is a Delone set Λ such that $\Lambda - \Lambda$ is a closed discrete set.
(ii) A *Meyer set* is a a Delone set Λ such that $\Lambda - \Lambda$ is a Delone set.

Delone sets of finite type are exactly those Delone sets that have a "finite number of local patterns", see [**La99a**, Theorem 2.2]. This property is also called "finite local complexity", see [**BH, S99**]. Meyer sets are an important subclass of these sets, introduced much earlier, whose properties are given in detail in Moody [**Mo97**]. They have several equivalent definitions, the one above being formulated in [**La95**].

DEFINITION 1.3. An *ideal crystal*(or *perfect crystal*) in $\mathbb{R}^n$ is any set Λ that consists of a finite number of translates of a full rank lattice L in $\mathbb{R}^n$. That is $\Lambda = L + F$, where F is a finite set.

Note that ideal crystals are Meyer sets, and we view them as a special kind of quasicrystal.

We consider three different concepts of pure point diffractive quasicrystal, all for Delone sets. The first concept, of Patterson set, is based on a mathematical analogue of X-ray diffraction developed by Hof ([**H92, H95a, H95b, H97**]). The second and third concepts, of Bohr almost periodic set and its extension to the concept of Besicovitch almost periodic set, are based on Fourier analysis, and each gives a notion of "spectrum" assigned to the Fourier transform of point masses at the points of Λ. These two concepts add "phase information" to the X-ray diffraction data.

The concept of *Patterson set* is studied in Section 2. It is based on the notion of autocorrelation measure (or Patterson function) associated to the difference set $\Lambda - \Lambda$ of the set Λ. The diffraction measure is the Fourier transform of the autocorrelation measure, and a Patterson set is a Delone set which has a unique diffraction measure which is a pure discrete measure. Our main new observation in Section 2 is to show that this concept has a precise relation with summation formulae in Fourier analysis (Theorem 2.9).

There are two general constructions of Delone sets of finite type Λ that are known to yield Patterson sets in special cases: model sets, which include as a special case cut-and-project sets, and certain Delone sets defined by self-similarity properties, which we call self-replicating Delone sets. A general method of proof that certain such sets are Patterson sets uses dynamical system methods described at the end of Section 2.

Cut and project sets are Delone sets in $\mathbb{R}^n$ constructed from a full rank lattice L in $\mathbb{R}^{n+m}$ for some $m \geq 0$, together with a compact set B in $\mathbb{R}^m$ which has nonempty interior. The space $\mathbb{R}^m$ is the "internal space" of the construction, and the set B in the "internal space" is called a mask or window. View $\mathbb{R}^{n+m} = \mathbb{R}^n \times \mathbb{R}^m$ with orthogonal projections $\pi_\|$ and $\pi_\perp$ onto the first n coordinates and last m

coordinates, respectively. The *cut and project set* $\Lambda = \Lambda(B, L)$ is defined by

$$\Lambda := \{\pi_\|(\boldsymbol{y}) : \boldsymbol{y} \in L \text{ and } \pi_\perp(\boldsymbol{y}) \in B\}. \tag{1.1}$$

Cut-and-project sets are always Meyer sets. Whenever a cut-and-project set is a Patterson set, its spectrum is contained in a finitely-generated $\mathbb{Z}$-module in $\mathbb{R}^n$, related to the dual lattice L^* of L.

Model Sets are Delone sets in $\mathbb{R}^n$ which generalize cut-and-project sets, and include them as a special case. The concept of model set was introduced in 1972 by Y. Meyer [**Me72**, p. 48]. They are produced by a similar construction in which the "internal space" is allowed to be an arbitrary locally compact Abelian group, see Schlottmann [**S98**] and Moody [**Mo99**]. The window set B is required to be a compact set with nonempty interior. A model set is said to be *regular* if the window B has a boundary $\partial B := B \setminus \text{Int}(B)$ of (Haar) measure zero. Schlottmann [**S99**] shows that regular model sets have a well-defined pure point diffraction measure. Model sets using a p-adic internal space occur in certain self-similar tiling constructions, see Baake, Moody and Schlottmann [**BMS**] and Lee and Moody [**LM**]. Model sets are always Meyer sets. The spectrum of a regular model set is not always contained in a finitely-generated $\mathbb{Z}$-module, as indicated by the example studied in [**BMS**].

Self-replicating Delone sets describe "control points" of associated self-affine tilings, see Gähler and Klitzing [**GK**] and Solomyak [**So97**, Section 5]. These sets are studied in Lagarias and Wang [**LaWa3**], and have a theory analogous to that for self-replicating tilings. At this point we observe only that such sets are Delone sets Λ that have a partition $\Lambda = \bigcup_{i=1}^{m} \Lambda_i$ in which the subsets Λ_i satisfy a system of functional equations

$$\Lambda_i = \bigcup_{j=1}^{m} (\phi(\Lambda_j) + \mathcal{D}_{ji}), \quad 1 \leq i \leq m, \tag{1.2}$$

in which

$$\phi(\boldsymbol{x}) = A\boldsymbol{x} + \boldsymbol{b} \tag{1.3}$$

is an expanding affine map, i.e., the matrix A has all eigenvalues $|\lambda| > 1$, and the *digit sets* $\mathcal{D}_{ij}$ are finite sets. The matrix A is called the *inflation matrix*. Associated to the functional equation (1.2) is a *substitution matrix* S which is a nonnegative integer matrix defined by

$$S_{ij} := |\mathcal{D}_{ij}|, \quad 1 \leq i, j \leq m. \tag{1.4}$$

We suppose that the substitution matrix is *primitive*, which means that some power S^k has strictly positive entries. Only special choices of the data $\{\phi, \mathcal{D}_{ij}\}$ yield functional equations (1.2) that have solutions which are self-replicating Delone sets of finite type. For example, the real matrix A must have algebraic integer eigenvalues, and if S is primitive then the largest eigenvalue of S must equal $|\det A|$. There exist self-replicating Delone sets which are Delone sets of finite type but are not Meyer sets. Some self-replicating Delone sets are Patterson sets, while others are not. All known examples of primitive self-replicating Delone sets that have been proved to have pure point diffraction spectrum are Meyer sets. The diffraction spectrum results in the literature are generally proved for the associated tiling models, as in Solomyak [**So97**], but Delone sets are explicitly considered in Solomyak [**So98b**]. For various results concerning self-affine and self-replicating tilings, see Gröchenig, Haas and Raugi [**GHR**], Kenyon[**Ke92, Ke94, Ke96**], Lagarias and Wang [**LaWa1, LaWa2**], Solomyak [**So97, So98a**] and Vince [**Vin1, Vin2**].

The concepts of *Bohr almost periodic set* and *Besicovitch almost periodic set* are presented in Section 3. We view these concepts as supplying "phase information" about f_Λ which is lost in passing to the autocorrelation measure. We associate to a Delone set Λ the Radon measure whose density function f_Λ consists of delta functions at the points of Λ. A Bohr almost periodic set is defined to be a distribution f_Λ whose Fourier transform is in a suitable weak sense a countable set of weighted delta functions. We formalize this concept using uniformly almost periodic functions and distributions, see Appendix B. However the concept of Bohr almost periodic set seems too narrow to include many sets regarded as quasicrystalline (including most cut-and-project sets), so we formulate a relaxed concept of *Besicovitch almost periodic set*, whose definition uses a wider class of almost periodic functions. This concept is expected to include cut-and-project sets; see the open problems in Section 4. We define more generally $\mathcal{B}$-*almost periodic sets*, where $\mathcal{B}$ is a suitable class of almost periodic distributions, and it remains to determine a good class $\mathcal{B}$ that gives a reasonable theory.

The inclusion relations between these three concepts of pure point diffraction quasicrystal are not known, except that Bohr almost periodic sets are Besicovitch almost periodic sets, which follows from the definition. It is natural to hope that a suitable class of $\mathcal{B}$-Besicovitch almost periodic sets will all be Patterson sets and have the consistent phase property given in (3.9), but this is an open question.

It is known that the information contained in the diffraction spectrum is not sufficient to reconstruct the set, up to translation. If Λ is a Patterson set and $\tilde{\Lambda}$ is a Delone set such that the symmetric difference

$$\Lambda \, \Delta \, \tilde{\Lambda} := (\Lambda \setminus \Lambda') \cup (\Lambda' \setminus \Lambda)$$

is a set of density zero, then $\tilde{\Lambda}$ is a Patterson set with the same spectrum. For similar reasons, the " phase information" obtained in a Besicovitch almost periodic set (of type B^2) is generally insufficient to reconstruct the set, because the "Fourier coefficients" are also unchanged by sets of density zero, see Section 3. However the narrower class of Bohr almost periodic sets (which include ideal crystals) are uniquely reconstructible from the "Fourier coefficients" of their spectrum.

The final section Section 4 lists a number of open problems raised by the topics above.

There are several other concepts of mathematical quasicrystal not considered here. The first models proposed for quasicrystalline structures were based on tilings of $\mathbb{R}^n$ using a finite number of different tile shapes, see Duneau and Katz [**DuKa**] and Levine and Steinhardt [**LSt**]. Later Lunnon and Pleasants [**LuP**] introduced a notion of *quasiperiodic tiling* in which $\mathbb{R}^n$ is tiled by tiles of a finite number of shapes $P_1, \ldots, P_k$, all polytopes, with the property that if any set of continuous functions $f_1, f_2, \ldots, f_k$ are assigned to these polytopes, and used to construct a function f on $L^\infty(\mathbb{R}^n)$ by replicating $f_i(\boldsymbol{x}+\boldsymbol{v})$ on each tile $P_i+\boldsymbol{v}$, then the Fourier transform $\hat{f}$ (in a suitable space of distributions) consists entirely of delta functions supported on a finitely-generated additive subgroup of $\mathbb{R}^n$ (a *quasilattice*), see also Le, Piunikhin, and Sadov [**LPS**]. There are also notions of quasicrystals as consisting of a collection of an uncountable number of tilings viewed as a dynamical system under the action of the group of translations of $\mathbb{R}^n$. These are called *tiling dynamical* systems. The eigenvalues of the $\mathbb{R}^n$-action of translation on tilings then play a role analogous to diffraction spectra, see Dworkin [**Dw**], and Hof [**H97**, p. 254]. Analogous dynamical systems for Delone sets play a role in proving certain sets are Patterson sets, see Section 2. A discussion of other mathematical concepts related to quasicrystals appears in Cahn and Taylor [**CT**] and Baake [**Ba**].

The final Section 4 lists a number of open problems.

Notation. The Euclidean inner product on $\mathbb{R}^n$ is
$$\langle \boldsymbol{\xi}, \boldsymbol{x} \rangle = \sum_{i=1}^{n} \xi_i x_i.$$

The Schwartz space of rapidly decreasing smooth functions on $\mathbb{R}^n$ is $\mathcal{S}(\mathbb{R}^n)$. The (normalized) Fourier transform $\hat{f}$ of $f \in \mathcal{S}(\mathbb{R}^n)$ is

$$\hat{f}(\boldsymbol{\xi}) := \int_{\mathbb{R}^n} e^{-2\pi i \langle \boldsymbol{\xi}, \boldsymbol{x} \rangle} f(\boldsymbol{x}) \, d\boldsymbol{x}. \tag{1.5}$$

The Fourier transform is defined for tempered distributions $f \in \mathcal{S}'(\mathbb{R}^n)$ in the usual fashion: $\langle \hat{f}, \psi \rangle = \langle f, \hat{\psi} \rangle$ for test functions $\psi \in \mathcal{S}(\mathbb{R}^n)$. The definition of spectrum in this paper removes a factor of 2π from the standard one, due to the 2π appearing in the definition of Fourier transform (1.5), see Appendix B.

2. Patterson Sets and Summation Formulas

The fundamental notion of a "quasicrystal" is a physical structure whose X-ray diffraction measure pattern consists of sharp spots. The concept of Patterson set is based on a mathematical concept of diffraction measure developed by A. Hof ([**H92, H95a, H95b, H97**]).

We will model sets of atoms[1] located at a discrete set Λ by the pure point measure μ_Λ which consists of unit masses at the points of Λ. The measure μ_Λ can be regarded either as a regular Borel measure on $\mathbb{R}^n$, or alternatively as a positive Radon measure (by the Riesz representation theorem). However we shall generally regard it as a distribution, also denoted μ_Λ, and written

$$\mu_\Lambda := \sum_{\boldsymbol{x} \in \Lambda} \delta_{\boldsymbol{x}}, \tag{2.1}$$

which is associated to the measure by

$$\langle \mu_\Lambda, g \rangle := \int_{\mathbb{R}^n} g(\boldsymbol{x}) \, d\mu_\Lambda(\boldsymbol{x}) = \sum_{\boldsymbol{x} \in \Lambda} g(\boldsymbol{x}). \tag{2.2}$$

for test functions g. If Λ is a Delone set, then μ_Λ is a tempered distribution. Given a distribution of the general type

$$g = \sum_{\boldsymbol{x} \in \Lambda} n(\boldsymbol{x}) \delta_{\boldsymbol{x}}, \tag{2.3}$$

in which Λ is a discrete set the weights $n(\boldsymbol{x})$ are complex numbers with $|n(\boldsymbol{x})| \leq C$, the associated regular Borel measure is uniquely defined via (2.2).

---

[1]An atomic structure may be more accurately represented as a measure obtained by convolving μ_Λ with a compactly supported nonnegative "bump function," see Hof [**H95b**]. Here we are concerned with the perfect idealization.

DEFINITION 2.1. A complex-valued regular Borel measure μ on $\mathbb{R}^n$ is called *translation-bounded* if there is a constant C such that

$$|\mu|(\boldsymbol{x} + [0,1]^n) \leq C, \quad \text{for all } \boldsymbol{x} \in \mathbb{R}^n. \tag{2.4}$$

A. Hof developed a mathematical formalization of a diffraction measure associated to an arbitrary translation-bounded measure μ. To define it we first need the notion of an autocorrelation measure.

DEFINITION 2.2. Given a translation-bounded measure μ an *autocorrelation measure* indexautocorrelation measure γ of μ is any measure that is a limit point in the vague topology of a sequence of measures $\{\nu_{j,\boldsymbol{w}_j} : j = 1, 2, 3, \dots\}$, where

$$\nu_{T,\boldsymbol{w}} := \frac{1}{T^n}(\mu|_{\boldsymbol{w}+T[0,1]^n} * \tilde{\mu}|_{\boldsymbol{w}+T[0,1]^n}), \tag{2.5}$$

in which $\tilde{\mu}$ is the complex-conjugate measure to μ, with the space direction reversed. Here convergence $\nu_j \to \nu$ as $n \to \infty$ in the vague topology means that for each compactly supported continuous function $\phi : \mathbb{R}^n \to \mathbb{C}$ one has $\int_{\mathbb{R}^n} \phi(\boldsymbol{x}) \, d\nu_j \to \int_{\mathbb{R}^n} \phi(\boldsymbol{x}) \, d\nu$.

A translation-bounded measure μ has at least one autocorrelation measure; in general it has many autocorrelation measures. We mainly consider cases where μ_Λ has a unique autocorrelation measure.

For a discrete measure μ_Λ an autocorrelation measure encodes information about the "two-point correlation function" of Λ, i.e., the difference set $\Lambda - \Lambda$, with elements counted with multiplicity. The notion of autocorrelation measure is translation-invariant: for each translate $\Lambda + \boldsymbol{x}$ of a set Λ the measure $\mu_\Lambda + \boldsymbol{x}$ has the same autocorrelation measures as μ_Λ.

LEMMA 2.3. *Let μ be a positive measure that is translation-bounded with constant C, and let γ be any autocorrelation measure for μ. Then:*

(1) *γ is a positive measure that is translation-bounded with constant C.*
(2) *γ is a positive-definite measure (in the sense of tempered distributions). That is, the Fourier transform $\widehat{\gamma}$ is a distribution of positive type.*
(3) *The Fourier transform $\widehat{\gamma}$ is a translation-bounded positive measure.*

PROOF. Properties (1) and (2) hold for all measures $\nu_{T,\boldsymbol{x}}$ and are inherited by γ. Part (3) follows from Hof [**H92**, Proposition 3.3], and is a result of Argabright and Gil de Lamadrid [**AGL**, Theorems 2.5 and 4.1]. The translation-boundedness constant C' for $\widehat{\gamma}$ generally differs from that of γ. See Appendix A for a discussion of positive-definite measures. □

DEFINITION 2.4. A *diffraction measure* for a Delone set Λ is the Fourier transform $\widehat{\gamma}$ of an autocorrelation measure γ of μ_Λ regarded as a tempered distribution.

The tempered distribution $\hat{\gamma}$ is identified with a translation-bounded positive measure by Lemma 2.3. We mainly consider cases in which the set Λ has a unique autocorrelation measure γ_Λ; in this case we call γ_Λ *the* autocorrelation measure of Λ and $\widehat{\gamma}_\Lambda$ *the* diffraction measure of Λ.

A diffraction measure $\widehat{\gamma}$ is a mathematical analogue of X-ray diffraction in the sense that values of the measure $\widehat{\gamma}$ evaluated on "bump functions" ("pixels") are analogues of physical X-ray diffraction pictures, see Gähler and Klitzing [**GK**].

DEFINITION 2.5. A *Patterson set* or *perfectly diffractive Delone set* s a Delone set Λ that has a unique autocorrelation measure γ_Λ whose associated diffraction measure $\widehat{\gamma}_\Lambda$ is a pure discrete measure. That is, there is a countable set $\sigma_P(X)$ such that

$$\widehat{\gamma}_\Lambda := \sum_{\boldsymbol{y} \in \sigma_P(\Lambda)} p(\boldsymbol{y})\delta_{\boldsymbol{y}}, \tag{2.6}$$

with all $p(\boldsymbol{y}) > 0$ for $\boldsymbol{y} \in \sigma_P(\Lambda)$. We call $\sigma_P(\Lambda)$ the *Patterson spectrum* of Λ.

The name 'Patterson set' reflects the fact that the autocorrelation is termed the *Patterson series* in X-ray crystallography, see Azároff [**Az**, p. 307], or [**Cow**, Chapter 5.3] for general background.

We note some elementary facts about Patterson sets. The positive-definiteness of the autocorrelation measure γ_Λ guarantees that $\sigma_P(\Lambda) = -\sigma_P(\Lambda)$ and

$$p(\boldsymbol{y}) = p(-\boldsymbol{y}) \geq 0. \tag{2.7}$$

Furthermore $p(\boldsymbol{y}) \leq C'$ follows from the translation-boundedness of μ_Λ with constant C'. In interesting examples the Patterson spectrum $Y = \sigma_P(\Lambda)$ is a dense set. However, for each $\epsilon > 0$ the set

$$Y_\epsilon := \{\boldsymbol{y} \in Y : p(\boldsymbol{y}) \geq \epsilon\}, \tag{2.8}$$

is a closed discrete set, as a consequence of the translation-boundedness of δ_Λ. In an actual X-ray diffraction spectrum only sufficiently large intensities will be detected above background levels, and the discreteness of Y_ϵ in (2.8) justifies how a Patterson set produces a pattern of discrete "bright spots."

The property of being a Patterson set is not affected by "small" changes in the set Λ. If Λ is a Delone set with a unique autocorrelation and Λ' is any Delone set, such that

$$\Lambda \, \Delta \, \Lambda' := (\Lambda \setminus \Lambda') \cup (\Lambda' \setminus \Lambda) \tag{2.9}$$

has density zero, in the sense that

$$\lim_{T \to \infty} \frac{1}{(2T)^n} \#(\Lambda \, \Delta \, \Lambda') \cap [-T, T]^n = 0, \tag{2.10}$$

then Λ' has the same autocorrelation measure as Λ. (See Hof [**H95a**].)

Many constructions of Patterson sets are based on the Poisson summation formula, which we state in the following form.

THEOREM 2.6 (Poisson summation formula). *For a full rank lattice L in $\mathbb{R}^n$ the tempered distribution*

$$\mu_L = \sum_{\boldsymbol{x} \in L} \delta_{\boldsymbol{x}}, \tag{2.11}$$

has Fourier transform

$$\hat{\mu}_L = \frac{1}{|\det(L)|}\mu_{L^*} = \frac{1}{|\det(L)|} \sum_{\boldsymbol{y} \in L^*} \delta_{\boldsymbol{y}}, \tag{2.12}$$

in which L^* is the dual lattice

$$L^* = \{\boldsymbol{y} \in \mathbb{R}^n : \langle \boldsymbol{y}, \boldsymbol{x} \rangle \in \mathbb{Z} \text{ for all } \boldsymbol{x} \in \Lambda\}. \tag{2.13}$$

The formula is equivalent to the assertion that for a Schwartz function $g \in \mathcal{S}(\mathbb{R}^n)$ one has

$$\sum_{\boldsymbol{x} \in L} g(\boldsymbol{x}) = \frac{1}{|\det(L)|} \sum_{\boldsymbol{y} \in L^*} \hat{g}(\boldsymbol{y}). \tag{2.14}$$

The Poisson summation formula is often stated in a more general form explicitly exhibiting the action of a translation $\boldsymbol{t} \in \mathbb{R}^n$,

$$\text{(2.15)} \quad \sum_{\boldsymbol{x} \in L} g(\boldsymbol{x} - \boldsymbol{t}) = \frac{1}{|\det(L)|} \sum_{\boldsymbol{y} \in L^*} \hat{g}(\boldsymbol{y}) e^{2\pi i \langle \boldsymbol{y}, \boldsymbol{t} \rangle}.$$

For a proof of the validity of (2.14), which applied to a wider class of test functions than the Schwartz class, see Katznelson [**K68**, p. 129] and Gröchenig [**Gr**]. A generalization of the Poisson summation formula to a class of unbounded measures on a general locally compact Abelian group appears in Argabright and Gil de Lamadrid [**AGL**, Theorem 3.3].

Using this formula it is easy to verify that all ideal crystals are Patterson sets.

THEOREM 2.7. *An ideal crystal $\Lambda = L + F$, in which L is a full rank lattice in $\mathbb{R}^n$ and F is a finite set, has a unique autocorrelation measure*

$$\text{(2.16)} \quad \gamma_\Lambda = \frac{1}{|\det(L)|} \sum_{\boldsymbol{f}_1 \in F} \sum_{\boldsymbol{f}_2 \in F} \left(\sum_{\boldsymbol{x} \in L} \delta_{\boldsymbol{x} + \boldsymbol{f}_1 - \boldsymbol{f}_2} \right).$$

Its Fourier transform $\widehat{\gamma}_\Lambda$ is given by

$$\text{(2.17)} \quad \widehat{\gamma}_\Lambda = \frac{1}{|\det(L)|^2} \sum_{\boldsymbol{y} \in L^*} \left(\sum_{\boldsymbol{f}_1 \in F} \sum_{\boldsymbol{f}_2 \in F} e^{2\pi i \langle \boldsymbol{f}_1 - \boldsymbol{f}_2, \boldsymbol{\xi} \rangle} \right) \delta_{\boldsymbol{y}}.$$

Thus Λ is a Patterson set with spectrum $\sigma_P(\Lambda)$ contained in the dual lattice L^*.

PROOF. To obtain (2.16) we use (2.5) and count elements of $(L+F) - (L+F)$ on a box $[-T, T]^n$ and let $T \to \infty$. These points fall in $L + (F - F)$ and the density yields a weight $1/|\det(L)|$; we omit the estimates. The Poisson summation formula yields (2.17). □

In the special case that Λ is a lattice L, its autocorrelation measure γ_L is equal to the measure μ_L up to a scale factor, namely

$$\text{(2.18)} \quad \gamma_L = \frac{1}{|\det(L)|} \mu_L = \frac{1}{|\det(L)|} \sum_{\boldsymbol{x} \in L} \delta_{\boldsymbol{x}}.$$

In this case the Patterson spectrum $\sigma_P(L) = L^*$. For a general ideal crystal $\Lambda = L + F$ one can have $\sigma_p(L) \neq L^*$. For the one-dimensional example

$$\Lambda := \mathbb{Z} \cup (\mathbb{Z} + a) \cup (\mathbb{Z} + b) \cup (\mathbb{Z} + c)$$

one can find irrational a, b, c such that

$$1 + e^{2\pi i n a} + e^{2\pi i n b} + e^{2\pi i n c} = 0$$

holds only for $n = \pm 1$, in which case $\sigma_P(\Lambda) = \mathbb{Z} \setminus \{\pm 1\}$.

There is a strong connection between Patterson sets and summation formulas. Recall that a Delone set of finite type is a Delone set Λ such that $\Lambda - \Lambda$ is a closed discrete set.

LEMMA 2.8. *If Λ is a Delone set of finite type then any autocorrelation measure γ of μ_Λ is a pure discrete measure of the form*

$$\text{(2.19)} \quad \gamma = \sum_{\boldsymbol{y} \in \Lambda - \Lambda} n(\boldsymbol{y}) \delta_{\boldsymbol{y}},$$

in which $n(\boldsymbol{y}) = n(-\boldsymbol{y}) \geq 0$.

PROOF. Each measure $\mu_{T,\boldsymbol{w}}$ in (2.5) is pure discrete, and has the form (2.19) with $n(\boldsymbol{y}) = n(-\boldsymbol{y}) \geq 0$. Since $\Lambda - \Lambda$ is a closed discrete set, any limit point in the vague topology of $\mu_{T,\boldsymbol{w}}$ inherits these properties. □

Thus we deduce:

THEOREM 2.9 (Quasicrystal summation formula). *Suppose that Λ is a Delone set of finite type in $\mathbb{R}^n$. If Λ is a Patterson set then its autocorrelation measure γ_Λ and Fourier transform $\widehat{\gamma}_\Lambda$ have the form*

$$(2.20) \qquad \gamma_\Lambda = \sum_{\boldsymbol{y} \in \Lambda - \Lambda} n(\boldsymbol{y})\delta_{\boldsymbol{y}} \quad \text{and} \quad \widehat{\gamma}_\Lambda = \sum_{\boldsymbol{z} \in \sigma_P(\Lambda)} p(\boldsymbol{z})\delta_{\boldsymbol{z}}.$$

Both γ_Λ and $\widehat{\gamma}_\Lambda$ are translation-bounded measures on $\mathbb{R}^n$. For each function g in the Schwartz space $\mathcal{S}(\mathbb{R}^n)$,

$$(2.21) \qquad \sum_{\boldsymbol{y} \in \Lambda - \Lambda} n(\boldsymbol{y})\hat{g}(\boldsymbol{y}) = \sum_{\boldsymbol{z} \in \sigma_P(\Lambda)} p(\boldsymbol{z})g(\boldsymbol{z}).$$

PROOF. This follows from Lemma 2.3 and the definition of Patterson set. For a test function $g \in \mathcal{S}(\mathbb{R}^n)$, the left side of (2.21) is $\langle \gamma_\Lambda, \hat{g} \rangle$ while the right side is $\langle \widehat{\gamma}_\Lambda, g \rangle$. □

The quasicrystal summation formula (2.21) may be valid for wider classes of functions $g(\boldsymbol{y})$ than just those functions in the Schwartz space $\mathcal{S}(\mathbb{R}^n)$. This is the case for the Poisson summation formula, see for example Gröchenig [**Gr**] and Kahane and Lemarié-Rieusset [**KLR**]. There are nontrivial limits to the range of validity of the Poisson summation formula, however. Katznelson ([**K67**], [**K68**, p. 155]) gives an example of a function $g \in L^1(\mathbb{R})$ such that $\hat{g} \in L^1(\hat{\mathbb{R}})$ and both sides of (2.21) converge absolutely but do not agree.

Theorem 2.9 applies more generally to sets Λ that are not Delone sets but retain the "finite local complexity" property that $\Lambda - \Lambda$ is a discrete closed set [2]. An interesting example of such a set having a pure point diffraction spectrum is the set of visible lattice points in $\mathbb{Z}^2$, as was recently shown by Baake, Moody and Pleasants [**BMP**].

There exist many interesting summation formulas known which are formally of the general type

$$\gamma_\Lambda = \sum_{\boldsymbol{y} \in Y} n(\boldsymbol{y})\delta_{\boldsymbol{y}} \quad \text{and} \quad \widehat{\gamma}_\Lambda = \sum_{\boldsymbol{z} \in Z(\Lambda)} p(\boldsymbol{z})\delta_{\boldsymbol{z}},$$

where Y and Z are countable sets, $n(\boldsymbol{y})$ and $p(\boldsymbol{z})$ are weights, which the formula applies to specific spaces of test functions (usually different from the Schwartz space), see Guinand [**Gu**, Section 10].

We now consider examples of Patterson sets. The most general method found so far for proving that certain Delone sets Λ in $\mathbb{R}^n$ are Patterson sets uses properties of an associated dynamical system $([[\Lambda]], \mathbb{R}^n)$.

DEFINITION 2.10. Given any Delone set Λ of finite type, the set $[[\Lambda]]$ is the collection of all Delone sets Λ' which are pointwise limits of some sequence of translates $\{\Lambda + \boldsymbol{x}_i : i = 1, 2, 3, \dots\}$ of Λ.

---

[2]Such sets must be uniformly discrete, but need not be relatively dense.

The *natural topology* on $[[\Lambda]]$ defines two sets Λ and Λ' as being within distance ϵ if there is a translation t with $\|t\| < \epsilon$ such that $\Lambda + t$ agrees with Λ' on a ball of radius $1/\epsilon$ around $\mathbf{0}$; the set $[[\Lambda]]$ is compact in this topology. More generally, for any Delone set one can define $[[\Lambda]]$ as the closure of the set of translates of $\Lambda + x$ in an appropriate topology, and $[[\Lambda]]$ is a compact set in this topology, see Solomyak [**So98b**]. The set $[[\Lambda]]$ is closed under translations, and we let $([[\Lambda]], \mathbb{R}^n)$ denote the (topological) dynamical system with this $\mathbb{R}^n$-action.

DEFINITION 2.11. (i) A topological dynamical system $\mathcal{X}$ with $\mathbb{R}^n$-action is *minimal* if every orbit of a point under translation by $\mathbb{R}^n$ is dense in $\mathcal{X}$.

(ii) A topological dynamical system is *uniquely ergodic* if it has a unique invariant measure μ; in this case we can regard it as the metrical dynamical system with measure μ.

(iii) A topological dynamical system is *strictly ergodic* if it is minimal and uniquely ergodic.

In the case of a topological dynamical system $\mathcal{X} = ([[\Lambda]], \mathbb{R}^n)$ these concepts have the following characterizations. $\mathcal{X}$ is minimal if and only if Λ is *repetitive*, which means that for each T-patch $\Lambda \cap B(x,T)$ of Λ there is a radius T' (depending only on T) such that Λ contains a translate of this patch inside any ball of radius T'. Such an $\mathcal{X}$ is uniquely ergodic if and only if any T-patch has a uniform limiting frequency of occurrence inside T'-patches, as $T' \to \infty$. Such an $\mathcal{X}$ is strictly ergodic if and only if it is uniquely ergodic and every T-patch has a uniform limiting frequency that is positive.

To any metrical dynamical system $([[\Lambda]], \mathbb{R}^n, \mu)$ we associate a family of commuting unitary operators $U(t): L^2([[\Lambda]], \mu) \to L^2([[\Lambda]], \mu)$ indexed by $t \in \mathbb{R}^n$, given by
$$U(t)f(\Lambda') = f(\Lambda' - t) \quad \text{for } \Lambda' \in [[\Lambda]].$$

DEFINITION 2.12. (i) A *measurable eigenfunction* $f \in L^2([[\Lambda]], \mu)$ with eigenvalue $\boldsymbol{\lambda} \in \mathbb{R}^n$ is one that satisfies
$$U(t)f(\Lambda) = e^{2\pi i \langle \lambda, t \rangle} f(\Lambda), \quad \text{for all } t \in \mathbb{R}^n.$$

(ii) A *continuous eigenfunction* is an eigenfunction which is continuous in the natural topology on $[[\Lambda]]$.

DEFINITION 2.13. (i) The *spectrum* of $([[\Lambda]], \mathbb{R}^n, \mu)$ is the joint spectrum of the family of commuting operators $\{U(t) : t \in \mathbb{R}^n\}$.

(ii) A dynamical system has *pure discrete spectrum* or *pure point spectrum* if the set of measurable eigenfunctions spans $L^2([[\Lambda]], \mu)$.

THEOREM 2.14. *If Λ is a Delone set of finite type such that the dynamical system $([[\Lambda]], \mathbb{R}^n)$ is strictly ergodic and has pure discrete spectrum, then every set Λ' in $[[\Lambda]]$ is a Patterson set.*

PROOF. The essential idea of this result appears in Dworkin [**Dw**]. A proof is sketched in Hof [**H97**, pp. 253–257]. □

It is known that if the dynamical system $([[\Lambda]], \mathbb{R}^n)$ is strictly ergodic and has purely continuous spectrum, then no set Λ' in $[[\Lambda]]$ is a Patterson set. If it has mixed spectrum—some discrete and some continuous—it is not known whether some Λ' in $[[\Lambda]]$ can be a Patterson set.

Essentially all model sets with a reasonable window set B have been proved to be Patterson sets by this method. The following result is due to Schlottmann [**S98**] [**S99**, Theorem 4.5].

THEOREM 2.15 (Schlottmann). *If Λ is a model set in $\mathbb{R}^n$ whose window set B is compact, with nonempty interior and with a boundary of Haar measure zero, then Λ is a Patterson set, whose spectrum $\sigma_p(\Lambda)$ has*

$$\sigma_P(\Lambda) \subseteq \pi_\|(L^*) \tag{2.22}$$

where L^* is the dual lattice of L.

Schlottmann proves that the associated dynamical system has a pure point spectrum, and Theorem 2.15 then follows from Theorem 2.14. That a result like this should hold was suggested by Meyer [**Me95**], and this result improves on an earlier result of Hof [**H97**, Section 4.4], whose proof did not use dynamical systems.

A number of self-replicating Delone sets have been proved to be Patterson sets using the associated dynamical system. For self-replicating Delone sets, Solomyak [**So97**] gives an algorithmic method for testing whether the dynamical system associated to a primitive self-replicating Delone set Λ of finite type in $\mathbb{R}^2$ has pure point spectrum. He applies this to several examples in Section 7 of his paper. His Example 7.2 implies that the set Λ of vertices of the "sphinx tiling" of Godrèche [**Go**] gives a dynamical system with purely discrete spectrum. Thus all elements Λ' of $[[\Lambda]]$ are Patterson sets. In this example the spectrum of the associated dynamical system $([[\Lambda]], \mathbb{R}^2)$ is not contained in any finite-dimensional $\mathbb{Z}$-module. This spectrum is contained in $\mathbb{Z}[\frac{1}{2}] \times \mathbb{Z}[\frac{1}{2}]$, and involves rationals with arbitrarily high powers of 2 in the denominator. This indicates that there is some set Λ' in $[[\Lambda]]$ which has a Patterson spectrum $\sigma_P(\Lambda')$ which is not contained in any finite-dimensional $\mathbb{Z}$-module. This would happen if Λ' had the same spectrum as that of the dynamical system, i.e., no coefficients were "extinguished." Note that this sort of spectrum differs from that of any cut-and-project set, because such sets have spectra $\sigma_P(\Lambda)$ contained in a finite-dimensional $\mathbb{Z}$-module by (2.22). However the sets constructed by this type of dynamical system may be model sets. The chair tiling in in $\mathbb{R}^2$ yields model sets based on a p-adic "internal space", as is shown in Baake, Moody and Schlottmann [**BMS**]. A similar result was established for the n-dimensional chair tiling by Lee and Moody [**LM**], who also showed that the sphinx tiling is a union of 36 model sets using such an "internal space." The dynamical systems associated to self-similar structures can be viewed as a generalization of substitution dynamical systems, the spectral properties of which have been extensively studied, see Queffélec [**Q**].

We say that a strictly ergodic dynamical system acting with an $\mathbb{R}^n$-action on a compact space Ω with invariant measure μ is *homogeneous* if $L^2(\Omega, \mu)$ has a basis of continuous eigenfunctions, see [**Ro94**, p. 494]. Such a dynamical system necessarily has pure discrete spectrum. Many of the constructions above yield Λ such that $([[\Lambda]], \mathbb{R}^n)$ is a homogeneous dynamical system. The potential relevance of such dynamical systems to the problem of defining "Fourier coefficients" for certain Delone sets Λ is discussed at the end of Section 3.

3. Fourier Quasicrystals

For an ideal crystal Λ it is well-known that the Fourier transform $\hat{\mu}_\Lambda$ is the density function of a measure which contains "phase information" that is lost in the

X-ray diffraction measure. Indeed, if

$$(3.1) \qquad \Lambda = \bigcup_{j=1}^{k} (L + \boldsymbol{f}_j),$$

where L is a full rank lattice on $\mathbb{R}^n$ having dual lattice L^*, then the Poisson summation formula gives

$$(3.2) \qquad \hat{\mu}_\Lambda = \sum_{\boldsymbol{y} \in L^*} c(\boldsymbol{y}) \delta_{\boldsymbol{y}}$$

in which

$$(3.3) \qquad c(\boldsymbol{y}) = \frac{1}{|\det(L)|} \sum_{j=1}^{k} \exp(2\pi i \langle \boldsymbol{f}_j, \boldsymbol{y} \rangle).$$

By Theorem 2.7 the autocorrelation measure γ_Λ of Λ has Fourier transform given by

$$(3.4) \qquad \widehat{\gamma}_\Lambda = \sum_{\boldsymbol{y} \in L^*} |c(\boldsymbol{y})|^2 \delta_{\boldsymbol{y}},$$

because

$$(3.5) \qquad |c(\boldsymbol{y})|^2 = \frac{1}{|\det(L)|^2} \sum_{i=1}^{k} \sum_{j=1}^{k} \exp(2\pi i \langle \boldsymbol{f}_i - \boldsymbol{f}_j, \boldsymbol{y} \rangle).$$

Knowledge of the Fourier coefficients $\{c(\boldsymbol{y}) : \boldsymbol{y} \in L^*\}$ suffices to uniquely reconstruct Λ, but it is well-known that knowledge of the intensities $\{|c(\boldsymbol{y})|^2 : \boldsymbol{y} \in L^*\}$ does not always uniquely determine the translation-equivalence class of Λ. This ambiguity is an important obstacle to the reconstruction of crystal structure from X-ray diffraction data.

This raises the problem:

PHASE PROBLEM. For which Patterson sets Λ can one define "phase information" $\{c(\boldsymbol{y}) : \boldsymbol{y} \in \sigma_P(\Lambda)\}$ such that the distribution $\hat{\mu}_\Lambda$ has a "formal δ-function expansion"

$$(3.6) \qquad \hat{\mu}_\Lambda \sim \sum_{\boldsymbol{y} \in Y} c(\boldsymbol{y}) \delta_{\boldsymbol{y}},$$

in which $Y = \sigma_P(\Lambda)$ and for which

$$(3.7) \qquad \widehat{\gamma}_\Lambda = \sum_{\boldsymbol{y} \in \sigma_P(\Lambda)} |c(\boldsymbol{y})|^2 \delta_{\boldsymbol{y}},$$

both hold?

The phase problem can be divided into two subproblems. The first problem is that of defining a "formal δ-function expansion" (3.6) for the distribution $\hat{\mu}_\Lambda$, for some countable spectrum Y. The second problem is obtaining conditions on a Patterson set Λ such that the coefficients $p(\boldsymbol{y})$ of the diffraction measure $\widehat{\gamma}_\Lambda$ given by

$$(3.8) \qquad \widehat{\gamma}_\Lambda = \sum_{\boldsymbol{y} \in \sigma_P(\Lambda)} p(\boldsymbol{y}) \delta_{\boldsymbol{y}}.$$

are related to the coefficients $c(\boldsymbol{y})$ of the formal δ-function expansion by
$$p(\boldsymbol{y}) = |c(\boldsymbol{y})|^2, \tag{3.9}$$
We call (3.9) the *consistent phase property*.

We first deal with the problem of defining a "formal δ-function expansion" (3.6). Here we do not assume that Λ is a Patterson set. The narrowest such definition is the following.

DEFINITION 3.1. A Delone set Λ is a *strongly almost periodic set* if the tempered distribution $\hat{\mu}_\Lambda$ is a translation-bounded measure that is a pure point measure.

In this case the Fourier transform of μ_Λ can be written
$$\hat{\mu}_\Lambda := \sum_{\boldsymbol{y} \in Y} c(\boldsymbol{y}) \delta_{\boldsymbol{y}}, \tag{3.10}$$
in which Y is a countable set. All such sets can be classified using the following result of Córdoba [**Co89**].

THEOREM 3.2 (Córdoba). *Suppose that $\Lambda = \bigcup_{i=1}^{k} \Lambda_i$ is a uniformly discrete set in $\mathbb{R}^n$, and let g_Λ denote the tempered distribution*
$$g_\Lambda = \sum_{i=1}^{k} w_i \left(\sum_{\boldsymbol{x} \in \Lambda_i} \delta_{\boldsymbol{x}} \right) \tag{3.11}$$
in which $\{w_1, \ldots, w_k\}$ are complex numbers. If the Fourier transform $\hat{g}_\Lambda$ is a translation-bounded measure which is pure point, i.e.
$$\hat{g}_\Lambda = \sum_{\boldsymbol{y} \in Y} m(\boldsymbol{y}) \delta_{\boldsymbol{y}}, \tag{3.12}$$
with
$$\sum_{\boldsymbol{y} \in \boldsymbol{z}+[0,1]^n} |m(\boldsymbol{y})| \leq C, \quad \text{for all } \boldsymbol{z} \in \mathbb{R}^n,$$
then Λ and each set Λ_i are a finite union of translates of some full rank lattice L in $\mathbb{R}^n$.

Córdoba's theorem as stated in [**Co89**] only concludes that each Λ_i is a finite disjoint union of translates of n-dimensional lattices. However the union of two such translates $(L_1 + \boldsymbol{a}_1) \cup (L_2 + \boldsymbol{a}_2)$ cannot be uniformly discrete unless the lattices L_1 and L_2 are commensurable, i.e., unless both can be written as a finite union of cosets of a common full-rank lattice L. This follows from Kronecker's theorem in Diophantine approximation. Since Λ is uniformly discrete, there must be a common refining lattice L for all these lattices simultaneously, which gives Theorem 3.2.

We immediately obtain:

COROLLARY 3.3. *A strongly almost periodic set is an ideal crystal and conversely.*

PROOF. Apply Theorem 3.2 with $\Lambda = \Lambda_1$ and $w_1 = 1$. □

The hypotheses of Theorem 3.2 cannot be relaxed to merely requiring that both g_Λ and $\hat{g}_\Lambda$ be translation-bounded pure discrete measures. Indeed, de Bruijn ([**B86**, Theorem 11.1]) gives examples of measures
$$\mu = \sum_{\boldsymbol{y} \in \Lambda} n(\boldsymbol{y}) \delta_{\boldsymbol{y}}$$

in which Λ is a Delone set, and μ and its Fourier transform $\hat{\mu}$ are both translation-bounded pure discrete measures, but Λ is not contained in a finite union of translates of a lattice. These examples are obtained from cut-and-project sets by a smoothing operation. In these examples the coefficients $m(\boldsymbol{y})$ necessarily assume infinitely many values.

For general Delone sets Λ the distribution $\hat{\mu}_\Lambda$ need not be a measure, so we cannot assign a direct meaning of "pure discrete measure" to $\hat{\mu}_\Lambda$. As an example, Hof [**H97**, p. 246] observes that any Delone set $\Lambda \subset \mathbb{Z}^n$ that is not fully periodic has $\hat{\mu}_\Lambda$ not a measure. To proceed, we observe that the existence of a Fourier transform $\hat{\mu}_\Lambda$ satisfying (3.2) can be rephrased as saying that μ_Λ has a "Fourier series"

$$(3.13) \qquad \mu_\Lambda \sim \sum_{\boldsymbol{y} \in L^*} c(\boldsymbol{y}) \exp(-2\pi i \langle \boldsymbol{y}, \cdot \rangle),$$

because the distributional Fourier transform of the function $\exp(-2\pi i \langle \boldsymbol{y}, \cdot \rangle)$ is $\delta_{\boldsymbol{y}}$. We therefore seek to directly define such a "Fourier series" associated to μ_Λ. To accomplish this, we consider various classes of almost periodic functions.

H. Bohr [**Bo1**] developed a theory of uniformly almost periodic functions on the real line, which was extended to $\mathbb{R}^n$ by S. Bochner. Uniformly almost periodic functions are those bounded continuous functions $h(\boldsymbol{x})$ that can be uniformly approximated on all of $\mathbb{R}^n$ by trigonometric polynomials. They have a well-defined "Fourier series"

$$(3.14) \qquad h(\boldsymbol{x}) \sim \sum_{\boldsymbol{y} \in Y} m(\boldsymbol{y}) \exp(-2\pi i \langle \boldsymbol{y}, \boldsymbol{x} \rangle)$$

in which Y is a countable set, and the coefficients are square-summable,

$$(3.15) \qquad \|h\|^2 := \sum_{\boldsymbol{y} \in Y} |m(\boldsymbol{y})|^2 < \infty.$$

The Fourier series data $\{m(\boldsymbol{y}) : \boldsymbol{y} \in Y\}$ permits unique reconstruction of the function $h(\boldsymbol{x})$. However not all countable sets Y and data $\{m(\boldsymbol{y}) : \boldsymbol{y} \in Y\}$ satisfying (3.15) give "Fourier series" of uniformly almost periodic functions. The condition

$$(3.16) \qquad \sum_{\boldsymbol{y} \in Y} |m(\boldsymbol{y})| < \infty,$$

is known to be a sufficient condition for (3.14) to be the Fourier series of a uniformly almost periodic function.

L. Schwartz [**Sch**, Section VI.9] introduced the following notion of uniformly almost periodic distribution based on uniformly almost periodic function.

DEFINITION 3.4. A tempered distribution f is a *uniformly almost periodic distribution* if for each compactly supported C^∞-function $g \in C_c^\infty(\mathbb{R}^n)$ the convolution $g * f$ is a uniformly almost periodic function on $\mathbb{R}^n$. (Here $g * f(\boldsymbol{y}) = \langle f, g_{-\boldsymbol{y}} \rangle$ where $g_{\boldsymbol{y}}(\boldsymbol{x}) = g(\boldsymbol{x} + \boldsymbol{y})$.) More generally, if $\mathcal{B}$ is a class of almost periodic functions, a $\mathcal{B}$-*almost periodic distribution* f is a tempered distribution f such that for each $g \in C_c^\infty(\mathbb{R}^n)$ the convolution $g * f \in \mathcal{B}$.

A uniformly almost periodic distribution f has a well-defined "Fourier series"

$$(3.17) \qquad f \sim \sum_{\boldsymbol{y} \in Y} m(\boldsymbol{y}) \exp(-2\pi i \langle \boldsymbol{y}, \cdot \rangle)$$

in which Y is a countable set. To construct it, given $\boldsymbol{y} \in \mathbb{R}^n$ take $g \in C_c^\infty(\mathbb{R}^n)$ to be a test function which has $\hat{g}(\boldsymbol{y}) \neq 0$, and if the uniformly almost periodic function $g * f$ has "Fourier series"

$$(3.18) \qquad g * f(x) \sim \sum_{\boldsymbol{z} \in \mathbb{R}^n} m_g(\boldsymbol{z}) \exp(-2\pi i \langle \boldsymbol{z}, \cdot \rangle)$$

where only countably many $m_g(\boldsymbol{z}) \neq 0$, then we set

$$(3.19) \qquad m(\boldsymbol{y}) := \frac{m_g(\boldsymbol{y})}{\hat{g}(\boldsymbol{y})}.$$

It can be checked that this definition is independent of the choice of test function g having $g(\boldsymbol{y}) \neq 0$. One can prove that the coefficients $m(\boldsymbol{y})$ are (uniformly) locally square-summable: there is a constant C such that for all $\boldsymbol{x} \in \mathbb{R}^n$,

$$(3.20) \qquad \sum_{\boldsymbol{y} \in \boldsymbol{x}+[0,1]^n} |m(\boldsymbol{y})|^2 < C.$$

However a drawback is that not all data $\{m(\boldsymbol{y}) : \boldsymbol{y} \in Y\}$ satisfying (3.20) are the "Fourier series" of a (uniformly) almost periodic distribution f. Burkill and Rennie [**BR**] develop a theory of almost periodic distributions extending that of Schwartz.

DEFINITION 3.5. A Delone set Λ is a *Bohr almost periodic set* Λ if its associated measure μ_Λ is a uniformly almost periodic distribution.

We view the Fourier transform $\hat{\mu}_\Lambda$ of a Bohr almost periodic set Λ as having a "formal δ-function expansion"

$$(3.21) \qquad \hat{\mu}_\Lambda \sim \sum_{\boldsymbol{y} \in Y} m(\boldsymbol{y}) \delta_{\boldsymbol{y}}$$

which is its "Fourier series" (3.18).

The unique reconstructability of uniformly almost periodic functions from their "Fourier series" has the following consequence for Bohr almost periodic sets:

(1) The "Fourier series" of a Bohr almost periodic set μ_Λ permits unique reconstruction of Λ.
(2) If Λ is a Bohr almost periodic set and G is a nonempty finite set disjoint from Λ, then $\Lambda \cup G$ is not a Bohr almost periodic set.

A simple sufficient condition for f to be a uniformly almost periodic distribution is the following.

LEMMA 3.6. *If a tempered distribution f and its Fourier transform $\hat{f}$ on $\mathbb{R}^n$ are both translation-bounded measures that are pure discrete, then f and $\hat{f}$ are both (uniformly) almost periodic distributions.*

PROOF. This is easy to verify using test functions $g \in C_c^\infty(\mathbb{R}^n)$ because (3.18) holds for $g * f$ and for $g * \hat{f}$. □

Lemma 3.6 implies that ideal crystals Λ are Bohr almost periodic sets. As mentioned earlier, de Bruijn ([**B86, B87**]) constructs a large number of distributions f which he calls "*Poisson combs*" that apparently satisfy the hypotheses of Lemma 3.6. (He works in the Gelfand-Shilov space $S_{1/2}^{1/2}$ of distributions, however, rather than with tempered distributions, see van Eijndhoven [**Eij**].) These sets are not Bohr almost periodic sets because points are assigned variable weights rather than having weight one at all points.

The concept of "Bohr almost periodic set" is so narrow as to exclude various Patterson sets. Hof [**H92**, p. 90] observes that the tempered distribution

$$f = \sum_{n \in \mathbb{Z}} w_n \delta_n \tag{3.22}$$

where $\{w_n : n \in \mathbb{Z}\}$ is a zero-one sequence that describes a "Fibonacci quasicrystal" is not a (uniformly) almost-periodic distribution. The same remains true even if the type of almost-periodicity used in defining the distribution is relaxed from that of Bohr to the wider classes of Stepanov or Wiener. The set

$$\Lambda_w := \{n \in \mathbb{Z} : w_n = 1\} \tag{3.23}$$

is a Meyer set and is known to be a Patterson set. Other examples are given by certain one-dimensional cut-and-project sets Λ do not have μ_Λ being an uniformly almost periodic measure, also due to Hof [**H97**, p. 257]

It seems to be unknown whether there exist any Bohr almost periodic sets that are not ideal crystals. A strong constraint on the nature of Bohr almost periodic sets arises from the restriction that

$$\mu_\Lambda = \sum_{\boldsymbol{x} \in \Lambda} n(\boldsymbol{x}) \delta_{\boldsymbol{x}} \tag{3.24}$$

has all coefficients $n(\boldsymbol{x}) = 1$. In regard to this property, we mention another result of Córdoba [**Co88**].

THEOREM 3.7 (Córdoba). *Suppose that X and Y are discrete sets in $\mathbb{R}^n$, that $\{p(\boldsymbol{y}) : \boldsymbol{y} \in Y\}$ are positive real numbers, and that the two distributions*

$$f_1 = \sum_{\boldsymbol{x} \in X} \delta_{\boldsymbol{x}} \quad \text{and} \quad f_2 = \sum_{\boldsymbol{y} \in Y} p(\boldsymbol{y}) \delta_{\boldsymbol{y}} \tag{3.25}$$

are tempered distributions. If $f_2 = \hat{f}_1$, then X is a full rank lattice L in $\mathbb{R}^n$ and Y is the dual lattice L^*, and all $p(\boldsymbol{y}) = 1/|\det(L)|$.

This result appears in [**Co88**, Theorem 2], except that Córdoba asserts that $|\det(L)| = 1$, which is too strong a conclusion. His method appears to establish the result above.

In Theorem 3.7 X is a discrete set, not necessarily a Delone set, while Y is required to be a discrete set, but translation-boundedness is not required, the growth on the sizes of the coefficients $|p(\boldsymbol{y})|$ being sufficient to give a tempered distribution. This result puts further restriction on any Bohr periodic set that is not an ideal crystal.

To obtain a wider class of sets Λ for which $\hat{\mu}_\Lambda$ has a well-defined δ-function expansion, we must relax the definition of "almost periodic distribution" to allow a wider class of almost periodic functions. We would like a definition that includes all cut-and-project sets which are Patterson sets. For this it seems that one needs a class $\mathcal{B} \subseteq L_2(\mathbb{R}^n)$ of almost periodic functions with the following three properties.

(1) *Translation-closure property.* If $f(\boldsymbol{x}) \in \mathcal{B}$ with "formal Fourier series"

$$f(\boldsymbol{x}) \sim \sum_{\boldsymbol{y} \in Y} c(\boldsymbol{y}) e^{2\pi i \langle \boldsymbol{y}, \boldsymbol{x} \rangle} \tag{3.26}$$

then for each translation $t \in \mathbb{R}^n$, $f_t(x) = f(x - t) \in \mathcal{B}$ and the "formal Fourier series" of f_t has the same spectrum Y as f with Fourier coefficients

(3.27) $$c_t(y) = c(y)e^{-2\pi i \langle y, t \rangle}.$$

(2) *Parseval property.* The "formal Fourier series" (3.26) of $f(x) \in \mathcal{B}$ satisfies

(3.28) $$\|f\|_2^2 = \sum_{y \in Y} |c(y)|^2.$$

(3) *Riesz-Fischer property.* For any countable set Y and set of coefficients $\{c(y) : y \in Y\}$ that are square-summable,

(3.29) $$\sum_{y \in Y} |c(y)|^2 < \infty,$$

there exists a function $f(x) \in \mathcal{B}$ which has "formal Fourier series"

(3.30) $$f(x) \sim \sum_{y \in Y} c(y) e^{2\pi i \langle y, x \rangle}.$$

In the one-dimensional case the Besicovitch class of B^2-almost periodic functions has these properties, see Appendix B. Definition 3.4 yields a notion of B^2-*almost periodic distribution* and we then also obtain an associated notion of *Besicovitch almost periodic set (of class B^2* analogous to Definition 3.5. Hof [**H97**, p. 258] observes that certain one-dimensional cut-and-project sets are Besicovitch almost periodic sets in this sense. The Besicovitch theory does not seem to have been extended to $\mathbb{R}^n$ for $n \geq 2$, but Følner [**Fø**] has developed a theory of almost periodic functions on $\mathbb{R}^n$ which has the Parseval and Riesz-Fischer properties.

DEFINITION 3.8. A Delone set Λ is a *Besicovitch almost periodic set* of class $\mathcal{B}$ if its associated measure μ_Λ is a uniformly almost periodic distribution of class $\mathcal{B}$.

This definition depends on the class $\mathcal{B}$, and one hopes that a suitable class $\mathcal{B}$ of functions define a concept of *Besicovitch almost periodic set* on $\mathbb{R}^n$ which will include all reasonable cut-and-project sets. Such a theory has not yet been worked out in any detail.

A price one pays in allowing larger classes of almost periodic functions with the Riesz-Fischer property is that a $\mathcal{B}$-almost periodic function cannot be reconstructed from its "formal Fourier series". For example, there are two B^2-almost periodic functions f and g on $\mathbb{R}$ which disagree on a set of infinite Lebesgue measures but have the same B^2-Fourier series. Thus we cannot hope to reconstruct a set Λ uniquely from "phase information" supplied by a "formal Fourier series" of this sort.

We conclude this section by describing results from another approach for associating "discrete spectrum" to the tempered distribution $\hat{\mu}_\Lambda$, which was originally explored by Bombieri and Taylor [**BT86**, **BT87**]. In some circumstances a tempered distribution $f(x)$ has a limit

(3.31) $$m_\xi := \lim_{T \to \infty} \frac{1}{(2T)^n} \int_{[-T,T]^n + a} e^{-2\pi i \langle \xi, x \rangle} f(x) \, dx,$$

which is independent of the translation $a \in \mathbb{R}^n$. We can view m_ξ as defining a "Fourier coefficient" of the distribution $f(x)$ at the frequency ξ. Hof ([**H95a**]) obtained the following result.

THEOREM 3.9 (Hof). *Let μ be a translation-bounded measure on $\mathbb{R}^n$ that has a unique autocorrelation measure γ and suppose that for some $\boldsymbol{\xi} \in \mathbb{R}^n$ the limit*

$$(3.32) \qquad m_{\boldsymbol{\xi}} = \lim_{T \to \infty} \frac{1}{(2T)^n} \int_{[-T,T]^n + \boldsymbol{a}} e^{-2\pi i \langle \boldsymbol{\xi}, \boldsymbol{x} \rangle} d\mu(\boldsymbol{x})$$

exists uniformly in $\boldsymbol{a}$. Then the pure discrete component $\widehat{\gamma}(\{\boldsymbol{\xi}\})$ of $\widehat{\gamma}$ at $\boldsymbol{\xi}$ has

$$(3.33) \qquad \widehat{\gamma}(\{\boldsymbol{\xi}\}) = |m_{\boldsymbol{\xi}}|^2.$$

The conclusion (3.33) asserts that the consistent phase property (3.9) holds at the point $\boldsymbol{\xi}$. The hypothesis (3.32) above is a uniformity condition which asserts that for each $\epsilon > 0$ there is a value T_ϵ such that for $T \geq T_\epsilon$,

$$(3.34) \qquad \left| m_{\boldsymbol{\xi}} - \frac{1}{(2T)^n} \int_{[-T,T]^n + \boldsymbol{a}} e^{2\pi i \langle \boldsymbol{\xi}, \boldsymbol{x} \rangle} d\mu(\boldsymbol{x}) \right| \leq \epsilon$$

holds for all $\boldsymbol{a} \in \mathbb{R}^n$. The uniformity condition (3.32) is known to hold for all $\boldsymbol{\xi} \in \mathbb{R}^n$ for cut-and-project sets with polytope masks B, see Hof [**H95a**, pp. 248–251] for precise results. The uniformity condition (3.32) for all $\boldsymbol{\xi} \in \mathbb{R}^n$ has also been verified for special classes of self-repetitive Delone sets, see Gähler and Klitzing [**GK**, Theorem 3.1], Hof [**H95a**, p. 247], and Solomyak [**So97**, Theorem 5.1].

Hof [**H97**, p. 247] presents an example based on Allouche and Mendès-France [**AM-F**, p. 336]) showing that some type of uniformity hypothesis is necessary for the conclusion (3.33) in Theorem 3.9 to hold. This example takes

$$(3.35) \qquad f_\mu := \sum_{m \in \mathbb{Z}} e^{2\pi i m^\alpha} \delta_m$$

with $\alpha = 1/(2k+1)$ for some integer $k \geq 1$. It has a well-defined "Fourier coefficient" (3.29) at $\xi = 0$, namely $m_0 = 0$, but $\widehat{\gamma}(\{0\}) = 1$. We also note that Theorem 3.9 does not provide any information regarding a possible continuous component of the measure $\widehat{\gamma}$, either singular continuous or absolutely continuous.

To conclude this section, we observe that Theorem 3.2 provides a mechanism to define "Fourier coefficients" for a sizeable class of aperiodic Delone sets Λ. This is evidenced by the examples above, and it may also apply to a class of Λ whose associated dynamical system has suitably strong properties. Suppose that Λ is a Delone set of finite type whose associated dynamical system $([[\Lambda]], \mathbb{R}^n)$ is minimal and uniquely ergodic. It is then expected that the uniformity condition (3.32) holds for those $\boldsymbol{\xi}$ not in the discrete spectrum of the dynamical system, and for those $\boldsymbol{\xi}$ for which the dynamical system has a continuous eigenfunction. An analogous theorem for a general uniquely ergodic transformation T on a compact space (with a $\mathbb{Z}$-action) was proved by E. A. Robinson, Jr. [**Ro94**, Theorem 1]. Assuming that a version of Robinson's result is valid for $\mathbb{R}^n$-actions, we could conclude that whenever the dynamical system $([[\Lambda]], \mathbb{R}^n)$ is homogeneous, i.e., $L^2(\Omega, \mu)$ has a basis of continuous eigenfunctions, then property (3.32) will hold for all $\boldsymbol{\xi} \in \mathbb{R}^n$. Theorem 3.2 then assigns "Fourier coefficients" at *every* $\boldsymbol{\xi} \in \mathbb{R}^n$, which satisfy the consistent phase property. Homogeneous dynamical systems have pure discrete spectrum, so that these "Fourier coefficients" would account for the entire spectrum. It follows that this class of sets Λ, which includes ideal crystals, would have a satisfactory definition of "phase information".

4. Open Problems

The first set of problems concerns Patterson sets and summation formulas. Aside from ideal crystals, all known constructions of Patterson sets Λ produce a Patterson spectrum $\sigma_P(\Lambda)$ that is a dense set in $\mathbb{R}^n$. What constraints does the assumption that $\sigma_P(\Lambda)$ is a discrete set put on $\Lambda - \Lambda$ and $\sigma_P(\Lambda)$? We first formulate a version of this question purely in terms of summation formulas.

PROBLEM 4.1. (a) Suppose that γ is a positive definite translation-bounded measure in $\mathbb{R}^n$ that is supported on a Delone set Λ in $\mathbb{R}^n$, with

$$\gamma = \sum_{\boldsymbol{x} \in \Lambda} n(\boldsymbol{x}) \delta_{\boldsymbol{x}}, \tag{4.1}$$

and that its Fourier transform $\widehat{\gamma}$ is also a discrete measure supported as a Delone set Y in $\mathbb{R}^n$

$$\widehat{\gamma} = \sum_{\boldsymbol{y} \in Y} p(\boldsymbol{y}) \delta_{\boldsymbol{y}}. \tag{4.2}$$

Is it true that there always exists a lattice L and a finite set F such that

$$X \subseteq L + F \quad \text{and} \quad Y \subseteq L^* \tag{4.3}$$

holds?

(b) If (a) is true, does the weaker hypothesis that X and Y are both discrete sets in $\mathbb{R}^n$ still imply that (4.3) holds?

An affirmative answer to this problem would significantly strengthen the result of Córdoba given as Theorem 3.7.

Since Problem 4.1 may be hard, we propose the following weaker version that involves Delone sets of finite type.

PROBLEM 4.2. (a) Let Λ be a Delone set of finite type in $\mathbb{R}^n$ that is a Patterson set and suppose that Patterson spectrum $\sigma_P(\Lambda)$ is a Delone set. Does there exist a lattice L such that

$$\sigma_P(\Lambda) \subseteq L^*, \tag{4.4}$$

holds?

(b) If (a) is true, does the weaker hypothesis that $\sigma_P(\Lambda)$ is a discrete set still imply (4.4)?

Next we ask a question concerning which substitution Delone sets are Patterson sets.

PROBLEM 4.3. Suppose that Λ is a Delone set of finite type that is a primitive self-replicating Delone set. If the dynamical system $([[\Lambda]], \mathbb{R}^n)$ has some continuous spectrum does it follow that every element of $[[\Lambda]]$ is not a Patterson set?

We next consider problems related to Bohr almost periodic sets.

PROBLEM 4.4. Is a Bohr almost periodic set necessarily an ideal crystal?

This problem was discussed in Section 3. In a related direction, one can ask for a classification of uniformly almost periodic measures whose Fourier transform is a uniformly almost periodic function.

PROBLEM 4.5. Characterize all translation-bounded measures μ in $\mathbb{R}^n$ that are uniformly almost periodic measures and whose (distributional) Fourier transform $\hat{\mu}$ is also a uniformly almost periodic measure.

A theory of Besicovitch almost periodic sets in $\mathbb{R}^n$ has not been worked out in any detail. At this point it is not clear what is the best class $\mathcal{B}$ of almost periodic distributions to take in order to get a good class of $\mathcal{B}$-*almost periodic Delone sets*. We will assume that the class of $\mathcal{B}$-almost periodic functions used necessarily satisfies properties (1)–(3) given in Section 3.

PROBLEM 4.6. Define a suitable class of $\mathcal{B}$-quasicrystals with the properties:

(1) Those Patterson sets that are $\mathcal{B}$-quasicrystals, have the consistent phase property (3.9).
(2) All cut-and-project sets that are Patterson sets are $\mathcal{B}$-quasicrystals.
(3) All self-replicating Delone sets that are Patterson sets are $\mathcal{B}$-quasicrystals.

More generally, we we may ask:

PROBLEM 4.7. Are all $\mathcal{B}$-quasicrystals necessarily Patterson sets? If so, do they all have the consistent phase property (3.12)?

We also consider the relation of the "phase information" determined by a $\mathcal{B}$-quasicrystal "formal Fourier expansion" to that determined by Theorem 3.9.

PROBLEM 4.8. Suppose that μ is a translation-bounded measure on $\mathbb{R}^n$ that is a $\mathcal{B}$-almost periodic measure with "formal Fourier series"

$$(4.5) \qquad \mu \sim \sum_{\boldsymbol{\xi} \in Y} c(\boldsymbol{\xi}) e^{2\pi i \langle \boldsymbol{\xi}, \boldsymbol{x} \rangle}.$$

If for a given $\boldsymbol{\xi} \in \mathbb{R}^n$ the limit

$$(4.6) \qquad m_{\boldsymbol{\xi}} := \lim_{T \to \infty} \frac{1}{(2T)^d} \int_{[-T,T]^d + \boldsymbol{a}} e^{-2\pi i \langle \boldsymbol{\xi}, \boldsymbol{x} \rangle} \, d\mu(\boldsymbol{x})$$

exists uniformly in $\boldsymbol{a} \in \mathbb{R}^n$, then does

$$(4.7) \qquad c(\boldsymbol{\xi}) = m_{\boldsymbol{\xi}}$$

always hold?

We have noted that the information contained in the spectrum of a Patterson set Λ does not suffice to reconstruct the set Λ up to a translation. Could this be done if extra information about the set Λ was known? Recall that a Delone set of finite type Λ is *repetitive* if for each radius T there is a finite bound $M_\Lambda(T)$ such that inside any patch of X of diameter $M_X(T)$ one can find a translate of each type of T-patch of Λ.

PROBLEM 4.9. (i) Suppose that Λ is a Delone set of finite type which is repetitive, and suppose that Λ is a Patterson set. Is it true that all repetitive Delone sets of finite type with the same autocorrelation measure as Λ are contained in the translation-closure $[[\Lambda]]$?

(ii) As an important special case, suppose further that $([[\Lambda]], \mathbb{R}^n)$ is uniquely ergodic. Is it true that the repetitive Delone sets of finite type with the same autocorrelation measure as Λ are exactly the translation closure Λ?

Note that reconstructing crystal structure from "phase information" is only possible by using the extra information that an ideal crystal is a fully periodic set, and in particular, that it is repetitive. An affirmative answer to this problem would extend the reconstruction results for crystals to some aperiodic sets.

We next consider a problem relating Patterson sets and Meyer sets.

PROBLEM 4.10. Suppose that Λ is a Delone set of finite type which is a repetitive. If Λ is a Patterson set, must Λ be a Meyer set?

In case this problem is too hard, one can ask it for special subclasses of sets. The following one is of particular interest.

PROBLEM 4.11. Suppose that Λ is a primitive self-replicating Delone set. If Λ is a Patterson set, must Λ be a Meyer set?

Note that a primitive self-replicating Delone set is necessarily a Delone set of finite type that is repetitive.

To conclude, a very general problem is to characterize summation formulae generalizing the Poisson summation formula. These would involve "weighted discrete sets" whose Fourier transform is also a "weighted discrete set", in an appropriate framework of almost periodic functions. This question extends far outside the framework of quasicrystals. The "explicit formulas" of prime number theory which relate the primes to the zeros of the Riemann zeta function can be viewed as a kind of summation formula. Guinand [**Gu**] has introduced a very general notion of uniform almost periodicity for weighted discrete sequences, with respect to a given family of test functions. In this framework he is able to show, for particular weights, that the truth of the Riemann hypothesis would say that the discrete set $\{m \log p : m \geq 0, p \text{ a prime}\}$ with weights $\log p / p^{m/2}$ has Fourier transform supported on $\{\gamma : \zeta(\frac{1}{2} + i\gamma) = 0\}$ with associated weight $1/(2\pi)$; see [**Gu**, p. 263].

Appendix A. Measures and Distributions

This appendix gives basic facts on measures and distributions. See also the appendices in Hof [**H97**]. This approach uses a theory of measures viewed as linear functionals on a suitable space of test functions.

DEFINITION A.1. A (complex-valued) *measure* μ is a continuous linear functional on the space $\mathcal{K}(\mathbb{R}^n)$ of compactly supported continuous functions on $\mathbb{R}^n$. Here continuity means that for each compact K there is a constant a_K such that

$$|\mu(f)| \leq a_K \|f\|_\infty$$

for all $f \in \mathcal{K}(\mathbb{R}^n)$ with support in K and $\|.\|_\infty$ is the supremum norm.

DEFINITION A.2. (i) A *positive measure* is a measure μ such that $f \in \mathcal{K}(\mathbb{R}^n)$ with $f \geq 0 \Rightarrow \mu(f) \geq 0$.

(ii) For every measure μ there is a smallest positive measure ρ such that $|\mu(f)| \leq \rho(|f|)$ for all $f \in \mathcal{K}(\mathbb{R}^n)$. This measure ρ is called the *absolute value* of μ, and is denoted $|\mu|$.

(iii) A measure is *bounded* if $|\mu|(\mathbb{R}^n)$ is finite, and is *unbounded* otherwise.

Every function $\phi(\boldsymbol{x})$ that is locally L^1 defines a measure μ_ϕ by

$$\mu_\phi(\boldsymbol{x}) = \int_{\mathbb{R}^n} f(\boldsymbol{x}) \phi(\boldsymbol{x}) \, d\boldsymbol{x} \quad \text{for } f \in \mathcal{K}(\mathbb{R}^n).$$

where $d\boldsymbol{x}$ is Lebesgue measure on $\mathbb{R}^n$. The convolution $\mu * \nu$ of two measures is given by

$$\mu * \nu(f) = \int_{\mathbb{R}^n \times \mathbb{R}^n} f(\boldsymbol{x} + \boldsymbol{y}) \, d\mu(\boldsymbol{x}) \, d\nu(\boldsymbol{y}),$$

and is well-defined if at least one of them has compact support.

A sequence of measures μ_n converges to a limit measure ν in the *vague topology* if for each test function $f \in \mathcal{K}(\mathbb{R}^n)$ the limit $\mu_n(f) \to \nu(f)$. (This is the weak-$*$ topology on the space of measures $\mathcal{M}(\mathbb{R}^n)$.)

This linear functional version of measure is related to concepts in classical measure theory via the Riesz-Markov representation theorem stated below.

DEFINITION A.3. (i) A *Borel measure* is a measure defined on the Borel sets $\mathcal{B}$ of $\mathbb{R}^n$ Such measures take values in $\mathbb{C}$, and may be unbounded.

(ii) A Borel measure is *positive* if it takes values in the nonnegative reals $\mathbb{R}_{\geq 0}$. Associated to a Borel measure μ is a measure $|\mu|$ which is the smallest positive measure such that
$$|\mu(X)| \leq |\mu|(X)$$
for all compact sets X.

DEFINITION A.4. A positive Borel measure μ is *regular* if it has the two properties:

(a) *Outer regular*. For each set A in $\mathbb{R}^n$,
$$\mu(A) = \inf\{\mu(U) : A \subset U, U \text{ open}\}.$$

(b) *Inner regular*. For each μ-measurable set $A \in \mathbb{R}^n$,
$$\mu(A) = \sup\{\mu(K) : K \subset A, K \text{ compact}\}.$$

A general Borel measure μ is *regular* if $|\mu|$ is.

If two regular Borel measures coincide on all open (resp. compact) sets, they are equal. If a positive Borel measure μ is not regular we can obtain a regularization by defining $\nu(A) = \inf\{\mu(U) : A \subset U : U \text{ open}\}$.

THEOREM A.5 (Lebesgue decomposition theorem). *Any regular Borel measure μ on $\mathbb{R}^n$ has a unique decomposition as*

(A.1) $$\mu = \mu_{\text{pp}} + \mu_{\text{ac}} + \mu_{\text{sc}}$$

where μ_{pp} is a pure point measure, μ_{ac} is absolutely continuous with respect to Lebesgue measure and μ_{sc} is singular continuous with respect to Lebesgue measure.

PROOF. See Reed and Simon [**RS**, Theorem I.14]. □

Here a *pure point measure* is a sum of weighted delta functions on a countable set X. There is no other restriction on the set X, which could be dense.

DEFINITION A.6. A Borel measure μ is a *Radon measure* if for each compact set $K \subset \mathbb{R}^n$, the measure $|\mu|(K)$ is finite.

The property that $|\mu|(K)$ is finite for all compact sets K in $\mathbb{R}^n$ implies that a Radon measure is a regular Borel measure, see Rudin [**Ru87**, Theorem 2.18] and Evans and Gariepy [**EG**, Theorem 1.1.4].

THEOREM A.7 (Riesz-Markov representation theorem). *There is a one to one correspondence between positive measures and positive Radon measures; for each positive measure ψ there is a unique Borel measure μ such that $\psi(f) = \int f \, d\mu$ for each $f \in \mathcal{K}(\mathbb{R}^n)$.*

PROOF. See Dieudonné [**Di**, Chapter XIII] or Rudin [**Ru87**, p. 42] or Reed and Simon [**RS**, Theorem IV.18]. □

To relate measures and distributions, note that *distributions* are continuous linear functionals on the space $\mathcal{D}(\mathbb{R}^n)$ of compactly supported smooth functions on $\mathbb{R}^n$, while *tempered distributions* are continuous linear functionals on the Schwartz space $\mathcal{S}$ of rapidly decreasing smooth functions on $\mathbb{R}^n$, with continuity with respect to an appropriate topology. Since $\mathcal{D}(\mathbb{R}^n) \subset \mathcal{K}(\mathbb{R}^n)$ we can associate to each measure μ a unique distribution, defined by

$$(\text{A.2}) \qquad \mu(g) = \int_{\mathbb{R}^n} g(x) d\mu(\boldsymbol{x}) \quad \text{for } g \in \mathcal{D}(\mathbb{R}^n).$$

We identify the measure with this distribution, noting that any measure is uniquely reconstructible from its associated distribution μ. If the distribution associated to a measure is a tempered distribution, we call it a *tempered measure*. Not all measures are tempered.

To have a well-defined Fourier transform, some restriction on the class of measures is required. There is an elegant theory of the Fourier transform for tempered distributions, described in Schwartz [**Sch**] and Rudin [**Ru91**, Chapter 7]. This theory applies to tempered measures, and the following subclass of measures are suitable for modeling diffraction questions.

DEFINITION A.8. A Radon measure μ on $\mathbb{R}^n$ is *translation-bounded* if there is a constant α such that

$$|\mu|([0,1]^n + \boldsymbol{y}) \leq \alpha \quad \text{for all } \boldsymbol{y} \in \mathbb{R}^n.$$

Translation-bounded measures are tempered measures. For tempered measures the Fourier transform $\hat{\mu}$ is well-defined as a tempered distribution, but in general it is not a measure.

DEFINITION A.9. A tempered distribution T on $\mathbb{R}^n$ is of *positive type* if $T(\overline{\phi(-\boldsymbol{x})} * \phi(\boldsymbol{x})) \geq 0$ for all compactly supported C^∞-test functions $\phi \in \mathcal{D}(\mathbb{R}^n)$.

L. Schwartz showed that a distribution is of positive type if and only if it is the Fourier transform of a positive measure μ of at most polynomial growth, see Reed and Simon [**RS**, p. 331]. This generalizes Bochner's theorem characterizing positive definite functions, and we therefore call a measure μ that is a distribution of positive type a *positive definite measure*. Such measures satisfy

$$\int_{\mathbb{R}^n} f(\boldsymbol{x}) * \overline{f(-\boldsymbol{x})} \, d\mu(\boldsymbol{x}) \quad \text{for } f \in \mathcal{K}(\mathbb{R}^n),$$

which is the definition of positive definite measure used in [**AGL**] below. Such measures are tempered measures.

THEOREM A.10. *If μ is a translation bounded positive definite measure then its Fourier transform $\hat{\mu}$ is a translation-bounded positive measure.*

PROOF. The first part follows from Hof [**H95a**, Proposition 3.3]. Finally $\hat{\mu}$ is a positive measure by Reed and Simon [**RS**, Theorem 1X.10, p. 331]; see also Berg and Forst [**BF**, Prop. 1.4.4]. □

A more general subclass of measures whose Fourier transforms are measures was introduced and studied by Argabright and Gil de Lamadrid [**AGL**].

DEFINITION A.11. A measure μ is a *transformable measure* if there exists a measure $\hat{\mu}$ defined on the character space $\hat{\mathbb{R}}^n := \{\boldsymbol{\xi} : \chi_{\boldsymbol{\xi}}(\boldsymbol{x}) = e^{2\pi i \langle \boldsymbol{\xi}, \boldsymbol{x} \rangle}\}$ such that

$$\int_{\mathbb{R}^n} f * f^*(\boldsymbol{x}) \, d\mu(\boldsymbol{x}) = \int_{\hat{\mathbb{R}}^n} |\check{f}(\boldsymbol{\xi})|^2 \, d\hat{\mu}(\boldsymbol{\xi}),$$

where $f^*(\boldsymbol{x}) = \overline{f(-\boldsymbol{x})}$ and $\check{f}(\boldsymbol{\xi})$ is the inverse Fourier transform of f. The measure $\hat{\mu}$ is called the *Fourier transform* of μ.

They show that transformable measures are necessarily tempered, and that on $\mathbb{R}^n$ their notion of Fourier transform agrees with that of tempered distributions, see [**AGL**, Theorem 7.2]. The Fourier transform $\hat{\mu}$ of a transformable measure is a translation-bounded measure [**AGL**, Theorem 2.5]. All positive definite measures are transformable [**AGL**, Theorem 4.1], and a transformable measure is positive definite if and only if $\hat{\mu}(x)$ is a positive measure. This class of measures is not symmetric under Fourier transform, i.e., if μ is transformable it need not be the case that $\hat{\mu}$ is transformable. There exist transformable measures that are not translation-bounded [**AGL**, Chapter 7].

Appendix B. Almost Periodic Functions and Almost Periodic Measures

This appendix describes various notions of almost periodic functions on $\mathbb{R}^n$ and uses them to define various notions of almost periodic measure μ. To such a measure μ one can associate a "formal Fourier series"

$$\mu \sim \sum_{\boldsymbol{\xi} \in \sigma(\mu)} c(\boldsymbol{\xi})e^{2\pi i \langle \boldsymbol{\xi}, \boldsymbol{x}\rangle},$$

in which $\sigma(\mu)$ is a countable set of frequencies is a spectrum of μ. In such a case the tempered distribution $\hat{\mu}$ will have a "formal δ-function expansion"

$$\hat{\mu} \sim \sum_{\boldsymbol{\xi} \in \sigma(\mu)} c(\boldsymbol{\xi})\delta_{\boldsymbol{\xi}}.$$

We first describe the theory of uniformly almost periodic functions as given in Bohr [**Bo1**].

DEFINITION B.1. A continuous function $f \in L^\infty(\mathbb{R}^n)$ is *uniformly almost periodic* (in the sense of Bohr) if for each $\epsilon > 0$ there exists a relatively dense set $\Lambda_\epsilon(f)$ of ϵ-almost periods of f. Here an ϵ-*almost period* is a value τ such that

(B.1) $$\sup_{x \in \mathbb{R}^n} |f(\boldsymbol{x} + \boldsymbol{\tau}) - f(\boldsymbol{x})| \leq \epsilon.$$

Let $AP(\mathbb{R}^n)$ denote the set of uniformly almost periodic functions on $\mathbb{R}^n$. It is closed under uniform limits: If $\{f_j\} \subseteq AP(\mathbb{R}^n)$ have $\|f - f_j\|_\infty \to 0$, then $f \in AP(\mathbb{R}^n)$. It is well-known that a function $f \in AP(\mathbb{R}^n)$ if and only if for each $\epsilon > 0$ there exists a finite trigonometric sum

$$P_\epsilon(\boldsymbol{x}) = \sum_{\boldsymbol{\lambda} \in F_\epsilon} c(\boldsymbol{\lambda}, \epsilon) e^{2\pi i \langle \boldsymbol{\lambda}, \boldsymbol{x}\rangle}$$

such that $\|f - P_\epsilon\|_\infty \leq \epsilon$.

There is a notion of *Fourier series* and a *Parseval relation* valid for uniformly almost periodic functions. For any almost periodic function f, for $\boldsymbol{x} \in \mathbb{R}^n$ the limit

(B.2) $$\mathcal{M}(f) := \lim_{R \to \infty} \frac{1}{R^n} \int_{C(\boldsymbol{x},R)} f(\boldsymbol{y})\, d\boldsymbol{y}$$

exists, where $C(\boldsymbol{x}, R) = R[0,1]^n + \boldsymbol{x}$. This limit is independent of $\boldsymbol{x}$, and is attained uniformly in $\boldsymbol{x}$ as $R \to \infty$. The function $e^{-2\pi i \langle \boldsymbol{\xi}, \boldsymbol{x}\rangle} f(\boldsymbol{x})$ is also uniformly almost periodic, and for $\boldsymbol{\xi} \in \mathbb{R}^n$ we define the Fourier coefficient $\gamma(\boldsymbol{\xi})$ of $f(\boldsymbol{x})$ by

(B.3) $$\gamma(\boldsymbol{\xi}) := \mathcal{M}(e^{-2\pi i \langle \boldsymbol{\xi}, \boldsymbol{x}\rangle} f(\boldsymbol{x})).$$

DEFINITION B.2. For a uniformly almost periodic function the set
$$\sigma_U(f) := \{\boldsymbol{\xi} : \gamma(\boldsymbol{\xi}) \neq 0\} \tag{B.4}$$
is called the *uap spectrum of f* or *Bohr spectrum of f*. (Note that Bohr's [**Bo1**] definition of the spectrum differs slightly, being $2\pi\sigma_U(f)$.)

THEOREM B.3. *If f is uniformly almost periodic on $\mathbb{R}^n$, then so is $|f|^2$ and*
$$\mathcal{M}(|f|^2) = \sum_{\boldsymbol{\xi} \in \sigma_U(f)} |\gamma(\boldsymbol{\xi})|^2. \tag{B.5}$$

This theorem implies that $\sigma_U(f)$ is at most a countable set. Thus f has a formal "Fourier expansion."
$$f(\boldsymbol{x}) \approx \sum_{\boldsymbol{\xi} \in \sigma_U(f)} \gamma(\boldsymbol{\xi}) e^{2\pi i \langle \boldsymbol{\xi}, \boldsymbol{x} \rangle}, \tag{B.6}$$
in the sense of mean values (Bohr [**Bo1**, p. 47]) and (B.5) can be viewed as a Parseval-type relation. Any Fourier expansion of an almost-periodic function satisfies
$$\sum_{\boldsymbol{\xi} \in S} |\gamma(\boldsymbol{\xi})|^2 < \infty \tag{B.7}$$
where $S = \sigma_U(f)$. However not all countable sets S and sequences (B.7) are "Fourier expansions" of some uniformly almost periodic functions. For any countable set S and any coefficient set $\{\gamma(\boldsymbol{\xi}) : \boldsymbol{\xi} \in S\}$ the condition
$$\sum_{\boldsymbol{\xi} \in S} |\gamma(\boldsymbol{\xi})| < \infty$$
is sufficient for there to exist a (unique) almost-periodic function f with "Fourier expansion" (B.6).

The Fourier expansion of a uniformly almost periodic function f uniquely determines f.

THEOREM B.4. *If f_1 and f_2 are uniformly almost periodic functions on $\mathbb{R}^n$ and if*
$$\mathcal{M}(e^{-2\pi i \langle \boldsymbol{\xi}, \boldsymbol{x} \rangle} f_1(\boldsymbol{x})) = \mathcal{M}(e^{-2\pi i \langle \boldsymbol{\xi}, \boldsymbol{x} \rangle} f_2(\boldsymbol{x}))$$
for all $\boldsymbol{\xi} \in \mathbb{R}^n$, then $f_1 \equiv f_2$.

The "Fourier expansion" (B.6) of a uniformly almost periodic function f can be used to reconstruct f using a Cesaro-like summation procedure.

THEOREM B.5. *Given any countable set S, there exists a family of weight functions $\{\beta_{S,\epsilon}(\boldsymbol{\xi})\}$ depending on: the parameter $\epsilon > 0$, such that:*
(1) $0 \leq \beta_{S,\epsilon}(\xi) \leq 1$,
(2) $\lim_{\epsilon \downarrow 0} \beta_{S,\epsilon}(\boldsymbol{\xi}) = 1$ *if* $\boldsymbol{\xi} \in S$,
(3) *For each ϵ there is a finite set $F_S(\epsilon)$ such that $\beta_{S,\epsilon}(\boldsymbol{\xi}) = 0$ if $\boldsymbol{\xi} \notin F_S(\epsilon)$.*
(4) *For every uniformly almost periodic function with spectrum contained in S, if*
$$P_{f,\epsilon}(x) := \sum_{\boldsymbol{\xi} \in S} \beta_{S,\epsilon}(\boldsymbol{\xi}) \gamma(\boldsymbol{\xi}) e^{2\pi i \langle \boldsymbol{\xi}, \boldsymbol{x} \rangle}$$
then $\|f - P_{f,\epsilon}\|_\infty \to 0$ as $\epsilon \to 0$.

The notion of almost periodic function was extended to measures by Schwartz [**Sch**, Section VI.9].

DEFINITION B.6. An (unbounded) Radon measure μ is a *uniformly almost periodic measure* if for each compactly supported C^∞-function g on $\mathbb{R}^n$ ($g \in \mathcal{D}(\mathbb{R}^n)$) the convolution

$$(\text{B.8}) \qquad g * \mu(\boldsymbol{x}) = \int_{\mathbb{R}^n} g(\boldsymbol{x} - \boldsymbol{y}) \, d\mu(\boldsymbol{y})$$

is a uniformly almost-periodic function.

We assign to a uniformly almost periodic measure μ the "formal Fourier expansion"

$$(\text{B.9}) \qquad \mu \sim \sum_{\boldsymbol{\xi} \in \sigma_U(\mu)} c(\boldsymbol{\xi}) e^{2\pi i \langle \boldsymbol{\xi}, \boldsymbol{x} \rangle},$$

in which the "Fourier coefficient" $c(\boldsymbol{\xi})$ is defined by observing that for each $g \in C_c^\infty(\mathbb{R}^n)$ the uniformly almost periodic function $\mu * g$ has the Fourier expansion

$$(\text{B.10}) \qquad \mu * g \approx \sum_{\boldsymbol{\xi} \in \sigma_U(\mu * g)} c(\boldsymbol{\xi}) \hat{g}(-\boldsymbol{\xi}) e^{2\pi i \langle \boldsymbol{\xi}, \boldsymbol{x} \rangle},$$

and $c(\boldsymbol{\xi})$ is determined using any $g(x)$ such that $\hat{g}(-\boldsymbol{\xi}) \neq 0$, and is well-defined. Here we set

$$(\text{B.11}) \qquad \sigma_U(\mu) := \bigcup_{g \in C_c^\infty} \sigma_U(\mu * g),$$

and $\sigma_U(\mu)$ can be proved to be a countable set.

Thus one may think of an uniformly almost periodic measure μ as having a fixed countable spectrum $\sigma_U(\mu)$, which contains the spectra of all $\mu * g$, and the formula (B.10) giving the "Fourier coefficients" of $\mu * g$ as a "weak summation formula."

The closed graph theorem implies that any uniformly almost periodic measure has

$$(\text{B.12}) \qquad \sup_{\boldsymbol{x} \in \mathbb{R}^n} \int_{C(\boldsymbol{x}, R)} |d\mu| = C(R) < \infty,$$

where $C(\boldsymbol{x}, R) = \boldsymbol{x} + R[0, 1]^n$ is a scaled unit cube. Thus any almost periodic measure is necessarily *translation-bounded*. This in turn implies that the convolution $\mu * g$ makes sense for any function $g \in \mathcal{S}(\mathbb{R}^n)$, the Schwartz space of rapidly-decreasing C^∞-functions. The tempered distribution $\hat{\mu}$ can be viewed as having a "formal δ-function expansion"

$$(\text{B.13}) \qquad \hat{\mu} \sim \sum_{\boldsymbol{\xi} \in \sigma_U(\mu)} c(\boldsymbol{\xi}) \delta_{\boldsymbol{\xi}}$$

given the term-by-term Fourier transform of the right side of (B.12).

For any full-rank lattice $L \in \mathbb{R}^n$, the measure

$$(\text{B.14}) \qquad \mu_L(\boldsymbol{x}) := \sum_{\boldsymbol{\lambda} \in L} \delta_{\boldsymbol{\lambda}}.$$

is an almost-periodic measure. For any compactly supported continuous function $f(\boldsymbol{x})$, the function

$$(\text{B.15}) \qquad \mu_L * f(\boldsymbol{y}) = \sum_{\boldsymbol{\lambda} \in L} f(\boldsymbol{y} - \boldsymbol{\lambda}),$$

is periodic with period lattice L, hence is uniformly almost-periodic.

The "Fourier expansion" of the almost periodic measure μ_L for a lattice L is related to the Poisson summation formula. This states (Theorem 2.6) that the (distributional) Fourier transform of μ_L is

(B.16) $$\hat{\mu}_L = \frac{1}{|det(L)|} \sum_{\boldsymbol{\xi} \in L^*} \delta(\boldsymbol{\xi}),$$

where L^* is the dual lattice

(B.17) $$L^* = \{\boldsymbol{\xi} \in \mathbb{R}^n : \langle \boldsymbol{\lambda}, \boldsymbol{\xi} \rangle \in \mathbb{Z} \text{ for all } \boldsymbol{\lambda} \in L\}.$$

However, taking the Fourier transform of (B.16) formally, term-by-term, gives

(B.18) $$\mu_L(\boldsymbol{x}) \sim \frac{1}{|det(L)|} \sum_{\boldsymbol{\xi} \in L^*} \exp(2\pi i \langle \boldsymbol{\xi}, \boldsymbol{x} \rangle).$$

This indicates that the spectrum $\sigma_U(\mu_L * f)$ of the periodic function $\mu_L * f$ is contained in L^*.

The space of uniformly almost periodic measures is not large enough to include measures μ_Λ associated to all regular cut-and-project sets. We can obtain a sufficiently large class of measures by relaxing the notion of almost periodic function used to define almost periodic measures. It was observed by Besicovitch and Bohr [BeBo] that various notions of almost periodicity are obtained by changing the topology of convergence used for approximation by finite trigonometric polynomials. In this way one can define the almost periodic functions of Stepanov, Wiener and others, see Besicovitch [Be]. We would like a class of almost periodic functions wide enough to permit as valid Fourier series expansions all expressions

(B.19) $$\sum_{\boldsymbol{\xi} \in S} c(\boldsymbol{\xi}) e^{2\pi i \langle \boldsymbol{\xi}, \boldsymbol{x} \rangle}$$

for any countable set S and any square-summable sequence

(B.20) $$\sum_{\boldsymbol{\xi} \in S} |c(\boldsymbol{\xi})|^2 < \infty,$$

and which satisfy a Parseval identity. The class $B^2(\mathbb{R})$ of Besicovitch almost periodic functions on $\mathbb{R}$ satisfy these conditions.

DEFINITION B.7. The *Besicovitch B^2-distance* between two functions $f, g \in L^2(\mathbb{R})$ is

(B.21) $$D_{B^2}[f,g] = \left(\limsup_{T \to \infty} \frac{1}{2T} \int_{-T}^{T} |f(x) - g(x)|^2 dx \right)^{1/2}.$$

The class $B^2(\mathbb{R})$ of *Besicovitch almost periodic functions* consists of all functions $f \in L^2(\mathbb{R})$ such that there is a sequence of trigonometric polynomials $\{f_n\}$ with

(B.22) $$D_{B^2}[f, f_n] \to 0 \quad \text{as} \quad n \to \infty.$$

Besicovitch's original definition of B^2-almost periodic functions incorporated a notion of ϵ-almost period which is parallel to definition (B.1) of uniformly almost periodic functions. Call a set P of real numbers *satisfactorily uniform* if there exists a number l such that the ratio of the maximum number of terms of P to the minimum number of terms in P in an interval of length l is less than 2. A function

in $L^2(\mathbb{R})$ is B^2-almost periodic if for each $\epsilon > 0$ there exists a satisfactorily uniform set
$$P_\epsilon = \{\tau_i : \cdots < \tau_{-1} < \tau_0 < \tau_1 < \tau_2 < \cdots\}$$
such that for each $i \in \mathbb{Z}$,
$$D_{B_2}[f, f_{\tau_i}] < \epsilon,$$
where f_{τ_i} is the translated function $f(\cdot - \tau_i)$, and for every positive $c > 0$,
$$\limsup_{x \to \infty}\left(\limsup_{i \to \infty} \frac{1}{c}\int_x^{x+c} |f(x+\tau_i) - f(x)|^2\, dx\right) < \epsilon^2.$$
This equivalence of this definition to the earlier one is given in Besicovitch [**Be**, p. 78 and p. 100].

We can associate to B^2-almost periodic function on the line a B^2-Fourier series
$$(B.23) \qquad f \sim_{B^2} \sum_{\xi \in \sigma_B(f)} c(\xi) e^{2\pi i \xi x}$$
which is supported on a countable set $\sigma_B(f)$ which we call the *Besicovitch spectrum* of f, cf. [**Be**, p. 104]. To do this we a mean value $\mathcal{M}(f)$ of a B^2-polynomial. Take a series of trigonometric polynomials
$$(B.24) \qquad s_n(x) = \sum_{\xi \in F_n} c(\xi) e^{2\pi i \xi x}$$
where F_n is a finite set, with $D_{B_2}[s_n(x) - f(x)] \to 0$. Each $s_n(x)$ has a mean value, and these have a limiting mean value, so we can define
$$(B.25) \qquad \mathcal{M}(f) := \lim_{n \to \infty} \mathcal{M}(s_n(x)).$$
Now $\{e^{2\pi i \xi x} s_n(x)\}_{n \to \infty}$ converges to $e^{2\pi i \xi x} f(x)$ in the B^2-sense, and we define
$$(B.26) \qquad c(\xi) \sim_{B^2} \mathcal{M}(f e^{-2\pi i \xi x}).$$

Theorem B.8. *If $f \in B^2(\mathbb{R})$ then f has a B^2-Fourier series supported on a countable set $\sigma_B(f)$,*
$$(B.27) \qquad f \sim_{B^2} \sum_{\xi \in \sigma_B(f)} c(\xi) e^{2\pi i \xi x}.$$

Also f satisfies the Parseval identity
$$(B.28) \qquad \|f\|^2 = D_{B_2}[f, 0] = \sum_{\xi \in \sigma_B(f)} |c(\xi)|^2.$$

Proof. See Besicovitch [**Be**, p. 109]. □

Theorem B.9. *If S is any countable set and $\{c(\xi) : \xi \in S\}$ is a square-summable sequence in $\mathbb{R}$, so that*
$$\sum_{\xi \in S} |c(\xi)|^2 < \infty,$$
then there exists $f \in B^2(\mathbb{R})$ which has B^2-Fourier series
$$(B.29) \qquad f \sim \sum_{\xi \in S} c(\xi) e^{2\pi i \xi x}.$$

Proof. See Besicovitch [**Be**, p. 110]. □

Finally, the uniqueness theorem for B^2-Fourier series is as follows.

THEOREM B.10. *The functions f, $g \in B^2(\mathbb{R})$ have the same B_2-Fourier series, if and only if*

(B.30) $$D_{B^2}[f,g] = 0.$$

PROOF. See Besicovitch [**Be**, p. 109]. □

One can find two functions f, $g \in B^2(\mathbb{R})$ which differ on a set of infinite Lebesgue measure and which have $D_{B^2}(f,g) = 0$, so that they have the same B^2-Fourier series. This ambiguity is similar to the ambiguity one encounters in the definition of the diffraction measure presented in Section 2, where the diffraction measure remains unaffected by "small" changes, cf. (2.9). However the Fourier coefficients of a B^2-almost periodic function f do contain "phase information" in the sense that the translated function $f_t(\boldsymbol{x}) = f(\boldsymbol{x} - t)$ has the same spectrum $\sigma_B(f_t) = \sigma_B(f)$ and its Fourier coefficients are

(B.31) $$c_t(\boldsymbol{\xi}) = c(\boldsymbol{\xi})e^{2\pi it\xi}.$$

We define B^2-almost periodic measures in exactly the same way as was done for uniformly almost periodic measures.

DEFINITION B.11. *An unbounded Radon measure μ is a Besicovitch almost periodic measure or B^2-almost periodic measure if for each compactly supported C^∞-function g on $\mathbb{R}$ the convolution $\mu * g(\boldsymbol{y})$ is a B^2-almost periodic function.*

One can now define a B^2-Fourier series and a B^2-almost periodic spectrum $\sigma_B(\mu)$ for a B^2-almost periodic measure, or distribution. Hof [**H97**, p. 257] indicates that certain one-dimensional cut-and-project sets Λ give B^2-almost periodic measures μ_Λ in this sense.

Besicovitch developed the theory of B^2-almost periodic functions on the real line $\mathbb{R}$, and apparently did not extend it to $\mathbb{R}^n$. In the 1950's Følner [**Fø**] developed an analogue of the Besicovitch theory which is valid on arbitrary infinite groups G, and in particular $\mathbb{R}^n$. He proves analogues of the Theorems B.4.–B.6. for his almost periodic functions. However, he notes that his function space does not agree with Besicovitch's class $B^2(\mathbb{R})$ on $\mathbb{R}$. Later Davis [**Da68**] gave an extension of B^p-almost periodic functions valid on an arbitrary locally compact Abelian group. His definition does extend Besicovitch's to $\mathbb{R}^n$ (choosing a suitable "complete homogeneous Bohr net" on $\mathbb{R}^n$). His functions have well-defined Fourier series with a countable spectrum, and he gives a sense in which the Fourier series recovers the function, cf. [**Da68**, Theorem 3.1]. However he does not work out the case of B^2-almost periodic functions in any detail.

The most interesting special case for this paper is that where the Besicovitch almost-periodic distribution f has a Fourier transform $\hat{f}$ that is itself a Besicovitch almost-periodic distribution. In this case both f and $\hat{f}$ have spectra—in the B^2-almost-periodic sense—that is supported in a countable set. Can one characterize such pairs of distributions? Is there some analogue of a "summation formula" associated to such a pair of distributions?

References

[AM-F] J.-P. Allouche and M. Mendès-France, *Automata and automatic sequences*, Beyond Quasicrystals (Les Houches, 1994) (F. Axel and D. Gratias, eds.) Springer, Berlin, 1995, pp. 293–367.

[AGL]   L. Argabright and J. Gil de Lamadrid, *Fourier analysis of unbounded measures on locally compact Abelian groups*, Mem. Amer. Math. Soc., vol. 145, Amer. Math. Soc., Providence, RI, 1974.

[AuM]   L. Auslander and Y. Meyer, *A generalized Poisson summation formula*, Appl. Comput. Harmon. Anal. **3** (1996), 372–376.

[Az]   L. D. Azároff, *Elements of X-ray crystallography*, McGraw-Hill, New York, 1968.

[Ba]   M. Baake, *A guide to mathematical quasicrystals*, Quasicrystals (J.-B. Suck, M. Schreiber, and P. Häussler, eds.) Springer, Berlin, 2001 (to appear); math-ph/9904014.

[BH]   M. Baake and M. Höffe, *Diffraction of random tilings: some rigorous results*, J. Statist. Phys. **99** (2000), 219–261; math-ph/9904005.

[BMP]   M. Baake, R. V. Moody, and P. A. B. Pleasants, *Diffraction from visible lattice points and k-th power free integers*, Discrete Math. **221** (2000), 3–42; math.MG/9906132.

[BMS]   M. Baake, R. V. Moody, and M. Schlottmann, *Limit-(quasi)periodic point sets as quasicrystals with p-adic internal spaces*, J. Phys. A **31** (1998), 5755–5765; math-ph/9901008.

[BF]   C. Berg and G. Forst, *Potential theory on locally compact Abelian groups*, Ergeb. Math. Grenzgeb., vol. 87, Springer, Berlin, 1975.

[Be]   A. S. Besicovitch, *Almost periodic functions*, Cambridge Univ. Press., Cambridge, 1932; reprint, Dover, New York, 1954.

[BeBo] A. S. Besicovitch and H. Bohr, *Almost periodicity and general trigonometric series*, Acta Math. **57** (1931), 203–292; reprint, [**Bo2**].

[Boc]   S. Bochner, *A generalization of Poisson's summation formula*, Duke Math. J. **6** (1940), 229–234.

[BocNe] S. Bochner and J. von Neumann, *Almost periodic functions in groups. II*, Trans. Amer. Math. Soc. **37** (1935), 21–50.

[Bo1]   H. Bohr, *Fastperiodische Funktionen*, Ergeb. Math., vol. 1, Springer, Berlin, 1932; English transl., *Almost periodic functions*, Chelsea, New York, 1947; reprint, [**Bo2**].

[Bo2]   _____, *Collected mathematical works. II. Almost periodic functions*, (E. Følner and B. Jessen, eds.) Dansk Matematisk Forening, Copenhagen, 1952.

[BT86] E. Bombieri and J. E. Taylor, *Which distributions of matter diffract? An initial investigation*, J. Physique **47** (1986), C3-19–C3-28.

[BT87] _____, *Quasicrystals, tilings and algebraic number theory: some preliminary connections*, The Legacy of Sonya Kovalevskaya (Cambridge, MA and Amherst, MA, 1985) (L. Keen, ed.), Contemp. Math., vol. 64, Amer. Math. Soc., Providence, RI, 1987, pp. 241–264.

[B81]   N. G. de Bruijn, *Algebraic theory of Penrose's nonperiodic tiling of the plane. I*, Nederl. Akad. Wetensch. Indag. Math. **43** (1981), 39–52; II, 53–66.

[B86]   _____, *Quasicrystals and their Fourier transform*, Nederl. Akad. Wetensch. Indag. Math. **48** (1986), 123–152.

[B87]   _____, *Modulated quasicrystals*, Nederl. Akad. Wetensch. Indag. Math. **49** (1987), 121–132.

[BR]   H. Burkill and B. C. Rennie, *Almost periodic generalized functions*, Math. Proc. Cambridge Philos. Soc. **94** (1983), 149–166.

[CT]   J. W. Cahn and J. E. Taylor, *An introduction to quasicrystals*, The Legacy of Sonya Kovalevskaya (Cambridge, MA and Amherst, MA, 1985) (L. Keen, ed.), Contemp. Math., vol. 64, Amer. Math. Soc., Providence, RI, 1987, pp. 287–297.

[Co88] A. Córdoba, *La formule sommatoire de Poisson*, C. R. Acad. Sci. Paris Sér. I Math. **306** (1988), 373–376.

[Co89] _____, *Dirac combs*, Lett. Math. Phys. **17** (1989), 191–196.

[Cow]   J. M. Cowley, *Diffraction physics*, 3rd ed., North-Holland, Amsterdam, 1995.

[Da68] H. W. Davis, *On completing the von Neumann almost periodic functions*, Duke Math. J. **35** (1968), 199–215.

[Da74] H. W. Davis, *Generalized almost periodicity in groups*, Trans. Amer. Math. Soc. **191** (1974), 329–352.

[D]   B. N. Delone, *Neue Darstellung der geometrischen Kristallographie*, Z. Kristallographie **84** (1932), 109–149.

[DDSG] B. N. Delone, N. P. Dolbilin, M. I. Štogrin, and R. V. Galiulin, *A local test for the regularity of a system of points*, Dokl. Akad. Nauk SSSR **227** (1976), 19–21 (Russian); English transl., Soviet Math. Dokl. **17** (1976), 319–322.
[Di] J. Dieudonné, *Treatise of Analysis*. II, Pure Appl. Math., vol. 10-II, Academic Press, New York, 1970.
[DuKa] M. Duneau and A. Katz, *Quasiperiodic structures*, Phys. Rev. Lett. **54** (1985), 2688–2691.
[DEK] A. L. Duran, R. Estrada, and R. P. Kanwal, *Extensions of the Poisson summation formula*, J. Math. Anal. Appl. **218** (1998), 581–606.
[Dw] S. Dworkin, *Spectral theory and X-ray diffraction*, J. Math. Phys. **34** (1993), 2965–2967.
[Eij] S. J. L. van Eijndhoven, *Functional analytic characterizations of the Gelfand-Shilov spaces S_α^β*, Nederl. Akad. Wetensch. Indag. Math. **49** (1987), 133–144.
[Els] V. Elser, *The diffraction pattern of projected structures*, Acta. Cryst. A **42** (1986), 36–43.
[EG] L. C. Evans and R. F. Gariepy, *Measure theory and fine properties of functions*, Stud. Adv. Math. CRC Press, Boca Raton, FL, 1992.
[Fø] E. Følner, *Besicovitch almost periodic functions in arbitrary groups*, Math. Scand. **5** (1957), 47–53.
[GK] F. Gähler and R. Klitzing, *The diffraction pattern of self-similar tilings*, The Mathematics of Long-Range Aperiodic Order (Waterloo, ON, 1995) (R. V. Moody, ed.), NATO ASI Ser. C: Math. Phys. Sci., vol. 489, Kluwer, Dordrecht, 1997, pp. 141–174.
[Go] C. Godrèche, *The sphinx: a limit periodic tiling of the plane*, J. Phys. A **22** (1989), L1163–L1168.
[Gr] K. Gröchenig, *An uncertainty principle related to the Poisson summation formula*, Studia Math. **121** (1996), 97–104.
[GHR] K. Gröchenig, A. Haas, and A. Raugi, *Self-affine tilings with several tiles. I*, Appl. Comput. Harmon. Anal. **7** (1999), 211–238.
[Gu] A. P. Guinand, *Concordance and the harmonic analysis of sequences*, Acta Math. **101** (1959), 235–271.
[H92] A. Hof, *Quasicrystals, aperiodicity, and lattice systems*, Ph.D. Thesis, Univ. Groningen, 1992.
[H95a] A. Hof, *On diffraction by aperiodic structures*, Commun. Math. Phys. **169** (1995), 25–43.
[H95b] A. Hof, *Diffraction by aperiodic structures at high temperatures*, J. Phys. A **28** (1995), 57–62.
[H97] A. Hof, *Diffraction by aperiodic structures*, The Mathematics of Long-Range Aperiodic Order (Waterloo, ON, 1995) (R. V. Moody, ed.), NATO ASI Ser. C: Math. Phys. Sci., vol. 489, Kluwer, Dordrecht, 1997, pp. 239–268.
[Jan] C. Janot, *Quasicrystals. A primer*, 2nd ed., Monographs on the Physics and Chemistry of Materials, Clarendon Press, Oxford, 1994.
[KLR] J.-P. Kahane and P.-G. Lemarié-Rieusset, *Remarques sur la formule sommatoire de Poisson*, Studia Math. **109** (1994), 303–316.
[K67] Y. Katznelson, *Une remarque concernant la formule de Poisson*, Studia Math. **29** (1967), 107–108.
[K68] _____, *An introduction to harmonic analysis*, Wiley, New York, 1968; 2nd ed., Dover, New York, 1976.
[Ke90] R. Kenyon, *Self-similar tilings*, Ph.D. Thesis, Princeton, 1990.
[Ke92] _____, *Self-replicating tilings*, Symbolic Dynamics and Its Applications (New Haven, CT, 1991) (P. Walters, ed.), Contemp. Math., vol. 135, Amer. Math. Soc, Providence, RI, 1992 pp. 239–264.
[Ke94] _____, *Inflationary tilings with self-similar structure*, Comm. Math. Helv. **69** (1994), 169–198.
[Ke96] _____, *The construction of self-similar tilings*, Geom. Funct. Anal. **6** (1996), 471–488.
[La95] J. C. Lagarias, *Meyer's concept of quasicrystal and quasiregular sets*, Commun. Math. Phys. **179** (1996), 365–376.
[La99a] _____, *Geometric models for quasicrystals. I. Delone sets of finite type*, Discrete Comput. Geom. **21** (1999), 161–191.
[La99b] _____, *Geometric models for quasicrystals. II. Local rules under isometries*, Discrete Comput. Geom. **21** (1999), 345–372.
[LaWa1] J. C. Lagarias and Y. Wang, *Self-affine tiles in $\mathbb{R}^n$*, Adv. Math. **121** (1996), 21–49.

[LaWa2] _____, *Integral self-affine tiles in $\mathbb{R}^n$. I. Standard and nonstandard digit sets*, J. London Math. Soc. (2) **54** (1996), 161–179.

[LaWa3] _____, *Substitution Delone sets* (in preparation).

[LPS] T. Q. T. Le, S. A. Piunikhin, and V. A. Sadov, *The geometry of quasicrystals*, Uspekhi Mat. Nauk **48** (1993), 41–102 (Russian); English transl., Russian Math. Surveys **48** (1993), 37–100.

[LM] J.-Y. Lee and R. V. Moody, *Lattice substitution systems and model sets*, Discrete Comput. Geom. (to appear); math.MG/0002019.

[LSt] D. Levine and P. J. Steinhardt, *Quasicrystals: a new class of ordered structures*, Phys. Rev. Lett. **53** (1984), 2477–2480.

[LuP] W. F. Lunnon and P. A. B. Pleasants, *Quasicrystallographic tilings*, J. Math. Pures Appl. (9) **66** (1987), 217–263.

[Me72] Y. Meyer, *Algebraic numbers and harmonic analysis*, North-Holland Math. Library, vol. 2, North-Holland, Amsterdam, 1972.

[Me95] _____, *Quasicrystals, Diophantine approximations and algebraic numbers*, Beyond Quasicrystals (Les Houches, 1994) (F. Axel and D. Gratias, eds.), Springer, Berlin, 1995, pp. 3–16.

[Mo95] R. V. Moody, *Meyer sets and the finite generation of quasicrystals*, Symmetries in Science. VIII (Bregenz, 1994) (B. Gruber, ed.), Plenum, New York, 1995, pp. 379–394.

[Mo97] _____, *Meyer sets and their duals*, The Mathematics of Long-Range Aperiodic Order (Waterloo, ON, 1995) (R. V. Moody, ed.), NATO ASI Ser. C: Math. Phys. Sci., vol. 489, Kluwer, Dordrecht, 1997, pp. 403–41.

[Mo99] _____, *Model sets: a survey*, From Quasicrystals to More Complex Systems (Les Houches, 1998) (F. Axel, F. Dénoyer, and J.-P. Gazeau, eds.), Centre de Physique des Houches, vol. 13, Springer, Berlin, 2000, pp. 145–166; math.MG/0002020.

[Q] M. Queffélec, *Substitution dynamical systems—spectral analysis*, Lecture Notes in Math., vol. 1294, Springer, Berlin, 1987.

[RS] M. Reed and B. Simon, *Methods of modern mathematical physics. I. Functional Analysis*, 2nd ed., Academic Press, San Diego, CA, 1980.

[Ren] B. C. Rennie, *On generalized functions. Essays in statistical science*, Z. Phys. C **10** (1981), 139–156.

[Ro94] E. A. Robinson, Jr., *On uniform convergence in the Wiener-Wintner theorem*, J. London Math. Soc. (2) **49** (1994), 493–501.

[Ro96a] _____, *The dynamical properties of Penrose tilings*, Trans. Amer. Math. Soc. **348** (1996), 4447–4464.

[Ro96b] _____, *The dynamical theory of tilings and quasicrystallography*, Ergodic Theory of $\mathbb{Z}^d$-Actions (Warwick, 1993–1994) (M. Pollicott and K. Schmidt, eds.) London Math. Soc. Lecture Note, vol. 228, Cambridge Univ. Press, Cambridge, 1996, pp. 451–473.

[Ru87] W. Rudin, *Real and complex analysis*, 3rd ed., McGraw-Hill, New York, 1987.

[Ru91] _____, *Functional Analysis*, 2nd ed., Internat. Ser. Pure Appl. Math., McGraw-Hill, New York, 1991.

[S98] M. Schlottmann, *Cut-and-project sets in locally compact Abelian groups*, Quasicrystals and Discrete Geometry (Toronto, ON, 1995) (J. Patera, ed.), Fields Inst. Monogr., vol. 10, Amer. Math. Soc., Providence, RI, 1998, pp. 247–264.

[S99] _____, *Generalized model sets and dynamical systems*, Directions in Mathematical Quasicrystals, CRM Monogr. Series, vol. 13, Amer. Math. Soc., Providence, RI, 2000, pp. 143–159, this volume.

[Sch] L. Schwartz, *Théorie des distributions*, 3rd ed., Hermann, Paris, 1998.

[Sen] M. Senechal, *Quasicrystals and geometry*, Cambridge Univ. Press, Cambridge,1995; corrected reprint, 1996.

[SBGC] D. Shechtman, I. Blech, D. Gratias, and J. W. Cahn, *Metallic phase with long-range orientational order and no translational symmetry*, Phys. Rev. Lett. **53** (1984), 1951–1953.

[So97] B. Solomyak, *Dynamics of self-similar tilings*, Ergodic Theory Dynam. Systems **17** (1997), 695–738.

[So98a] _____, *Nonperiodicity implies unique composition for self-similar translationally finite tilings*, Discrete Comput. Geom. **20** (1998), 265–279.

[So98b] _____, *Spectrum of dynamical systems arising from Delone sets*, Quasicrystals and Discrete Geometry (Toronto, ON, 1995) (J. Patera, ed.), Fields Inst. Monogr., vol. 10, Amer. Math. Soc., Providence, RI, 1998, pp. 265–275.
[Vin1] A. Vince, *Replicating tesselations*, SIAM J. Discrete Math. **6** (1993), 501–521.
[Vin2] _____, *Digit tiling of Euclidean space*, Directions in Mathematical Quasicrystals, CRM Monogr. Series, vol. 13, Amer. Math. Soc., Providence, RI, 2000, pp. 327–368, this volume.

AT&T LABS–RESEARCH, FLORHAM PARK, NJ 07932-0971, USA
E-mail address: jcl@research.att.com

Designer Quasicrystals: Cut-and-Project Sets with Pre-Assigned Properties

Peter A. B. Pleasants

ABSTRACT. We investigate what conditions on the lattice, projections and window in the cut-and-project construction are required for various properties of the resulting model set, such as uniformity, diffraction, Ammann bars, symmetry, inflation and local rules. In particular, we show that if such a model set with a "nice" window has a local refining inflation then the multiplier must be a quadratic irrational. We describe a natural way of obtaining a lattice and projections from a module over an algebraic number field and show that these necessarily satisfy the conditions for all quasicrystal properties. A consequence is that there exist model sets in any dimension with arbitrary symmetry group that have all quasicrystal properties except a local refining inflation. From these we show how to derive by iterated substitution, without reducing the symmetry group, point sets that have a local refining inflation and retain all the quasicrystal properties with the possible exception of diffraction, which we are unable to establish. Ammann bars play a central rôle. They turn out to be a nongeneric feature of cut-and-project sets but arise naturally from the algebraic number field construction. We use them to set up the iterated substitution just mentioned and to prove the existence of weak local rules. Further to the subject of constructing quasicrystals with given symmetries, we distinguish a set of "overt" symmetries of a quasicrystal and ask whether this is necessarily the full symmetry group or whether indeed it need be a group at all.

Introduction

In this paper we investigate in detail which aspects of the cut-and-project construction give rise to which features of the resulting quasicrystal or model set with the aim of elucidating to what extent quasicrystals can be constructed by this means with prescribed properties and of understanding the interrelations between the properties. We concentrate on what we call the "plain" cut-and-project construction, where the lattice lies in Euclidean space and the window is Riemann measurable. Riemann measurability of the window (rather than Lebesgue measurability) seems naturally suited to ensuring the uniformity of the model set, and it

---

2000 *Mathematics Subject Classification.* Primary 52C23, 11P21, 11R04, 11R06; Secondary 52C07, 52C22, 82D25.

I should like to thank Michael Baake for his encouragement and many helpful comments and Takashi Soma for some comments and help in preparing the figure.

implies, in particular, that the boundary of the window has measure zero. We also describe in detail a canonical way of deriving cut-and-project sets from modules over real algebraic number fields, introduced in [**P2**]. To a large extent this paper is a reworking of [**LuP**], which built up higher dimensional quasicrystals from 1-dimensional model sets (there called "prism patterns"), in the light of this more direct way of obtaining higher dimensional model sets. We take this opportunity to correct some of the mistakes in [**P1, LuP, P2**] and introduce some new ones. As a vestige of the prism patterns, Ammann bars play a central rôle. Unlike most of the features we study, they do not occur in "generic" cut-and-projectsets, but they do always occur in sets derived from the algebraic number field construction where they correspond to submodules defined over the number field (but not to submodules defined over $\mathbb{Q}$). In order to study them we introduce the concepts of submodels and quotient models of a model set. We make use of Ammann bars to establish weak local rules and local inflation for the quasicrystals we construct.

Section 1 lists 17 properties of quasicrystals, among them such things as uniformity, diffraction, symmetry, inflation, Ammann bars and local rules (also known as "matching rules") and notes some of the simpler relations between them. I have tried to make these definitions as wide as possible so as to capture all recognizable instances of the concepts they embody. In particular, the definition of inflation (or "flation" as I call it here, to encompass deflation as well) is more general than the definition in [**P2**]. What was called an inflation in [**P2**] is here called a "refining inflation". The definitions rely heavily on the concept of "patches", which are defined up to translation equivalence throughout. We do not discuss isometry equivalence of patches. Proposition 1.6, that noncrystallographic repetitive sets have uncountable LI-class, even modulo congruence, is of particular interest. A version of it proved in the more difficult context of isometry patches is given in [**DD**] (another proof of this version will appear in [**P3**]) and sheds light on the intriguing result of Dolbilin [**Do**] that if at most countably many different tilings of $\mathbb{R}^n$ satisfy a given finite set of local rules then at least one of them is crystallographic.

Section 2 describes what I call the "plain" cut-and-project construction and attempts a systematic analysis of which features of the construction contribute to which properties of the resulting model set. Mostly we show only that some condition on the construction is sufficient for a given property of the resulting model set (the appropriate direction if we wish to construct a model set with the property) but sometimes we can show that a condition is also necessary. Instead of the usual closed window we define a concept of "half-openness" and use half-open windows as a natural way of treating singular model sets which largely avoids having to regard them as special cases. Also the concepts of submodel and quotient model are introduced in order to explain Ammann bars. Some properties of the model set, among them diffraction, place no constraints on the construction. But most properties require what we call "orientation conditions" which say something about how the lattice is positioned relative to the physical and internal spaces. Most of the orientation conditions (such as our "V-Condition" and "W-Condition") are "generic", meaning that they are satisfied by a "random" positioning of the lattice and the two subspaces, but one (our "Submodels Condition") is "anti-generic" and not satisfied by random positioning. In spite of this it is possible for all the orientation conditions to be satisfied simultaneously. A common requirement for model sets is that the lattice should project densely into internal space. In Corollary 2.12 we show that this is equivalent to our main generic orientation condition, the V-Condition, which is purely algebraic. The fact that our generic orientation conditions imply uniformity of the model set is equivalent to a standard result of Diophantine approximation—a higher dimensional generalization of the uniform distribution of $\{n\theta\}$ mod 1 when θ is irrational. Lemma 2.9 gives a proof of this result, in keeping with the geometrical approach of this paper, which to the best of my knowledge is novel. The anti-generic Submodels Condition leads to alignment (of which Ammann bars are a strong form): a "lattice-like" property of the model set. The symmetry and flation properties (closely related, since flation can be regarded as a form of symmetry) require linear operators that stabilize the lattice (often automorphisms of the lattice) in addition to orientation conditions. We derive local rules from Ammann hyperplanes, using the treatment of [**LuP**], which unfortunately leads only to weak local rules which we cannot prove perfect. Of the results in this section we mention particularly the Diophantine approximation result Lemma 2.9 and its corollary, Lemma 2.18 which slightly generalizes a local derivability criterion of Baake, Schlottmann and Jarvis [**BSJ**], and Proposition 2.35 which shows that a local refining inflation cannot be obtained by these methods unless the multiplier is a quadratic algebraic integer.

Section 3 describes the canonical way, introduced in [**P2**], of deriving a cut-and-project scheme from a module in K^n, an n-dimensional vector space over an algebraic number field K. Cut-and-project schemes derived in this way satisfy all the orientation conditions, both generic and anti-generic. In particular, they have a plentiful supply of submodels (potential Ammann bars) which correspond to K-subspaces of K^n (but not to $\mathbb{Q}$-subspaces of K^n). The linear operators for symmetry and inflation derive from linear operators defined over K on K^n. Consequently K must be chosen so that the symmetry group G has a representation over K. When this is done, model sets with symmetry group containing G can be constructed having all the quasicrystal properties with the inevitable exception, when $\deg K > 2$, of local refining inflation. The key results of this section are Lemma 3.2, which shows that submodels of the cut-and-project scheme correspond to K-subspaces of K^n, Lemma 3.3, which shows that the algebraic number field construction automatically satisfies all the orientation conditions, and Theorem 3.5, which shows that for an arbitrary finite group G of isometries a model set can be constructed with all these symmetries and with all the quasicrystal properties except local refining inflation.

Reverting to general cut-and-project sets, Section 4 investigates the result of iterating the effect of a nonlocal refining inflation on a finite segment of a model set, much as in [**LuP**, Section 6]. It turns out that the inflation can be made local in this way, while retaining many of the properties of the model set. (When $\dim K > 2$ the new set defined in this way cannot be a plain model set, however.) The key result is Proposition 4.3 which shows that if the model set has a "spanning set" of Ammann hyperplanes then a set with local inflation can be derived from it that retains this feature. A consequence of this is Theorem 4.4, which says that it is possible to construct a set with all the symmetries of a given finite group G of isometries, with a local refining inflation, and with all the main quasicrystal properties except possibly diffraction (which I have not been able to establish for these sets).

Theorems 3.5 and 4.4 assert that for every group G of isometries there are point sets with G-symmetry that exhibit a large collection of quasicrystal properties (a somewhat different collection for each theorem). In Section 5 we contrast this situation with the crystallographic restriction (which says that for a given dimension n the finite isometry groups that are point groups of crystals are a very limited subset of the set of all finite isometry groups) and view these theorems as saying that there is no quasicrystallographic restriction (though such a statement is necessarily vague in the light of the vagueness as to constitutes a mathematical quasicrystal). This is not the end of the matter however, as the natural habitat of a quasicrystal is in an LI-class and the symmetry group of its LI-class may be larger than the symmetry group of any particular point set, as defined in Property 11. In the examples we construct, the LI-class has no extra symmetries not already present in the point set, but questions remain of to what extent we can specify that the symmetry group of the LI-class is larger than the symmetry group of any of its members, or perhaps larger than the union of the symmetry groups of all its members, or perhaps even whether we can specify that the union of the symmetry groups of the members of the LI-class is not a group.

For the purpose of proving the existence theorems (Theorems 3.5 and 4.4) it would be much simpler and more straightforward to start from the K-module construction of Section 3 and derive the various properties directly, since proofs and definitions are considerably more troublesome in the context of a general cut-and-project scheme. I have opted for a two stage approach—first, in Section 2, establishing conditions for properties of a model set in the general situation then, in Section 3, showing that these conditions are satisfied by cut-and-project schemes derived from K-modules—to clarify what the various properties depend on and how they are related, indicating, for example, that Ammann bars and alignment are nongeneric for cut-and-project schemes.

An alternative very simple and elegant construction of quasicrystals with a prescribed group of symmetries, without the use of algebraic number fields, has been given by Cotfas and Verger-Gaugry [**CV-G, Cot**]. Their method is based on symmetric clusters. Unlike the method we describe, which always leads to sets having a refining inflation by a sufficiently large Pisot number in K, the cluster method can lead to quasicrystals both with inflation and without inflation.

I have not touched on local inflation which is not refining. That would seem to be a useful area for further study.

1. Discrete Point Sets

Notation and Definitions. By a *discrete* point set we mean a set $\mathcal{D} \subset \mathbb{E}^n$ ($n \geq 1$) such that for every $\boldsymbol{x}$ in $\mathcal{D}$ there is a distance $\delta = \delta(\boldsymbol{x}) > 0$ such that $|\boldsymbol{y} - \boldsymbol{x}| > \delta$ for every $\boldsymbol{y} \neq \boldsymbol{x}$ in $\mathcal{D}$. In this section we list some quasicrystal-related properties that a discrete point set may have and discuss the relationships between them. Before doing that we introduce some general concepts about discrete point sets.

A point set $\mathcal{D}$ is *uniformly discrete* if there is a *packing radius* $r > 0$ such that every open ball of radius r in $\mathbb{E}^n$ contains at most one point of $\mathcal{D}$. (That is, the single value $\delta = 2r$ does for every $\boldsymbol{x}$ in $\mathcal{D}$.) A point set $\mathcal{D}$ is *relatively dense* if there is a *covering radius* $R > 0$ such that every closed ball of radius R in $\mathbb{E}^n$ contains at least one point of $\mathcal{D}$. A set that is both uniformly discrete and relatively dense is called a *Delone* set.

The *difference set* of $\mathcal{D}$ is
$$\mathcal{D} - \mathcal{D} = \{\boldsymbol{x} - \boldsymbol{y} \mid \boldsymbol{x}, \boldsymbol{y} \in \mathcal{D}\}.$$

If $\mathcal{D}$ is relatively dense and $\mathcal{D} - \mathcal{D}$ is uniformly discrete, $\mathcal{D}$ is called a *Meyer set*.[1]

A strong form of relative denseness (at least when $\nu > 0$) is the following.

DEFINITION 1.1. The discrete point set $\mathcal{D}$ has *uniform frequency* ν if for every ball B in $\mathbb{E}^n$ the number of points of $\mathcal{D}$ in B is
$$\nu \operatorname{vol}(B) + o\bigl(\operatorname{vol}(B)\bigr),$$
where $\operatorname{vol}(B)$ is the volume of B. To be more precise, for every $\varepsilon > 0$ there is an N such that
$$(\nu - \varepsilon)\operatorname{vol}(B) < |\mathcal{D} \cap B| < (\nu + \varepsilon)\operatorname{vol}(B)$$
for every ball B with $\operatorname{vol}(B) > N$.

The set $\mathcal{D}$ has *upper uniform frequency* ν if
$$\limsup\left(\frac{|\mathcal{D} \cap B|}{\operatorname{vol}(B)}\right) = \nu \quad \text{as } \operatorname{vol}(B) \to \infty.$$

Several of the properties we shall describe are framed in terms of patches of $\mathcal{D}$. Given $\boldsymbol{x} \in \mathcal{D}$ and $\rho > 0$ the ρ-*patch* of $\boldsymbol{x}$, $P_\rho(\boldsymbol{x}) = P_\rho(\boldsymbol{x}, \mathcal{D})$, is the configuration (up to translation) of points of $\mathcal{D}$ within a distance ρ of $\boldsymbol{x}$. More explicitly,
$$P_\rho(\boldsymbol{x}) = (\mathcal{D} - \boldsymbol{x}) \cap B_\rho(\mathbf{0})$$
where $B_\rho(\mathbf{0})$ is the ball centre $\mathbf{0}$ radius ρ. Thus only patches centred on points of $\mathcal{D}$ itself are considered and they are identified only up to translation. We denote the set of distinct ρ-patches of $\mathcal{D}$ by $\mathcal{A}(\rho) = \mathcal{A}_\mathcal{D}(\rho)$ and call it the ρ-*atlas* of $\mathcal{D}$. The *atlas* $\mathcal{A}_\mathcal{D}$ of $\mathcal{D}$ is the union of the $\mathcal{A}_\mathcal{D}(\rho)$'s for $\rho > 0$ (in other words, the set of all patches of $\mathcal{D}$). We denote the number of distinct ρ-patches by
$$p(\rho) = p_\mathcal{D}(\rho) = |\mathcal{A}_\mathcal{D}(\rho)|.$$

Clearly $p(\rho)$ is a nondecreasing function of ρ. For a general discrete point set it could well be infinite for large enough ρ, but we shall consider here only sets for which $p(\rho)$ is finite for all ρ (a property known as "finite local complexity").

Note that these definitions are in terms of translation equivalence classes of patches, we shall not deal with isometry equivalence of patches.

DEFINITION 1.2. Two discrete point sets $\mathcal{D}_1$ and $\mathcal{D}_2$ are *locally isomorphic*[2] if every patch of each also occurs as a patch of the other, that is, if $\mathcal{A}_{\mathcal{D}_1} = \mathcal{A}_{\mathcal{D}_2}$. Thus it is impossible to distinguish $\mathcal{D}_1$ from a translate of $\mathcal{D}_2$ by observing the two sets over any bounded region. The class of all sets locally isomorphic to $\mathcal{D}$ is the *local isomorphism class* (or *LI-class*) of $\mathcal{D}$.

Clearly the LI-class of $\mathcal{D}$ contains all translates of $\mathcal{D}$, and when $\mathcal{D}$ is crystallographic the translates account for the entire LI-class. But sets like the Penrose set (illustrated in Figure 1) have LI-classes that are uncountable even when translates are regarded as equal.

---

[1] Lagarias [**La**] characterized Meyer sets $\mathcal{D}$ as sets for which both $\mathcal{D}$ and $\mathcal{D} - \mathcal{D}$ are Delone. Our definition is trivially equivalent to this since $\mathcal{D} - \mathcal{D}$ is relatively dense when $\mathcal{D}$ is and $\mathcal{D}$ is uniformly discrete when $\mathcal{D} - \mathcal{D}$ is.

[2] Another term for this is *locally indistinguishable*

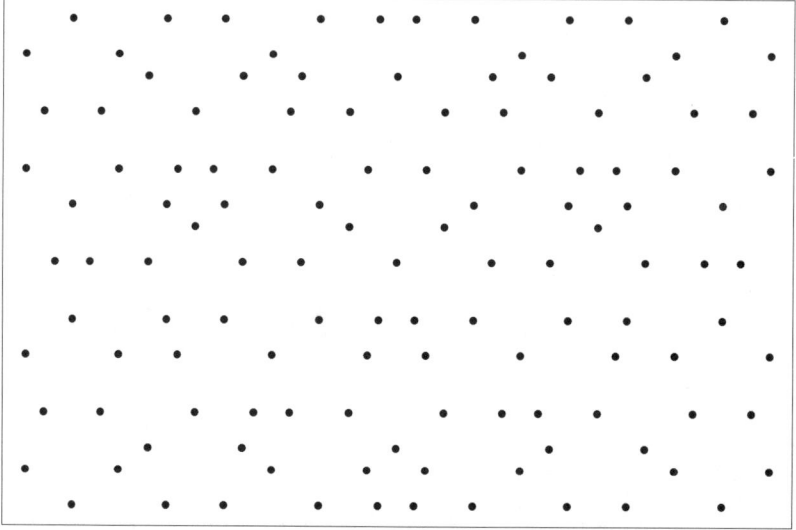

FIGURE 1. The Penrose point set $\mathcal{P}$, derived form the "cartwheel" Penrose tiling. There is approximate D_{10} symmetry about the centre. A refining inflation with multiplier $\tau = \frac{1}{2}(\sqrt{5}+1)$ is clearly visible. Ammann lines are clearly visible horizontally, nearly parallel to the diagonals of the box, and at two steeper slopes.

DEFINITION 1.3. Let $\mathcal{D}_1$ and $\mathcal{D}_2$ be relatively dense discrete point sets in $\mathbb{E}^n$ with covering radii R_1 and R_2. Then $\mathcal{D}_2$ is said to be *locally derivable* from $\mathcal{D}_1$ (see [**BSJ, BSchl**]) if there are radii $\rho, R > 0$ and a function

$$f : \mathcal{A}_{\mathcal{D}_1}(\rho) \to \mathcal{A}_{\mathcal{D}_2}(R_1 + R) \times B_R(\mathbf{0})$$

such that, for every $\boldsymbol{x} \in \mathcal{D}_1$, if $f(P_\rho(\boldsymbol{x})) = (P, \boldsymbol{t})$ then $\boldsymbol{x} + \boldsymbol{t} \in \mathcal{D}_2$ and $P_{R_1+R}(\boldsymbol{x} + \boldsymbol{t}, \mathcal{D}_2) = P$.

This definition ensures that $\mathcal{D}_2$ is determined in its entirety by $\mathcal{D}_1$. Local derivability is therefore transitive: if $\mathcal{D}_2$ is locally derivable from $\mathcal{D}_1$ and $\mathcal{D}_3$ from $\mathcal{D}_2$ then $\mathcal{D}_3$ is locally derivable from $\mathcal{D}_1$. The size of ρ is a measure of how local the derivability is. If $\mathcal{D}_2$ is locally derivable from $\mathcal{D}_1$ then clearly so is every translate of $\mathcal{D}_2$, but other sets in the LI-class of $\mathcal{D}_2$ are not in general locally derivable from $\mathcal{D}_1$. However, for each set in the LI-class of $\mathcal{D}_1$ the same function f enables us to derive a set in the LI-class of $\mathcal{D}_2$, and for this reason it is usual to think of local derivability as a relation between LI-classes.

Quasicrystal Properties. We now make a list of properties that some quasicrystals (in the form of discrete point sets) have. I am not suggesting that a set needs to have all these properties to be regarded as a mathematical quasicrystal, but they represent the sort of behaviour we have in mind when we think of quasicrystals. When interpreting them it is helpful to bear in mind the Penrose point set $\mathcal{P}$, illustrated in Figure 1, which possesses all the properties. This set can be derived from the Penrose kite and dart tiling, which has edges of two lengths in the ratio $\tau : 1$, where $\tau = (\sqrt{5}+1)/2$. The set $\mathcal{P}$ is obtained by placing a point on each long edge of the tiling, nearer the tip of the dart (for dart edges) and further from the base of the kite (for kite edges), to divide the edge in the ratio $\tau : 1$. The set illustrated comes from the "cartwheel" tiling, which has approximate D_{10} symmetry about the origin. An alternative description of this set is that it is the set of integers of the 5th cyclotomic field $\mathbb{Q}(\xi)$ (where $\xi^5 = 1$) in the complex plane whose conjugates by a generator of the Galois group lie in the regular decagon whose vertices are the 10th roots of 1. (This decagon should be regarded as half-open, with the direction $\boldsymbol{h}$ of Definition 2.2 equal to $-i$.) But we frame the properties quite generally, in particular making them applicable to sets in space of any dimension. We also note that any mathematical crystal possesses all the properties except Properties 5, 15, and 16, in accord with the principle that any definition of quasicrystal should include crystals as a special case. In describing these properties we denote by $\mathcal{Q}$ the discrete point set which is the putative quasicrystal.

PROPERTY 1 (Finite local complexity). $p(\rho)$ is finite for all $\rho > 0$.

The connexion between periodicity and the size of $p(\rho)$ is illustrated by the following easily verified results (see [**LaP2**]):

$$p(\rho) = 1, \ \forall \rho > 0 \iff \mathcal{Q} \text{ is a translate of a lattice,}$$
$$p(\rho) \text{ is bounded} \iff \mathcal{Q} \text{ is crystallographic.}$$

In fact, if the maximum value of $p(\rho)$ is p then $\mathcal{Q}$ is the union of precisely p distinct translates of a lattice. (The term "lattice" here must be interpreted as including lattices confined to a subspace not of the full dimension n. Also "crystallographic" means having n linearly independent periods, but again, in addition to ordinary periods, "infinite periods", which have the effect of confining $\mathcal{Q}$ between a pair of hyperplanes perpendicular to the period, are counted.) It is conjectured in [**LaP1**] (Conjecture 2.2) that, for $\mathcal{Q} \in \mathbb{E}^n$,

$$p(\rho) = o(\rho^k) \implies \mathcal{Q} \text{ has } n - k + 1 \text{ linearly independent periods.}$$

Requiring that $p(\rho)$ be finite seems a simple way of staking out another level beyond crystallographic sets in a hierarchy of discrete point sets.

PROPERTY 2 (Repetitive). For every $\rho > 0$ there is a radius $R(\rho)$ such that for every ρ-patch P of $\mathcal{Q}$ and every ball B of radius $R(\rho)$ there is an $\boldsymbol{x}$ in B with $P_\rho(\boldsymbol{x}) = P$.

When $\mathcal{Q}$ is uniformly discrete with packing radius r, $R(\rho) = R$, the covering radius of $\mathcal{Q}$, for $\rho < 2r$. A repetitive set $\mathcal{Q}$ has the crystal-like quality that every finite configuration that occurs is ubiquitous throughout the set. It also has the minimality property (analogous to lack of defects in a crystal) that no "simpler" point set can be formed whose atlas is a proper subset of $\mathcal{A}_\mathcal{Q}$. (Because every $(R(\rho) + \rho)$-patch of $\mathcal{Q}$ contains copies of all ρ-patches of $\mathcal{Q}$.) We might want to restrict the growth rate of the function $R(\rho)$. Sets with such restrictions, and in particular *linearly repetitive* sets, where the growth rate is at most linear, are studied in [**LaP1**].

PROPERTY 3 (Uniform). Every patch $P_\rho(\boldsymbol{x})$ of $\mathcal{Q}$ occurs with positive uniform frequency. That is, there is a $\nu > 0$ (depending only on $P_\rho(\boldsymbol{x})$) such that

$$\{\boldsymbol{y} \in \mathcal{Q} \mid P_\rho(\boldsymbol{y}) = P_\rho(\boldsymbol{x})\}$$

has uniform frequency ν.

A fuller description would be to say that $\mathcal{D}$ has "uniform patch frequencies", but since we refer to this property frequently it is convenient to use the single word "uniform".

PROPERTY 4 (Diffracting). The set has a pure point diffraction spectrum. By this we mean that the distribution
$$\sum_{\boldsymbol{x}\in\mathcal{Q}} \delta_{\boldsymbol{x}},$$
consisting of a δ-peak with amplitude 1 at each point of $\mathcal{Q}$, has a unique tempered autocorrelation distribution γ, given by
$$\gamma = \lim_{R\to\infty} \frac{1}{\mathrm{vol}(B_R)} \sum_{\boldsymbol{x},\boldsymbol{y}\in\mathcal{Q}\cap B_R} \delta_{\boldsymbol{x}-\boldsymbol{y}},$$
where B_R is the ball with centre $\mathbf{0}$ and radius R, and that the Fourier transform $\hat{\gamma}$ of γ consists of a countable, locally convergent sum of δ-peaks with positive intensity. The distribution $\hat{\gamma}$ is called the *diffraction spectrum* of $\mathcal{Q}$. The locations of the peaks of $\hat{\gamma}$ will not in general be discrete, but for any threshold $\varepsilon > 0$ the peaks with intensity $> \varepsilon$ are discrete.

Some background to diffraction of noncrystallographic point sets can be found in [**H1, H2, B, BMP**].

PROPERTY 5 (Nonperiodic). There is no period $\boldsymbol{p}$ such that $\mathcal{Q} + \boldsymbol{p} = \mathcal{Q}$.

PROPERTY 6 (Near-crystallographic). For every ε, $\delta > 0$ and unit vector $\boldsymbol{u}$ there is a vector $\boldsymbol{p}$ with $\boldsymbol{u}\cdot\boldsymbol{p}/|\boldsymbol{p}| > 1 - \delta$ such that $\mathcal{Q}\triangle(\mathcal{Q}+\boldsymbol{p})$ has upper uniform frequency $< \varepsilon$, where $\triangle$ denotes the symmetric difference.

This says that there is a "near period" close to any given direction $\boldsymbol{u}$.

PROPERTY 7 (Lattice-like). Let $\mathcal{X}$ be any subset of $\mathcal{Q}$, $A = \mathrm{aff}(\mathcal{X})$ the minimal affine subspace of $\mathbb{E}^n$ containing $\mathcal{X}$, and $A^\perp$ an orthogonal complement of A in $\mathbb{E}^n$. Then the set of vectors $\boldsymbol{t}$ in $A^\perp$ for which $\mathcal{Q}\cap(A+\boldsymbol{t}) \neq \emptyset$ is a Meyer set of positive uniform frequency in $A^\perp$ and, for all such $\boldsymbol{t}$ with the possible exception of a set of uniform frequency 0 in $A^\perp$, $\mathcal{Q}\cap(A+\boldsymbol{t})$ is a Meyer set of positive uniform frequency in $A+\boldsymbol{t}$.

PROPERTY 8 (Ammann bars). An *Ammann grid* or system of *Ammann bars* of $\mathcal{Q}$ is a set of translates $\{A + \boldsymbol{t} \mid \boldsymbol{t} \in \mathcal{T} \subset A^\perp\}$ of some linear subspace A of dimension k, say, in $\mathbb{E}^n$ that satisfies the following for some numbers $\delta, \Delta > 0$:

(1) $\mathcal{T}$ is a Meyer set with positive uniform frequency in $A^\perp$.
(2) Every $\boldsymbol{x} \in \mathcal{Q}$ lies in $A + \boldsymbol{t}$ for some $\boldsymbol{t} \in \mathcal{T}$.
(3) For every $\boldsymbol{t} \in \mathcal{T}$, $\mathcal{Q}\cap(A+\boldsymbol{t})$ is a Meyer set in $A+\boldsymbol{t}$ with uniform frequency $> \delta$ and covering radius $< \Delta$.

Particular cases:

(i) If $\mathcal{Q}$ is a Meyer set with positive uniform frequency then $\mathcal{Q}$ itself is a singleton Ammann n-space system.
(ii) If $\mathcal{Q}$ is a Meyer set with positive uniform frequency then the points of $\mathcal{Q}$ are a single system of Ammann points.
(iii) In addition to these, the Penrose set $\mathcal{P}$ of Figure 1 has five systems of Ammann lines, one horizontal, two close to the diagonals of the bounding rectangle of the figure, and two others of steeper slope.

Clearly this property is closely related to the previous one. The difference is that there is a uniform lower bound for the frequency of points and a uniform upper bound for the covering radius on the bars of an Ammann system and that there are no exceptional bars. This makes Ammann bars an easily recognizable feature which can be used to formulate local rules and define local inflation.

PROPERTY 9 (Finitely generated). There is a finite set of vectors $\boldsymbol{g}_1, \ldots, \boldsymbol{g}_N$ such that the additive group (or $\mathbb{Z}$-module) they generate contains $\mathcal{Q} - \mathcal{Q}$ as a subset.

What we really have in mind is that $\mathcal{Q} - \boldsymbol{x}_0$ is contained in a finitely generated subgroup of $\mathbb{E}^n$, where $\boldsymbol{x}_0 \in \mathcal{Q}$ is chosen as an origin, but these two notions are clearly equivalent and the definition using $\mathcal{Q} - \mathcal{Q}$ avoids singling out an origin.

PROPERTY 10 (Meyer). $\mathcal{Q}$ is a Meyer set.

PROPERTY 11 (Symmetry). We call an isometry g of $\mathbb{E}^n$ that fixes the origin a *symmetry* of $\mathcal{Q}$ if $\mathcal{Q} \triangle g\mathcal{Q}$ has uniform frequency 0. Clearly the symmetries of $\mathcal{Q}$ form a group, which we call the *symmetry group* of $\mathcal{Q}$.

The reason for allowing such approximate symmetries is that two model sets that agree except on a set of frequency zero are "singular" (we explain this in the next section) and in this case it is better, for topological consistency, to regard the fundamental object as being the collection of related singular sets rather than the individual sets themselves. The Penrose set $\mathcal{P}$, for example, is singular and the fact that we do not require symmetries to be exact enables us to recognize D_{10} as its symmetry group even though there are six straight lines on which $\mathcal{P}$ fails to match its rotation through $\pi/5$. In fact the only exact symmetry of $\mathcal{P}$ is reflexion in a vertical axis, but the collection consisting of the 10 rotations of $\mathcal{P}$ through multiples of $\pi/5$ has symmetry group D_{10}. This accords with our feeling that $\mathcal{P}$ is the most symmetric set in its LI-class, even though there are other sets in the class with exact D_5 symmetry.

PROPERTY 12 (Flation). There is a real number $\mu \neq \pm 1$ such that $(\mathcal{Q}+\boldsymbol{t}) \cap \mu\mathcal{Q}$ is relatively dense for some $\boldsymbol{t} \in \mathbb{R}^n$.

This ascribes a type of self-similarity or scale symmetry to $\mathcal{Q}$. The Penrose set has this property for any $\mu \neq \pm 1$ in $\mathbb{Q}(\tau)$ and $\boldsymbol{t} = \boldsymbol{0}$. It is usual to call it an *inflation* when $|\mu| > 1$ and a *deflation* when $|\mu| < 1$. In terms of a tiling corresponding to $\mathcal{Q}$, an inflation ensures that after dilation by a factor μ the original tiling can be recovered by subdividing each of the dilated tiles and perhaps glueing together some of the pieces of neighbouring dilated tiles. However, it may not be possible to use such an inflation to generate the tiling from a finite patch because identical tiles in different locations may need to be subdivided differently and there may be no rule, based on the configuration of nearby tiles, for which subdivision to use. For an inflation that can be used to generate $\mathcal{Q}$ we need Property 14 below.

The translation $\boldsymbol{t}$ has been introduced to exhibit this property as a more general form of Property 14 and also because there are some advantages in regarding a flation as a property of translation classes.

We note that flation comes in inflation-deflation pairs. If $\mu, \boldsymbol{t}$ give a flation then so do $\mu^{-1}, -\mu^{-1}\boldsymbol{t}$.

In [**P2**] I used the simpler definition $\mu\mathcal{Q} \subset \mathcal{Q}$ for inflation, but that excludes, for example, the subset $\mathbb{Z} \cup (\mathbb{Z} + \theta)$ of $\mathbb{R}$ with θ irrational, even though this is a periodic set. The property in its present form is possessed by every crystal for every rational value of μ. Nevertheless, it is worth stating the simpler form of the property as a special case.

PROPERTY 13 (Refining inflation). There is a real number μ with $|\mu| > 1$ such that $\mu \mathcal{Q} \subset \mathcal{Q} + \boldsymbol{t}$ for some $\boldsymbol{t} \in \mathbb{R}^n$.

Apart from the translation $\boldsymbol{t}$, which has been introduced for the reasons described above, this is the definition of inflation given in [**P2**]. The Penrose set has this property with $\mu = \tau$ (or any other Pisot number in $\mathbb{Q}(\tau)$) and $\boldsymbol{t} = \boldsymbol{0}$. There is a case for allowing an exceptional set of frequency zero (requiring that $(\mathcal{Q}+\boldsymbol{t}) \setminus \mu\mathcal{Q}$ has frequency zero instead of $\mu\mathcal{Q} \subset \mathcal{Q} + \boldsymbol{t}$), as in the definition of symmetry in Property 11, since inflation is best regarded as a kind of symmetry. We say more about this later.

PROPERTY 14 (Local flation). There is a real number $\mu \neq \pm 1$ such that $\mathcal{Q}$ is locally derivable from $\mu\mathcal{Q}$.

This clearly implies Property 12, when $\mathcal{Q}$ is repetitive, with $\boldsymbol{t}$ equal to any of the translation vectors associated with the local derivability function. When $|\mu| > 1$ the property enables $\mathcal{Q}$ to be generated iteratively, starting from any sufficiently large finite patch. Every crystal has this property for every nonzero integer μ, but not for any μ with $|\mu| < 1$. The Penrose set has this property with $\mu \neq 1$ any power of $-\tau$ and $\boldsymbol{t} = \boldsymbol{0}$.

For a local flation to exist $\mathcal{Q}$ must be relatively dense so that Definition 1.3 is applicable.

Local flations do not necessarily come in pairs since there may be local derivability in one direction only. The case of reversible local flation, when $\mathcal{Q}$ and $\mu\mathcal{Q}$ are mutually locally derivable, is of particular interest, however.

PROPERTY 15 (Uncountable LI-class). There are $2^{\aleph_0}$ pairwise incongruent point sets in the LI-class of $\mathcal{Q}$.

PROPERTY 16 (Weak local rules). For some $\rho > 0$ there is no periodic point set all of whose ρ-patches are ρ-patches of $\mathcal{Q}$.

PROPERTY 17 (Perfect local rules). For some $\rho > 0$ every point set all of whose ρ-patches are ρ-patches of $\mathcal{Q}$ is locally isomorphic to $\mathcal{Q}$.

In the context of tilings these last two properties are often referred to as "matching rules".

We list some relations between these properties whose proofs are clear.

PROPOSITION 1.4. *The following relations hold between the above properties.*

(i) *Property 2 $\Rightarrow$ Property 1 $\Rightarrow$ $\mathcal{Q}$ is uniformly discrete.*
(ii) *Properties 1 and 3 $\Rightarrow$ Property 2 $\Rightarrow$ $\mathcal{Q}$ is relatively dense.*
(iii) *Property 10 $\Rightarrow$ Property 1.*
(iv) *Properties 2 and 13 $\Rightarrow$ Property 12.*
(v) *Properties 2 and 14 $\Rightarrow$ Property 12.*
(vi) *Properties 5 and 17 $\Rightarrow$ Property 16 $\Rightarrow$ Property 5.*

Proposition 1.6 (or rather its corollary, which puts the result in the format of Proposition 1.4) gives a slightly less obvious implication. Before proving it we state a lemma which puts on record the comment following the description of Property 2 that repetitivity is equivalent to minimality of atlas (cf. [**LaP1**, Theorem 3.2]).

LEMMA 1.5. *If $\mathcal{D}_1$ and $\mathcal{D}_2$ are discrete point sets with $\mathcal{D}_1$ repetitive and $\mathcal{A}_{\mathcal{D}_2} \subseteq \mathcal{A}_{\mathcal{D}_1}$ then $\mathcal{D}_2$ is locally isomorphic to $\mathcal{D}_1$.*

PROOF. Let P be any ρ-patch of $\mathcal{D}_1$. Since $\mathcal{D}_1$ is repetitive there is a radius R such that every R-patch of $\mathcal{D}_1$ contains a point whose ρ-patch is P. Hence every R-patch of $\mathcal{D}_2$ contains a point with ρ-patch P, so $P \in \mathcal{A}_{\mathcal{D}_2}$. Thus $\mathcal{A}_{\mathcal{D}_2} = \mathcal{A}_{\mathcal{D}_1}$ and $\mathcal{D}_1, \mathcal{D}_2$ are locally isomorphic. □

PROPOSITION 1.6. *Let $\mathcal{D}$ be a repetitive but noncrystallographic point set. Then the number of pairwise incongruent point sets locally isomorphic to $\mathcal{D}$ is $2^{\aleph_0}$ (the cardinal of the continuum).*

PROOF. We construct inductively an infinite binary rooted tree whose nodes are labelled with patches $P_\rho(\boldsymbol{x})$ of $\mathcal{D}$. The label for the root is chosen arbitrarily. For each node with label $P_\rho(\boldsymbol{x})$ we label its offspring as follows. Choose two points $\boldsymbol{x}_1, \boldsymbol{x}_2 \in \mathcal{D}$, with $\boldsymbol{x}_1 - \boldsymbol{x}_2$ not a period of $\mathcal{D}$, whose ρ-patches are $P_\rho(\boldsymbol{x})$. (This is possible because the points of $\mathcal{D}$ with ρ-patch $P_\rho(\boldsymbol{x})$ are relatively dense but the periods of $\mathcal{D}$ lie in a proper subspace of $\mathbb{E}^n$.) Since $\boldsymbol{x}_1 - \boldsymbol{x}_2$ is not a period of $\mathcal{D}$ there is some $\rho' > 0$ with $P_{\rho'}(\boldsymbol{x}_1) \neq P_{\rho'}(\boldsymbol{x}_2)$. Choose such a $\rho' > 2\rho$ and take the resulting two patches to be the labels of the offspring of the node with label $P_\rho(\boldsymbol{x})$. Then we have

(1.1) $$P_\rho(\boldsymbol{x}) \subseteq P_{\rho'}(\boldsymbol{x}_1), P_{\rho'}(\boldsymbol{x}_2)$$

but

(1.2) $$P_{\rho'}(\boldsymbol{x}_1) \neq P_{\rho'}(\boldsymbol{x}_2).$$

The nestings (1.1) show that every branch of the tree gives a unique limiting point set containing $\boldsymbol{0}$, and the fact that the value of ρ tends to infinity along each branch shows that each patch of the limiting set is a patch of a patch of $\mathcal{D}$, hence a patch of $\mathcal{D}$ itself. So by Lemma 1.5 each limiting point set is locally isomorphic to $\mathcal{D}$. On the other hand, the bifurcations (1.2) show that every limiting set is different. Since an infinite binary tree has $2^{\aleph_0}$ branches, there are $2^{\aleph_0}$ point sets in the LI-class of $\mathcal{D}$ containing $\boldsymbol{0}$. Each such set is discrete, and hence countable, so is congruent to at most countably many others. It follows that there are $2^{\aleph_0}$ incongruent sets locally isomorphic to $\mathcal{D}$. Since there are only $2^{\aleph_0}$ countable subsets of $\mathbb{E}^n$ altogether there are exactly $2^{\aleph_0}$ such sets. □

COROLLARY 1.7.
Properties 2 and 5 $\Rightarrow \mathcal{Q}$ is repetitive and noncrystallographic $\Rightarrow$ Property 15.

2. The Plain Cut-And-Project Construction

The Basic Construction. The ingredients for the cut-and-project construction for a model set $\mathcal{M}$ are:
- a high dimensional space $\mathbb{E}^N$ with a pair of complementary subspaces V (the *physical space*, to contain $\mathcal{M}$) and W (the *internal space*), so that
$$\mathbb{E}^N = V \oplus W;$$
- a lattice $\mathcal{L} \subset \mathbb{E}^N$; and
- a bounded *window* or *acceptance domain* $\Omega \subset W$ with the properties described below.

We denote by π_V and π_W the associated projections of $\mathbb{E}^N$ onto V and W with kernels W and V respectively. It is also convenient to consider the *strip* Σ in $\mathbb{E}^N$ given by
$$\Sigma = \pi_W^{-1}(\Omega) = V + \Omega.$$
Except where stated otherwise we shall assume that the window has the following properties:

> The window Ω is bounded and Riemann measurable, all sections of Ω by affine subspaces are Riemann measurable, all projections of the interior of Ω onto affine subspaces are Riemann measurable, and Ω is half-open in the sense of Definition 2.2 below.

The measurability conditions are satisfied by all polytopes, for example. The half-open condition ensures that $\mathcal{M}$ is repetitive in most cases and also makes some of our discussion clearer. However, because half-openness does not always survive taking sections or projections, we shall need to waive this assumption for some of our results. An example in the literature of an explicitly half-open window is [**BKSZ**, Figure 4.2]. The window being Riemann measurable means that its characteristic function is Riemann integrable and is equivalent to its boundary having Riemann measure zero. This assumption is needed to establish the uniformity properties of $\mathcal{M}$. We restrict attention to "nice" windows because our concern is with using cut-and-project to construct model sets, and a nice window makes it easy to study the properties of the set we construct. There are numerous examples in the literature of windows with fractal boundary, but they usually arise from the interpretation as model sets of quasicrystals originally constructed by inflation. We do not assume that Ω is connected, however. Commonly V and W will be chosen as invariant subspaces of a group of operators on $\mathcal{L}$.

DEFINITION 2.1. The *model set* $\mathcal{M}$ derived from V, W, $\mathcal{L}$ and Ω is
$$\mathcal{M} = \{\pi_V(l) \mid l \in \mathcal{L} \cap \Sigma\}.$$

Thus $\mathcal{M}$ is the π_V-projection of the set of lattice points in Σ. The construction can be pictured like this, with all the arrows surjective mappings.

$$\begin{array}{ccccc} V & \xleftarrow{\pi_V} & \mathbb{E}^N & \xrightarrow{\pi_W} & W \\ \cup & & \cup & & \cup \\ \mathcal{M} & \xleftarrow{\pi_V} & \overbrace{\mathcal{L} \cap \Sigma, \Sigma} & \xrightarrow{\pi_W} & \Omega \end{array}$$

We call this the *plain cut-and-project construction*, where the word "plain" is meant to indicate that the containing space is ordinary Euclidean space (general locally compact spaces have been considered, for example in [**M, Schl1, Schl2**]) and that the window is nonpathological.

With Definition 2.1 as we have it, a translate of a model set is not in general a model set itself. Nevertheless, the properties of model sets we consider are translation invariant and it is perhaps best to regard $\mathcal{M}$ as the representative of a translation equivalence class.

The definition does not include the so-called "multicomponent" quasicrystals, for which the cut-and-project construction uses an individual window for each point of $\mathcal{L}$, depending on which coset of a sublattice $\mathcal{L}'$ it belongs to. All we can say about those is that they are finite unions of plain model sets. Perhaps the nicest way to represent them as model sets (though not plain model sets) is to replace $\mathbb{E}^N$ by the locally compact group $\mathbb{E}^N \times F$, where $F = \{\boldsymbol{f}_1, \ldots, \boldsymbol{f}_k\}$ is the finite quotient group $\mathcal{L}/\mathcal{L}'$, and W by $W \times F$. The window Ω is then $\bigcup_{i=1}^{k}(\Omega_i \times \{\boldsymbol{f}_i\})$, where Ω_i is the window for the ith coset.

We now make precise the definition of half-open.

DEFINITION 2.2. A subset H of Euclidean space is *half-open in the direction* $\boldsymbol{h}$ (where $\boldsymbol{h}$ is a unit vector) if it has the property that

(i) $\boldsymbol{x} \in H \Rightarrow \exists \delta > 0$ such that $\boldsymbol{x} + d\boldsymbol{h}$ is in the interior of H for all d with $0 < d < \delta$, and

(ii) $\boldsymbol{x} \notin H \Rightarrow \exists \delta > 0$ such that $\boldsymbol{x} + d\boldsymbol{h}$ is in the exterior of H for all d with $0 < d < \delta$.

(The *exterior* of H means the interior of the complement of H.)

We note that this agrees with the kind of half-openness that is appropriate for a fundamental region. For example, a parallelogram that is half-open in this sense contains only one of its vertices and the two edges incident with this vertex. The lattice tiling of the plane by such half-open parallelograms covers each point exactly once, consistent with the fact that each edge of the tiling is adjacent to two tiles and each vertex is adjacent to four tiles. Similarly a half-open convex decagon contains four of its vertices and five of its edges. An illustration of this is the inner shell of four points around the origin in Figure 1, which arise from the vertices of a half-open decagonal window. The next three shells have ten points each, conforming to the decagonal symmetry.

While it can matter which boundary points are included in a set, it can also be convenient to distinguish sets less finely.

DEFINITION 2.3. Two subsets of $\mathbb{E}^m$ are *geometrically equal* if they have the same closure and the same interior.

PROPOSITION 2.4.

(i) If H is half-open in the direction $\boldsymbol{h}$ then it is contained in the closure of its interior and contains the interior of its closure.

(ii) If H is half-open in the direction $\boldsymbol{h}$ then so are all translates of H.

(iii) If H is half-open in the direction $\boldsymbol{h}$ then so is its complement.

(iv) If H_1 and H_2 are half-open in the direction $\boldsymbol{h}$ then so is $H_1 \cap H_2$.

(v) There are only finitely many half-open polytopes geometrically equal to a given polytope $\mathcal{P}$.

PROOF. All except (v) are straightforward. For (v) we observe that if $\boldsymbol{x}$ lies on a face Φ of $\mathcal{P}$, where Φ is to be understood in the strict sense of not including its lower dimensional boundaries, and $\boldsymbol{x} + d\boldsymbol{h}$ is in the interior of $\mathcal{P}$ for d sufficiently small then the same is true of every point of Φ. On the other hand, if $\boldsymbol{x} + d\boldsymbol{h}$ is in the exterior of $\mathcal{P}$ for d sufficiently small then again the same holds for all points of Φ. Consequently if H is a half-open polytope geometrically equal to $\mathcal{P}$ then each face Φ of $\mathcal{P}$ is either included in H in its entirety or excluded from H in its entirety, and the number of possibilities for H is bounded above by 2^ϕ, where ϕ is the total number of faces of $\mathcal{P}$ of all dimensions. □

REMARK 2.5. Compared to openness and closedness, half-openness has the obvious drawback that it depends on a direction $\boldsymbol{h}$. However parts (iii) and (iv) of Proposition 2.4 show that, for given $\boldsymbol{h}$, half-openness is preserved by all Boolean set operations, in contrast to openness and closedness which are preserved by unions and intersections but not by taking complements. This can be a considerable advantage. On the other hand, openness and closedness survive taking lower dimensional sections and projections, which half-openness unfortunately does not. In the cut-and-project construction the choice of the direction $h \in W$ to give the half-openness of Ω affects only the boundaries of Ω and Σ, so *nonsingular* model sets, with no lattice points on the boundary of Σ, are unaffected by it. When there are lattice points on the boundary of Σ, however, different but closely related *singular* model sets can be obtained by different choices of h. Part (v) of Proposition 2.4 shows that when Ω is a polytope any such collection of related singular model sets is finite. The same is true for the spherical windows that have been used in [**MP**], for example; in that case because only finitely many lattice points fall on the boundary of Σ.

Terminology. For the lattice $\mathcal{L}$ we use the notation dens($\mathcal{L}$) to denote the reciprocal of the volume of its fundamental region (which is also its uniform frequency) to avoid the ambiguity[3] of whether det($\mathcal{L}$) refers to the the square of this volume or the volume itself. By a *lattice subspace* or $\mathcal{L}$-*subspace* we mean a subspace L of $\mathbb{E}^N$ spanned by vectors of $\mathcal{L}$. We write $\langle l_1, \ldots, l_r \rangle$ for the lattice subspace spanned by the lattice vectors $\{l_1, \ldots, l_r\}$. In particular, an $\mathcal{L}$-subspace of dimension 1 is a *lattice line* and an $\mathcal{L}$-subspace of codimension 1 is a *lattice hyperplane*. (Note that lattice lines and hyperplanes in this sense are linear subspaces of $\mathbb{E}^N$ containing the origin—we are not including the affine lines and hyperplanes that result from translating them to other lattice points.)

Submodels and Quotient Models. If an $\mathcal{L}$-space L has $\pi_V(L) = V \cap L$ then also $\pi_W(L) = W \cap L$ (since for any $u \in L$ we have $\pi_W(u) = u - \pi_V(u) \in L$) and L has the direct sum decomposition

$$L = V' \oplus W',$$

where $V' = V \cap L$ and $W' = W \cap L$. Also, since L is an $\mathcal{L}$-subspace, $\mathcal{L}' = \mathcal{L} \cap L$ is a lattice in L. If we take $\Omega' = \Omega \cap W'$ then the model set $\mathcal{M}'$ derived from $\mathcal{L}', V', W', \Omega'$ by the cut-and-project construction is a subset of $\mathcal{M}$ that we call a *submodel* of $\mathcal{M}$. The new strip, Σ', is $\Sigma \cap (V' \oplus W')$. (We note that the new window Ω' may not be half-open. This is something to be borne in mind when using submodels.)

Not every $\mathcal{L}$-subspace gives a submodel, because in general $\pi_V(L) \neq V \cap L$. Nor is it necessarily true that every subspace of V' of V spanned by vectors in $\mathcal{M}$ is a submodel, since there may be no $\mathcal{L}$-subspace with $L \cap V = V'$.

For a given subspace V' of V there may be more than one $\mathcal{L}$-subspace L with

(2.1) $$\pi_V(L) = V \cap L = V',$$

but since the sum of two $\mathcal{L}$-subspaces satisfying (2.1) also satisfies (2.1) there is a unique maximal L satisfying (2.1). When L is maximal in this sense we call the submodel a *full submodel*. For a full submodel $\mathcal{M}'$ we have $\mathcal{M}' = \mathcal{M} \cap V'$.

Given a submodel derived from an $\mathcal{L}$-subspace L we can factor out L and observe that $\mathcal{L}/\mathcal{L}'$ is a lattice in the quotient space $\mathbb{E}^N/L$, which has the direct sum decomposition

(2.2) $$\mathbb{E}^N/L = (V/V') \oplus (W/W').$$

---

[3] See, for example, [**Con**, pp. 3 and 48].

(In fact $\mathbb{E}^N/L$ can be explicitly represented as an $\mathcal{L}$-subspace complementary to L.) We define a window Ω'' as the projection of Ω on W/W' (or, to describe it another way, as the set of orbits of Ω under the action of W'). The model set in $\mathbb{E}^N/L$ given by the direct sum decomposition (2.2), the lattice $\mathcal{L}/\mathcal{L}'$, and the window Ω'' is the *quotient model* of $\mathcal{M}$ by $\mathcal{M}'$. Physically, it is essentially the set of translates of V' in V that have nonempty intersection with $\mathcal{M}$, but it may also include some translates of V' that do not meet $\mathcal{M}$ (corresponding to points $\boldsymbol{l} \in \mathcal{L}$ with $(\pi_W(\boldsymbol{l})+W')\cap\Omega \neq \emptyset$ but for which there is no $\boldsymbol{l}' \in \mathcal{L}'+\boldsymbol{l}$ with $\pi_W(\boldsymbol{l}') \in \Omega$). We note that if Ω is nonempty and half-open in the direction $\boldsymbol{h}$ then Ω'' is nonempty and is half-open provided that $\boldsymbol{h} \notin W'$.

Unconditional Quasicrystal Properties. The rest of this section is devoted to investigating what extra conditions need to be placed on the plain cut-and-project construction to obtain the various quasicrystal properties. We begin with the properties common to all model sets.

PROPOSITION 2.6. *Every plain model set has Properties* 1, 4 *and* 9.

PROOF. Property 4 for model sets is established in [**H1**] and also in the articles by Bernuau and Duneau [**BD**] and Schlottmann [**Schl2**] in this volume, under the less restrictive assumption that the boundary of Ω has Lebesgue measure zero.

For Property 1 we note that

$$\boldsymbol{y} \in \mathcal{M} - \mathcal{M} \implies \exists \boldsymbol{l}_1, \boldsymbol{l}_2 \in \mathcal{L} \text{ with } \boldsymbol{y} = \pi_V(\boldsymbol{l}_1) - \pi_V(\boldsymbol{l}_2) \text{ and } \pi_W(\boldsymbol{l}_1), \pi_W(\boldsymbol{l}_2) \in \Omega$$
$$\implies \exists \boldsymbol{l} \in \mathcal{L} \text{ with } \boldsymbol{y} = \pi_V(\boldsymbol{l}) \text{ and } \pi_W(\boldsymbol{l}) \in \Omega - \Omega.$$

Since there are only finitely many lattice points $\boldsymbol{l}$ in the bounded region $B_\rho(\boldsymbol{0}) + (\Omega - \Omega)$ of $\mathbb{E}^N$ it follows that $(\mathcal{M} - \mathcal{M}) \cap B_\rho(\boldsymbol{0})$ is finite. Hence there are only finitely many possibilities for vectors $\boldsymbol{y} - \boldsymbol{x}$ with $\boldsymbol{x} \in \mathcal{M}$ and $\boldsymbol{y} \in P_\rho(\boldsymbol{x})$, so only finitely many patches $P_\rho(\boldsymbol{x})$.

For Property 9, we note that every $\boldsymbol{y} \in \mathcal{M} - \mathcal{M}$ is an integer linear combination of $\pi_V(\boldsymbol{l}_1), \ldots, \pi_V(\boldsymbol{l}_N)$, where $\{\boldsymbol{l}_1, \ldots, \boldsymbol{l}_N\}$ is a basis of $\mathcal{L}$. $\square$

REMARK 2.7. This proposition does not require Ω to be half-open.

Orientation Conditions. Many of the remaining properties require conditions on the orientation of V or W relative to $\mathcal{L}$. Four kinds of condition on the orientation of $\mathcal{L}$ relative to a subspace U, or pair of complementary subspaces U_1, U_2, of $\mathbb{E}^N$ are:

(1) For every $\mathcal{L}$-subspace L, $\operatorname{codim}(U \cap L) = \min(\operatorname{codim} U + \operatorname{codim} L, N)$.
(2) $U \cap \mathcal{L} = \{\boldsymbol{0}\}$; that is, U contains no $\mathcal{L}$-line.
(3) U is contained in no $\mathcal{L}$-hyperplane.
(4) For every $\mathcal{L}$-subspace L there is a subspace U_2' of U_2 such that

(2.3) $\qquad (L + U_1) \cap U_2 \subseteq U_2' \subseteq U_2$ and $L + U_2'$ is an $\mathcal{L}$-subspace.

Condition (1) says that U has "general orientation" with respect to $\mathcal{L}$. It is satisfied by almost all proper subspaces U of any given dimension.

Conditions (2) and (3) are particular cases of Condition (1) that are dual to each other. It is easy to see by simple examples that they are independent of each other when U has dimension and codimension both > 1.

We shall make Condition (4) more recognizable shortly. Unlike the other conditions, it is restrictive. When $L = \{\mathbf{0}\}$ we can always take $U_2' = \{\mathbf{0}\}$ (remembering that U_1 and U_2 are complementary). When U_2 satisfies Condition (1) and $\dim L \geq \dim U_1$ we can take $U_2' = U_2$, giving $L + U_2' = \mathbb{E}^N$. However, when $1 \leq \dim L < \dim U_1$ (2.3) is incompatible with U_1 and U_2 both satisfying Condition (1). Nevertheless, the restrictive Condition (4) is compatible with Conditions (2) and (3) being satisfied by both U_1 and U_2, and indeed we shall see later that for the most commonly studied quasicrystals V and W both satisfy Conditions (2) and (3) and (V, W) and (W, V) both satisfy Condition (4).

We now state some special cases of the above conditions that arise in the cut-and-project context.

THE W-CONDITION. $W \cap \mathcal{L} = \{\mathbf{0}\}$; that is, Condition (2) holds for $\mathcal{L}$ and W.

THE V-CONDITION. V is contained in no $\mathcal{L}$-hyperplane; that is, Condition 3 holds for $\mathcal{L}$ and V.

THE SUBMODELS CONDITION. Condition 4 holds for $\mathcal{L}$ with $U_1 = V$ and $U_2 = W$.

Equivalent formulations of the submodels condition are the following (the last of which explains its name).

FORMULATION 2. For every $\mathcal{L}$-subspace L there is a subspace W' of W with $W' \supseteq \pi_W(L)$ and $L + W'$ an $\mathcal{L}$-subspace.

FORMULATION 3. For every $\mathcal{L}$-subspace L there is an $\mathcal{L}$-subspace L' such that $L' \cap V = \pi_V(L') = \pi_V(L)$.

FORMULATION 4. For every $\mathcal{L}$-subspace L there is a submodel $\mathcal{L}', V', W'$ with $V' = \pi_V(L)$.

The equivalence of the Submodels Condition to Formulation 2 is clear, since $\pi_W(L) = (L + V) \cap W$. Formulation 3 follows from Formulation 2 by taking $L' = L + W'$. In the converse direction, we can assume (replacing L' by $L + L'$, if necessary) that $L' \supseteq L$ in Formulation 3 and then it can be shown that L' can be expressed in the form $L' = L + W'$ and that this W' satisfies Formulation 2. Finally, Formulations 3 and 4 are clearly equivalent.

We shall also need two further conditions.

THE STRONG V-CONDITION. Every full submodel satisfies the V-Condition.

THE STRONG W-CONDITION. Every quotient model by a full submodel satisfies the W-Condition.

Note that, in contrast to the other conditions, which are "generic", the Submodels Condition is restrictive. Also note that the orientation conditions depend only on $\mathbb{Q}\mathcal{L}$ (a vector space over $\mathbb{Q}$), not on $\mathcal{L}$ itself, and do not depend on Ω. Lattices $\mathcal{L}_1, \mathcal{L}_2$ with $\mathbb{Q}\mathcal{L}_1 = \mathbb{Q}\mathcal{L}_2$ are called *commensurate*, and are characterized by $\mathcal{L}_1 \cap \mathcal{L}_2$ having finite index in $\mathcal{L}_1 + \mathcal{L}_2$.

Consequences of Orientation Conditions. First, a preliminary lemma.

LEMMA 2.8. *For every $\varepsilon > 0$ and $\boldsymbol{x} \in \mathbb{E}^N$ there is a $q \in \mathbb{Z}^+$ and an $\boldsymbol{l} \in \mathcal{L}$ such that $|q\boldsymbol{x} - \boldsymbol{l}| < \varepsilon$.*

PROOF. This is an easy application of Dirichlet's Schubfach-Schluss (Box Principle). Cover a fundamental region $\mathcal{F}$ of $\mathcal{L}$ with balls of diameter ε, say Q of them, and reduce each of the $Q+1$ vectors $\mathbf{0}, \boldsymbol{x}, 2\boldsymbol{x}, \ldots, Q\boldsymbol{x}$ modulo $\mathcal{L}$ so that the remainders lie in $\mathcal{F}$. Then two of these remainders must lie in the same ball, say

$$|(q_i\boldsymbol{x} - \boldsymbol{l}_i) - (q_j\boldsymbol{x} - \boldsymbol{l}_j)| < \varepsilon$$

with $0 \le q_j < q_i \le Q$ and $\boldsymbol{l}_i, \boldsymbol{l}_j \in \mathcal{L}$. This gives the result with $q = q_i - q_j$ and $\boldsymbol{l} = \boldsymbol{l}_i - \boldsymbol{l}_j$. $\square$

Our next lemma is a result in Diophantine approximation equivalent to [**C**, Theorem 1, Chapter IV]. Because the equivalence is slightly troublesome to establish and because a proof using completely different methods from [**C**] seems of interest, we include a proof (generalizing and completing the proof of Lemma 3 of [**LuP**]) that is in tune with the geometrical nature of this article.

LEMMA 2.9. *If the V-Condition is satisfied (that is, V lies in no $\mathcal{L}$-hyperplane), $\dim V = n$ and $B_R(\boldsymbol{x})$ is a ball in V with radius R then the number of points of $\mathcal{L}$ in $B_R(\boldsymbol{x}) + \Omega$ is*

(2.4) $$\operatorname{dens}(\mathcal{L})\operatorname{vol}(B_R(\boldsymbol{x}) + \Omega) + o(R^n)$$

uniformly in $\boldsymbol{x}$.

REMARK 2.10. Although we shall need this lemma only for balls, it clearly remains true for an arbitrary Riemann measurable region of V. We also note that the proof of this lemma only requires Ω to be Riemann measurable—it need not be half-open.

PROOF. It is enough to prove this when Ω is an $(N-n)$-dimensional cube and with $B_R(\boldsymbol{x})$ replaced by C_R, an n-dimensional cube of side R, and to ensure that our error estimates are independent of the position of C_R. (Then the result for $B_R(\boldsymbol{x})$ can be obtained by covering it with cubes of side $\sqrt{R}$, say, and the result for a general Riemann measurable Ω by approximating it as a union of small cubes.)

We use the idea (based on estimating the number of lattice cells that meet its boundary) that the number of lattice points in a polytope $\mathcal{P}$ (in this case a cubic prism) is

(2.5) $$\operatorname{dens}(\mathcal{L})\operatorname{vol}(\mathcal{P}) + O\left(\sum_F \operatorname{vol}(F)\right),$$

where the sum is over all proper[4] faces F of $\mathcal{P}$ and $\operatorname{vol}(F)$ denotes d-dimensional volume, where d is the dimension of F. Applying this directly to $C_R + \Omega$ does not give a small enough error term, since the volumes of its faces parallel to V are of the order R^n, so we construct a linear operator T that maps $\mathcal{L}$ onto itself but reduces volumes in V, then apply (2.5) to $T(C_R + \Omega)$.

Let $\{\boldsymbol{b}_1, \ldots, \boldsymbol{b}_N\}$ be a basis of $\mathcal{L}$ and let $\boldsymbol{v}$ be a vector in V that lies in no $\mathcal{L}$-subspace. (Such vectors $\boldsymbol{v}$ exist because there are only countably many $\mathcal{L}$-subspaces to be avoided and, in view of the V-Condition, each of them meets V in a proper subspace.) Given an $\varepsilon > 0$, we construct a basis $\{\boldsymbol{l}_1, \ldots, \boldsymbol{l}_N\}$ of $\mathcal{L}$ with

$$|\boldsymbol{l}_i - q_i\boldsymbol{v}| < \varepsilon \quad \text{for } i = 1, \ldots, N,$$

---

[4]There is an oversight in the proof of Lemma 3 of [**LuP**] in that faces that are not facets are ignored. This is what I meant by "completing" the proof as well as generalizing it.

where each $q_i \in \mathbb{R}$. We do this inductively, and for the purpose of the induction satisfy the stronger inequalities

(2.6) $$|l_i - q_i v| < \varepsilon/2^{N-i}.$$

Suppose this has been done for $i = 1, \ldots, k$ and we have linearly independent vectors $\{l_1, \ldots l_k\}$ satisfying (2.6) which are a basis for the lattice $\mathcal{L} \cap \langle l_1, \ldots l_k \rangle$. Apply Lemma 2.8 to find an $l' \in \mathcal{L}$ with $|l' - q'v| < \varepsilon'$, where $\varepsilon' < \varepsilon/2^N$ is chosen less than the distance of v from $\langle l_1, \ldots, l_k \rangle$ (which is positive, because v lies in no $\mathcal{L}$-subspace). If we extend $\{l_1, \ldots, l_k\}$ to a basis $\{l_1, \ldots, l_k, l''\}$ of $\mathcal{L} \cap \langle l_1, \ldots, l_k, l' \rangle$ then

$$l' = a_1 l_1 + \cdots + a_k l_k + a l''$$

with $a_1, \ldots, a_k, a \in \mathbb{Z}$. We now take

$$l_{k+1} = l'' + \lfloor a_1/a \rfloor l_1 + \cdots + \lfloor a_k/a \rfloor l_k$$

and

$$q_{k+1} = q'/a - \{a_1/a\} q_1 - \cdots - \{a_k/a\} q_k,$$

where $\lfloor x \rfloor$ and $\{x\}$ are the integer and fractional parts of x. Then $\{l_1, \ldots, l_k, l_{k+1}\}$ is a basis of $\mathcal{L} \cap \langle l_1, \ldots, l_k, l_{k+1} \rangle$ and

$$|l_{k+1} - q_{k+1} v| < \frac{\varepsilon}{2^N} + \frac{\varepsilon}{2^{N-1}} + \cdots + \frac{\varepsilon}{2^{N-k}} < \frac{\varepsilon}{2^{N-k-1}}.$$

We now define the linear operator T by

$$T l_i = b_i \quad (i = 1, \ldots, N).$$

Then T maps $\mathcal{L}$ onto itself and hence $|\det T| = 1$. The key property of T for our purposes is that there is a constant c, independent of ε, such that for every $x \in \mathbb{E}^N$

(2.7) $$|\pi_{V^\perp}(T^{-1} x)| < c\varepsilon |x|,$$

where we temporarily use $\pi_{V^\perp}$ for the projection onto the orthogonal complement of V. This is because x can be written as $x = \sum x_i b_i$ with each $|x_i| < c|x|/N$, for some c, and the component of $T^{-1} b_i = l_i$ orthogonal to V has length $< \varepsilon$. We can now estimate the effect of T on volumes of proper faces F of $C_R + \Omega$. These have the form $C + \Phi$, where C is a face of C_R and Φ a face of Ω. First consider a parallel-to-V, $(N-d)$-dimensional face F of the form $C_R + \Phi$, where Φ is a proper face of Ω. Take a unit cube Γ in $\mathbb{E}^N$ with an $(N-d)$-dimensional face $\Gamma_1 \subseteq T(F)$ and complementary d-dimensional face Γ_2 (that is, $\Gamma = \Gamma_1 \oplus \Gamma_2$). Since T is unimodular, $T^{-1}(\Gamma)$ has volume 1. But since $T^{-1}(\Gamma_1) \subseteq F$ we can also compute this volume as the product of $\text{vol}(T^{-1}(\Gamma_1))$ with the volume of the projection of $T^{-1}(\Gamma_2)$ orthogonal to F. Since Γ_2 has diameter $\sqrt{d}$ this latter volume is $< (c\sqrt{d}\varepsilon)^d$, by (2.7), so $\text{vol}(T^{-1}(\Gamma_1)) > (c\sqrt{d}\varepsilon)^{-d}$ and hence T reduces the volume of F by a factor smaller than $(c\sqrt{d}\varepsilon)^d$. For a not-parallel-to-V, $(N-d)$-dimensional face F of $C_R + \Omega$ of the form $C + \Phi$, where C is a proper face of C_R and Φ any face of Ω, we have

$$\text{vol}(T(C + \Phi)) \leq \|T\|^{N-d} \text{vol}(C + \Phi) \leq c_1 \|T\|^{N-d} R^{n-1},$$

where c_1 is the maximum volume of Ω or any of its faces. For the sum of the volumes of all proper faces of $T(C_R + \Omega)$ we have (when $\varepsilon \leq 1$) the upper bound

(2.8) $$c_2 \varepsilon R^n + c_3 \|T\|^{N-1} R^{n-1},$$

where the constants depend only on $\mathcal{L}$ and Ω.

To complete the proof we note that, since T is unimodular and maps $\mathcal{L}$ to itself, the number of lattice points in $C_R + \Omega$ is given by (2.5) with $T(C_R + \Omega)$ in place of $\mathcal{P}$. For any $\varepsilon' > 0$ we choose $\varepsilon = \varepsilon'/2c_2c_4$, where c_4 is the constant implied by the O-notation in (2.5), and construct the corresponding T. Then, for R large enough as a function of T, the error term (2.8) is $< \varepsilon' R^n$, as required. $\square$

REMARK 2.11. It is known (see the first note on p. 74 of [**C**]) that the error term in this lemma cannot be universally improved to any explicit function of R. The convergence rate depends on the Diophantine character of V with respect to $\mathcal{L}$.

COROLLARY 2.12. *The V-Condition is necessary and sufficient for $\pi_W(\mathcal{L})$ to be dense in W.*

PROOF. If the V-Condition holds then, by the lemma, for any small ball Ω in W there is a point l of $\mathcal{L}$ in $B_R(\mathbf{0}) + \Omega$, when R is large enough. Since $\pi_W(l) \in \Omega$, $\pi_W(\mathcal{L})$ is dense in W. If the V-Condition does not hold then $V \subseteq L$, for some $\mathcal{L}$-hyperplane L, and $\mathcal{L} = \mathcal{L}' + \mathbb{Z}l$ with $\mathcal{L}' \subset L$. Since V and W are complementary, $\pi_W(L)$ has codimension 1 in W and $\pi_W(\mathcal{L}) \subset \pi_W(L) + \mathbb{Z}\pi_W(l)$, an evenly spaced grid of codimension 1 affine subspaces of W. So $\pi_W(\mathcal{L})$ is not dense in W. $\square$

REMARK 2.13. All accounts of the cut-and-project method require $\pi_W(\mathcal{L})$ to be dense in W but few point out that this is equivalent to the algebraic V-Condition. In [**H3**] Hof shows that when V and W are orthogonal the denseness of $\pi_W(\mathcal{L})$ in W is equivalent to W containing no nonzero point of the dual lattice of $\mathcal{L}$—clearly the same as the V-Condition in this case.

LEMMA 2.14. *The W-Condition is necessary and sufficient for the restriction of π_V to $\mathcal{L}$ to be one-one.*

PROOF. The W-Condition holds

$$\iff l_1 - l_2 \in W \text{ implies } l_1 = l_2 \text{ when } l_1, l_2 \in \mathcal{L}$$
$$\iff \pi_V(l_1) = \pi_V(l_2) \text{ implies } l_1 = l_2 \text{ when } l_1, l_2 \in \mathcal{L}$$
$$\iff \pi_V \text{ is one-one on } \mathcal{L}. \qquad \square$$

LEMMA 2.15. *If the V-Condition and W-Condition hold then $\mathcal{M}$ has uniform frequency equal to $\mathrm{dens}(\mathcal{L})$ times the Riemann measure of the projection of Ω onto the orthogonal complement of V.*

PROOF. This follows from Lemma 2.9, Remark 2.10 and Lemma 2.14. $\square$

REMARK 2.16. This lemma does not require Ω to be half-open since this was not a requirement for Lemma 2.9.

Quasicrystal Properties Needing Orientation Conditions. Since patches feature in many of the properties, we begin with a lemma that relates the ρ-patch of a point $x \in \mathcal{M}$ to the position of the corresponding point, or points, of $\mathcal{L}$.

LEMMA 2.17. *For any $\rho > 0$ let $P_1, P_2, \ldots, P_p$ be the distinct ρ-patches of the plain model set $\mathcal{M}$. Then Σ can be partitioned into strips $\Sigma_1, \Sigma_2, \ldots, \Sigma_p$, each a finite Boolean combination of $\mathcal{L}$-translates of Σ, so that, for each $l \in \mathcal{L} \cap \Sigma$, $\pi_V(l)$ has ρ-patch P_i if and only if $l \in \Sigma_i$.*

PROOF. The only points of $\mathcal{L}$ that can contribute to the ρ-patch of $\pi_V(l)$ are those of the form $l + m$ with $m \in B_\rho(0) + (\Omega - \Omega)$, and the ρ-patch of $\pi_V(l)$ is determined by which of these belong to Σ. This is a finite set of m's, call it S, and to obtain a given patch P_i there is a certain subset $E(i)$ of S (not containing $\mathbf{0}$) which must be excluded. There are also finitely many subsets $I(i,1), \ldots, I(i,k)$ of S (all containing $\mathbf{0}$) such that the patch P_i is obtained if and only if, as well as the m's in $E(i)$ being excluded, the m's in at least one of the $I(i,j)$'s are included. (Although the exclusion set $E(i)$ is uniquely determined by P_i there may be multiple possibilities for the inclusion set when the W-Condition is not satisfied, so that different lattice points m can give the same point $\pi_V(l+m)$.) Thus Σ_i is given by

$$\Sigma_i = \bigcup_{j=1}^{k} \bigcap_{m \in I(i,j)} (\Sigma - m) \setminus \bigcup_{m \in E(i)} (\Sigma - m),$$

a subset of Σ and a Boolean combination of $\mathcal{L}$-translates of Σ. □

We note that the partition of the strip Σ partitions the window Ω into sets $\Omega_i = \Sigma_i \cap W$ which we call the *windowpanes* for the ρ-patches of $\mathcal{M}$. In view of Proposition 2.4, these windowpanes are half-open for the same direction h as Σ.

The next lemma is a slight generalization of a result of Baake, Schlottmann and Jarvis [**BSJ**] (see also [**BSchl**]).

LEMMA 2.18. *Let $\mathcal{M}_1$ and $\mathcal{M}_2$ be plain model sets sharing the same physical and internal spaces V and W, with commensurate lattices $\mathcal{L}_1$ and $\mathcal{L}_2$ that satisfy the V-Condition and the W-Condition, and with strips Σ_1 and Σ_2. If $\mathcal{M}_2$ is locally derivable from $\mathcal{M}_1$ then*

(i) $\mathcal{L}_1 \subseteq \mathcal{L}_2$, *and*
(ii) Σ_2 *is geometrically equal to a finite Boolean combination of $\mathcal{L}_2$-translates of Σ_1.*

If $\mathcal{L}_1 = \mathcal{L}_2$ then, nearly conversely, $\mathcal{M}_2$ is locally derivable from $\mathcal{M}_1$ provided Σ_2 is exactly equal to a finite Boolean combination of $\mathcal{L}_2$-translates of Σ_1.

REMARK 2.19. The lack of an exact converse when $\mathcal{L}_1 = \mathcal{L}_2$ is an inevitable consequence of the fact that distinct singular model sets formed from geometrically equal windows with a different direction h of half-openness are not locally derivable from each other. Another way of stating this lemma would be to explicitly allow singular model sets as exceptional rather than distinguish different kinds of equality of window.

PROOF. By Lemma 2.15, $\mathcal{M}_1$ and $\mathcal{M}_2$ are relatively dense.

We first assume that $\mathcal{M}_2$ is locally derivable from $\mathcal{M}_1$ and show that (i) holds. Let $x = \pi_V(l)$ be a point of $\mathcal{M}_1$, let P be its ρ-patch and Σ' the corresponding substrip of Σ_1, and let $x + t = \pi_V(l + m)$ be the corresponding point in $\mathcal{M}_2$, where $l \in \mathcal{L}_1$, $l + m \in \mathcal{L}_2$ and ρ and t are as in Definition 1.3. By the W-Condition the vector m is the same for all points x such that $P_\rho(x) = P$. Suppose that $\mathcal{L}_1 \not\subseteq \mathcal{L}_2$ and let l' be a vector in $\mathcal{L}_1 \setminus \mathcal{L}_2$. By the V-Condition and Lemma 2.9 there is an l'' in $\mathcal{L}_1 \cap \mathcal{L}_2$ such that $l + l' + l'' \in \Sigma'$. Then $\pi_V(l + l' + l'')$ is a point of $\mathcal{M}_1$ whose ρ-patch is P but $y = \pi_V(l + l' + l'') + t \notin \mathcal{M}_2$ (because $l + l' + l'' + m$ does not belong to $\mathcal{L}_2$ but, in view of the W-Condition, is the only point of $\mathcal{L}_1 + \mathcal{L}_2$ to project to y). Since this is inconsistent with the derivability of $\mathcal{M}_2$ from the ρ-patches of $\mathcal{M}_1$, $\mathcal{L}_1 \subseteq \mathcal{L}_2$.

Next we show that if $\mathcal{M}_2$ is locally derivable from $\mathcal{M}_1$ then (ii) also holds. By Lemma 2.14 the restriction of π_V to $\mathcal{L}_2$ has an inverse λ. By Definition 1.3, $\boldsymbol{v}_2 \in \mathcal{M}_2$ if and only if $\boldsymbol{v}_2 = \boldsymbol{v}_1 + \boldsymbol{t} + \boldsymbol{u}$ for some $\boldsymbol{v}_1 \in \mathcal{M}_1$ with $f(P_\rho(\boldsymbol{v}_1)) = (P', \boldsymbol{t})$ and $\boldsymbol{u} \in P'$. Let Σ_2' be a union of translates of strips corresponding to the patches of $\mathcal{M}_1$ (as in Lemma 2.17), the strip corresponding to the patch P being translated by the set of vectors $\{\lambda(\boldsymbol{t} + \boldsymbol{u}) \mid \boldsymbol{u} \in P'\}$, where $f(P) = (P', \boldsymbol{t})$. (Since each $\boldsymbol{t} + \boldsymbol{u}$ has the form $\boldsymbol{v}_1 - \boldsymbol{v}_2$ with $\boldsymbol{v}_1 \in \mathcal{M}_1$ and $\boldsymbol{v}_2 \in \mathcal{M}_2$, it lies in $\pi_V(\mathcal{L}_2)$.) Then for a vector $\boldsymbol{l} \in \mathcal{L}_2$

$$(2.9) \qquad \boldsymbol{l} \in \Sigma_2' \iff \boldsymbol{l} \in \Sigma_2.$$

Also, by Lemma 2.17, each strip for a ρ-patch of $\mathcal{M}_1$ is a Boolean combination of $\mathcal{L}_1$-translates of Σ_1. Hence, since $\mathcal{L}_1 \subseteq \mathcal{L}_2$, Σ_2' is a Boolean combination of $\mathcal{L}_2$-translates of Σ_1. By Corollary 2.12, $\pi_W(\mathcal{L}_2)$ is dense in W, so it follows from (2.9) that the interior of Σ_2' is contained in the closure of Σ_2 and vice versa. Hence Σ_2 and Σ_2' are geometrically equal.

For the near converse, we start by observing that the union of two sets, $\mathcal{D}_1$ and $\mathcal{D}_2$, locally derivable from $\mathcal{M}_1$ is also locally derivable from $\mathcal{M}_1$. This can be verified by choosing for the new ρ the largest of the two values for $\mathcal{D}_1$ and $\mathcal{D}_2$ plus the largest of the two covering radii of these sets and defining the new f so that $\boldsymbol{t}$ is the shortest vector from $\boldsymbol{x}$ to any point of $\mathcal{D}_1$ or $\mathcal{D}_2$. Consequently we can suppose that the disjunctive normal form of the Boolean expression for Σ_2 in terms of translates of Σ_1 has only one term, so that

$$\Sigma_2 = \bigcap_{i=1}^{j}(\Sigma_1 + \boldsymbol{l}_i) \setminus \bigcup_{i=j+1}^{k}(\Sigma_1 + \boldsymbol{l}_i),$$

with $\boldsymbol{l}_i \in \mathcal{L}_1$ for $i = 1, \ldots, k$, an intersection of $\mathcal{L}_1$-translates of Σ_1 and complements of $\mathcal{L}_1$-translates of Σ_1. Moreover $j \geq 1$, since the cross-section of Σ_1 is bounded. Hence

$$\boldsymbol{y} \in \mathcal{M}_2 \iff \begin{cases} \boldsymbol{y} + \pi_V(\boldsymbol{l}_i) \in \mathcal{M}_1 & \text{for } 1 \leq i \leq j; \\ \boldsymbol{y} + \pi_V(\boldsymbol{l}_i) \notin \mathcal{M}_1 & \text{for } j+1 \leq i \leq k. \end{cases}$$

Thus the points $\boldsymbol{y} \in \mathcal{M}_2$ all have the form $\boldsymbol{y} = \boldsymbol{x} - \pi_V(\boldsymbol{l}_1)$, for some $\boldsymbol{x} \in \mathcal{M}_1$, and whether $\boldsymbol{x} - \pi_V(\boldsymbol{l}_1)$ is in $\mathcal{M}_2$ is determined by the ρ'-patch of $\boldsymbol{x}$ in $\mathcal{M}_1$, where

$$\rho' > 2 \max_{1 \leq i \leq k} |\pi_V(\boldsymbol{l}_i)|.$$

If we now take $\rho = 2\rho' + R_1 + 2R$, for some $R > R_2$, then for each $\boldsymbol{x} \in \mathcal{M}_1$ we can identify a shortest $\boldsymbol{t}$ with $\boldsymbol{x} + \boldsymbol{t} \in \mathcal{M}_2$ (by the ρ'-patch of $\boldsymbol{x} + \boldsymbol{t} + \pi_V(\boldsymbol{l}_1)$ in $\mathcal{M}_1$) and determine the $(R_1 + R)$-patch of $\boldsymbol{x} + \boldsymbol{t}$ in $\mathcal{M}_2$ (by the ρ'-patches of the points of $\mathcal{M}_1$ within a distance $R_1 + R$ of $\boldsymbol{x} + \boldsymbol{t}$). Hence $\mathcal{M}_2$ is locally derivable from $\mathcal{M}_1$. $\square$

REMARK 2.20. It is possible for $\mathcal{M}_2$ to be locally derivable from $\mathcal{M}_1$ when $\mathcal{L}_2 \supset \mathcal{L}_1$, but (ii) of Lemma 2.18 (even with exact equality) is not sufficient for local derivability in general.

As examples, take V and W as the x- and y-axes in $\mathbb{E}^2$ and $\mathcal{L}_1$ as a lattice in $\mathbb{E}^2$ with no point except $\boldsymbol{0}$ on either axis (so that the V- and W-Conditions are satisfied). Choose a primitive point $\boldsymbol{l} = (l, m) \in \mathcal{L}_1$ and take for $\mathcal{L}_2$ the lattice $\mathcal{L}_1 \cup (\mathcal{L}_1 + \frac{1}{2}\boldsymbol{l})$ that contains $\mathcal{L}_1$ with index 2.

If we take for $\mathcal{M}_1$ and $\mathcal{M}_2$ the model sets derived from $\mathcal{L}_1$ and $\mathcal{L}_2$ with the half-open interval $[0, 3m/2)$ on the y-axis as window (the same window for both sets) then $\mathcal{M}_2$ is locally derivable from $\mathcal{M}_1$. If fact $\mathcal{M}_2 \setminus \mathcal{M}_1$ consists of all points x such that at least one of $x \pm l/2$ belongs to $\mathcal{M}_1$ but neither of $x \pm 3l/2$ belongs to $\mathcal{M}_1$, and any $\rho > 2l$ will do in Definition 1.3.

If we take for $\mathcal{M}_1$ and $\mathcal{M}_2$ the model sets derived from $\mathcal{L}_1$ and $\mathcal{L}_2$ with identical windows $[0, m)$ instead of $[0, 3m/2)$ then $\mathcal{M}_2$ is not locally derivable from $\mathcal{M}_1$. This is because, since now no two points on the boundary of the strip Σ differ by a vector in $\mathcal{L}_1 + \frac{1}{2}l$, for every $\rho > 0$ it is possible to find points $l_1, l_2 \in \mathcal{L}_1$ with $l_1 + \frac{1}{2}l$ and $l_2 + \frac{1}{2}l$ on opposite sides of the boundary of Σ but with $(\mathcal{L}_1 - l_1) \cap \Sigma$ and $(\mathcal{L}_1 - l_2) \cap \Sigma$ agreeing up to at least a distance ρ from the origin.

PROPOSITION 2.21.
  (i) *If the V-Condition holds then $\mathcal{M}$ has Properties 2, 6 and 10.*
  (ii) *If the V-Condition and W-Condition both hold then $\mathcal{M}$ has Property 3.*
  (iii) *If the V-Condition and W-Condition both hold and $W \neq \{\mathbf{0}\}$ then $\mathcal{M}$ has Property 15.*

PROOF. We assume throughout that the V-Condition is satisfied. Let P_i be a given ρ-patch of $\mathcal{M}$, Σ_i the corresponding strip given by Lemma 2.17, and $\Omega_i = \Sigma_i \cap W \subseteq \Omega$ the windowpane of P_i. Being nonempty and half-open, Ω_i has nonempty interior, so by Lemma 2.9, if R is large enough then for every $\mathbf{v} \in V$ there is at least one point l of $\mathcal{L}$ in $B_R(\mathbf{v}) + \Omega_i$. Then $\pi_V(l)$ lies in $\mathcal{M} \cap B_R(\mathbf{v})$ and has ρ-patch P_i. Taking the largest R that is needed for any ρ-patch shows that $\mathcal{M}$ has Property 2.

If $l(i, R, \mathbf{v})$ is the number of points of $\mathcal{L}$ in $B_R(\mathbf{v}) + \Omega_i$ then Lemma 2.9 also shows that
$$l(i, R, \mathbf{v})/R^n \to \mathrm{dens}(\mathcal{L}) \, \mathrm{vol}(B_1(\mathbf{v}) + \Omega_i) > 0 \quad \text{as } R \to \infty$$
uniformly in $\mathbf{v}$. If the W-Condition holds then by Lemma 2.14 there is a one-one correspondence between the points of $\mathcal{L}$ in $B_R(\mathbf{v}) + \Omega_i$ and the points of $\mathcal{M}$ in $B_R(\mathbf{v})$ with ρ-patch P_i. So $\mathcal{M}$ has Property 3.

For Property 10 we note that $\mathcal{M} - \mathcal{M}$ is itself essentially a model set (derived from the window $\Omega - \Omega$, which is bounded but not half-open) so, by Proposition 2.6 and Remark 2.7, $\mathcal{M} - \mathcal{M}$ is of finite local complexity and hence uniformly discrete. The relative denseness of $\mathcal{M}$ follows from Property 2.

For Property 6, first choose ε' so small that $\mathrm{dens}(\mathcal{L}) \, \mathrm{vol}(\Omega \triangle (\Omega + \mathbf{w})) < \varepsilon$ for every $\mathbf{w} \in W$ with $|\mathbf{w}| < \varepsilon'$, then (using Lemma 2.9) choose ρ so large that, for every $\mathbf{v} \in V$, $B_\rho(\mathbf{v}) + B_{\varepsilon'}(\mathbf{0})$ contains a point l of $\mathcal{L}$. Finally, choose D so large that $\mathbf{u} \cdot \mathbf{v}/|\mathbf{v}| > 1 - \delta$ for every $\mathbf{v} \in B_\rho(D\mathbf{u})$. Then there is a point l of $\mathcal{L}$ in $B_\rho(D\mathbf{u}) + B_{\varepsilon'}(\mathbf{0})$ and if we take $\mathbf{p} = \pi_V(l)$ then $\mathbf{u} \cdot \mathbf{p}/|\mathbf{p}| > 1 - \delta$. Also
$$\mathbf{x} \in \mathcal{M} + \mathbf{p} \iff \exists l' \in \mathcal{L} \cap \Sigma \text{ with } \pi_V(l') = \mathbf{x} - \mathbf{p}$$
$$\iff \exists l'' \in \mathcal{L} \cap (\Sigma + l) \text{ with } \pi_V(l'') = \mathbf{x}.$$

(Here $l'' = l' + l$.) Consequently the number of points of $\mathcal{M} \triangle (\mathcal{M} + \mathbf{p})$ in $B_R(\mathbf{v})$, for any $\mathbf{v} \in V$ and $R > 0$, is at most equal to the number of points of $\mathcal{L}$ in $B_R(\mathbf{v}) + \Omega \triangle (\Omega + \pi_W(l))$. By Lemma 2.9 and our choice of l this is $< \varepsilon \, \mathrm{vol}(B_R(\mathbf{v})) + o(R)$. Letting R tend to infinity shows that the upper uniform frequency of $\mathcal{M} \triangle (\mathcal{M} + \mathbf{p})$ is $< \varepsilon$.

For Property 15 we note that translating the window Ω in W gives locally isomorphic model sets, because if Ω_i is the windowpane in Ω for a ρ-patch P_i then $\Omega_i + \boldsymbol{w}$ is the windowpane for the identical patch P_i in the model set obtained from the window $\Omega + \boldsymbol{w}$. Also the model sets obtained from the windows $\Omega + \boldsymbol{w}$ are all different, because $\Omega \triangle (\Omega + \boldsymbol{w})$ has nonempty interior when $\boldsymbol{w} \neq \boldsymbol{0}$, so by Lemma 2.9 there is a point $\boldsymbol{l}$ of $\mathcal{L}$ in $\Sigma \triangle (\Sigma + \boldsymbol{w})$ and then $\pi_V(\boldsymbol{l})$ belongs to one of the model sets obtained from Ω and $\Omega + \boldsymbol{w}$ but not the other. As $\boldsymbol{w}$ varies this gives $2^{\aleph_0}$ model sets in the LI-class of $\mathcal{M}$, all of them subsets of $\pi_V(\mathcal{L})$. Since each isometry class of these model sets is countable (because an isometry is uniquely determined by the images of any $n+1$ points not lying on a hyperplane and there are only countably many $(n+1)$-element subsets of $\pi_V(\mathcal{L})$) this gives $2^{\aleph_0}$ incongruent sets in the LI-class of $\mathcal{M}$. There are only $2^{\aleph_0}$ countable subsets of $\mathbb{E}^n$ in all, hence exactly this number of incongruent sets in then LI-class of $\mathcal{M}$. □

REMARK 2.22. Properties 6, 10 and 15, like Lemma 2.15, merely require the window Ω to have nonempty interior. This is not true of Properties 2 and 3, however, which rely on the half-openness of Ω to ensure that all windowpanes of patches have nonempty interior.

PROPOSITION 2.23. *If the W-Condition holds then $\mathcal{M}$ has Property 5 (that is, $\mathcal{M}$ is nonperiodic) if and only if $\mathcal{L} \cap V = \{\boldsymbol{0}\}$.*

PROOF. Clearly any nonzero vector $\boldsymbol{p}$ of $\mathcal{L} \cap V$ is a period of $\mathcal{M}$. Conversely, if $\boldsymbol{p} \neq \boldsymbol{0}$ is a period of $\mathcal{M}$ and $\boldsymbol{x}$ is any point of $\mathcal{M}$ then $\boldsymbol{x}$ lifts to a point $\boldsymbol{l} \in \mathcal{L}$ and $\boldsymbol{x} + \boldsymbol{p}$ to a point $\boldsymbol{l} + \boldsymbol{l}' \in \mathcal{L}$. In view of the W-Condition the projection π_V is one-one on $\mathcal{L}$, by Lemma 2.14, so the point $\boldsymbol{x} + j\boldsymbol{p}$ of $\mathcal{M}$, for $j \in \mathbb{N}$, lifts uniquely to the point $\boldsymbol{l} + j\boldsymbol{l}' \in \mathcal{L}$. Hence $\pi_W(\boldsymbol{l}) + j\pi_W(\boldsymbol{l}') \in \Omega$, for every $j \in \mathbb{N}$, and since Ω is bounded $\pi_W(\boldsymbol{l}') = \boldsymbol{0}$. Thus $\boldsymbol{l}' = \boldsymbol{p} \in \mathcal{L} \cap V$. □

REMARK 2.24. It is possible to obtain a necessary and sufficient condition for $\mathcal{M}$ to be periodic without assuming the W-Condition, but it is too complicated to be of interest.

PROPOSITION 2.25. *If the Strong V-Condition, the Strong W-Condition and the Submodels Condition hold then $\mathcal{M}$ has Property 7. If, in addition, $\mathcal{M}$ is nonsingular then Property 7 holds with no exceptional vectors $\boldsymbol{t}$. If $\dim W \leq 2$, $\mathcal{L} \cap V = \{\boldsymbol{0}\}$ and Ω is convex then, for each exceptional vector $\boldsymbol{t}$, $\mathcal{M} \cap (A + \boldsymbol{t})$ consists of only a single point.*

REMARK 2.26. The last sentence is included because it shows that many of the well-known 2-dimensional quasicrystals with a quadratic inflation multiplier (for example, the Penrose set and the Ammann-Beenker set) have the property that any line joining two points of the set contains infinitely many points of the set, even when the set is a singular one of its type.

PROOF. Let $\mathcal{X}$ be any subset of $\mathcal{M}$ and $\boldsymbol{x}_0, \ldots, \boldsymbol{x}_d$ points in $\mathcal{X}$ that span $A = \mathrm{aff}(\mathcal{X})$. Let $\boldsymbol{l}_0, \ldots, \boldsymbol{l}_d$ be points of $\mathcal{L}$ that project to $\boldsymbol{x}_0, \ldots, \boldsymbol{x}_d$ and put $L = \langle \boldsymbol{l}_1 - \boldsymbol{l}_0, \ldots, \boldsymbol{l}_d - \boldsymbol{l}_0 \rangle$, an $\mathcal{L}$-subspace. By Formulation 4 of the Submodels Condition, there is a submodel $\mathcal{L}'$, V', W' of $\mathcal{M}$ with $V' = \mathrm{aff}(\mathcal{X}) - \boldsymbol{x}_0$, which we can take to be the full submodel with this V'. This full submodel satisfies the W-Condition because $\mathcal{M}$ does and the V-Condition because $\mathcal{M}$ satisfies the Strong V-Condition. Since the submodel is full $\pi_V(\boldsymbol{l}) \in V'$, for $\boldsymbol{l} \in \mathcal{L}$, if and only if $\boldsymbol{l} \in \mathcal{L}'$ and hence $\pi_V(\boldsymbol{l})$ belongs to the translate $V' + \boldsymbol{t}$ of V' (where $\boldsymbol{t}$ can be chosen in the orthogonal complement of V' in V) if and only if there is some $l(t) \in \mathcal{L}$ with $\pi_V\bigl(l(t)\bigr) \in V' + t$ and l belongs to the coset $\mathcal{L}' + l(t)$ of $\mathcal{L}'$ in $\mathcal{L}$. It follows that $\mathcal{M} \cap (V' + t)$ is a translate of the model set in V' derived from $\mathcal{L}'$, V', W' and the strip $\bigl(\Sigma - l(t)\bigr) \cap (V' \oplus W')$. Hence, by Lemma 2.15, Remark 2.16 and Proposition 2.21(i), $\mathcal{M} \cap (V' + t)$ is a Meyer set in $V' + t$ with positive uniform frequency whenever the window has nonempty interior.

The t's with $\mathcal{M} \cap (V' + t)$ nonempty all belong to the quotient model of $\mathcal{M}$ by $\mathcal{M} \cap V'$ and, except possibly when $\pi_W\bigl(l(t)\bigr)$ is on the boundary of the projection of the interior of Ω, the strip $\bigl(\Sigma - l(t)\bigr) \cap (V' \oplus W')$ for $\mathcal{M} \cap (V'+t)$ has nonempty interior. The quotient model (which can be regarded as a set of translation vectors t in V orthogonal to V') inherits the V-Condition from $\mathcal{M}$ and satisfies the W-Condition because $\mathcal{M}$ satisfies the Strong W-Condition. Hence, by Lemma 2.15, Remark 2.16 and Proposition 2.21(i), it is a Meyer set with positive uniform frequency. The t's for which $\mathcal{M} \cap (V' + t)$ is not a Meyer set of positive uniform frequency are among those that project onto the boundary of the projection of the interior of the quotient model window, so by Lemma 2.15 again have uniform frequency zero.

For the last two sentences of the proposition, we note that when $\mathcal{M}$ is nonsingular l is in the interior of Σ for every $l \in \mathcal{L}$, so whenever $\bigl(\Sigma - l(t)\bigr) \cap (V' \oplus W')$ is nonempty it has nonempty interior (as a subset of $V' \oplus W'$). Also, for any two distinct points $l', l'' \in \mathcal{L}' + l(t)$ with $\pi_V(l'), \pi_V(l'') \in \mathcal{M}$, $\pi_W(l')$ and $\pi_W(l'')$ are distinct points of Ω when $\mathcal{L} \cap V = \mathbf{0}$. So when $\dim W' = 1$ and Ω is convex the window in W' contains an interval and thus has nonempty interior. When $\dim W = 2$ and $\dim W' = 2$ the window in W' is Ω itself, which again has nonempty interior. $\square$

PROPOSITION 2.27. *Suppose that the model set $\mathcal{M}$ derived from $\mathcal{L}$, V, W, Ω satisfies the Strong V-Condition and the Strong W-Condition and has a full submodel given by $\mathcal{L}', V', W'$. Suppose also that there is a $\psi \geq 0$ such that, for every translate $W' + t$ of W' in W, $(W' + t) \cap \Omega$ is either empty or contains a $(\dim W')$-dimensional ball of radius ψ. Then $\mathcal{M}$ has a system of Ammann bars that are translates of V'.*

PROOF. This follows from the proof of Proposition 2.25. The lower bound for the uniform frequency arises from Lemma 2.15 and the upper bound for the covering radius from Lemma 2.9, since the submodel on each Ammann bar $V' + t$ is obtained from the same lattice $\mathcal{L}'$ and from a window that contains a ball of radius ψ. $\square$

REMARK 2.28. A simple way to construct a window satisfying the hypothesis of Proposition 2.27 for several subspaces W' simultaneously is to use a zonotope. A *zonotope* is a region of the form

(2.10) $$\{x_1 \boldsymbol{w}_1 + \cdots + x_k \boldsymbol{w}_k : 0 \leq x_1, \ldots, x_k \leq 1\}$$

for some finite set of vectors $\boldsymbol{w}_1, \ldots, \boldsymbol{w}_k$ that span W. (Equivalently, it is a projection of a k-dimensional parallelotope.) Clearly a zonotope satisfies the hypothesis of the proposition for every subspace W' that is spanned by any subset of $\{\boldsymbol{w}_1, \ldots, \boldsymbol{w}_k\}$. Given a finite collection of subspaces we can choose a spanning set for each of them and take $\{\boldsymbol{w}_1, \ldots, \boldsymbol{w}_k\}$ to be the union of these spanning sets augmented by extra vectors, if necessary, to give a spanning set of W. Then the zonotope (2.10) satisfies the hypothesis for every subspace in the collection.

Quasicrystal Properties Needing Operators. These are the symmetry and flation properties.

PROPOSITION 2.29. *If $\mathcal{M}$ satisfies the V-Condition and there is a linear operator T on $\mathbb{E}^N$ such that V and W are T-invariant, $T\mathcal{L} = \mathcal{L}$ and $T\overline{\Omega} = \overline{\Omega}$, where $\overline{\Omega}$ is the closure of Ω, then $\mathcal{M} \triangle T\mathcal{M}$ has uniform frequency 0. If $\mathcal{M}$ is nonsingular then $T\mathcal{M} = \mathcal{M}$.*

PROOF. Since V and W are T-invariant, T commutes with π_V and π_W, so $\boldsymbol{x} \in T\mathcal{M}$ if and only if $\boldsymbol{x} = \pi_V(\boldsymbol{l})$ for some $\boldsymbol{l} \in \mathcal{L}$ with $\pi_W(\boldsymbol{l}) \in T\Omega$. Hence if $\boldsymbol{x} \in \mathcal{M} \triangle T\mathcal{M}$ then $\boldsymbol{x} = \pi_V(\boldsymbol{l})$ for some $\boldsymbol{l} \in \mathcal{L}$ with $\pi_W(\boldsymbol{l}) \in \Omega \triangle T\Omega$. Since $T\overline{\Omega} = \overline{\Omega}$, $\Omega \triangle T\Omega$ is contained in the boundary of Ω, which has Riemann measure zero since Ω is Riemann measurable. So, by Lemma 2.9, $\mathcal{M} \triangle T\mathcal{M}$ has frequency zero. When $\mathcal{M}$ is nonsingular no points of $\mathcal{L}$ have $\pi_W(\boldsymbol{l}) \in \partial\Omega$ so $T\mathcal{M} = \mathcal{M}$. □

PROPOSITION 2.30. *If $\mathcal{M}$ satisfies the V-Condition and there is a linear operator T on $\mathbb{E}^N$ such that T acts on V as multiplication by a scalar $\mu \neq \pm 1$, W is T-invariant, $\mathcal{L} \cap T\mathcal{L}$ is a sublattice of $\mathcal{L}$ of full dimension N and $\Omega \cap T\Omega$ has nonempty interior then $\mathcal{M}$ has Property 12.*

PROOF. Again T commutes with π_V and π_W, so $\boldsymbol{x} \in \mathcal{M} \cap \mu\mathcal{M}$ if and only if $\boldsymbol{x} = \pi_V(\boldsymbol{l})$ for some $\boldsymbol{l} \in \mathcal{L} \cap T\mathcal{L}$ with $\pi_W(\boldsymbol{l}) \in \Omega \cap T\Omega$. Since $\mathcal{L} \cap T\mathcal{L}$ is commensurate with $\mathcal{L}$, and hence inherits the V-Condition, Lemma 2.9 shows that there is an R such that $\mathcal{M} \cap \mu\mathcal{M} \cap B_R(\boldsymbol{v})$ is nonempty for every $\boldsymbol{v} \in V$, so that $\mathcal{M} \cap \mu\mathcal{M}$ is relatively dense. □

PROPOSITION 2.31. *If V, W, $\mathcal{L}$, Ω, $\mathcal{M}$ and T are as in Proposition 2.30 and satisfy both the V-Condition and the W-Condition, and Σ is the strip associated with Ω, then μ gives a refining inflation of $\mathcal{M}$ if $T\mathcal{L} \subseteq \mathcal{L}$ and $T\Sigma \subseteq \Sigma + \boldsymbol{l}$ for some $\boldsymbol{l} \in \mathcal{L}$. If $T\mathcal{L} \nsubseteq \mathcal{L}$ or $T\Sigma \setminus (\Sigma + \boldsymbol{l})$ has nonempty interior for every $\boldsymbol{l} \in \mathcal{L}$ then μ does not give a refining inflation.*

PROOF. Clearly if $T\mathcal{L} \subseteq \mathcal{L}$ and $T\Sigma \subset \Sigma + \boldsymbol{l}$ for some $\boldsymbol{l} \in \mathcal{L}$ then $\mu\mathcal{M} \subset \mathcal{M} + \boldsymbol{t}$, where $\boldsymbol{t} = \pi_V(\boldsymbol{l})$, and μ gives a refining inflation. Conversely, if $T\mathcal{L} \nsubseteq \mathcal{L}$ then, by Corollary 2.12 and the fact that $T\mathcal{L} \setminus (\mathcal{L} + \boldsymbol{l})$ is a union of translates of the lattice $T\mathcal{L} \cap \mathcal{L}$, for every $\boldsymbol{l} \in \mathcal{L} + T\mathcal{L}$ there is an $\boldsymbol{m} \in T\mathcal{L} \setminus (\mathcal{L} + \boldsymbol{l})$ with $\boldsymbol{m} \in T\Sigma$. Then $\pi_V(\boldsymbol{m}) \in \mu\mathcal{M}$ but $\pi_V(\boldsymbol{m}) \notin \mathcal{M} + \pi_V(\boldsymbol{l})$, by Lemma 2.14. If $T\mathcal{L} \subseteq \mathcal{L}$ and $T\Sigma \setminus (\Sigma + \boldsymbol{l})$ has nonempty interior for every $\boldsymbol{l} \in \mathcal{L}$ then, by Corollary 2.12 again, for every $\boldsymbol{l} \in \mathcal{L}$ there is an $\boldsymbol{m} \in T\mathcal{L} \cap T\Sigma$ with $\boldsymbol{m} \notin \Sigma + \boldsymbol{l}$. Hence $\mu\mathcal{M} \nsubseteq \mathcal{M} + \pi_V(\boldsymbol{l})$. (We have used the fact that if $\mu\mathcal{M} \subseteq \mathcal{M} + \boldsymbol{t}$ then $\boldsymbol{t}$ must be of the form $\pi_V(\boldsymbol{l})$ with $\boldsymbol{l} \in \mathcal{L} + T\mathcal{L}$.) □

To obtain necessary and sufficient conditions for the existence of a strip Σ with $T\Sigma \subseteq \Sigma + \boldsymbol{l}$ we first need a lemma.

LEMMA 2.32. *If V, W and $\mathcal{L}$ satisfy the W-Condition and T is an operator on $V \oplus W$ such that V is an eigenspace of T with eigenvalue μ, W is T-invariant and $T\mathcal{L} \cap \mathcal{L}$ is commensurate with $\mathcal{L}$ then the characteristic polynomial of T has the form ϕ^m, where ϕ is the minimum polynomial of μ over $\mathbb{Q}$ and $m \in \mathbb{Z}^+$.*

PROOF. Since $T\mathcal{L}$ is commensurate with $\mathcal{L}$ the characteristic polynomial has rational coefficients and hence has the form $\phi^m \psi$, where ψ has rational coefficients and $\gcd(\phi, \psi) = 1$. Now $\ker \psi(T)$ is an $\mathcal{L}$-subspace and, since $\psi(\mu) \neq 0$, $V \cap$

$\ker \psi(T) = \{\mathbf{0}\}$ so $\ker \psi(T) \subseteq W$. In view of the W-Condition, $\ker \psi(T) = \{\mathbf{0}\}$ and ψ is a constant. □

Being a refining inflation obtained from an operator T with a polytopic window puts a severe restriction on the multiplier μ.

PROPOSITION 2.33. *If V, W, $\mathcal{L}$ and T are as in Proposition 2.30, satisfy the V-Condition and the W-Condition and have $T\mathcal{L} \subseteq \mathcal{L}$ then a necessary and sufficient condition for the existence of a polytope $\Omega \subset W$ such that T gives a refining inflation on the model set $\mathcal{M}$ obtained from $\mathcal{L}$ and Ω is that μ is a Pisot number (that is, an algebraic integer > 1 all of whose conjugates have absolute value < 1) and V is the entire eigenspace of T for the eigenvalue μ.*

PROOF. If the condition is satisfied then, by Lemma 2.32, the characteristic polynomial of T on $\mathbb{E}^N$ is a power of ϕ, the minimum polynomial of μ. In fact, since V is an eigenspace and μ is not an eigenvalue of T on W, the characteristic polynomial is precisely ϕ^n. Hence W, as a vector space over $\mathbb{C}$, is the direct sum of the $d-1$ algebraic conjugates of V, where $d = \deg \phi$, so has a basis of eigenvectors and therefore, as a vector space over $\mathbb{R}$, can be written as a direct sum of 1-dimensional subspaces $\langle \mathbf{e}_j \rangle$ (where the $\mathbf{e}_j$'s are eigenvectors of real eigenvalues of T) and 2-dimensional subspaces E_j (the real subspaces generated by the real and imaginary parts of pairs of complex conjugate eigenvectors, $\mathbf{e}_j$, $\bar{\mathbf{e}}_j$, of T). If for the first type of summand we define ϖ_j to be the interval $\{x\mathbf{e}_j \mid x \in \mathbb{R}, |x| \leq 1\}$ and for the second type we define ϖ_j to be the ellipse $\{\Re(z\mathbf{e}_j) \mid z \in \mathbb{C}, |z| \leq 1\}$ then the direct product of the ϖ_j's is a bounded region Ω' of W with $T\Omega' \subseteq M\Omega'$, where $M < 1$ is the largest of the absolute values of the conjugates of μ (other than μ itself). By replacing any ellipses ϖ_j by polygons that approximate them sufficiently closely, we obtain a polytope Ω with $T\Omega \subseteq \Omega$.

Conversely, if T has an eigenvalue μ_j on W with $|\mu_j| > 1$ let $M > 0$ be maximal such that $M\mathbf{e}_j$ is in the closure of $\Omega - \Omega$, where $\mathbf{e}_j$ is an eigenvector for μ_j. Then $|\mu_j|M\mathbf{e}_j$ is in the closure of $T\Omega - T\Omega$ but not in the closure of $\Omega - \Omega$, so there is no $\mathbf{w}$ with $T\Omega \subseteq \Omega + \mathbf{w}$. Thus if T gives a refining inflation then μ is not an eigenvalue of T on W and all the conjugates μ_j of μ have $|\mu_j| \leq 1$. We still need to show that $|\mu_j| = 1$ is impossible. If $|\mu_j| = 1$ then μ_j is complex (since $\mu_j = \pm 1$ is impossible). Let E_j, as before, be a 2-dimensional real subspace of W generated by the real and imaginary parts of an eigenvector for μ_j and let $\mathbf{w} = \Re(z\mathbf{e}_j)$ be a point of E_j in the closure of $\Omega - \Omega$ with $|z|$ ($= f(\mathbf{w})$, say) maximal. Then the points $T^k\mathbf{w} = \Re(\mu_j^k z\mathbf{e}_j)$, $k \in \mathbb{N}$, have the same maximal value $|\mu_j^k z| = |z|$ of f and also belong to E_j and to the closure of $\Omega - \Omega$. Since μ_j is not a root of unity this is an infinite set of points (its closure is an ellipse, in fact) whereas when Ω is a polytope $E_j \cap (\Omega - \Omega)$ is a polygon and can achieve the maximal value of f at only a finite set of points (a subset of its vertices). □

REMARK 2.34. If we allow nonpolytopic windows then refining inflations with Salem number multipliers become possible by taking the ellipses that occur in the proof of the proposition to be cross-sections of the window. (Salem numbers are defined like Pisot numbers except that conjugates of absolute value 1 are allowed.) This will fail for some singular sets in each LI-class, however, unless an exceptional set of frequency zero is allowed to the requirement that $\mu \mathcal{Q} \subset \mathcal{Q} + \mathbf{t}$ in the definition of refining inflation.

Being a local refining inflation obtained from an operator T puts an even stronger restriction on μ.

PROPOSITION 2.35. *With the notation and assumptions of Proposition 2.33, if there is a polytope $\Omega \subset W$ such that T gives a refining inflation that is local then either $\mathcal{M}$ is a lattice or $\dim W = \dim V$ and μ is a quadratic Pisot number. If, in addition, the Submodels Condition and the Strong V-Condition hold then, conversely, if μ is a quadratic Pisot unit a polytope $\Omega \subset W$ can be chosen so that either T or $-T$ gives a local refining inflation.*

PROOF. If the inflation given by T is refining and local then $T\mathcal{L} \subseteq \mathcal{L}$ and $\mathcal{M}$ is locally derivable from $\mu\mathcal{M}$ so, by Lemma 2.18, Ω is a Boolean combination of $\pi_W(\mathcal{L})$-translates of $T\Omega$. In particular, every point on the boundary $\partial\Omega$ of Ω is a $\pi_W(\mathcal{L})$-translate of a point on the boundary $\partial(T\Omega)$ of $T\Omega$. This provides a mapping from $\partial\Omega$ to $\partial(T\Omega)$, in spite of the latter having smaller measure than the former when all eigenvalues μ_j of T on W have $|\mu_j| < 1$. Our strategy is to show that this can happen only in very exceptional circumstances.

A consequence of Ω being a Boolean combination of translates of $T\Omega$ is that for every facet F_1 of Ω there is a parallel facet, TF_2 say, of $T\Omega$, where F_2 is a facet of Ω, and then a facet TF_3 of $T\Omega$ parallel to F_2, and so on. Since Ω has finitely many facets, there are positive integers l, m such that $T^m F_l$ is parallel to F_l, so the subspace H of codimension 1 in W that is parallel to F_l is T^m-invariant. The proof of Proposition 2.33 shows that (over $\mathbb{C}$) W has a basis of eigenvectors for T. If $\{\mu_j \mid j = 2, \ldots, d\}$ are the eigenvalues of T on W then, for $2 \leq k \leq d$, $\prod_{j \neq k}(T^m - \mu_j^m)/(\mu_k^m - \mu_j^m)$ is an operator on H (defined over $\mathbb{C}$) which, for each $\boldsymbol{h} \in H$ picks out the component of $\boldsymbol{h}$ in the eigenspace for μ_k. (No ratio of conjugates of a Pisot number can be a root of unity.) Hence H is a direct sum of eigenspaces of T over $\mathbb{C}$ and, in particular, is T-invariant. We also see that the original facet F_1 is parallel to $T^l F_l$ which is parallel to H. So every facet of Ω is parallel to some T-invariant codimension 1 subspace of W.

To return to the putative mapping from $\partial\Omega$ to $\partial(T\Omega)$, for every $\boldsymbol{w} \in \partial\Omega$ there is a $\boldsymbol{k} \in \pi_W(\mathcal{L})$ such that $\boldsymbol{w} + \boldsymbol{k} \in \partial(T\Omega)$ and if $\boldsymbol{w}$ lies on a facet of Ω parallel to a hyperplane $H \subseteq W$ then $\boldsymbol{k}$ can be chosen so that $\boldsymbol{w} + \boldsymbol{k}$ lies on a facet of $T\Omega$ parallel to H. However, the mapping

$$(2.11) \qquad \boldsymbol{w} \to \boldsymbol{w} + \boldsymbol{k}$$

is in general neither single-valued nor one-one: different values of $\boldsymbol{k}$ may be possible for a given $\boldsymbol{w}$ and different $\boldsymbol{w}$'s may give the same $\boldsymbol{w} + \boldsymbol{k}$. This difficulty can be overcome by regarding some points on $\partial\Omega$ as equivalent. Let D be the union of the facets of Ω that are parallel to a given codimension 1 subspace $H \subseteq W$. We have seen that H is T-invariant, so the union of the facets of $T\Omega$ parallel to H is TD. Denote by L the maximal $\mathcal{L}$-subspace in $V \oplus H$. Clearly L is T-invariant. We regard two points $\boldsymbol{w}_1, \boldsymbol{w}_2 \in \partial\Omega$ as equivalent if $\boldsymbol{w}_1 - \boldsymbol{w}_2 \in \pi_W(L) + \pi_W(\mathcal{L})$, and similarly for pairs of points on TD. Since L and $\mathcal{L}$ are T-invariant (2.11) defines a one-one mapping from equivalence classes on D to equivalence classes on TD. There is a natural r-dimensional measure on the set of equivalence classes, where $r = N - n - 1 - \dim \pi_W(L)$, which is preserved by the mapping (2.11). (The set of equivalence classes on H is given by $H/\pi_W(L)$, and the equivalence classes on the individual facets of D and TD are similar. Factoring out projections of lattice vectors not in L as well allows part of one facet to be identified with part of another one when the parts differ by the projection of a lattice vector.) On the other hand, by Proposition 2.33, all eigenvalues of T on W (and hence on $H/\pi_W(L)$) are < 1 in absolute value and hence the operator T reduces the measure by a factor $\leq M^r$, where M is the largest absolute value of any eigenvalue of T on W. This is incompatible with the existence of a one-one measure-preserving map from the equivalence classes of D to the equivalence classes of TD unless $r = 0$. To bound r below we note from the proof of Proposition 2.33 that the characteristic polynomial of T on $V \oplus W$ has the form ϕ^n, with ϕ irreducible over $\mathbb{Q}$, from which it follows that the characteristic polynomial of T on L, a proper T-invariant $\mathcal{L}$-subspace, is ϕ^m for some $m \leq n - 1$. Hence each eigenvalue of T on $\pi_W(L)$ has multiplicity $\leq n - 1$ so $\dim \pi_W(L) \leq (d-1)(n-1)$, where d is the degree of ϕ. Thus $r \geq (d-1)n - 1 - (d-1)(n-1) = d - 2$ and can be 0 only when $d = 1$ or 2.

For the converse direction, let V' be any linear subspace of V spanned by vectors of $\pi_V(\mathcal{L})$. By Formulation 4 of the Submodels Condition there is an $\mathcal{L}$-subspace L' such that $V', W', \mathcal{L} \cap L'$ give a full submodel, where $W' = \pi_W(L') = W \cap L'$. Since V' is T-invariant and T commutes with π_V and π_W, the $\mathcal{L}$-subspace TL' gives the same submodel. As the submodel is full, we have $TL' \subseteq L'$, so that L' is T-invariant. The characteristic polynomial of T on L' has rational coefficients (because T stabilizes the lattice $\mathcal{L} \cap L'$ that spans L') so is ϕ^k, for some k, where ϕ is the minimum polynomial of μ. Hence $\dim V' = \dim W' = k$. Now suppose that V'' is another subspace of V that is the image of an $\mathcal{L}$-subspace and the corresponding full submodel is given by $V'', \mathcal{L} \cap L'', W''$. Then $V' \cap V''$ is also the image of an $\mathcal{L}$-subspace and the corresponding full submodel is given by $V' \cap V'', \mathcal{L} \cap L' \cap L''$ and $W \cap L' \cap L'' = (W \cap L') \cap (W \cap L'') = W' \cap W''$. Hence $\dim(W' \cap W'') = \dim(V' \cap V'')$.

Now choose n codimension 1 subspaces of V, say $V_1, \ldots, V_n$, each spanned by vectors of $\pi_V(\mathcal{L})$ and with $V_1 \cap \cdots \cap V_n = \{\mathbf{0}\}$. This is possible since, in view of the V-Condition, $\pi_V(\mathcal{L})$ is relatively dense. If $W_1, \ldots, W_n$ are the corresponding subspaces of W then every intersection of W_i's is the internal space for a full submodel and has the same dimension as the corresponding intersection of V_i's. In particular, $\dim(W_1 \cap \cdots \cap W_n) = \dim(V_1 \cap \cdots \cap V_n) = 0$ and the W_i's have trivial intersection. By the Strong V-Condition and Corollary 2.12, if W' is any intersection of W_i's then $\pi_W(\mathcal{L}) \cap W'$ is dense in W'. In particular, if we denote by $\widehat{W}_j$ the intersection of all the W_i's except W_j (a 1-dimensional subspace of W) then we can pick a vector $\boldsymbol{w}_j \neq \mathbf{0}$ in $\pi_W(\mathcal{L}) \cap \widehat{W}_j$, and $\boldsymbol{w}_j \notin W_j$. Now let Ω be the centrally symmetric convex polytope with nonempty interior enclosed by the hyperplanes $W_i \pm \boldsymbol{w}_i$ ($i = 1, \ldots, n$). Then $\pm T\Omega = \pm \mu'\Omega \subset \Omega$, where $\mu' = (\operatorname{norm} \mu)/\mu = \pm 1/\mu$ is the conjugate of μ and has $|\mu'| < 1$, so by Proposition 2.31 both T and $-T$ give a refining inflation with this Ω. To show that $(\operatorname{norm} \mu)T$ gives a local inflation we shall show that Ω is a union of $\pi_W(\mathcal{L})$-translates of $|\mu'|\Omega$ and appeal to Lemma 2.18, bearing in mind that since μ is a unit $\det T = \pm 1$ and $T\mathcal{L} = \mathcal{L}$. Let $\boldsymbol{w}$ be any point in the closure $\overline{\Omega}$ of Ω and F the lowest dimensional face of $\overline{\Omega}$ that $\boldsymbol{w}$ lies in. Then $\operatorname{aff}(F)$ has the form $W' + \boldsymbol{w}'$, where W' is an intersection of some of the hyperplanes $W_1, \ldots, W_n$, and $\boldsymbol{w}'$ is a signed sum of the corresponding $\boldsymbol{w}_j$'s. Since $\pi_W(\mathcal{L}) \cap W'$ is dense in W' and $\boldsymbol{w}' \in \pi_W(\mathcal{L})$, $\pi_W(\mathcal{L}) \cap F$ is dense in F, so there are points $\boldsymbol{w}'' \in \pi_W(\mathcal{L}) \cap F$ arbitrarily close to $\boldsymbol{w}$. (When $F = \boldsymbol{w}$ is a vertex of Ω, $\boldsymbol{w} = \boldsymbol{w}'$ is itself a point of $\pi_W(\mathcal{L})$.) If $\boldsymbol{w}''$ is close enough to $\boldsymbol{w}$ then for the

contraction of Ω towards $\boldsymbol{w}''$ by a factor $|\mu'|$ we have
$$|\mu'|\Omega + (1-|\mu'|)\boldsymbol{w}'' \subset \Omega$$
and
$$\bigl(|\mu'|\Omega + (1-|\mu'|)\boldsymbol{w}''\bigr) \cap B_\varepsilon(\boldsymbol{w}) = \Omega \cap B_\varepsilon(\boldsymbol{w})$$
when the ball $B_\varepsilon(\boldsymbol{w})$ is small enough (where ε depends on $\boldsymbol{w}$). Also $(1-|\mu'|)\boldsymbol{w}'' = \boldsymbol{w}'' \pm T\boldsymbol{w}'' \in \pi_W(\mathcal{L})$, since $\mathcal{L}$ is T-invariant. Being compact, $\overline{\Omega}$ is covered by finitely many of the balls $B_\varepsilon(\boldsymbol{w})$, and Ω is the union of the corresponding sets
$$(\operatorname{norm}\mu)T\Omega + (1-|\mu'|)\boldsymbol{w}'',$$
which are translates of $(\operatorname{norm}\mu)T\Omega$ by vectors of $\pi_W(\mathcal{L})$. So by Lemma 2.18 $\mathcal{M}$ is locally derivable from $(\operatorname{norm}\mu)\mu\mathcal{M}$. □

REMARK 2.36. In the light of this proof we note that the usual local inflation with multiplier τ on the Penrose tiling does not strictly take $\mathcal{P}$ to itself, for example the horizontal Ammann line just above the x-axis moves below the x-axis. The exact local inflation of $\mathcal{P}$ has multiplier $-\tau$. Perhaps there is a case for allowing exceptions of frequency zero to the definition of local derivability, as suggested for refining inflation in Remark 2.34.

Local Rules. It remains to consider the quasicrystal Properties 16 and 17, namely, weak and perfect local rules, for model sets. I know of no sufficient conditions for a model set to have perfect local rules, but weak local rules are a consequence of the existence of Ammann hyperplanes together with a sufficiently rich symmetry group.

DEFINITION 2.37. For an isometry g acting on $\mathbb{E}^n$ we denote by $\delta(g)$ the degree of the minimum polynomial of g.

DEFINITION 2.38. Let G be a group of isometries acting on $\mathbb{E}^n$. We call a set $\mathcal{U}$ of unit vectors in $\mathbb{E}^n$ *appropriate for* G if it is G-invariant and there is a set of vectors $\boldsymbol{v}_1,\ldots,\boldsymbol{v}_n$ such that each $\boldsymbol{v}_i$ is orthogonal to $n-1$ linearly independent vectors of $\mathcal{U}$ and for each $g \in G$ the $\langle g \rangle$-orbit of $\boldsymbol{v}_i$ spans a subspace of dimension $\delta(g)$.

LEMMA 2.39. *If the discrete point set $\mathcal{Q} \subset \mathbb{E}^n$ has finite local complexity and a finite group G of symmetries such that the subgroup $H = \langle g^{\delta(g)} \mid g \in G \rangle$ is not the symmetry group of any crystal in $\mathbb{E}^n$ and there are n sets of unit vectors, $\mathcal{U}_1,\ldots,\mathcal{U}_n$, each appropriate for G, such that*

(i) *every set of n vectors $\{\boldsymbol{u}_i \mid \boldsymbol{u}_i \in \mathcal{U}_i, i=1,\ldots,n\}$ is linearly independent and*

(ii) *for each $\boldsymbol{u} \in \bigcup \mathcal{U}_i$, $\mathcal{Q}$ has a system of Ammann hyperplanes normal to $\boldsymbol{u}$,*

then $\mathcal{Q}$ has weak local rules.

PROOF. By (ii), there are constants $D, \Delta > 0$ such that $\mathcal{Q}$ has the property that for each $\boldsymbol{u} \in \mathcal{U} = \bigcup \mathcal{U}_j$ the hyperplanes normal to $\boldsymbol{u}$ that contain points of $\mathcal{Q}$ have maximum separation $< D$ and covering radius $< \Delta$. If we take $\rho = D + 3\Delta$ then every discrete point set $\mathcal{Q}'$ all of whose ρ-patches are ρ-patches of $\mathcal{Q}$ also has this property. Consequently the hyperplanes normal to vectors in $\mathcal{U}$ that contain points of $\mathcal{Q}'$ determine what is called in [**LuP**] a "$\mathcal{U}$-grid tiling" (because the vectors in $\mathcal{U}$ span $\mathbb{E}^n$). The result now follows from the Second Grid Restriction Theorem of [**LuP**, p.262], which depends on the existence of a set of vectors $\mathcal{U}$ satisfying the hypotheses of the lemma and also having the property that all dot products of the vectors in $\mathcal{U}$ are algebraic numbers. This latter property is used in the proof only to establish that the $\mathcal{U}$-grid tiling has "finite type" (i.e., finitely many different shapes of tiles), which in the present context is guaranteed by the fact that $\mathcal{Q}'$ has finite local complexity (because $\mathcal{Q}$ does). □

PROPOSITION 2.40. *Let $V = \mathbb{E}^n$ and let $\mathcal{L} \subset V \oplus W = \mathbb{E}^N$ satisfy the Submodels Condition, the Strong V-Condition and the Strong W-Condition. Let G be a finite group of operators acting on $\mathbb{E}^N$ such that V, W and $\mathcal{L}$ are G-invariant and the restriction of G to V is a group of isometries. If the restriction to V of $H = \langle g^{\delta(g)} \mid g \in G \rangle$ is not the symmetry group of a crystal in $\mathbb{E}^n$ then it is possible to find a polytope $\Omega \subset W$ such that the resulting model set $\mathcal{M}$ has all the of symmetries in G and has weak local rules.*

PROOF. This depends on constructing sets of unit vectors $\mathcal{U}_1, \ldots, \mathcal{U}_n$ in $\mathbb{E}^n$, appropriate for G, that satisfy the hypotheses of Lemma 2.39 and are such that each unit vector $\boldsymbol{u} \in \mathcal{U} = \bigcup \mathcal{U}_i$ is normal to a subspace of codimension 1 in $\mathbb{E}^n$ that is spanned by points of $\pi_V(\mathcal{L})$. Such a construction is described in the proof of Lemma 17 of [**LuP**], except now, instead of the dot products of the unit vectors being algebraic (requirement (ii) of Lemma 17 of [**LuP**]) we require these vectors to be normal to hyperplanes spanned by vectors of $\pi_V(\mathcal{L})$. In [**LuP**], requirements (i), (iii) and (iv) of Lemma 17 are satisfied by choosing the unit vectors in turn to avoid certain finite collections of hyperplanes and by including all vectors in the G-orbit of each vector chosen. These choices are consistent with the vectors being normal to hyperplanes spanned by by vectors of $\pi_V(\mathcal{L})$, since $\pi_V(\mathcal{L})$ is relatively dense in V (because of the V-Condition) and is G-invariant (because $\mathcal{L}$ is). Now for each $\boldsymbol{u} \in \mathcal{U}$ the hyperplane $V' \subset V$ normal to $\boldsymbol{u}$ is spanned by $\pi_V(\boldsymbol{l}_1), \ldots, \pi_V(\boldsymbol{l}_{n-1})$, with $\boldsymbol{l}_1, \ldots, \boldsymbol{l}_{n-1} \in \mathcal{L}$. By taking L to be the $\mathcal{L}$-subspace spanned by $\boldsymbol{l}_1, \ldots, \boldsymbol{l}_{n-1}$ in Formulation 4 of the Submodels Condition we see that $\mathcal{L}'$, V', W' is a full submodel for some sublattice $\mathcal{L}' \subset \mathcal{L}$ and $W' \subset W$. Finally we choose a basis $\mathcal{B}$ for each $\mathcal{L}'$ compatible with G (i.e., if the basis $\mathcal{B}$ is chosen for $\mathcal{L}'$ then $g(\mathcal{B})$ should be chosen for $g\mathcal{L}'$) and take for the window Ω the zonotope generated by the W-projections of the union of all these bases, which will therefore be G-invariant. Then the resulting model set has all the symmetries in G, by Proposition 2.29, and has a system of Ammann hyperplanes normal to each vector $\boldsymbol{u}$, by Proposition 2.27. Hence, by Lemma 2.39 it has weak local rules. □

REMARK 2.41. The subgroup H of Lemma 2.39 and Proposition 2.40 is often larger than might be expected: even having $g^{\delta(g)} = e$ does not necessarily exclude g from H, since it may be a product of other generators of H. For the actions of the dihedral groups D_m and their cyclic subgroups C_m on $\mathbb{E}^2$ we have

$$H = \begin{cases} C_m & \text{if } m \text{ is not a multiple of 4,} \\ C_{m/2} & \text{if } m \text{ is a multiple of 4.} \end{cases}$$

So apart from the crystallographic groups C_m, D_m ($m = 1, 2, 3, 4, 6$) the only other finite isometry groups on $\mathbb{E}^2$ to which the local rules criterion of Proposition 2.40 does not apply are C_8, D_8, C_{12}, D_{12}. These happen to be the most commonly studied quasicrystal symmetries apart from the symmetry group D_{10} of the Penrose tiling. Although our weak local rules criterion does not apply to D_8 or D_{12} there are quasicrystals with these symmetries that have local rules. The quasicrystals with these symmetries constructed by Baake, Joseph and Schlottmann in [**BJS**], for example, were later shown by the authors to have perfect local rules.

For the action of the icosahedral group Y_h on $\mathbb{E}^3$ (which is known to be non-crystallographic) we have $H = Y_h$.

3. The Algebraic Number Field Construction

Modules Over Algebraic Number Fields. Let K be a real algebraic number field of degree d; that is, K is a subfield of $\mathbb{R}$ that has dimension d as a vector space over $\mathbb{Q}$. Then K^n has dimension nd as a vector space over $\mathbb{Q}$.

DEFINITION 3.1 (See [**BS**, Chapter 2, Section 1.3] for the case $n=1$). A *full module* $\mathcal{Z}$ in K^n is a $\mathbb{Z}$-module generated by nd vectors in K^n that are linearly independent over $\mathbb{Q}$.

For example, it is known that the ring of integers $\mathcal{O}$ of K has an "integral basis" $\{\omega_1, \ldots, \omega_d\}$ such that every $\alpha \in \mathcal{O}$ can be uniquely written as $\alpha = a_1\omega_1 + \cdots + a_d\omega_d$ with each $a_i \in \mathbb{Z}$; consequently $\mathcal{O}^n$ is a full module in K^n. In particular, $\mathcal{O}$ is a full module in K and the mapping

$$(3.1) \qquad \alpha \mapsto (a_1, \ldots, a_d)$$

shows that $\mathcal{O}$ is isomorphic to a d-dimensional lattice.

It is clear that the intersection of two full modules in K^n is again a full module in K^n.

Models From Modules. There is a more canonical way than (3.1) of constructing a lattice isomorphic to a full module, which is not dependent on choosing a basis for the module and is also more suited to constructing model sets.

There are d distinct homomorphisms of K into the complex numbers $\mathbb{C}$ which can be ordered as

$$\sigma_1, \sigma_2, \ldots, \sigma_r, \sigma_{r+1}, \bar{\sigma}_{r+1}, \ldots, \sigma_{r+s}, \bar{\sigma}_{r+s},$$

where $r + 2s = d$, $\sigma_1, \ldots, \sigma_r$ are homomorphisms into $\mathbb{R}$ and $\sigma_{r+i}, \bar{\sigma}_{r+i}$ (for $1 \leq i \leq s$) are obtained from each other by complex conjugation (which is an automorphism of $\mathbb{C}$). (It is always possible to write K as $\mathbb{Q}(\theta)$, where θ satisfies an equation of degree d over $\mathbb{Q}$, and the d homomorphisms arise from mapping θ to the distinct roots of this equation.) The mapping

$$(3.2) \qquad \kappa(\boldsymbol{\alpha}) = \big(\sigma_1(\boldsymbol{\alpha}), \ldots, \sigma_r(\boldsymbol{\alpha}), \sigma_{r+1}(\boldsymbol{\alpha}), \ldots, \sigma_{r+s}(\boldsymbol{\alpha})\big),$$

where $\boldsymbol{\alpha} = (\alpha_1, \ldots, \alpha_n)$, takes K^n into

$$(3.3) \qquad (\mathbb{E}^n)^r \oplus (\mathbb{C}^n)^s \cong \mathbb{E}^{nd}$$

and is clearly one-one. Because the product of the conjugates of a nonzero algebraic integer is a nonzero rational integer (hence ≥ 1 in absolute value), the image of $\mathcal{O}^n$ under this mapping is discrete—in fact the open unit ball centred on a point of the image contains no other point of the image. Hence the image of $\mathcal{O}^n$ is a lattice (of full dimension) in $\mathbb{E}^{nd}$. For any other full module $\mathcal{Z}$, $\mathcal{Z} \cap \mathcal{O}^n$ has finite index in $\mathcal{Z}$, and consequently the image of $\mathcal{Z}$ is also a lattice $\mathcal{L}$ in $\mathbb{E}^{nd}$. (The case $n=1$ of this construction is described in detail in [**BS**, Chapter 2, Section 3.1].)

Given a full module $\mathcal{Z}$ in K^n we can now obtain a model set $\mathcal{M}$ by taking for V the first $\mathbb{E}^n$ factor in (3.3), for W the remaining factors, namely

$$(3.4) \qquad W = (\mathbb{E}^n)^{r-1} \oplus \mathbb{C}^{ns} \cong \mathbb{E}^{n(d-1)}$$

(we shall name the individual factors $W_2,\ldots,W_{r+s}$) and for $\mathcal{L}$ the image $\kappa(\mathcal{Z})$ of $\mathcal{Z}$ under the mapping (3.2). The projection π_V consists of selecting the component in the first factor of (3.3) and the projection π_W of selecting the remaining components. Each $\mathcal{L}$-subspace in $V \oplus W$ is the $\mathbb{R}$-subspace generated by the image of a $\mathbb{Q}$-subspace of K^n, and this in fact gives a one-one correspondence between $\mathcal{L}$-subspaces of $V \oplus W$ and $\mathbb{Q}$-subspaces of K^n.

LEMMA 3.2.
   (i) If the $\mathcal{L}$-subspace L corresponds to a $\mathbb{Q}$-subspace S of K^n then $\dim_{\mathbb{R}} \pi_V(L)$ is equal to $\dim_K KS$, the dimension, as a vector space over K, of the K-subspace of K^n spanned by S.
   (ii) The $\mathcal{L}$-subspaces L of $V \oplus W$ that satisfy $\pi_V(L) = V \cap L$ are precisely those that correspond to K-subspaces of K^n.

PROOF. For (i), we note that if $\boldsymbol{\alpha} \in K^n$ and $\beta \in K$ then the V-component of $\kappa(\beta\boldsymbol{\alpha})$ is $\sigma_1(\beta)\sigma_1(\boldsymbol{\alpha})$, a scalar multiple of the V-component $\sigma_1(\boldsymbol{\alpha})$ of $\kappa(\boldsymbol{\alpha})$. So if L' is the $\mathcal{L}$-subspace corresponding to KS then $\pi_V(L') = \pi_V(L)$. Let $\{\boldsymbol{\epsilon}_1,\ldots,\boldsymbol{\epsilon}_k\}$ be a K-basis of KS. Then $\pi_V(L')$ is spanned by $\{\sigma_1(\boldsymbol{\epsilon}_1),\ldots,\sigma_1(\boldsymbol{\epsilon}_k)\}$ and these vectors are linearly independent. (If they were linearly dependent over $\mathbb{R}$ then they would be linearly dependent over $\sigma_1(K)$, implying a corresponding linear dependence between $\boldsymbol{\epsilon}_1,\ldots,\boldsymbol{\epsilon}_k$ over K.) So $\dim \pi_V(L) = \dim \pi_V(L') = k$.

For (ii), in one direction, we need to show that if the $\mathcal{L}$-subspace L corresponds to a K-subspace S of K^n then, for each $\boldsymbol{\alpha} \in S$, $(\sigma_1(\boldsymbol{\alpha}), 0, \ldots, 0) \in \mathcal{L}$. Let $\{1, \omega, \ldots, \omega^{d-1}\}$ be a power basis of K over $\mathbb{Q}$. Then each $\omega^i \boldsymbol{\alpha}$ is in S so

$$\boldsymbol{u}_i = \big(\sigma_1(\omega^i)\sigma_1(\boldsymbol{\alpha}), \sigma_2(\omega^i)\sigma_2(\boldsymbol{\alpha}), \ldots, \sigma_{r+s}(\omega^i)\sigma_{r+s}(\boldsymbol{\alpha})\big) \in L.$$

Now the matrix $\big(\sigma_j(\omega^i)\big)$ (where $\sigma_{r+k} = \bar{\sigma}_{r+k-s}$, for $s < k \leq 2s$) is nonsingular since if a nontrivial linear combination of its rows vanished then the conjugates $\sigma_1(\omega),\ldots,\sigma_d(\omega)$ would all satisfy a polynomial of degree $d-1$, which is impossible since they are distinct. So there is a solution $x_1,\ldots,x_d$ ($x_i \in \mathbb{R}$) to the system of equations

$$\sum_{i=1}^{d} x_i \sigma_j(\omega^i) = \begin{cases} 1 & \text{for } j = 1, \\ 0 & \text{for } j > 1, \end{cases}$$

and hence

$$(\sigma_1(\boldsymbol{\alpha}), 0, \ldots, 0) = \sum_{i=1}^{d} x_i \boldsymbol{u}_i \in L.$$

For the other direction of (ii), let S be a general $\mathbb{Q}$-subspace of K^n with basis $\{\boldsymbol{\epsilon}_1,\ldots,\boldsymbol{\epsilon}_l\}$ and let L be the corresponding $\mathcal{L}$-subspace of $V \oplus W$. For each $\boldsymbol{\alpha} \in K^n$ all the coordinates of $\kappa(\boldsymbol{\alpha})$ lie in the compositum field $F = \sigma_1 K \sigma_2 K \cdots \sigma_{r+s} K$. Each mapping $\sigma_j \sigma_i^{-1} : \sigma_i K \to \sigma_j K$ extends to an automorphism of F and F is normal over $\mathbb{Q}$. We can find a basis of $V \cap L$ of the form $\{\boldsymbol{e}_1,\ldots,\boldsymbol{e}_k\}$ with

$$\boldsymbol{e}_j = \sum_{i=1}^{l} x_{ij} \kappa(\boldsymbol{\epsilon}_i) \quad (j = 1,\ldots,k),$$

and each $x_{ij} \in F$, since the x_{ij}'s are solutions of a system of linear equations with coefficients in F. Let σ be the automorphism of F that takes $\sigma_1 K$ to $\sigma_h K$. Then σ permutes the coordinates of each $\kappa(\boldsymbol{\epsilon}_i)$, and $\sigma(\boldsymbol{e}_j) \in W_h \cap L$. Since $\boldsymbol{e}_1,\ldots,\boldsymbol{e}_k$ have coefficients in $F \cap \mathbb{R}$ and are linearly independent over $\mathbb{R}$ they are linearly independent over F, and hence $\sigma(\boldsymbol{e}_1), \ldots, \sigma(\boldsymbol{e}_k)$ are also linearly independent over F and therefore over $\mathbb{C}$. Consequently

$$\dim_{\mathbb{R}}(W_h \cap L) \geq \begin{cases} k & \text{if } W_h = \mathbb{E}^n, \\ 2k & \text{if } W_h = \mathbb{C}^n. \end{cases}$$

(There is actually equality here: by considering the effect of σ^{-1} too we find that $\{\sigma(\boldsymbol{e}_1), \ldots, \sigma(\boldsymbol{e}_k)\}$ is an $\mathbb{R}$-basis of $W_h \cap L$ in the real case and a $\mathbb{C}$-basis in the complex case.) We now have

$$\dim_{\mathbb{Q}} S = \dim_{\mathbb{R}} L \geq \sum_{i=1}^{r+s} \dim_{\mathbb{R}}(W_i \cap L) \geq dk.$$

On the other hand, by (i),

$$\dim_{\mathbb{Q}} S \leq \dim_{\mathbb{Q}} KS = d \dim_K KS = d \dim_{\mathbb{R}} \pi_V(L).$$

If $\pi_V(L) = V \cap L$ then $\dim \pi_V(L) = k$ and hence $\dim_{\mathbb{Q}} S = \dim_{\mathbb{Q}} KS = dk$. Thus $S = KS$ and is a K-subspace of K^n. □

We next show that a model set $\mathcal{M}$ derived from a module $\mathcal{Z}$ in K^n satisfies all the orientation conditions of the previous section.

LEMMA 3.3. *Let V, W and $\mathcal{L}$ be derived from a full module $\mathcal{Z}$ in K^n as described above. Then*:
  (i) *An $\mathcal{L}$-subspace $L \subseteq V \oplus W$ gives a submodel if and only if L corresponds to a K-subspace of K^n.*
  (ii) *Every submodel is full.*
  (iii) *There are full submodels of every dimension from 0 to n.*
  (iv) *V, W, $\mathcal{L}$ satisfy the Submodels Condition, the Strong V-Condition and the Strong W-Condition.*
  (v) *If $K \neq \mathbb{Q}$ then $\mathcal{L} \cap V = \{\mathbf{0}\}$.*

PROOF. Clearly (i) is equivalent to Lemma 3.2(ii). For (ii), let L be an $\mathcal{L}$-subspace that gives a submodel and S the corresponding K-subspace of K^n. If $L' \supseteq L$ is another $\mathcal{L}$-subspace with corresponding subspace S' of K^n and $\pi_V(L') = \pi_V(L)$ then $S' \supseteq S$ and, by Lemma 3.2(i), the dimension of the K-subspace spanned by S' is equal to the dimension of S, so $S' = S$ and hence $L' = L$. Thus L gives a full submodel. Since there are K-subspaces of K^n of every dimension from 0 to n, (iii) follows from (i) and (ii). For (iv), Formulation 3 of the Submodels Condition follows from Lemma 3.2 by taking S' to be the K-subspace of K^n spanned by S, where S and S' are the subspaces of K^n corresponding to L and L'. Now suppose that we have a submodel derived from an $\mathcal{L}$-space L' with $\pi_V(L') = V'$ and that S' is the corresponding K-subspace of K^n. If $L \subseteq L'$ is an $\mathcal{L}$-subspace with $L \supseteq \pi_V(L')$ then $L \cap V = \pi_V(L) = V'$ so L corresponds to a K-subspace $S \subseteq S'$, by Lemma 3.2(ii), and, by Lemma 3.2(i), $\dim S = \dim V' = \dim S'$. So $S = S'$ and $L = L'$. Thus no proper $\mathcal{L}$-subspace of L' contains V' and the submodel derived from L' satisfies the V-Condition, whence V, W, $\mathcal{L}$ satisfy the Strong V-Condition. Next suppose that $\boldsymbol{\alpha} \in K^n$ has $\sigma_1(\boldsymbol{\alpha}) \in V'$. If L'' is the $\mathcal{L}$-subspace corresponding to the K-subspace of K^n spanned by S' and $\boldsymbol{\alpha}$ then $\pi_V(L'') = \pi_V(L')$ and hence $\boldsymbol{\alpha} \in S'$. Consequently $(\sigma_2(\boldsymbol{\alpha}), \ldots, \sigma_d(\boldsymbol{\alpha})) \in \pi_W(L') = W'$, the internal space for the submodel given by L'. In the notation of (2.2), this shows that $\mathbf{0}$ is the only point of the quotient lattice to belong to the quotient internal space W/W', so that

V, W, $\mathcal{L}$ satisfy the Strong W-Condition. For (v), we have $d \geq 2$ when $K \neq \mathbb{Q}$ so that $\big(\sigma_1(\boldsymbol{\alpha}), \ldots, \sigma_{r+s}(\boldsymbol{\alpha})\big) \in \mathcal{L} \cap V$ implies $\sigma_{r+s}(\boldsymbol{\alpha}) = \mathbf{0}$ and hence $\boldsymbol{\alpha} = \mathbf{0}$. $\square$

Linear Operators. There is a natural way of obtaining linear operators on $V \oplus W$ that stabilize V, W and $\mathcal{L}$ from linear operators on K^n that stabilize the module $\mathcal{Z}$. A linear operator T on K^n can be represented with respect to a K-basis of K^n by a matrix M with entries in K. From M we obtain the linear operator on $(\mathbb{E}^n)^r \oplus (\mathbb{C}^n)^s$ given by the map

$$(3.5) \quad (\boldsymbol{x}_1, \ldots, \boldsymbol{x}_r, \boldsymbol{z}_1, \ldots, \boldsymbol{z}_s)$$
$$\mapsto (\sigma_1(M)\boldsymbol{x}_1, \ldots, \sigma_r(M)\boldsymbol{x}_r, \sigma_{r+1}(M)\boldsymbol{z}_1, \ldots, \sigma_{r+s}(M)\boldsymbol{z}_s)$$

We shall denote this operator by T too. It is clear from (3.5) that the individual factors $\mathbb{E}^n$ and $\mathbb{C}^n$ of $(\mathbb{E}^n)^r \oplus (\mathbb{C}^n)^s$ are invariant subspaces of T.

If $\mathcal{Z}$ is a T-invariant full module of K^n then

$$T(\mathcal{Z}) \subseteq \mathcal{Z} \implies \sigma_i\big(M(\mathcal{Z})\big) \subseteq \sigma_i(\mathcal{Z}) \quad (i = 1, \ldots, d)$$
$$\implies \sigma_i(M)\big(\sigma_i(\mathcal{Z})\big) \subseteq \sigma_i(\mathcal{Z}) \quad (i = 1, \ldots, d)$$
$$\implies T(\mathcal{L}) \subseteq \mathcal{L},$$

so $\mathcal{L}$ is T-invariant too.

Flation. The simplest linear operators on K^n consist of multiplication by a scalar $\mu \in K$ (so that M is a scalar matrix). When μ is not an integer of K there is no module $\mathcal{Z}$ with $\mu \mathcal{Z} \subseteq \mathcal{Z}$, but the lattice $\mu \mathcal{Z}$ is always commensurate with $\mathcal{Z}$. The effect on $(\mathbb{E}^n)^r \oplus (\mathbb{C}^n)^s$ of multiplying by μ is nontrivial:

$$(\boldsymbol{x}_1, \ldots, \boldsymbol{x}_r, \boldsymbol{z}_1, \ldots, \boldsymbol{z}_s) \mapsto (\sigma_1(\mu)\boldsymbol{x}_1, \ldots, \sigma_r(\mu)\boldsymbol{x}_r, \sigma_{r+1}(\mu)\boldsymbol{z}_1, \ldots, \sigma_{r+s}(\mu)\boldsymbol{z}_s).$$

It acts on V as a scaling with multiplier $\sigma_1(\mu)$ and its eigenvalues on W are the conjugates $\sigma_2(\mu), \ldots, \sigma_d(\mu)$ of $\sigma_1(\mu)$. (The eigenvalues on $\mathbb{C}$ of multiplication by z are z and $\bar{z}$.)

Constructing a G-Invariant Lattice. For this we need to choose a number field K appropriate to the group G.

PROPOSITION 3.4. *For every finite isometry group G acting on $\mathbb{E}^n$ there is a plain model set $\mathcal{M} \subset \mathbb{E}^n$ such that every $g \in G$ is a symmetry of $\mathcal{M}$.*

PROOF. In view of Proposition 2.29 we need to find an internal space W with a G-action and a G-invariant lattice $\mathcal{L} \subset \mathbb{E}^n \oplus W$ that satisfies the V-Condition.

The first step is to find a real algebraic number field K over which G can be represented. It is an elementary result of representation theory that for every representation of a finite group G there is an equivalent representation over an algebraic number field. It is less elementary, but true [**Ser**], that for every real representation of G there is an equivalent representation over a real algebraic number field K. Consequently, given a finite group G acting on $\mathbb{E}^n$ there exists a real algebraic number field K such that, with a suitable choice of basis of $\mathbb{E}^n$, the matrices of the elements of G have all their entries in K. Using this same basis as a basis for K^n, we have an action of G on K^n and an embedding, σ_1, of K^n in $\mathbb{E}^n$ that commutes with the G-actions.

We next find a G-invariant full module in K^n. Let $\mathcal{Y}$ be any full module in K^n (say $\mathcal{Y} = \mathcal{O}^n$) and put

$$\mathcal{Z} = \bigcap_{g \in G} g(\mathcal{Y}).$$

Then $\mathcal{Z}$ is a full module in K^n and is clearly G-invariant. If now κ is the canonical map (3.2) and the action of each $g \in G$ on $V \oplus W$ is as defined in (3.5) then $\mathcal{L} = \kappa(\mathcal{Z})$ is a G-invariant lattice in $V \oplus W$ and the action of G on $V = \mathbb{E}^n$ agrees with the given action. If we choose a window Ω whose closure $\overline{\Omega}$ is G-invariant then, by Proposition 2.29, the resulting model set $\mathcal{M}$ will have each $g \in G$ as a symmetry in the sense of Property 11. $\square$

THEOREM 3.5. *Let G be an arbitrary finite isometry group acting on $\mathbb{E}^n$ and $K \neq \mathbb{Q}$ any real algebraic number field over which G has an equivalent representation. Then there is a plain model set $\mathcal{M}$ that has Properties 1–7, 9, 10 and 15, has each $g \in G$ as a symmetry, has a refining inflation with a multiplier μ such that $\mathbb{Q}(\mu) = K$, and has systems of Ammann bars of all dimensions from 0 to n such that the normals to the $(n-1)$-dimensional systems span $\mathbb{E}^n$. If K is quadratic then such an $\mathcal{M}$ can be constructed so that the refining inflation, or its negative, is local. If there is no crystal in $\mathbb{E}^n$ having $\langle g^{\delta(g)} \mid g \in G \rangle$ as symmetries then there exists such an $\mathcal{M}$ which has weak local rules.*

PROOF. We first construct V, W and $\mathcal{L}$ as in the proof of Proposition 3.4. Next we choose a spanning set $\mathcal{S} = \{\boldsymbol{\alpha}_1, \ldots, \boldsymbol{\alpha}_m\}$ of K^n over K and take for Ω a half-open set whose closure is the zonotope generated by

$$\mathcal{W} = \{\pi_W(\kappa(g(\omega_i \boldsymbol{\alpha}_j))) \mid g \in G, i = 1, \ldots, d, j = 1, \ldots, m\},$$

where $\{\omega_1, \ldots, \omega_d\}$ is a basis of K over $\mathbb{Q}$. By Lemma 3.3(ii), (iii) and (iv) and Proposition 2.27, $\mathcal{M}$ has a system of Ammann bars of every dimension. (Clearly Ω meets the requirements of Proposition 2.27 and, in particular, Ω has nonempty interior.) It is possible to find n subsets of $\mathcal{S}$ that span $(n-1)$-dimensional K-subspaces of K^n with trivial intersection, and then the normals to the corresponding systems of $(n-1)$-dimensional Ammann bars span V.

For the symmetry, $\overline{\Omega}$ is clearly G-invariant (because $\mathcal{W}$ is) so, by Proposition 2.29, $\mathcal{M}$ has every $g \in G$ as a symmetry.

For the inflation we need a Pisot number μ such that $\mathbb{Q}(\mu) = K$ and $\mu \mathcal{Z} \subseteq \mathcal{Z}$. The condition $\mu \mathcal{Z} \subseteq \mathcal{Z}$ confines μ to a certain order in K known as the *coefficient ring* of $\mathcal{Z}$, where an *order* is a module in K that contains 1 and is a ring (see [**BS**, Chapter 2, Section 2.2]), and it is known that there are Pisot numbers μ with $\mathbb{Q}(\mu) = K$ in every module of a real algebraic number field. (This follows from [**BS**, Theorem 4, Chapter 2], for example, by choosing $c_2, \ldots, c_{s+t} < 1$.) By replacing μ by a sufficiently large power of μ, if necessary, we can suppose that the absolute value of every conjugate of μ is less than the ratio of the inradius to the circumradius of Ω. Then if T is the linear operator on $V \oplus W$ derived from multiplication by μ on K, we have $T\mathcal{L} \subseteq \mathcal{L}$ and $T\Omega \subset \Omega$ so, by Proposition 2.31, μ gives a refining inflation on $\mathcal{M}$.

When K is quadratic $\pi_W(\kappa(\omega \boldsymbol{\alpha})) = \sigma_2(\omega) \pi_W(\kappa(\boldsymbol{\alpha}))$, so the subspace of W spanned by any subset of $\mathcal{W}$ is the internal space of some submodel derived from a K-subspace of K^n according to Lemma 3.2(ii). Also $\pm T\Omega = (1/\mu)\Omega$ is a scaled reduction of Ω when K is quadratic. If we choose the $\boldsymbol{\alpha}_j$'s in $\mathcal{Z}$, the ω_i's in the coefficient ring of $\mathcal{Z}$ and μ a sufficiently large unit in K then $\pi_W(\mathcal{L})$ is dense in each face of Ω and Ω is a union of $\pi_W(\mathcal{L})$-translates of $(1/\mu)\Omega$, as in the proof of Proposition 2.35, so that $\mathcal{M}$ is locally derivable from $(\text{norm }\mu)\mu\mathcal{M}$.

The last sentence of the proposition follows from the proof of Proposition 2.40 on taking $\mathcal{S}$ to be the pre-image of the union of all the bases $\mathcal{B}$ constructed in that proof. (It is clearly possible to choose these bases so that each $\boldsymbol{\alpha}_j$ lies in $\mathcal{Z}$.)

Finally, by Lemma 3.3 and Propositions 2.6, 2.21, 2.23 and 2.25, $\mathcal{M}$ has Properties 1–7, 9, 10 and 15. □

REMARK 3.6. By choosing as a basis of $\mathbb{E}^2 = \mathbb{C}$ the vectors 1 and a primitive mth root of 1 (or $\{1, i\}$ when $m = 2$) we see that the dihedral group D_m acting on $\mathbb{E}^2$ has a representation over the field $\mathbb{Q}(\cos(2\pi/m))$ (the maximal real subfield of the mth cyclotomic field) which has degree $\phi(m)/2$ over $\mathbb{Q}$, where ϕ is Euler's totient function. In fact this is the unique minimal field over which D_m can be represented. Our construction gives $N = \dim(V \oplus W) = \phi(m)$ in this case. It has been shown by Baake, Joseph and Schlottmann in [**BJS**, Appendix A] that $\phi(m)$ is indeed the minimal embedding dimension for obtaining quasicrystals with a symmetry of order m by the cut-and-project construction.

Likewise the application of the construction to the icosahedral group in the example below gives $N = 6$, the known minimal embedding dimension for cut-and-project icosahedral quasicrystals (see, for example, [**Du**]).

Examples. Given the group G, the choices to be made in carrying out the construction implicit in Theorem 3.5 are, in order:

(1) A basis of $\mathbb{E}^n$ with respect to which the matrices of elements of G are algebraic (generating a number field K, say).
(2) A G-invariant module $\mathcal{Z}$ in K^n.

Different choices at this stage may lead to different model sets. The next choice makes no difference to the model set but is needed to fix the lattice and the action of G on W.

(3) A basis of each of the $r + s - 1$ components of W in (3.4) for use in interpreting the coordinates given by the canonical mapping (3.2).

Different choices have the effect of distorting W by different linear operators without changing the configuration in V. It is convenient to make a choice so that G acts as an isometry on W and the lattice $\mathcal{L}$ is easily recognizable (a help in identifying the construction as equivalent to some other known construction, for example).

(4) A G-invariant window Ω.
(5) An inflation multiplier μ in the coefficient ring of $\mathcal{Z}$ with $\mu\Omega \subseteq \Omega + \boldsymbol{t}$.

The first of these again affects the model set. In actual calculations there is a further choice to be made.

(6) A $\mathbb{Z}$-basis for $\mathcal{Z}$.

Different choices merely lead to different bases for the same lattice $\mathcal{L}$, but a good choice can again be helpful for recognizing the lattice.

We now describe the same two examples discussed in [**P2**] and use the opportunity to try to correct the treatment of the icosahedral group given there, where the fact that the rotation matrix was wrong led to the description being quite garbled and incorrect.

The pentagonal group D_5. This acts on $\mathbb{E}^2$ and is generated by the rotation g about $\mathbf{0}$ through $2\pi/5$ and the reflexion h in the x-axis. The matrices of these generators with respect to the standard basis have all entries in the field $\mathbb{Q}(\sin(2\pi/5))$ of degree 4 over $\mathbb{Q}$. So we could use this field as the K in our construction. But better is to choose a basis $\{(1,0)', g(1,0)'\}$ (where $\boldsymbol{x}'$ denotes the column vector corresponding to the row vector $\boldsymbol{x}$), giving matrices

(3.6)
$$\begin{pmatrix} 0 & -1 \\ 1 & 1/\tau \end{pmatrix} \text{ and } \begin{pmatrix} 1 & 1/\tau \\ 0 & -1 \end{pmatrix}$$

for g and h and allowing us to take $K = \mathbb{Q}(\tau)$ of degree 2 over $\mathbb{Q}$.

The ring of integers of K is $\mathbb{Z}[\tau]$, and we can choose $\mathcal{Z} = \mathbb{Z}[\tau]^2$. Since $1/\tau$ is an integer this is D_5-invariant. The smallest choice for μ as a Pisot unit in $\mathbb{Q}(\tau)$ is $\mu = \tau$.

The choice of basis $\{(1,0)', g^2(1,0)'\}$ for W (and regarding W as orthogonal to V) makes $\mathcal{L}$ a scalar multiple of the root lattice A_4^*. Choosing for the window Ω the regular decagon inscribed in the unit circle with one pair of sides parallel to the x-axis gives the Penrose point set $\mathcal{P}$, from which the Penrose tiling can be derived. Choosing for Ω the rotation of this regular decagon with one pair of sides parallel to the y-axis gives a point set $\mathcal{T}$ from which the Tübingen tiling can be derived. Both point sets actually have D_{10} symmetry.

We notice that the matrices (3.6) both have column sums congruent to 1 mod $\sqrt{5}$ (a prime in $\mathbb{Q}(\tau)$). As a result, the module

$$\mathcal{Z}' = \{(\alpha, \beta) \mid \alpha, \beta \in \mathbb{Z}[\tau], \alpha + \beta \equiv 0 \pmod{\sqrt{5}}\}$$

is also D_5-invariant. This leads to a lattice $\mathcal{L}$ that is a scalar multiple of A_4, but the model sets derivable from this are scaled versions of what we already have.

The icosahedral rotation group Y. If we position a regular icosahedron with centre at the origin of $\mathbb{E}^3$ so that each coordinate axis passes through the mid-point of an edge then, with respect to the standard basis, the matrices

$$\begin{pmatrix} -1 & 0 & 0 \\ 0 & -1 & 0 \\ 0 & 0 & 1 \end{pmatrix} \text{ and } \frac{1}{2} \begin{pmatrix} \tau & -1 & 1/\tau \\ 1 & 1/\tau & -\tau \\ 1/\tau & \tau & 1 \end{pmatrix}$$

give order 2 and order 5 rotations, g_2 and g_5 say, which generate its rotation group Y. Each of these matrices has $(\tau, 1, 1/\tau)'$ and $(-1, 1, 1)'$ as right eigenvectors mod 2 and $(-\tau, 1, 1/\tau)$ as a left eigenvector mod 2 (with unit eigenvalues in each case), so

$$\{(\alpha, \beta, \gamma) \in \mathbb{Z}[\tau]^3 \mid \tau\alpha + \beta + (1/\tau)\gamma \equiv 0 \pmod{2}\}$$

is a Y-invariant $\mathbb{Z}[\tau]$-module. (This module can also be expressed as

$$\mathbb{Z}[\tau](\tau, 1, 1/\tau)' + \mathbb{Z}[\tau](-1, 1, 1)' + 2\mathbb{Z}[\tau]^3,$$

so the right eigenvectors show that g_5 maps the module into $\mathbb{Z}[\tau]^3$, even though its matrix has denominator 2. The left eigenvector then shows that g_2 and g_5 map the module into itself.) Since $(1, 1, -1)$ is also a left eigenvector mod 2 with eigenvalue 1 for both matrices, we have additional G-invariant modules

$$\{(\alpha, \beta, \gamma) \in \mathbb{Z}[\tau]^3 \mid \tau\alpha + \beta + (1/\tau)\gamma \equiv 0 \pmod{2}, \alpha + \beta + \gamma \equiv 0 \text{ or } 1 \pmod{2}\}$$

and

$$\{(\alpha, \beta, \gamma) \in \mathbb{Z}[\tau]^3 \mid \tau\alpha + \beta + (1/\tau)\gamma \equiv 0 \pmod{2}, \alpha + \beta + \gamma \equiv 0 \pmod{2}\}.$$

Each of these three modules has index 2 in the preceding one. (In $\mathbb{Z}[\tau]$ there are 4 residue classes mod 2, represented by 0, 1, τ and $1/\tau$.) The first and last are $\mathbb{Z}[\tau]$-modules so have coefficient ring $\mathbb{Z}[\tau]$, allowing the choice of Pisot unit multiplier $\mu = \tau$. The second is only a $\mathbb{Z}$-module and has coefficient ring $\mathbb{Z}[2\tau]$. For this the smallest possible Pisot unit inflation multiplier is τ^3. With an appropriate choice of basis for W the corresponding lattices are scaled versions of D_6^*, $\mathbb{Z}^6$ and D_6 from which the three classes of icosahedral model set, as described in [**RMW**], can be obtained. When the window Ω is chosen to be invariant under the full icosahedral symmetry group Y_h (for example, Ω can be a regular icosahedron, dodecahedron or rhombic triacontahedron) the resulting model set also has the full group of symmetries Y_h. In both these examples the extra symmetry is given by the matrix $-I$, which we have not explicitly mentioned but which is always a symmetry of every module.

4. Modifying Plain Cut-And-Project Sets

We would like to have quasicrystals with a local refining inflation, particularly as this gives an efficient way of generating the entire set from some finite segment of it. However, Proposition 2.35 shows that this is impossible for a model set whose window is a polytope when the inflation multiplier is not a quadratic algebraic integer. Nevertheless, it is possible to use the nonlocal inflation of a model set as a springboard to define a local inflation that generates a set retaining most of the properties of the original model set.

PROPOSITION 4.1. *Let $\mathcal{M}$ be a model set derived from V, W, $\mathcal{L}$ and Ω, which satisfy the V-Condition and the W-Condition and have $\mathbf{0} \in \Omega$. Suppose also that $\mathcal{M}$ has a refining inflation by a Pisot number μ with $\mathbf{t} = \mathbf{0}$, derived from an operator T, as in Proposition 2.31, and has a group G of symmetries, derived from operators T_g $(g \in G)$, as in Proposition 2.29. Then for every $\varepsilon > 0$ there is a repetitive Meyer set $\mathcal{Q}$ that has all the symmetries in G, has a local refining inflation, and is such that $\mathcal{M} \bigtriangleup \mathcal{Q}$ has upper uniform frequency $< \varepsilon$.*

PROOF. We may assume that $\mathcal{M}$ is nonsingular by replacing Ω with $(1+\delta)\Omega$, if necessary, for some sufficiently small $\delta > 0$. The resulting model set retains the symmetries and refining inflation of $\mathcal{M}$, but the G-invariance is now strict. The frequency of the extra points introduced in this way can be kept arbitrarily small by choosing δ small enough. By taking a high power of the inflation we may also assume that μ is as large as we like and its conjugates as small as we like. Then $\mathcal{Q}$ is constructed like this. Let $\mathcal{Q}_1$ be the patch $P_{q_1}(\mathbf{0})$ of $\mathcal{M}$ for sufficiently large q_1 and divide $\mu\mathcal{Q}_1$ into Voronoï cells.[5] In each closed Voronoï cell $V(\boldsymbol{x})$ corresponding to $\boldsymbol{x} \in \mu\mathcal{Q}_1$ that does not meet the boundary of $B_{q_1}(\mathbf{0})$ place the points $(\mathcal{M}+\boldsymbol{x}) \cap V(\boldsymbol{x})$, and call the union of all these points $\mathcal{V}_2$. These Voronoï cells completely cover $B_{q_2}(\mathbf{0})$, where $q_2 = \mu(q_1 - R)$ and R is an upper bound for the covering radius of $\mathcal{M}$. Now put $\mathcal{Q}_2 = \mathcal{V}_2 \cap B_{q_2}(\mathbf{0})$ and repeat the process to obtain a point set $\mathcal{V}_3$ contained in a union of Voronoï cells of $\mu\mathcal{Q}_2$ and a radius q_3 such that $B_{q_3}(\mathbf{0})$ is contained in the union of these Voronoï cells, and put $\mathcal{Q}_3 = \mathcal{V}_3 \cap B_{q_3}(\mathbf{0})$. Continuing in this way we get a sequence of point sets $\{\mathcal{Q}_i\}$. If there is a uniform

---

[5]The *Voronoï cell* of $\boldsymbol{x}$ in $\mathcal{D}$ is the polytope consisting of all points in $\mathbb{E}^n$ that are closer to $\boldsymbol{x}$ than to any other point of $\mathcal{D}$. Other names for it are the *Dirichlet region*, the *Brillouin zone* or the *Wigner-Seitz cell* of $\boldsymbol{x}$ in $\mathcal{D}$.

radius R such that, for each i, there is at least one point of $\mathcal{V}_i$ in each ball of radius R that is contained in the union of the Voronoï cells used to form $\mathcal{V}_i$, then we can take $q_{i+1} = \mu(q_i - R)$ for each i, so that if $q_1 > 2R$ and $\mu > 2$ then q_i is increasing and tends to infinity. If we also choose μ large enough so that the Voronoï cell of $\mathbf{0}$ in $\mathcal{M}$ contains $B_{q_1/\mu}(\mathbf{0})$ then $\mathcal{Q}_i \subset \mathcal{Q}_{i+1}$ so that there is a limiting set $\mathcal{Q}$. In this case too, $\mu \mathcal{Q}_i \subset \mathcal{Q}_{i+2}$ for each i, so $\mu \mathcal{Q} \subset \mathcal{Q}$ and μ gives a refining inflation on $\mathcal{Q}$. We shall establish the existence of the radius R shortly. Since $\mathcal{M}$ is strictly G-invariant each $\mathcal{Q}_i$ is G-invariant and hence so is $\mathcal{Q}$.

Since each point $\boldsymbol{x}$ of $\mu \mathcal{Q}_1$ has the form $\pi_V(\boldsymbol{l})$ with $\boldsymbol{l} \in T\mathcal{L}$ and $\pi_W(\boldsymbol{l}) \in T\Omega$, each $\boldsymbol{x} \in \mathcal{V}_2$ has the form $\pi_V(\boldsymbol{l}')$ with $\boldsymbol{l}' \in T\mathcal{L} + \mathcal{L} = \mathcal{L}$ (since $T\mathcal{L} \subseteq \mathcal{L}$) and $\pi_W(\boldsymbol{l}') \in T\Omega + \Omega$. In view of the W-Condition π_V is one-one on $\mathcal{L}$ so there is a unique lifting $\lambda \colon \pi_V(\mathcal{L}) \to \mathcal{L}$ and we have

$$\pi_W(\lambda(\mathcal{V}_2)) \subset T\Omega + \Omega.$$

Similarly $\pi_W(\lambda(\mu \mathcal{Q}_2)) \subseteq \pi_W(\lambda(\mu \mathcal{V}_2)) \subset T^2\Omega + T\Omega$ and hence

$$\pi_W(\lambda(\mathcal{V}_3)) \subset T^2\Omega + T\Omega + \Omega,$$

and so on. Since all eigenvalues of T on W have absolute value < 1, the diameter of $T^k\Omega$ tends exponentially to zero as $k \to \infty$, so for any $\eta > 0$ we can arrange (by replacing T by a sufficiently large power of T if necessary) that

$$(4.1) \qquad \pi_W(\lambda \mathcal{Q}) \subset \sum_{k=0}^{\infty} T^k \Omega \subset \Omega + B_\eta(\mathbf{0}).$$

In the converse direction, each point $\boldsymbol{l} \in \mathcal{L}$ with $\pi_V(\boldsymbol{l})$ in the Voronoï cells of $\mu \mathcal{Q}_1$ and $\pi_W(\boldsymbol{l}) \in \Omega - \boldsymbol{w}$ for all $\boldsymbol{w} \in T\Omega$ has $\pi_V(\boldsymbol{l}) \in \mathcal{V}_2$, then each $\boldsymbol{l} \in \mathcal{L}$ with $\pi_V(\boldsymbol{l})$ in the Voronoï cells of $\mu \mathcal{Q}_2$ and $\pi_W(\boldsymbol{l}) \in \Omega - \boldsymbol{w}$ for all $\boldsymbol{w} \in T^2\Omega + T\Omega$ has $\pi_V(\boldsymbol{l}) \in \mathcal{V}_3$, and so on. In the limit we have

$$(4.2) \qquad \pi_W(\lambda \mathcal{Q}) \supset \pi_W(\mathcal{L}) \cap \left(\Omega \setminus (\partial \Omega + B_\eta(\mathbf{0}))\right).$$

One consequence of this is that if we choose η small enough for $\Omega \setminus (\partial \Omega + B_\eta(\mathbf{0}))$ to have positive Riemann measure then Lemma 2.9 ensures the existence of the radius R, the same for all the sets $\mathcal{V}_i$, required above. Another consequence is that, by (4.1), (4.2) and Lemma 2.9, $\mathcal{M} \triangle \mathcal{Q}$ has upper frequency $< \varepsilon$ if η is small enough. Also (4.1) shows that $\mathcal{Q} - \mathcal{Q}$ is uniformly discrete, and (4.2) that $\mathcal{Q}$ is relatively dense, so $\mathcal{Q}$ is a Meyer set.

Now choose μ large enough so that $\mu r > q_1$, where r is the packing radius of the model set derived from the window $\Omega + B_\eta(\mathbf{0})$. (There is no circularity about this choice, since increasing μ only decreases η.) This ensures that a single inflation of every Voronoï cell contains a copy of $\mathcal{Q}_1$ (whose iterated inflation generates the whole of $\mathcal{Q}$) and hence that $\mathcal{Q}$ is repetitive.

Finally we need to show that $\mathcal{Q}$ is locally derivable from $\mu \mathcal{Q}$. If we take $\rho = 3\mu R$ then the ρ-patch of $\boldsymbol{x}$ in $\mu \mathcal{Q}$ determines all the Voronoï cells of $\mu \mathcal{Q}$ that meet $B_{2\mu R}(\boldsymbol{x})$ and hence determines the $2\mu R$-patch of $\boldsymbol{x}$ in $\mathcal{Q}$. This gives a function from $\mathcal{A}_{\mu \mathcal{Q}}(\rho)$ to $\mathcal{A}_\mathcal{Q}(2\mu R)$ (we take all $\boldsymbol{t}$'s to be $\mathbf{0}$ in Definition 1.3) which establishes the local derivability, since the covering radius of $\mu \mathcal{Q}$ is $\leq \mu R$. □

REMARK 4.2. The assumptions of Proposition 4.1 imply that the operator T commutes with T_g for each $g \in G$. To see this, note that since the linear operator $T - T_g^{-1} T T_g$ annihilates V its image lies in W. But it also maps $\mathcal{L}$ to itself and hence annihilates $\mathcal{L}$ too because of the W-Condition. Since $\mathcal{L}$ spans $V \oplus W$ the operator is identically zero.

Although the set $\mathcal{Q}$ provided by Proposition 4.1 is repetitive there seems a difficulty in proving that it is uniform, related to the fact that a Voronoï cell of $\mathcal{Q}$ may overlap more than one Voronoï cell of $\mu\mathcal{Q}$, so that the cells of $\mu\mathcal{Q}$ do not normally partition into cells of $\mathcal{Q}$. This difficulty can be circumvented when $\mathcal{M}$ has Ammann hyperplanes. If $\mathcal{M}$ has systems of Ammann hyperplanes that partition $\mathbb{E}^n$ into bounded polytopes, or "Ammann cells", and μ gives a refining inflation on $\mathcal{M}$ then the Ammann cells of $\mu\mathcal{M}$ do partition into Ammann cells of $\mathcal{M}$ and can be used to define a substitution-generated tiling of $\mathbb{E}^n$. Ammann cells provide an alternative to Voronoï cells as a way of deriving a tiling from a point set. The partitions of the individual Ammann cells to be used for the substitution can be derived from substitutions on each system of Ammann hyperplanes which in turn can be got from the 1-dimensional case of Proposition 4.1. The argument follows very closely Section 6 of [**LuP**], which deals with tilings rather than point sets and makes heavy use of "prism patterns" which are 1-dimensional model sets.

PROPOSITION 4.3. *Let $\mathcal{M}$ be as in Proposition 4.1 and suppose that it satisfies the Strong V-Condition and the Strong W-Condition and that the window Ω is convex and centrally symmetric about $\mathbf{0}$. Suppose also that $\mathcal{M}$ has k systems of Ammann hyperplanes, whose normals span $\mathbb{E}^n$, each derived from a full $(n-1)$-dimensional submodel $\mathcal{M}'$ of $\mathcal{M}$, as in Proposition 2.27, whose corresponding sublattice $\mathcal{L}'$ is T-invariant. Then there is a uniform Meyer set $\mathcal{Q}$ that has all the symmetries in G, has a local refining inflation, and has k systems of Ammann hyperplanes parallel to the Ammann systems of $\mathcal{M}$.*

PROOF. Again we may assume that $\mathcal{M}$ is nonsingular. We may also assume that the k systems of Ammann hyperplanes are closed under the action of G, by including, with each Ammann system, all its distinct images under the action of T_g for $g \in G$. These images are clearly Ammann systems and since T_g commutes with T the corresponding sublattices $T_g \mathcal{L}'$ are T-invariant. The central symmetry of Ω implies that each individual Ammann system is centrally symmetric, so that any image which is parallel to its original system is in fact identical to its original. We assume too that $n \geq 2$, since when $n = 1$ the fact that the set $\mathcal{Q}$ given by Proposition 4.1 is repetitive and is generated from two central intervals by iterating a locally determined subdivision process ensures that it is uniform, by the argument used for higher dimensional tilings below, and Ammann points are automatic for uniform Meyer sets.

First consider one particular system of hyperplanes given by the full submodel V', W', $\mathcal{L}'$, $\Omega \cap W'$, where $\dim V' = n - 1$ and $\mathcal{L}'$ is T-invariant. According to the proof of Proposition 2.25 the set $\mathcal{T}$ consisting of the intersection points $\mathbf{t}$ of these Ammann hyperplanes with the line $(V')^\perp$ orthogonal to V' is given by the quotient model V/V', W/W', $\mathcal{L}/\mathcal{L}'$, Ω'', where Ω'' is the image of Ω in W/W'. The quotient inherits the V-Condition from $\mathcal{M}$ and satisfies the W-Condition because of the Strong W-Condition on $\mathcal{M}$. Since $\mathcal{L}'$ (and hence $V' \oplus W'$) is T-invariant, T acts on $\mathcal{L}/\mathcal{L}'$, V/V' and W/W'. On V/V' it acts as scalar multiplication by μ, its eigenvalues on W/W' are the conjugates of μ, and from $T\Omega \subseteq \Omega$ it follows that $T\Omega'' \subseteq \Omega''$. We can now apply Proposition 4.1 to this quotient model to obtain a Meyer set $\mathcal{U} \subset V/V'$ that has a local refining inflation by μ^a, for some $a \in \mathbb{N}$, and is such that $\mathcal{T} \triangle \mathcal{U}$ has upper uniform frequency $< \varepsilon$. Also the proof of Proposition 4.1 shows that each point of $\mathcal{U}$ is the (V/V')-projection of some point of $\mathcal{L}/\mathcal{L}'$ whose (W/W')-projection lies in the bounded region $(1+\eta)\Omega''$ and, in the converse direction, that the (V/V')-projection of every point of $\mathcal{L}/\mathcal{L}'$ whose (W/W')-projection lies in $(1-\eta)\Omega''$ is in $\mathcal{U}$, where η can be made as small as we like by choosing a large enough. (The η here is not the same as the η in the proof of Proposition 4.1, but is related to it. The sets $\Omega'' + B(\mathbf{0})$ and $\Omega'' \setminus (\partial\Omega'' + B(\mathbf{0}))$ of Proposition 4.1 are sandwiched between sets of the form $(1 \pm \eta)\Omega''$ because of the convexity and central symmetry of Ω''.) The 1-dimensional set $\mathcal{T}$ divides the line into intervals and the boundaries of the Voronoï cells for $\mathcal{T}$ are the mid-points of these intervals. So the derivation of $\mathcal{U}$ from $\mu^a \mathcal{T}$ just consists of filling the left half of each interval of $\mu^a \mathcal{T}$ by the segment of $\mathcal{T}$ of the same length starting from the origin and moving right and the right half of the interval by the mirror image segment of $\mathcal{T}$ to the left of the origin. Since Ω is symmetric about $\mathbf{0}$ so is Ω'', and hence $\mathcal{T}$ has mirror symmetry about the origin which is preserved by the construction of $\mathcal{U}$.

We now consider the tiling of V got by slicing along the hyperplanes parallel to V' through the points of $\mathcal{U}$, for all the sets $\mathcal{U}$ obtained from the different systems of Ammann hyperplanes. Since the normals to these systems span V and the spacings between the points of each set $\mathcal{U}$ are bounded, the tiles are of bounded size. If C is the cluster of tiles with the origin as a vertex in the tiling got from $\mathcal{T}$ instead of $\mathcal{U}$ (i.e., slicing by the Ammann hyperplanes of $\mathcal{M}$) then the uniformity of $\mathcal{M}$ implies that C occurs with positive uniform frequency in the $\mathcal{T}$-tiling. Since the frequency of each $\mathcal{T} \triangle \mathcal{U}$ is $< \varepsilon$, the frequency of the clusters C that meet a slicing hyperplane of the $\mathcal{U}$-tiling that is not a slicing hyperplane of the $\mathcal{T}$-tiling (or vice versa) is $< c_1 \varepsilon$, where c_1 depends only on $\mathcal{M}$. So by choosing ε small enough we can ensure that the cluster C occurs with positive upper uniform frequency in the $\mathcal{U}$-tiling. We note that, by the way $\mathcal{U}$ was constructed, the cluster C occurs in particular at the origin of the $\mathcal{U}$-tiling. The $\mathcal{U}$-tiling is G-invariant because the sets $\mathcal{T}$ are closed under the action of G, each $\mathcal{U}$ is centrally symmetric, and if $\mathcal{T}$ gives rise to $\mathcal{U}$ then $g\mathcal{T}$ gives rise to $g\mathcal{U}$.

To show that the $\mathcal{U}$-tiling has finite local complexity consider its vertex set $\mathcal{V}$. This consists of the intersection points of all those sets of slicing hyperplanes that meet in a single point. Each point $\boldsymbol{x}$ of $\mathcal{V}$ has

(4.3) $$\boldsymbol{x} \in V'_i + \pi_V(\boldsymbol{l}_i) \quad (i = 1, \ldots, n),$$

where $V'_1, \ldots, V'_n$ are n of the $(n-1)$-dimensional subspaces V' with $V'_1 \cap \cdots \cap V'_n = \{\mathbf{0}\}$ and the $\boldsymbol{l}_i$'s are in $\mathcal{L}$. So $\boldsymbol{x} = \pi_V(\boldsymbol{\lambda})$ where

(4.4) $$\boldsymbol{\lambda} = \boldsymbol{\lambda}_i + \boldsymbol{l}_i \quad \text{and} \quad \boldsymbol{\lambda}_i \in V'_i \oplus W'_i \quad (i = 1, \ldots, n).$$

Now
$$\pi_V\big((V'_1 \oplus W'_1) \cap \cdots \cap (V'_n \oplus W'_n)\big) = V'_1 \cap \cdots \cap V'_n = \{\mathbf{0}\}$$

so
$$(V'_1 \oplus W'_1) \cap \cdots \cap (V'_n \oplus W'_n) \subseteq W$$

which in view of the W-Condition gives

(4.5) $$(V'_1 \oplus W'_1) \cap \cdots \cap (V'_n \oplus W'_n) = \{\mathbf{0}\}$$

since this subspace is an $\mathcal{L}$-subspace. When expressed in terms of a basis of $\mathcal{L}$, (4.4) can be written as a set of linear equations for $\boldsymbol{\lambda}$ with rational coefficients, which by (4.5) have a unique solution. Hence there is an $m \in \mathbb{N}$, independent of the constant terms $\boldsymbol{l}_i$, such that $\boldsymbol{\lambda} \in m^{-1}\mathcal{L}$. We can also choose m to be independent of the particular set of hyperplanes $V'_1, \ldots, V'_n$ by taking a common multiple of the m's for the individual sets. The point x of (4.3) also has the image of $\pi_W(l_i)$ lying in the bounded region $(1+\eta)\Omega''_i$ of W/W'_i for $i = 1, \ldots, n$. Since $W'_i = \pi_W(V'_i \oplus W'_i)$ for each i, (4.5) gives $W'_1 \cap \cdots \cap W'_n = \{0\}$ and it follows that $\pi_W(l_i)$ lies in a bounded region of W. Thus $\mathcal{V}$ is a subset of the cut-and-project set with this bounded region as window and $m^{-1}\mathcal{L}$ as lattice, so has finite local complexity by Proposition 2.6. Since its vertex set has finite local complexity, so does the tiling.

We now consider how inflation works on the $\mathcal{U}$-tiling. Its scale is expanded by a factor μ^a and then each tile is subdivided in a manner that depends on its position relative to the neighbouring slicing hyperplanes on either side of it in each of the k systems. Tiles of the same shape may have different types for the purpose of subdividing, but the fact that the tiling has finite local complexity ensures that there are finitely many types of tile, even taking this into account. It is also clear that the type of a tile determines the types of all the tiles it subdivides into. Consequently the process of iterated inflation starting from a single tile is a finite Markov process, the numerical proportions of the various types of tile in an inflated tile corresponding to the transition probabilities.[6]

Such inflation processes for tilings are discussed in Section 3 of [**LuP**], for example. The $\mathcal{U}$-tiling itself is the limit tiling that results from starting with the central cluster C and iterating this inflation. Because the cluster C does not have uniform frequency zero in the tiling no tile occurs with uniform frequency zero, so by Lemma 15 of [**LuP**] the inflation is irreducible in the sense that no proper subset of the tiles is closed under the inflation operation. (In the language of Markov processes, there are no transient states.) In fact, because the inflation of every tile in C contains a copy of itself, the inflation operation is primitive (i.e., every power of the operation is irreducible). By Lemma 2 of [**LuP**] the irreducibility of the inflation operation implies that the tiling is uniform. (Strictly speaking, Lemma 2 of [**LuP**] applies to tilings generated from a single tile, whereas the cluster C contains several tiles. However, since C occurs with positive frequency it is possible to generate a locally isomorphic tiling from a single tile, whose uniformity implies the uniformity of the original tiling.)

We notice that the same argument shows that the sets $\mathcal{U}$ themselves are uniform and that so is the $(n-1)$-dimensional tiling on any slicing hyperplane through $\mathbf{0}$ got by slicing with the hyperplanes of the other systems. In the latter case not only may tiles of the same shape have different inflation types but even tiles of the same shape similarly situated with respect to all neighbouring $(n-2)$-dimensional subspaces, since parallel $(n-2)$-dimensional subspaces may arise from nonparallel slicing hyperplanes in $\mathbb{E}^n$. Nevertheless, there is only a finite number of inflation types of tiles and the types of the subdivisions of a tile depend only on the type of the tile. But now it may not be possible to identify the type of a tile from the

---

[6]The transition matrix described here differs slightly from the counting matrix used in [**LuP**], whose entries are the numbers of the various kinds of tile in an inflated tile instead of the proportions. What we describe here is equivalent to conjugating the counting matrix by the diagonal matrix whose entries are the volumes of the tiles to get a matrix that is stochastic.

This is a convenient place to correct an error in [**LuP**]. The statement there (Remark 2 on p. 234) that the frequencies of tiles are proportional to their volumes is wrong and results from the howler of confusing left and right eigenvectors: for the counting matrix the right eigenvector of the maximum eigenvalue consists of the frequencies of the tiles and the left eigenvector consists of their volumes.

geometry of the tiling: identifying the type may require knowledge of which slicing hyperplane in $\mathbb{E}^n$ intersects the tiling in a given $(n-2)$-dimensional cut.

We now define the point set $\mathcal{Q}$ as the points that lie on k slicing hyperplanes of the $\mathcal{U}$-tiling (one hyperplane of each system), so that $\mathcal{Q} \subseteq \mathcal{V}$. Since $\mathcal{Q}$ is uniquely determined by the $\mathcal{U}$-tiling, which is G-invariant, every $g \in G$ is a symmetry of $\mathcal{Q}$. Also $\mathcal{Q}$ inherits a refining inflation with multiplier μ^a from the 1-dimensional sets $\mathcal{U}$ and inherits finite local complexity and uniformity from the corresponding properties of the $\mathcal{U}$-tiling.

We need to show that the slicing hyperplanes of the $\mathcal{U}$-tiling are Ammann hyperplanes of $\mathcal{Q}$. First note that if V_1', V_2' are $(n-1)$-dimensional subspaces of V parallel to two different systems of Ammann hyperplanes then the corresponding subspaces W_1', W_2' of W satisfy $W_1' + W_2' = W$. This is because $V_1' + V_2' = V$ (since $n \geq 2$) so $(V_1' \oplus W_1') + (V_2' \oplus W_2') \supseteq V$, which in view of the V-Condition and the fact that $(V_1' \oplus W_1') + (V_2' \oplus W_2')$ is an $\mathcal{L}$-subspace gives $(V_1' \oplus W_1') + (V_2' \oplus W_2') = V + W$ and hence $W_1' + W_2' = W$. Now let $A = V_1' + \pi_V(l)$ ($l \in \mathcal{L}$) be one of the slicing hyperplanes of the $\mathcal{U}$-tiling, $W_1' + \pi_W(l)$ the corresponding affine subspace of W and Ω_1'' the image of Ω in W/W_1'. Then $W_1' + \pi_W(l) \in (1+\eta)\Omega_1''$, so $W_1' + \pi_W\bigl((1-2\eta)l\bigr)$ has nonempty intersection with $(1-\eta)\Omega$ and by the hypotheses of Proposition 2.27 (and the convexity and central symmetry of Ω) this intersection contains the ball $B_{(1-\eta)\psi}(\mathbf{0})$ in $W_1' + \pi_W\bigl((1-2\eta)l\bigr)$. From the fact that $W_1' + W_2' = W$ it follows that there is a constant c_2, depending only on the directions of the Ammann systems, such that any subspace $W_2' + \mathbf{w}$ of W (corresponding to an Ammann hyperplane not parallel to V_1) which meets the ball $B_{(1-c_2\eta)\psi}(\mathbf{0})$ in $W_1' + \pi_W(l)$ also meets the ball $B_{(1-\eta)\psi}(\mathbf{0})$ in $W_1' + \pi_W\bigl((1-2\eta)l\bigr)$. Hence any point $l' \in \mathcal{L} \cap (V_1' + W_1') + l$ with $\pi_W(l') \in B_{(1-c_2\eta)\psi}(\mathbf{0})$ has $\pi_V(l') \in A$ and also $\pi_V(l')$ belonging to a slicing hyperplane from each of the other systems in the $\mathcal{U}$-tiling. So $\pi_V(l') \in \mathcal{Q}$. When η is chosen less than $1/2c_2$ the Strong V-Condition shows that the points of $\mathcal{Q}$ on A have a subset with uniform frequency bounded away from 0, so the covering radius of $\mathcal{Q}$ on A is bounded. Having finite covering radius on A and being a subset of the Meyer set $\mathcal{V}$, $\mathcal{Q} \cap A$ is a Meyer set on A. To show that $\mathcal{Q} \cap A$ is uniform we relate it to the $\mathcal{U}$-tiling on A_0, the hyperplane through $\mathbf{0}$ parallel to A. There is a constant c_3, depending on the distance of A from A_0 and the angles between the systems of hyperplanes, such that for any $\mathbf{a} + \mathbf{x} \in A$ (where $\mathbf{a} \perp A$ and $\mathbf{x} \in A_0$) and $\rho > 0$ the ρ-patch of $\mathcal{Q} \cap A$ about $\mathbf{a} + \mathbf{x}$ is determined by the region of the tiling of A_0 within a distance $\rho + c_3$ of $\mathbf{x}$. (Regions are equivalent for this purpose only when the inflation types of the tiles in the regions agree, as well as the tiles themselves.) The uniformity of $\mathcal{Q} \cap A$ on A now follows from the uniformity of the $\mathcal{U}$-tiling on A_0. To complete the proof that the slicing hyperplanes are Ammann hyperplanes of $\mathcal{Q}$ we note that each set $\mathcal{U}$ is a uniform Meyer set and that, by the definition of $\mathcal{Q}$, every point of $\mathcal{Q}$ lies on a slicing hyperplane of each system.

It remains to verify that the inflation on $\mathcal{Q}$ is local. Let R be a strict upper bound for the covering radii of the sets $\mathcal{U}$ and for the covering radii of the intersections of $\mathcal{Q}$ with each of the slicing hyperplanes of the $\mathcal{U}$-tiling. Then $\mathcal{Q}$ itself has covering radius $< 2R$. Also let ρ' be the largest of the ρ-parameters in the local derivation of $\mathcal{U}$ from $\mu^a \mathcal{U}$, for all the sets $\mathcal{U}$. If we take $\rho = \rho' + 4\mu^a R$ then the ρ-patch in $\mu^a \mathcal{Q}$ of any $\mathbf{x}$ in $\mu^a \mathcal{U}$ determines all the hyperplanes of the $\mathcal{U}$-tiling magnified by a factor μ^a that meet $B_{\rho' + 2\mu^a R}(\mathbf{x})$. These in turn determine the hyperplanes of the $\mathcal{U}$-tiling itself that meet $B_{2\mu^a R}(\boldsymbol{x})$, showing that $\mathcal{Q}$ is locally derivable from $\mu^a \mathcal{Q}$. □

THEOREM 4.4. *Let G be an arbitrary finite isometry group acting on $\mathbb{E}^n$ and $K \neq \mathbb{Q}$ any real algebraic number field over which G has an equivalent representation. Then there is a discrete point set $\mathcal{Q}$ that has Properties 1–3, 5, 9, 10 and 15, has each $g \in G$ as a symmetry, has a local refining inflation with a multiplier μ such that $\mathbb{Q}(\mu) = K$, and has systems of Ammann hyperplanes whose normals span $\mathbb{E}^n$. If there is no crystal in $\mathbb{E}^n$ having $\langle g^{\delta(g)} \mid g \in G \rangle$ as symmetries then there exists such a $\mathcal{Q}$ which has weak local rules.*

PROOF. Except for Properties 5 and 15 this follows from Theorem 3.5 and Proposition 4.3. For Property 5, if $\boldsymbol{p}$ is a period of $\mathcal{Q}$ then $\mu\boldsymbol{p}$ is a period of $\mu\mathcal{Q}$ and hence also of $\mathcal{Q}$, since $\mathcal{Q}$ is locally derivable from $\mu\mathcal{Q}$. It follows that $(a\mu+b)\boldsymbol{p}$ is a period of $\mathcal{Q}$ for all $a, b \in \mathbb{Z}$. Since the set of periods of $\mathcal{Q}$ is necessarily discrete and μ is irrational this implies that $\boldsymbol{p} = \boldsymbol{0}$. Property 15 is now a consequence of Corollary 1.7. □

5. Is There a Quasicrystallographic Restriction?

In Section 1 we made a list of desirable properties for quasicrystals and in Section 2 investigated, for each property individually, what conditions are required of the cut-and-project construction for the model set to possess that property. A negative result in Section 2 was that a local refining inflation is not possible for a noncrystallographic plain model set with polytopic window unless the multiplier is quadratic. In Section 3 we described how to construct model sets from modules over algebraic number fields and showed that they satisfy all the conditions required in Section 2. In Section 4 we used model sets with a refining inflation as seeds to generate sets for which the inflation is local and which retain most of the properties of the model set: indeed, Theorem 4.4 shows that it is possible to construct point sets that possess all the main quasicrystal properties simultaneously, including local refining inflation but with the possible exception of diffraction.

The only parameters in our list of properties are the symmetry group G and the inflation multiplier μ, and to judge from examples μ appears to be constrained by G. So Theorem 4.4 (and perhaps Theorem 3.5 too, for those who don't mind if inflation is nonlocal but value diffraction) amounts to asserting that for any group G of isometries there is a point set having all the symmetries in G and all the main quasicrystal properties. A succinct way of expressing this is to say that there is no "quasicrystallographic restriction". The *crystallographic restriction* tells us that, for a given dimension n, not every finite subgroup of $O(n)$ is the point group of a crystal. It is usually expressed as a restriction on the possible orders of crystal symmetries. For example, the only orders possible for crystal symmetries in dimensions ≤ 3 are 1, 2, 3, 4 and 6, so that C_5, C_m for $m \geq 7$ and the icosahedral group Y_h are not point groups of crystals. A version of the crystallographic restriction for general n is given in [**Schw**, Chapter 3, Section 3].[7] Theorems 3.5 and 4.4

---

[7]This has an error, pointed out to me by the author, which is corrected in [**P1**].

show that, with fairly tight interpretations of what a quasicrystal[8] is, there is no such restriction on the point groups of quasicrystals.

There is still a question remaining about the possible symmetries of quasicrystals, however. The symmetries of Property 11 are symmetries of an individual point set $\mathcal{Q}$. But the other members of the LI-class of $\mathcal{Q}$ are closely related to it and in a strong sense indistinguishable from it, so a symmetry should be a property of an entire LI-class. The usual definition of the point symmetry group of a point set of finite local complexity (but not necessarily crystallographic) is:.

DEFINITION 5.1. The *point symmetry group* G of $\mathcal{D} \subset \mathbb{E}^n$ is the set of all isometries g of $\mathbb{E}^n$ such that, for any patch P of $\mathcal{D}$, $g\mathcal{D}$ is also a patch of $\mathcal{D}$.

Being defined entirely in terms of the atlas $\mathcal{A}_\mathcal{D}$ of patches of $\mathcal{D}$, G depends only on the LI-class of $\mathcal{D}$. It is a group (closed under taking inverses because the number of patches of a given size is finite) and is in fact the symmetry group of $\mathcal{A}_\mathcal{D}$. For an isometry $g \in G$, however, there may be no large region of $\mathcal{D}$ that is invariant, or even approximately invariant, under g. For this reason it seems natural to define "overt" symmetries of a Delone set $\mathcal{D}$ of finite local complexity, also in an LI-invariant manner.

DEFINITION 5.2. The set S of *overt point symmetries* of $\mathcal{D} \subset \mathbb{E}^n$ is the set of all g in $O(n, \mathbb{R}$ that satisfy either of the following two equivalent conditions:

(i) g is a symmetry (in the sense of Property 11) of some point set in the LI-class of $\mathcal{D}$;

(ii) $\min_{\boldsymbol{x} \in \mathbb{E}^n} |((B_\rho(\boldsymbol{x}) \cap \mathcal{D}) - \boldsymbol{x}) \triangle g((B_\rho(\boldsymbol{x}) \cap \mathcal{D}) - \boldsymbol{x})| = o(\rho^n)$ as $\rho \to \infty$.

Version (ii) of this definition says that for arbitrarily large radii ρ there is a ball of radius ρ centred at a point $\boldsymbol{x}$ (not necessarily in $\mathcal{D}$) within which g conjugated to $\boldsymbol{x}$ is an approximate symmetry of the points of $\mathcal{D}$. So overt symmetries show up in every set of the LI-class as large symmetrical regions. When $\mathcal{D}$ is repetitive, $S \subseteq G$. The set S is clearly closed under taking inverses and powers, but the example of the 2-dimensional crystal type $P2mg$ shows that S is not always equal to G and not always a group. For a crystal of this symmetry type the point group G is D_2, but S consists of only 3 of the 4 elements of G: the mirror symmetry associated with the glide reflexion is not an overt symmetry. For the LI-class of the set $\mathcal{M}$ of Theorem 3.5, $S = G$ (though these may be larger than the group G used in the construction).

For crystals there are finitely many types in each dimension (see [**Z**]) which have been listed for low dimensions, so the corresponding pairs G, S can be computed. For quasicrystals we have the question of which pairs G, S are possible and, in particular, whether there is any restriction on the relationship of S to G apart from inclusion and S being closed under taking inverses and powers? It is likely that this question for low dimensional quasicrystals is related to the same question for high dimensional crystals.

References

[B]   M. Baake, *A guide to mathematical quasicrystals*, Quasicrystals (J.-B. Suck, M. Schreiber, and P. Häussler, eds.), Springer, Berlin, 2001 (to appear); math-ph/9904014.

---

[8]As far as I am aware there is currently no generally agreed mathematical definition of "quasicrystal". In view of the wide variety of objects studied under this name it is probably still too soon to lay one down.

[BJS]   M. Baake, D. Joseph, and M. Schlottmann, *The root lattice D_4 and planar quasilattices with octagonal and dodecagonal symmetry*, Internat. J. Modern Phys. B **5** (1991), 1927–1953; mp_arc/00-255.

[BKSZ]   M. Baake, P. Kramer, M. Schlottmann, and D. Zeidler, *Planar patterns with fivefold symmetry as sections of periodic structures in 4-space*, Internat. J. Modern Phys. B **4** (1990), 2217–2268.

[BMP]   M. Baake, R. V. Moody, and P. A. B. Pleasants, *Diffraction from visible lattice points and kth power free integers*, Discrete Math. **221** (2000), 3–42; math.MG/9906132.

[BSchl] M. Baake and M. Schlottmann, *Geometric aspects of tilings and equivalence concepts*, Quasicrystals. Proceedings of the 5th International Conference on Quasycrystals (Avignon, 1995) (C. Janot and R. Mosseri, eds.), World Scientific, Singapore, 1995, pp. 15–21.

[BSJ]   M. Baake, M. Schlottmann, and P. D. Jarvis, *Quasiperiodic tilings with tenfold symmetry and equivalence with respect to local derivability*, J. Phys. A **24** (1991), 4637–4654.

[BD]   G. Bernuau and M. Duneau, *Fourier analysis of deformed model sets*, Directions in Mathematical Quasicrystals (M. Baake and R. V. Moody, eds.), CRM Monogr. Ser., vol. 13, Amer. Math. Soc., Providence, RI, 2000, pp. 43–60, this volume.

[BS]   Z. I. Borevich and I. R. Shafarevich, *Number theory*, Academic Press, New York, 1966.

[C]   J. W. S. Cassels, *An introduction to Diophantine approximation*, Cambridge Tracts in Math. and Math. Phys., vol. 45, Cambridge Univ. Press, Cambridge, 1957.

[Con]   J. H. Conway, *The sensual (quadratic) form*, Carus Math. Monogr., vol. 26, Math. Assoc. Amer., Washington, DC, 1997.

[Cot]   N. Cotfas, *Quasiperiodic patterns defined by G-clusters and their self-similarities*, 1999.

[CV-G]   N. Cotfas and J.-L. Verger-Gaugry, *A mathematical construction of n-dimensional quasicrystals starting from G-clusters*, J. Phys. A **30** (1997), 4283–4291.

[DD]   L. Danzer and N. Dolbilin, *On Delone graphs and certain species of such*, The Mathematics of Long-Range Aperiodic Order (Waterloo, ON, 1995) (R. V. Moody, ed.), NATO ASI Ser. C: Math. Phys. Sci., vol. 489, Kluwer, Dordrecht, 1997, pp. 85–114.

[Do]   N. Dolbilin, *The countability of a tiling family and the periodicity of a tiling*, Discrete Comput. Geom. **13** (1995), 405–414.

[Du]   M. Duneau, *N-dimensional crystallography and the icosahedral group*, Lectures on Quasicrystals (F. Hippert and D. Gratias, eds.), EDP Sciences, Les Ulis, 1994, pp. 153–186.

[H1]   A. Hof, *On diffraction by aperiodic structures*, Commun. Math. Phys. **169** (1995), 25–43.

[H2]   ———, *Diffraction by aperiodic structures*, The Mathematics of Long-Range Aperiodic Order (Waterloo, ON, 1995) (R. V. Moody, ed.), NATO ASI Ser. C: Math. Phys. Sci., vol. 489, Kluwer, Dordrecht, 1997, pp. 239–268.

[H3]   ———, *Uniform distribution and the projection method*, Quasicrystals and Discrete Geometry (Toronto, ON, 1995) (J. Patera, ed.), Fields Inst. Monogr., vol. 10, Amer. Math. Soc., Providence, RI, 1998, pp. 201–206.

[La]   J. C. Lagarias, *Meyer's concept of quasicrystal and quasiregular sets*, Commun. Math. Phys. **179** (1996), 365–376.

[LaP1]   J. C. Lagarias and P. A. B. Pleasants, *Repetitive Delone sets and quasicrystals*, Ergodic Theory & Dynam. Systems (to appear); math.DS/9909033.

[LaP2]   ———, *Local complexity characterizations of ideal crystals* (in preparation).

[LuP]   W. F. Lunnon and P. A. B. Pleasants, *Quasicrystallographic tilings*, J. Math. Pures Appl. (9) **66** (1987), 217–263.

[M]   R. V. Moody, *Meyer sets and their duals*, The Mathematics of Long-Range Aperiodic Order (Waterloo, ON, 1995) (R. V. Moody, ed.), NATO ASI Ser. C: Math. Phys. Sci., vol. 489, Kluwer, Dordrecht, 1997, pp. 403–441.

[MP]   R. V. Moody and J. Patera, *Quasicrystals and icosians*, J. Phys. A **26** (1993), 2829–2853.

[P1]   P. A. B. Pleasants, *Quasicrystallography: some interesting new patterns*, Elementary and Analytic Theory of Numbers (Warsaw, 1982) (H. Iwaniec, ed.), Banach Center Publ., vol. 17, PWN, Warsaw, 1985, pp. 439–461.

[P2] _____, *The construction of quasicrystals with arbitrary symmetry group*, Quasicrystals. Proceedings of the 5th International Conference on Quasicrystals (Avignon, 1995) (C. Janot and R. Mosseri, eds.), World Scientific, Singapore, 1995, pp. 22–30.

[P3] _____, *Abstract patch calculus* (in preparation).

[RMW] D. S. Rokhsar, N. D. Mermin, and D. C. Wright, *Rudimentary quasicrystallography: the icosahedral and decagonal reciprocal lattices*, Phys. Rev. B **35** (1987), 5487–5495.

[Schl1] M. Schlottmann, *Cut-and-project sets in locally compact Abelian groups*, Quasicrystals and Discrete Geometry (Toronto, ON, 1995) (J. Patera, ed.), Fields Inst. Monogr., vol. 10, Amer. Math. Soc., Providence, RI, 1998, pp. 247–264.

[Schl2] _____, *Generalized model sets and dynamical systems*, Directions in Mathematical Quasicrystals (M. Baake and R. V. Moody, eds.), CRM Monogr. Ser., vol. 13, Amer. Math. Soc., Providence, RI, 2000, pp. 143–159, this volume.

[Schw] R. L. E. Schwarzenberger, *N-dimensional crystallography*, Res. Notes Math., vol. 41, Pitman, London, 1980.

[Sen] M. Senechal, *Quasicrystals and geometry*, Cambridge Univ. Press, Cambridge, 1995; corrected reprint, 1996.

[Ser] J.-P. Serre, *Linear representations of finite groups*, Grad. Texts in Math., vol. 42, Springer, New York, 1977; 4th corrected printing, 1993.

[Z] H. Zassenhaus, *Neuer Beweis der Endlichkeit der Klassenzahl bei unimodularer Äquivalenz endlicher ganzzahliger Substitutionsgruppen*, Abh. Math. Sem. Hamburg **12** (1938), 276–288.

DEPARTMENT OF MATHEMATICS AND COMPUTING SCIENCE, UNIVERSITY OF THE SOUTH PACIFIC, SUVA, FIJI

E-mail address: `pleasants_p@usp.ac.fj`

Generalized Model Sets and Dynamical Systems

Martin Schlottmann

ABSTRACT. It is shown that the dynamical systems approach to the diffraction properties of model sets can be generalized to regular model sets in arbitrary σ-compact Abelian groups with arbitrary locally compact Abelian groups as internal spaces. It is then shown that these regular model sets possess pure point diffraction spectra.

Introduction

Model sets have been of interest for physicists since the observation of crystallographically anomalous Bragg diffraction spectra with characteristic symmetries in connection with so-called quasi-crystals. Occurring in different disguise as the cut-and-project method, the klotz construction, and several other brands, which have been developed independently [**KN, DK, KKL**], it turned out to be possible to consider all these methods as special cases of a construction introduced by Meyer [**Me**], dubbed "model sets".

Although the understanding of the quasicrystal diffraction spectra has been the primary goal of the application of model sets in physics, it took until the late nineties (in comparison with the discovery of quasicrystals in 1984) before the Bragg character of the diffraction of model sets was rigorously established under fairly general conditions by Hof [**H**]. Hof used a beautiful trick invented in a different context by Dworkin [**D**] which connects the diffraction spectrum with the spectrum of a specific dynamical system constructed from the point sets.

The present paper is devoted to a further generalization of this scheme which has already proved fruitful in the study of substitution tilings [**BMS**]. The class of model sets used in physical applications is usually restricted to the case of Euclidean space as "internal space". This restriction seems, in the spirit of Meyer's book, unnecessary and even unnatural, and, according to a suggestion by Moody [**M**], it is immediately clear that the theory should extend to general locally compact Abelian groups as internal spaces. This impression is confirmed by the observation

---

2000 *Mathematics Subject Classification.* Primary 11K70, 37A30; Secondary 43A25, 52C23.

This work was supported by the Pacific Institute of Mathematical Sciences (PIMS) and the Natural Sciences and Engineering Research Council of Canada (NSERC).

It is my pleasure to express my thanks to M. Baake and R. V. Moody for valuable discussions and encouragement.

©2000 Martin Schlottmann

that model sets with these general internal spaces can be characterized by a natural criterion which is formulated exclusively in terms of the "direct space" [**Sch**]. The latter paper gives some preliminary results mostly concerning the geometrical properties of general model sets and the existence of relative frequencies of local patterns. Here we continue the development by extending Hof's analysis to the general case, the main result (Theorem 4.5) establishing the pure point nature of the dynamical spectrum, and hence also of the diffraction spectrum, of regular model set in the complete generality of model sets over locally compact Abelian groups.

The paper is organized as follows. In Section 1, we collect some notation and facts on translationally bounded measures and establish two technical lemmas which seem to fit nowhere else. We decided to generalize the direct space to locally compact groups, too, so some of the functional analysis used at this point becomes a little bit abstract; the reader uncomfortable with this additional generalization may safely assume an Euclidean space as direct space and will recognize the cited constructions when restricted to the Euclidean case. Section 2 describes the construction of a topology, more precisely a uniform structure, on sets of point sets which gives a precise meaning of two point sets being close. The study of the dynamical system generated by the action of the group on a local indistinguishability (LI) class is carried out in Section 3; here, Dworkin's argument is introduced. Finally, in Section 4, we will introduce Pleasant's "torus parametrization" of LI classes [**BHP**] and apply the preceding results to model sets.

1. Translationally Bounded Measures

In this section, we collect some technicalities and notations concerning translationally bounded measures on locally compact Abelian groups (LCAG).

Throughout this paper, we assume that $\mathfrak{G}$ is an LCAG which is σ-compact. As usual, the commutative group operation will be written additively. The group $\mathfrak{G}$ is supposed to be equipped with an invariant (Haar) measure which will be fixed throughout the following. We denote integration with respect to this Haar measure by $\int \ldots dt$; the Haar measure of a subset D of $\mathfrak{G}$ will be written as $|D|$.

We identify the set $\mathfrak{M}(\mathfrak{G})$ of complex regular Borel measures in $\mathfrak{G}$ with the space $\mathfrak{C}_c(\mathfrak{G})^*$ of complex continuous linear functionals on the space $\mathfrak{C}_c(\mathfrak{G})$ of complex continuous functions with compact support in $\mathfrak{G}$ which is equipped with the inductive locally convex limit topology induced by the canonical embeddings $\mathfrak{C}_K(\mathfrak{G}) \to \mathfrak{C}_c(\mathfrak{G})$ where $\mathfrak{C}_K(\mathfrak{G})$ is the space of complex continuous functions with support in K for $K \subseteq \mathfrak{G}$ compact. The vague topology on $\mathfrak{M}(\mathfrak{G})$ equals the weak* topology on $\mathfrak{C}_c(\mathfrak{G})^*$, i.e., the topology $\sigma(\mathfrak{C}_c(\mathfrak{G})^*, \mathfrak{C}_c(\mathfrak{G}))$ which is the weakest topology making all the functionals $\nu \mapsto \nu(\varphi)$ continuous ($\varphi \in \mathfrak{C}_c(\mathfrak{G})$).

The identification of $\mathfrak{M}(\mathfrak{G})$ and $\mathfrak{C}_c(\mathfrak{G})^*$ generates some ambiguity in the functional notation of measures: for a Borel set A, $\nu(A)$ is the measure of A, whereas, for a function φ, $\nu(\varphi)$ is the integral of φ with respect to ν. In the following, the intended meaning will always be clear from the context, and we will get rid of a lot of integral signs this way.

For $t \in \mathfrak{G}$, $\varphi \in \mathfrak{C}_c(\mathfrak{G})$, we denote by τ_t the translate of φ by t, i.e., $\tau_t\varphi(x) = \varphi(x-t)$ ($x \in \mathfrak{G}$); similarly, if $\nu \in \mathfrak{M}(\mathfrak{G})$, we let $\tau_t\nu(\varphi) = \nu(\tau_{-t}\varphi)$ ($\varphi \in \mathfrak{C}_c(\mathfrak{G})$). Analogously, $\varphi^*(x) := \overline{\varphi(-x)}$ (the complex conjugate of $\varphi(-x)$), and $\nu^*(\varphi) := \overline{\nu(\varphi^*)}$. The convolution of a function $\varphi \in \mathfrak{C}_c(\mathfrak{G})$ with a measure ν is the continuous function $(\varphi * \nu)(t) = \nu(\tau_t\overline{\varphi^*})$. This matches the usual definition of convolution if

ν is absolutely continuous (with respect to Haar measure) and represented by its Radon-Nikodym density.

A measure ν is defined to be translationally bounded iff, for every $\varphi \in \mathfrak{C}_c(\mathfrak{G})$, the set $\{\tau_t \nu(\varphi) \mid t \in \mathfrak{G}\}$ is bounded. By the Banach-Steinhaus theorem [**T**, Theorem 33.2], this is equivalent to the condition that the set $\{\tau_t \nu \mid t \in \mathfrak{G}\}$ is relatively compact in the vague topology; it is, furthermore, equivalent to the condition that the set $\{\tau_t \nu \mid t \in \mathfrak{G}\}$ is strongly bounded, i.e., that, for every bounded subset B of $\mathfrak{C}_c(\mathfrak{G})$, the set $\{\tau_t \nu(\varphi) \mid t \in \mathfrak{G}, \varphi \in B\}$ is bounded. Because we assume $\mathfrak{G}$ to be σ-compact, we can finally conclude that ν is translationally bounded iff, for every compact $K \subseteq \mathfrak{G}$, the set $\{|\nu|(t+K) \mid t \in \mathfrak{G}\}$ is bounded, where $|\nu|$ is the variation of ν. We will use these different characterizations of translational boundedness interchangeably.

Several results of the following sections will deal with averaging of translationally bounded measures. For this, we need "nice" sequences of compact sets over which the averaging is carried out. We choose generalized van Hove sequences as defined below for this purpose.

If C and K are nonempty compact subsets of $\mathfrak{G}$, we denote

$$(1.1) \qquad \partial^K C := (C+K) \backslash C^\circ \cup ((\overline{\mathfrak{G} \backslash C} - K) \cap C).$$

The compact set $\partial^K C$ can be superficially interpreted as the set of points which have a distance measured by K from the "boundary of C"; however, this "boundary" cannot, in general, be taken in the topological sense, as in the example of a discrete $\mathfrak{G}$ the topological boundary would be always empty.

A sequence $(D_n)_{n \in \mathbb{N}}$ of compact subsets of $\mathfrak{G}$ of positive volume is a generalized *van Hove sequence* iff, for every compact $K \subseteq \mathfrak{G}$,

$$(1.2) \qquad \lim_{n \to \infty} \frac{|\partial^K D_n|}{|D_n|} = 0.$$

The meaning of (1.2) is that the surface/bulk ratio becomes sufficiently negligible in the limit $n \to \infty$.

The existence of van Hove sequences in compactly generated Abelian groups is easily seen from the structure theorem for these groups which states that every such group is of the form $\mathfrak{G} = \mathbb{Z}^d \times \mathbb{R}^{d'} \times \mathfrak{K}$ for nonnegative integers d, d', and a compact group $\mathfrak{K}$, see [**HR**, Theorem 9.6]; then, a van Hove sequence is given by

$$(1.3) \qquad D_n = \{(\vec{m}, \vec{x}, k) \in \mathbb{Z}^d \times \mathbb{R}^{d'} \times \mathfrak{K} \mid |m_i| \leq n, |x_j| \leq n\}.$$

In the case of a general σ-compact LCAG $\mathfrak{G}$, the group $\mathfrak{G}$ can be written as a union of countably many relatively compact open subsets V_n. Because every $\overline{V_n}$ generates a compactly generated open subgroup $\mathfrak{G}_n$ of $\mathfrak{G}$, we can find a compact subset $D_n \subseteq \mathfrak{G}$ of positive volume such that $|\partial^{\overline{V_n}} D_n|/|D_n| < 1/n$. Then, the sequence $(D_n)_{n \in \mathbb{N}}$ is a van Hove sequence in $\mathfrak{G}$.

The benefit of van Hove sequences lies in the following lemma.

LEMMA 1.1. *Let $(D_n)_{n \in \mathbb{N}}$ be a van Hove sequence in $\mathfrak{G}$, and ν a translationally bounded measure.*

(1) *For every compact K,*

$$(1.4) \qquad \lim_{n \to \infty} \frac{|\nu|(\partial^K D_n)}{|D_n|} = 0$$

(2) *The sequence $(|\nu|(D_n)/|D_n|)_{n \in \mathbb{N}}$ is bounded.*

(3) If φ is a continuous function with compact support, then

(1.5) $$\lim_{n\to\infty} \frac{1}{|D_n|}\left((\varphi * \nu)(D_n) - \int \varphi(t)dt \cdot \nu(D_n)\right) = 0$$

PROOF. (1) There is a compact subset C of $\mathfrak{G}$ and a discrete closed $L \subseteq \mathfrak{G}$ such that $\mathfrak{G} = L + C$ and such that different $t + C$ ($t \in L$) overlap at most in a set of zero Haar measure. For example, take D_1 from (1.3) for C and an appropriate discrete subgroup for L in the case of a compactly generated $\mathfrak{G}$; the general case is handled by doing this for an open compactly generated subgroup and translating into cosets.

Because ν is translationally bounded, there is a constant $c \geq 0$ such that $|\nu|(t + C) \leq c \cdot |t + C|$ independently of $t \in L$. For every n, let $W_n := \bigcup \{t + C \mid t \in L, (t+C) \cap \partial^K D_n \neq \varnothing\}$; this set is the union of finitely many $t + C$ and $\partial^K D_n \subseteq W_n \subseteq \partial^{K+C-C} D_n$. Then $|\nu|(W_n) \leq c \cdot |W_n|$, therefore, $|\nu|(\partial^K D_n) \leq c \cdot |\partial^{K+C-C} D_n|$, and (1.4) follows because $(D_n)_{n\in\mathbb{N}}$ is a van Hove sequence.

(2) In the same fashion, we find W_n with $|\nu|(W_n) \leq c \cdot |W_n|$ and $D_n \subseteq W_n \subseteq D_n \cup \partial^{C-C} D_n$; the assertion follows from (1).

(3) One calculates:

$$\left|(\varphi * \nu)(D_n) - \int \varphi(t)\,dt \cdot \nu(D_n)\right|$$
$$= \left|\int\int_{D_n - s} \varphi(t)\,dt\,\nu(ds) - \int_{D_n}\int \varphi(t)\,dt\,\nu(ds)\right|$$
$$= \left|\int_{\mathfrak{G}\setminus D_n}\int_{D_n - s}\varphi(t)\,dt\,\nu(ds) + \int_{D_n}\int_{\mathfrak{G}\setminus(D_n - s)}\varphi(t)\,dt\,\nu(ds)\right|$$
$$\leq \int_{(D_n + K)\setminus D_n}\int|\varphi(t)|\,dt\,|\nu|(ds) + \int_{(\mathfrak{G}\setminus D_n - K)\cap D_n}\int|\varphi(t)|\,dt\,|\nu|(ds)$$
$$\leq \int|\varphi(t)|\,dt \cdot |\nu|(\partial^K D_n)$$

for K containing the support of φ. $\square$

The point in introducing van Hove sequences is that later on, when considering autocorrelation measures, we would like to form the convolution of translationally bounded measures with themselves, a usually ill-defined operation. Instead, one does that with finite portions of these measures located on members of van Hove sequences and normalizes by the volume of the van Hove set. More precisely: for a measure ν, we denote the trace of ν on a measurable subset D by ν_D which is defined by the expression

(1.6) $$\nu_D(\varphi) := \int_D \varphi\,d\nu.$$

A measure γ is said to be the (unique) *autocorrelation* of the translationally bounded measure ν if and only if, for every van Hove sequence $(D_n)_{n\in\mathbb{N}}$, the sequence $((\nu_{D_n})^* * \nu_{D_n})/|D_n|$ converges vaguely to γ. Because ν_{D_n} has compact support, the formation of this convolution does not pose a problem. One major problem will be to establish that certain translationally bounded measures arising from certain point sets in $\mathfrak{G}$ have such autocorrelation measures. As a tool for this, we note

LEMMA 1.2. *If μ, ν are translationally bounded and $(D_n)_{n\in\mathbb{N}}$ is a van Hove sequence, then we have, in the vague topology,*

$$\frac{1}{|D_n|}(\mu_{D_n} * \nu_{D_n} - \mu * \nu_{D_n}) \to 0 \quad (n \to \infty). \tag{1.7}$$

PROOF. A short calculation shows that, for $\varphi \in \mathfrak{C}_c(\mathfrak{G})$ with support in a compact subset K,

$$|(\mu_{D_n} * \nu_{D_n})(\varphi) - (\mu * \nu_{D_n})(\varphi)| \leq \int_{\partial^K D_n} (|\varphi| * |\mu|)(t) |\nu|(dt). \tag{1.8}$$

Because μ is translationally bounded, the expression on the right side is smaller than $c \cdot |\nu|(\partial^K D_n)$ for a suitable constant c. Now, the lemma follows from (1.5). □

If a measure ν has an autocorrelation measure γ, then the latter is clearly positive definite (cf. [**AG**, p. 23]). In particular, it has a well-defined Fourier transform $\widehat{\gamma}$ which is a positive measure on the dual group $\widehat{\mathfrak{G}}$ (for the definition of Fourier transform for unbounded measures the reader may refer to [**AG**, p. 8]). Because in diffraction physics the latter describes the kinematic diffraction pattern of a structure modeled with a measure ν, we will call this Fourier transform $\widehat{\gamma}$ the *diffraction measure* of ν. (The reader interested in a more detailed description of the connection between diffraction and Fourier transform may consult any textbook on crystallography, e.g, [**C**].) Note that in our general setup there is no well-defined possibility to build the Fourier transform of ν itself even if ν is fairly regular.

As a positive measure, $\widehat{\gamma}$ can be decomposed according to Lebesgue's decomposition theorem. We are particularly interested in the point part of $\widehat{\gamma}$ because this is the part which corresponds with so-called Bragg diffraction pattern, an important tool in the analysis of ordered structures [**C**].

2. Uniform Structure on Sets of Point Sets

In the following, we denote by $\mathfrak{D}$ the set of all closed and discrete subsets of $\mathfrak{G}$. Our first task is to define a topology, more specifically a uniformity (cf. [**K**]) on $\mathfrak{D}$ which makes it a complete uniform Hausdorff space. For this, we have to define what it means for two point sets, P_1, P_2, to be close; basically P_1 will be close to P_2 if, after a small translation, these two coincide on a large compact region.

In the following, we will use the following notations for operations on sets of ordered pairs:

$$U^{-1} := \{(x,y) \mid (y,x) \in U\} \tag{2.1}$$

$$U \circ V := \{(x,z) \mid \exists y : (x,y) \in U \text{ and } (y,z) \in V\} \tag{2.2}$$

$$U[x] := \{y \mid (x,y) \in U\}. \tag{2.3}$$

Now, we set, for compact $K \subseteq \mathfrak{G}$ and a neighborhood V of 0:

$$U_{K,V} := \{(P_1,P_2) \in \mathfrak{D} \times \mathfrak{D} \mid \exists s \in V : P_1 \cap K = (s+P_2) \cap K\}. \tag{2.4}$$

It is immediately verified that $(P,P) \in U_{K,V}$, $U_{K_1 \cup K_2, V_1 \cap V_2} \subseteq U_{K_1,V_1} \cap U_{K_2,V_2}$; furthermore, for a compact neighborhood $W \subseteq -V$ of 0, $U_{K+W,W} \subseteq (U_{K,V})^{-1}$; if we take a compact neighborhood W of 0 such that $W + W \subseteq V$, we calculate $U_{K-W,W} \circ U_{K-W,W} \subseteq U_{K,V}$. This shows that the set of $U_{K,V}$ as defined in (2.4) forms a base of a uniform structure [**K**] on $\mathfrak{D}$; the topology on $\mathfrak{D}$ induced by this uniformity is given by taking the sets $U_{K,V}[P]$ as a base of neighborhoods of $P \in \mathfrak{D}$.

This uniform structure separates points: if $(P_1, P_2) \in U_{K,V}$ for all K, V, then, for any K: $P_1 \cap K \subseteq \overline{P_2} = P_2$, hence $P_1 \subseteq P_2$. By symmetry, $P_1 = P_2$. Hence $\mathfrak{D}$ becomes a uniform Hausdorff space.

From the definition of the topology on $\mathfrak{G}$, it is easy to conclude that the action

(2.5) $$\mathfrak{G} \times \mathfrak{D} \to \mathfrak{D}, \quad (t, P) \mapsto t + P$$

is continuous (in both variables simultaneously). What is more, its restriction to $K \times \mathfrak{D}$ is uniformly continuous for every compact subset K of $\mathfrak{G}$ (one calculates that, if K, $K' \subseteq \mathfrak{G}$ compact, V, V' are neighborhoods of 0 with $V' + V' \subseteq V$, then, if t_1, $t_2 \in K$ with $t_1 - t_2 \in V'$ and P_1, $P_2 \in \mathfrak{D}$ with $(P_1, P_2) \in U_{K'-K, V'}$, then $(t_1 + P_1, t_2 + P_2) \in U_{K', V}$).

PROPOSITION 2.1. $\mathfrak{D}$ is complete in the uniformity defined by eq. (2.4).

PROOF. Let $(P_i)_{i \in I}$ be a Cauchy net in $\mathfrak{D}$ for some index set I directed by a relation $\leq$; we show that a subnet converges in $\mathfrak{D}$. Without loss of generality, we may assume that there is a compact $K_0 \subseteq \mathfrak{G}$ such that $P_i \cap K_0 \neq \varnothing$, for all $i \in I$, because otherwise we can find a subnet converging to $\varnothing \in \mathfrak{D}$. We pick one $t_i \in P_i \cap K_0$ for each $i \in I$; passing to a subnet, we can assume that t_i converge to some t. It is clear that $(-t_i + P_i)_{i \in I}$ form a Cauchy net in $\mathfrak{D}$ and convergence of $(-t_i + P_i)_{i \in I}$ entails convergence of $(P_i)_{i \in I}$. Hence, we may assume that $0 \in P_i$ for all $i \in I$.

We may, furthermore, assume that, for some compact neighborhood V_0 of 0 and some $i_0 \in I$, every $P_i \cap V_0$ equals some $(s_i + P_{i_0}) \cap V_0 = s_i + P_{i_0} \cap (-s_i + V_0)$ for suitably chosen $s_i \in V_0$. As $0 \in P_i \cap V_0$, this s_i is in $(-P_{i_0}) \cap V_0$, therefore all $P_i \cap V_0 \subseteq P_{i_0} \cap (V_0 - V_0) - P_{i_0} \cap (V_0 - V_0)$ which is a finite set. We can, therefore, choose some neighborhood V_1 of 0 such that $P_i \cap V_1 = \{0\}$ for all $i \in I$.

As a consequence, if $K \subseteq \mathfrak{G}$ is compact and contains 0, and if $(P_i, P_j) \in U_{K, V_1}$, then necessarily $P_i \cap K = P_j \cap K$. The Cauchy property of the net $(P_i)_{i \in I}$ can therefore be strengthened: for every compact $K \subseteq \mathfrak{G}$, there is some $i_K \in I$ such that, if $i, j \geq i_K$, then $P_i \cap K = P_j \cap K$.

It is now clear that $(P_i)_{i \in I}$ converges to the closed discrete set P as defined by $P := \bigcup \{P_{i_K} \cap K \mid K \subseteq \mathfrak{G} \text{ compact}\}$. □

Of particular interest are those point sets whose $\mathfrak{G}$-orbits are precompact and which therefore, after closure, form well-behaved dynamical systems. These sets have a nice geometrical characterization.

DEFINITION 2.2. A point set P in $\mathfrak{G}$ is of *finite local complexity* (FLC) iff $P - P$ is closed and discrete.

The term "finite local complexity" is justified by property (3) in the following proposition which basically states that, up to translations, there are only finitely many patterns of any given finite size in P.

PROPOSITION 2.3. Let $P \in \mathfrak{D}$. The following are equivalent:
(1) P is of finite local complexity,
(2) The $\mathfrak{G}$-orbit $[P]_\mathfrak{G} := \{t + P \mid t \in \mathfrak{G}\}$ of P is precompact.
(3) For every compact $K \subseteq \mathfrak{G}$, there is a compact $K' \subseteq \mathfrak{G}$ such that, for every $t \in \mathfrak{G}$, there is some $t' \in K'$ with $(t + P) \cap K = (t' + P) \cap K$.

PROOF. (1) $\Rightarrow$ (2) For any compact $K \subseteq \mathfrak{G}$ and any neighborhood V of 0, the set $F_K := (P - P) \cap (K - K)$ is finite, because $P - P$ is closed and discrete.

We can, furthermore, find a finite set $E_K \subseteq \mathfrak{G}$ such that $K \subseteq E_K + V$. A simple calculation shows that $t + P \in \bigcup \{U_{K,V}^{-1}[s+F] \mid s \in E_K, F \subseteq F_K\}$, for any $t \in \mathfrak{G}$: if $(t+P) \cap K = \varnothing$, there is nothing to show; otherwise, choose $p \in P$ such that $t + p \in K$, find $s \in E_K$ and $r \in V$ such that $r + s = t + p$ and observe $(t+P) \cap K = (r+s+(P-p)) \cap K = r + s + (P-p) \cap (K - (s+r)) = (r+s+F) \cap K$ for $F := (P-p) \cap (K-(s+r)) \subseteq (P-P) \cap (K-K)$, i.e., $t + P \in U_{K,V}^{-1}[s+F]$. The assertion follows.

(2) $\Rightarrow$ (3) If $[P]_\mathfrak{G}$ is precompact, for given K we choose some compact neighborhood V of 0 and $t_1, \ldots, t_l \in \mathfrak{G}$ such that, given $t \in \mathfrak{G}$, we have $t + P \in U_{K,V}[t_i + P]$, for some suitable t_i. This means $(t+P) \cap K = (s + t_i + P) \cap K$, for some $s \in V$. Then, $K' := \bigcup_{i=1}^{l}(t_i + V)$ fulfills the requirement of (3).

(3) $\Rightarrow$ (1) Let K be compact, K' chosen for K according to (3). We may assume that $0 \in K$. Let $q \in (P-P) \cap K$, for example $q_1, q_2 \in P$ with $q = q_1 - q_2$. Then, we find $t' \in K'$ such that $q_1 - q_2 \in (-q_2 + P) \cap K = (t' + P) \cap K = P \cap (K - t') + t'$, and, because $0 \in (-q_2 + P) \cap K$, $-t'$ is an element of P, and therefore $q = q_1 - q_2 \in P \cap (K - K') - P \cap (-K')$, which is a finite set. $\square$

In the rest of this paper, unless stated otherwise, we will deal only with sets $P \subseteq \mathfrak{G}$ of finite local complexity (FLC sets). The closure of the $\mathfrak{G}$-orbit $[P]_\mathfrak{G}$ in $\mathfrak{D}$ will be denoted by X_P. A consequence of the previous proposition is that, for an FLC set P, the topological space X_P is a compact Hausdorff space, because it is the closure of a precompact set in a complete uniform Hausdorff space.

Proposition 2.3 is an example of the intimate connection of geometrical properties of P with certain topological properties of the space X_P; further examples of this will show up in the sequel. In order to study the topology of X_P, it is useful to determine the algebra $\mathfrak{C}(X_P)$ of complex continuous functions on X_P. We note that, under the present assumptions, P may have arbitrarily large holes, i.e., compact regions of arbitrary size in $\mathfrak{G}$ which contain no points of P. We don't want to exclude this class of sets from the start, because it contains interesting examples like the set of visible points in a lattice (cf. [**A**]). In these cases, the empty set $\varnothing$ will be an element of X_P. In the following description of $\mathfrak{C}(X_P)$, this turns out to be rather inconvenient, so we consider the locally compact space $X_P \setminus \{\varnothing\}$ and the algebra $\mathfrak{C}_0(X_P \setminus \{\varnothing\})$ of complex continuous functions vanishing at ∞ on $X_P \setminus \{\varnothing\}$ instead. Of course, if $\varnothing \notin X_P$, then $\mathfrak{C}(X_P) = \mathfrak{C}_0(X_P \setminus \{\varnothing\})$; on the other hand, if $\varnothing \in X_P$, then $\mathfrak{C}_0(X_P \setminus \{\varnothing\})$ can be identified with the algebra of those continuous functions on X_P which vanish at $\varnothing$.

In the following, we fix some FLC set $P_0 \in \mathfrak{D}$. First, we introduce some notation.

DEFINITION 2.4. For $t \in \mathfrak{G}$ and compact $K \subseteq \mathfrak{G}$ with $(t + P_0) \cap K \neq \varnothing$, we set

(2.6) $$T_{K,t}(P) := \{s \in \mathfrak{G} \mid (s+P) \cap K = (t + P_0) \cap K\},$$

for any $P \in X_{P_0}$. Furthermore, if $\varphi \in \mathfrak{C}_c(\mathfrak{G})$ (continuous functions on $\mathfrak{G}$ with compact support), we define the following function on X_{P_0}:

(2.7) $$\varphi_{K,t}(P) := \sum_{s \in T_{K,t}(P)} \varphi(-s) \quad (P \in X_{P_0})$$

Later on, the case $T_{\{0\},-t} = -P$ for $t \in P$ will be of particular interest. Obviously, $T_{K,t}$ is a subset of a translate of $-P$, therefore, $\varphi_{K,t}$ is well-defined as, for each P, the sum in (2.7) has only finitely many nonzero terms. Furthermore, if

$\varnothing \in X_{P_0}$, then $\varphi_{K,t}(\varnothing) = 0$. In view of the definition of the topology on X_{P_0} we see easily that $\varphi_{K,t}$ is continuous, because, if P_1 and P_2 coincide on a large enough compact set, then $T_{K,t}(P_1)$ and $T_{K,t}(P_2)$ coincide on the support of φ; on the other hand, if $P_1 = s + P_2$ for a small s then $\varphi_{K,t}(P_1)$ and $\varphi_{K,t}(P_2)$ can differ only by a small amount (see (2.7)).

The promised description of $\mathfrak{C}_0(X_{P_0}\setminus\{\varnothing\})$ is given in the following proposition.

PROPOSITION 2.5. *The set of all functions of the form* (2.7) *forms a total subset of* $\mathfrak{C}_0(X_{P_0}\setminus\{\varnothing\})$, *i.e., every complex continuous function on* X_{P_0} *vanishing at* $\varnothing$ *can be uniformly approximated by linear combinations of suitable* $\varphi_{K,t}$.

PROOF. We use the Stone-Weierstraß approximation theorem [**La**, Chapter III, §1, Theorem 1.4]. For this, we have to show that, (1) if $\varnothing \neq P_1 \neq P_2$, there are φ, K, and t such that $\varphi_{K,t}(P_1) = 1$ and $\varphi_{K,t}(P_2) = 0$ (separation property); and (2) the set of linear combinations of $\varphi_{K,t}$'s form an subalgebra of $\mathfrak{C}(X_{P_0})$.

(1) We choose $q_0 \in P_1 \Delta P_2$ (symmetric difference), $q_1 \in P_1$ and a neighborhood V of 0 in G such that $(q_1 - V) \cap (P_1 \cup P_2) = \{q_1\}$. We take a $\varphi \in \mathfrak{C}_c(\mathfrak{G})$ with support in V and $\varphi(0) = 1$. Let $K = \{q_0, q_1\}$ and choose $t \in \mathfrak{G}$ such that $P_1 \cap K = (t + P_0) \cap K$. Now, if $s \in V$ and $(s + P_i) \cap K = P_1 \cap K$ for $i = 1$ or $i = 2$, then necessarily $s = 0$ (by the choice of V). By construction, $P_1 \cap K \neq P_2 \cap K$. Therefore, $\varphi_{K,t}$ has the asserted properties.

(2) We have to show that the product of two functions of the form (2.7) is a linear combination of suitable $\varphi_{K,t}$'s. Let, for $i = 1, 2$,

$$(2.8) \qquad \Phi^i(P) = \sum_{s \in T_{K_i,t_i}(P)} \varphi^i(s) \quad (P \in X_{P_0})$$

(skipping the sign under φ which is irrelevant for the moment). Setting $F_i := (t_i + P_0) \cap K_i$ ($i = 1, 2$) which are finite sets, we calculate:

$$(2.9) \qquad \Phi^1(P)\Phi^2(P) = \sum_{s_1, s_2} \varphi^1(s_1)\varphi^2(s_2)$$

where the sum is taken over all s_1, s_2 fulfilling $(s_i + P) \cap K_i = F_i$ ($i = 1, 2$). Note that such a pair (s_1, s_2) necessarily fulfills $s_1 - s_2 \in F_1 - F_2 + P_0 - P_0$.

Let K be compact containing the supports of both, φ^1 and φ^2, in its interior; then, $F := (F_1 - F_2 + P_0 - P_0) \cap (K - K)$ is finite, and we find a compact neighborhood V of 0 with $(F - F) \cap (V + V) \subseteq \{0\}$.

Because we can always decompose φ^i into a sum of continuous functions with arbitrarily small supports in K, we may assume without loss of generality that $\mathrm{supp}(\varphi^i) - \mathrm{supp}(\varphi^i) \subseteq V$; then $V' - V' \subseteq V + V$ for $V' := \mathrm{supp}(\varphi^1) - \mathrm{supp}(\varphi^2)$ and, hence, $V' \cap F \subseteq \{s_0\}$ for some $s_0 \in \mathfrak{G}$. As a consequence, any pair (s_1, s_2) contributing at all to the sum (2.9) must fulfill $s_1 - s_2 = s_0$.

If we now set $F_0 := F_1 \cup (s_0 + F_2)$ and $K_0 := K_1 \cup (s_0 + K_2)$, then $s_1 - s_2 \in V'$ and $(s_i + P) \cap K_i = F_i$ for $i = 1, 2$ if and only if $s_1 - s_2 = s_0$ and $(s_1 + P) \cap K_0 = F_0$. Now, either there is no $t_0 \in \mathfrak{G}$ such that $(t_0 + P_0) \cap K_0 = F_0$, and then $\Phi^1 \Phi^2$ vanishes, or we can choose such a t_0 and see that $\Phi^1 \Phi^2 = \psi_{K_0,t_0}$ setting $\psi(s) = \varphi^1(-s)\varphi^2(-s - s_0)$. $\square$

The functions of the form (2.7) can be expressed in terms of convolution. For this, we associate to every discrete closed point set $P \subseteq \mathfrak{G}$ the uniquely determined measure ν^P defined by

$$(2.10) \qquad \nu^P(\varphi) := \sum_{t \in P} \varphi(t) \quad (\varphi \in \mathfrak{C}_c(\mathfrak{G})),$$

i.e., $\nu^P = \sum_{t \in P} \delta_t$ where δ_t is the unit point measure concentrated at t. It is clear that ν^P is translationally bounded for FLC sets P. For $K \subseteq \mathfrak{G}$ compact, $t, t' \in \mathfrak{G}$, $\varphi \in \mathfrak{C}_c(\mathfrak{G})$, and $P \in X_{P_0}$, we calculate,

$$(2.11) \qquad \varphi_{K,t}(t' + P) = \sum_{s \in T_{K,t}(P)} \varphi(t' - s) = \nu^{T_{K,t}(P)}(\tau_{t'} \overline{\varphi^*}) = (\varphi * \nu^{T_{K,t}(P)})(t'),$$

a formula which will come in handy when $\varphi_{K,t}$ is studied on a single translation orbit.

3. The Dynamical System of an FLC Set

We now take a closer look at the dynamical system $(X_P, \mathfrak{G})$ given by the canonical action (2.5) of $\mathfrak{G}$ on X_P. As usual, we assume P to be an FLC set.

Let us first give a geometric description of the elements of X_P. By definition, $P' \in X_P$ iff P' can be approximated by translates $t + P$ in the uniform topology defined in (2.4). Expanding the definition, we see that $P' \in X_P$ iff, for every compact $K \subseteq \mathfrak{G}$, there is a $t \in \mathfrak{G}$ such that $P' \cap K = (t + P) \cap K$.

This is one half of the condition of "local indistinguishability" of P' and P. That is, P' and P are locally indistinguishable iff $P' \in X_P$ and $P \in X'_P$, i.e., $X_P = X'_P$. In terms of dynamical systems: $(X_P, \mathfrak{G})$ is minimal iff every $P' \in X_P$ is locally indistinguishable from P, or, equivalently, iff the equivalence class of P with respect to local indistinguishability, the so-called LI class $[P]_{LI}$ of P, coincides with X_P. The last condition clearly is equivalent to compactness of $[P]_{LI}$.

Now, there is a nice geometric interpretation of compactness of $[P]_{LI}$. For this, we introduce the simplified notation

$$(3.1) \qquad T_K(P) := \{t \in \mathfrak{G} \mid (t + P) \cap K = P \cap K\}.$$

(This is a special case of (2.6) for $P = P_0$ and $t_0 = 0$.)

Recall that a subset S of $\mathfrak{G}$ is relatively dense iff there is a compact $K \subseteq \mathfrak{G}$ such that $S + K = \mathfrak{G}$. We have now the following equivalences:

PROPOSITION 3.1. *For an FLC set P, the following statements are equivalent*:
  (1) $T_K(P)$ *is relatively dense for every compact K*
  (2) $X_P = [P]_{LI}$
  (3) P *is repetitive, i.e., for every compact $K \subseteq \mathfrak{G}$, there is a compact $K' \subseteq \mathfrak{G}$ such that, for every $t_1, t_2 \in \mathfrak{G}$, there is an $s \in K'$ with $(t_1 + P) \cap K = (s + t_2 + P) \cap K$.*

PROOF. For any compact neighborhood V of 0, one calculates that $T_K(P) \subseteq \{t \in \mathfrak{G} \mid (t + P, P) \in U_{K,V}\} \subseteq T_K(P) - V$. Therefore, (1) is true iff P is an almost periodic point of the dynamical system $(X_P, \mathfrak{G})$. As already noted, (2) holds iff $(X_P, \mathfrak{G})$ is minimal. Therefore, (1) $\Leftrightarrow$ (2) follows from general theorems on dynamical systems [**E**, Proposition 2.5, p. 10], sometimes referred to as Gottschalk's theorem.

Furthermore, (1) is a special case of (3), so (3) $\Rightarrow$ (1).

Assume now that (1) holds and let $K \subseteq \mathfrak{G}$ be compact. Because P is an FLC set, with Proposition 2.3(3) we obtain a compact K_1 such that for all $t_1 \in \mathfrak{G}$, there is $t' \in K_1$ with $(t+P) \cap K = (t'+P) \cap K$. According to assumption (1) we find a compact K_2 such that, for every $t_2 \in \mathfrak{G}$, there is some $t'' \in K_2$ with $(t''+t_2+P) \cap (K-K_1) = P \cap (K-K_1)$. Because $-t_1 + K \subseteq K - K_1$, we calculate $(t_2 + t' + t'' + P) \cap K = (t_1 + P) \cap K$. As $t_1 + t_2 \in K_1 + K_2$, (3) is fulfilled for $K' := K_1 + K_2$. □

We now move from the topological aspects of the dynamical system $(X_{P_0}, \mathfrak{G})$ to the study of measure theoretic properties. The following theorem connects unique ergodicity of $(X_{P_0}, \mathfrak{G})$ to a property which admits a direct geometric interpretation. Let $(D_n)_{n \in \mathbb{N}}$ be a fixed van Hove sequence in $\mathfrak{G}$.

THEOREM 3.2. *The dynamical system $(X_{P_0}, \mathfrak{G})$ is uniquely ergodic iff, for every compact $K \subseteq \mathfrak{G}$ and every $t_0 \in \mathfrak{G}$, there is a constant c_{K,t_0} such that*

$$(3.2) \qquad \lim_{n \to \infty} \frac{1}{|D_n|} \nu^{T_{K,t_0}(P)}(D_n) = c_{K,t_0},$$

in the vague topology, independently of $P \in X_{P_0}$.

In this case, (3.2) holds for every van Hove sequence in $\mathfrak{G}$, the constants c_{K,t_0} being independent of the choice of the sequence.

PROOF. Because of (1.5) in connection with (2.11), the condition (3.2) is equivalent to

$$(3.3) \qquad \lim_{n \to \infty} \frac{1}{|D_n|} \int_{D_n} \varphi_{K,t_0}(t+P)\, dt = c_{K,t_0} \int \varphi(t)\, dt,$$

for all $\varphi \in \mathfrak{C}_c(\mathfrak{G})$.

Because the set of φ_{K,t_0} is total in $\mathfrak{C}(X_{P_0})$ (Proposition 2.5), by a 3ϵ argument we see that the existence and independence of P of the limit in (3.3) for all φ_{K,t_0} is equivalent to the existence and independence of P of

$$(3.4) \qquad \lim_{n \to \infty} \frac{1}{|D_n|} \int_{D_n} \Phi(t+P)\, dt,$$

for every $\Phi \in \mathfrak{C}(X_{P_0})$. By the pointwise ergodic theorem for continuous functions, this is equivalent to unique ergodicity of the dynamical system $(X_{P_0}, \mathfrak{G})$. (For the pointwise ergodic theorem for continuous functions, see, e.g., [**F**, Theorem 3.5]; the theorem given there is immediately generalizable to the present situation).

In particular, in the uniquely ergodic case, the left side of (3.3) equals $\mu(\varphi_{K,t_0})$ independently of the choice of the van Hove sequence, where we write μ for the uniquely determined $\mathfrak{G}$-invariant probability measure on X_{P_0}. □

The expression $\nu^{T_{K,t_0}(P)}(D_n)$ is just the number of points of $T_{K,t_0}(P)$ which are located inside D_n. Theorem 3.2, therefore, states that $(X_{P_0}, \mathfrak{G})$ is uniquely ergodic iff all $T_{K,t_0}(P)$ have a well-defined average density (or frequency) independently of $P \in X_{P_0}$ when averaged over a van Hove sequence. In view of the definition (2.6), the frequency of $T_{K,t_0}(P)$ can be interpreted as the frequency of the finite pattern $(t_0 + P_0) \cap K$ in P. Hence, informally stated,

COROLLARY 3.3. *The dynamical system $(X_{P_0}, \mathfrak{G})$ is uniquely ergodic iff the frequencies of finite patterns exist in every $P \in X_{P_0}$ and are independent of the particular choice of P.*

One can show that frequencies of finite patterns exist and coincide for arbitrary $P \in X_{P_0}$ for a given van Hove sequence iff they exist and coincide only in P_0, but for arbitrary van Hove sequences. This way, one can connect this Corollary with similar results in [**LP**].

In the uniquely ergodic case, we can prove the existence of the autocorrelation measure by an argument due to Dworkin [**D**]. We denote the uniquely determined $\mathfrak{G}$-invariant probability measure on X_{P_0} by μ. The action of $\mathfrak{G}$ on X_{P_0} induces a strongly continuous unitary representation u_t of $\mathfrak{G}$ on $L^2(\mu)$ by

$$(3.5) \qquad (u_t\xi)(s) := \xi(s-t) \quad (\xi \in L^2(\mu), s, t \in \mathfrak{G}).$$

We denote the scalar product on $L^2(\mu)$ by $\langle \cdot, \cdot \rangle_\mu$.

THEOREM 3.4. *If $(X_{P_0}, \mathfrak{G})$ is uniquely ergodic, the autocorrelation of $P \in X_{P_0}$,*

$$(3.6) \qquad \gamma^P := \lim_{n \to \infty} \frac{1}{|D_n|}((\nu^P{}_{D_n})^* * \nu^P{}_{D_n}),$$

exists in the vague sense and is independent of the particular choice of P and the van Hove sequence $(D_n)_{n \in \mathbb{N}}$.

Furthermore, for $\varphi, \psi \in \mathfrak{C}_c(\mathfrak{G})$, the equation

$$(3.7) \qquad (\psi^* * \varphi * \gamma^P)(t) = \langle \psi_{\{0\}, -t_0}, u_{-t}\varphi_{\{0\}, -t_0}\rangle_\mu$$

holds for any $t_0 \in P_0$, where $\psi_{\{0\}, -t_0}$ and $\varphi_{\{0\}, -t_0}$ are continuous functions on X_{P_0} according to (2.7).

PROOF. Choose some $P \in X_{P_0}$. In order to make the notation more readable, we let $\nu := \nu^P$. Using (1.7) three times, we see that we have only to show that

$$\frac{1}{|D_n|}((\nu^*)_{D_n} * \nu)$$

converges vaguely to some limit independently of P.

Let φ and ψ be some elements of $\mathfrak{C}_c(\mathfrak{G})$. We want to show that

$$\frac{1}{|D_n|}(\psi^* * \varphi * (\nu^*)_{D_n} * \nu))(t) = \frac{1}{|D_n|}((\nu^*)_{D_n} * (\varphi * \nu))(\tau_t \overline{\psi})$$

converges pointwise to a function on $\mathfrak{G}$ which is independent of P. Using (1.7) again, we are reduced to show that

$$\frac{1}{|D_n|}(\nu^* * (\varphi * \nu)_{D_n})(\tau_t \overline{\psi}) = \frac{1}{|D_n|}((\psi * \nu)^* * (\varphi * \nu)_{D_n})(t)$$

converges (independently of P). We choose some $t_0 \in P_0$; let $\Phi := \varphi_{\{0\}, -t_0}$ and $\Psi := \psi_{\{0\}, -t_0}$ and observe that $\Phi(t+P) = (\varphi * \nu)(t)$ and $\Psi(t+P) = (\psi * \nu)(t)$ according to (2.11).

$$(3.8) \qquad \frac{1}{|D_n|}((\psi*\nu)^* * (\varphi*\nu)_{D_n})(t) = \frac{1}{|D_n|}\int_{D_n}(\psi*\nu)^*(t-s)(\varphi*\nu)(s)\,ds$$

$$= \frac{1}{|D_n|}\int_{D_n}\overline{\Psi(s-t+P)}\Phi(s+P)\,ds$$

$$\xrightarrow{n\to\infty} \int \overline{\Psi(-t+P')}\Phi(P')\,\mu(dP')$$

$$= \langle \Psi, u_{-t}\Phi\rangle_\mu$$

using the pointwise ergodic theorem for continuous functions.

In particular, $((\nu_{D_n})^* * \nu)/|D_n|$ converges pointwise on the subspace of $\mathfrak{C}_c(\mathfrak{G})$ generated by functions of the form $\varphi * \psi$ which is a total set in $\mathfrak{C}_c(\mathfrak{G})$; furthermore, the limit is independent of $P \in X_{P_0}$. A short calculation shows that, for arbitrary

$\varphi \in \mathfrak{C}_c(\mathfrak{G})$, $|((\nu_{D_n})^* * \nu)(\varphi)| \leq \nu(D_n) \cdot \sup_{t \in \mathfrak{G}}(|\varphi| * \nu)(t)$. With Lemma 1.1, we can conclude that $((\nu_{D_n})^* * \nu)/|D_n|$ is weakly and, therefore, strongly bounded. Hence, convergence on a total set entails convergence on all of $\mathfrak{C}_c(\mathfrak{G})$. It follows also that the vague limit is independent of $P \in X_{P_0}$.

Equation (3.7) is now a simple consequence of (3.8). □

Equation (3.7) connects the unitary representation u_t of $\mathfrak{G}$ and the autocorrelation γ^P. In particular, we can say more about the Fourier transform of γ^P, i.e., the diffraction properties of P_0 in terms of (3.7). By Stone's (or SNAG) theorem [**RN**, §140], there is a projection valued measure E (the spectral resolution of u_t) on the dual group $\widehat{\mathfrak{G}}$ of $\mathfrak{G}$ such that, for every $\zeta, \xi \in L^2(\mu)$, the function $t \mapsto \langle \zeta, u_t \xi \rangle_\mu$ is the Fourier transform of the measure $A \mapsto \langle \zeta, E(A)\xi \rangle_\mu$ (we identify $\mathfrak{G}$ and the dual of $\widehat{\mathfrak{G}}$). If we allow for the sign in (3.7) we obtain that, for $\varphi, \psi \in \mathfrak{C}_c(\mathfrak{G})$, the inverse Fourier transform of $A \mapsto \langle \psi_{\{0\},-t_0}, E(A)\varphi_{\{0\},-t_0} \rangle_\mu$ is the function $\psi^* * \varphi * \gamma^P$. Hence (cf. [**AG**]),

$$(3.9) \qquad \overline{\widehat{\psi}}\widehat{\varphi}\widehat{\gamma}^P = \langle \psi_{\{0\},-t_0}, E(\cdot)\varphi_{\{0\},-t_0} \rangle_\mu$$

Because in general the functions of the form $\varphi_{\{0\},-t_0}$ do not form a total set in the Hilbert space $L^2(\mu)$ (cf. 2.5), we do not get a one-to-one correspondence between the diffraction measure $\widehat{\gamma}^P$ and the dynamical spectrum E. For example, it is not clear that $\widehat{\gamma}^P$ has a nontrivial discrete component solely on the grounds that E has such a component, if the latter turns out to be the case. But, from (3.9), we can certainly conclude that $\widehat{\gamma}^P$ is pure point if E is pure point; this will be used in the next section. Of course, it is equally possible to conclude that $\widehat{\gamma}^P$ has no nontrivial discrete component if E has none.

4. The Torus Parametrization

In the present section we show how, for a particular class of FLC sets, the diffraction spectrum can be analyzed more deeply. This class encompasses the so-called model sets (also known as cut-and-project sets).

Let us assume that $P_0 \neq \varnothing$ is a repetitive FLC subset of $\mathfrak{G}$, fixed throughout this section. The subgroup of $\mathfrak{G}$ generated by $P_0 - P_0$ will be denoted by L. Without loss of generality, we may assume that $P_0 \subset L$. We assume, furthermore, (with loss of generality) that there is a LCAG $\mathfrak{G}_{\text{int}}$ and an algebraic homomorphism

$$(4.1) \qquad L \to \mathfrak{G}_{\text{int}}, \quad t \mapsto t^*$$

with the properties

S1 The subgroup $\tilde{L} := \{(t, t^*) \in \mathfrak{G} \times \mathfrak{G}_{\text{int}} \mid t \in L\}$ of $\mathfrak{G} \times \mathfrak{G}_{\text{int}}$ is a lattice in $\mathfrak{G} \times \mathfrak{G}_{\text{int}}$ (i.e., $\tilde{L}$ is discrete and $(\mathfrak{G} \times \mathfrak{G}_{\text{int}})/\tilde{L}$ is compact),

S2 P_0^* is relatively compact.

If such a map $*$ exists, we can, without loss of generality, assume that, additionally, the following conditions are fulfilled:

S3 L^* is dense in $\mathfrak{G}_{\text{int}}$,

S4 $W_0 := \overline{P_0^*}$ (closure of P_0^* in $\mathfrak{G}_{\text{int}}$) has no nontrivial translation invariance (i.e., $\{c \in \mathfrak{G}_{\text{int}} \mid c + W_0 = W_0\} = \{0\}$),

(otherwise, form suitable subgroups and quotients). A map $*$ with these properties, following [**M**], will be called a *star map* for P_0. Such a star map will be called *regular* iff, additionally,

S5 The boundary of W_0 as defined in **S4** has zero Haar measure.

We assume that we have fixed some star map for P_0 fulfilling S1–S4. We now construct a $\mathfrak{G}$-map β from the LI class onto $\mathfrak{T} := (\mathfrak{G} \times \mathfrak{G}_{\text{int}})/\tilde{L}$ (the so-called "torus parametrization" [**BHP**] of X_{P_0}). Let

$$\pi \colon \mathfrak{G} \times \mathfrak{G}_{\text{int}} \to \mathfrak{T} \tag{4.2}$$

be the canonical projection.

LEMMA 4.1. *If $P \in X_{P_0}$ such that $P \subseteq L$, then the set*

$$\bigcap \{t^* - W_0 \mid t \in P\} \tag{4.3}$$

contains exactly one element $c_P \in \mathfrak{G}_{int}$, and $\overline{P^*} = c_P + W_0$.

PROOF. Let $q_0 \in P$ be fixed, $K_0 \subseteq \mathfrak{G}$ compact such that $q_0 \in P \cap K_0$. For every compact $K \supseteq K_0$, we can find a $t_K \in L$ such that $P \cap K = (t_K + P_0) \cap K$. Then, $q_0 \in t_K + P_0$, therefore $t_K \in q_0 - P_0$, hence $t_K^* \in q_0^* - W_0$. As $q_0^* - W_0$ is compact, we can find a subnet $(t_{K_i})_{i \in I}$ of $(t_K)_{(K\text{compact})}$ such that its image under $*$ converges to some $c \in \mathfrak{G}_{\text{int}}$. Any $q \in P$ is contained in some K_i, therefore $q \in t_{K_i} + P_0$, i.e., $q^* \in t_{K_i}^* + W_0$. Going to the limit we see that $q^* \in c + W_0$ which implies that c is an element of the intersection (4.3). It is clear that any such c must fulfill $P^* \subseteq c + W_0$.

Let $W := \overline{P^*}$. Because $W \subseteq c + W_0$, it is compact. Up to now, we have not used the fact that W_0 fulfills condition S4, therefore, by symmetry, we can conclude that, for some $c' \in \mathfrak{G}_{\text{int}}$, we have $P_0^* \subseteq c' + W$. Therefore, $W_0 \subseteq c' + W \subseteq c' + c + W_0$, hence $W_0 \subseteq c' + c + W_0$. Because W_0 is compact, this can only be if $W_0 = c' + c + W_0$. As a consequence, $W = c + W_0$. We can now use condition S4 in order to conclude that c is uniquely determined. $\square$

Obviously, for $P \in X_{P_0}$ with $P \subseteq L$, and for $t \in L$, we have $c_{t+P} = t^* + c_P$. For every $P \in X_{P_0}$, there is some $s \in \mathfrak{G}$ with $s + P \subseteq L$; if $s + P \subseteq L$ and $s' + P \subseteq L$, then $s - s'$ must be in L. What we have just shown is that the following is well-defined:

DEFINITION 4.2. Define

$$\beta \colon X_{P_0} \to \mathfrak{T}, \quad P \mapsto \beta(P) := \pi((s, c_{s+P})) \tag{4.4}$$

for s chosen such that $s + P \subseteq L$.

It is clear that $\beta(t + P) = t + \beta(P)$ for every $t \in \mathfrak{G}$, hence β is a $\mathfrak{G}$-map (note that we have a canonical action of $\mathfrak{G}$ on $\mathfrak{T}$).

PROPOSITION 4.3. *β is a continuous map of X_{P_0} onto $\mathfrak{T}$.*

PROOF. Let $P \in X_{P_0}$ and choose $s \in \mathfrak{G}$ with $s + P \subseteq L$. Let, furthermore, V' be an open neighborhood of c_{s+P} in $\mathfrak{G}_{\text{int}}$. Lemma 4.1 now shows us that $\{c_{s+P}\} = \bigcap \{t^* - W_0 \mid t \in s + P\}$, thus $\bigcap \{(t^* - W_0) \backslash V' \mid t \in s + P\} = \varnothing$. As every $(t^* - W_0) \backslash V'$ is compact, there is a finite subset F of $s + P$ such that $\bigcap \{(t^* - W_0) \backslash V' \mid t \in F\} = \varnothing$. In particular, there is a compact set $K \subseteq \mathfrak{G}$ with $\bigcap \{(t^* - W_0) \backslash V' \mid t \in (s + P) \cap K\} = \varnothing$, so $\bigcap \{t^* - W_0 \mid t \in (s + P) \cap K\} \subseteq V'$. From this it follows that, if $t \in \mathfrak{G}$, $P' \in X_{P_0}$, and $(s + t + P') \cap K = (s + P) \cap K$, then necessarily $c_{s+t+P'} \in V'$.

For any neighborhood V of 0 in $\mathfrak{G}$, we can conclude that, if $(P', P) \in U_{-s+K,V}$ (cf. (2.4)), then $(s + t, c_{s+t+P'}) \in (s + V) \times (c_{s+P} + V')$, where we have chosen

$t \in V$ such that $(t + P') \cap (-s + K) = P \cap (-s + K)$. The continuity of β follows from the observation that $\beta(P) = \pi((s, c_{s+P}))$ and $\beta(P') = \pi((s + t, c_{s+t+P'}))$.

Because X_{P_0} is compact, in order to show that β is onto it suffices to show that its image is dense in $\mathfrak{T}$. This follows if we note that the orbit of P_0 under $\mathfrak{G}$ is mapped onto the orbit of 0 in $\mathfrak{T}$ under the action of $\mathfrak{G}$. The latter is given by $\pi(\mathfrak{G} \times \{0\}) = \pi(\mathfrak{G} + \tilde{L})$ which is easily seen to be dense in $\mathfrak{T}$ in view of S3. $\square$

Before we continue to study the dynamical system $(X_{P_0}, \mathfrak{G})$, we would like to note a nice consequence of the previous proposition.

COROLLARY 4.4. *If defined as in* S4, W_0 *equals the closure of its interior.*

PROOF. Because L is generated by $P_0 - P_0$, L^* is generated by $(W_0 - W_0) \cap L^*$. Then, $\mathfrak{G}_{\text{int}}$ must be compactly generated because of S3. As a consequence, $\mathfrak{G} \times \mathfrak{G}_{\text{int}}$ is σ-compact, because $\mathfrak{G}$ was supposed to be σ-compact from the beginning. Because $\tilde{L}$ is discrete, it must be countable. Therefore, L^* is countable.

By Baire's category theorem, we can choose an element c of $\mathfrak{G}_{\text{int}}$ which is not in $L^* - \partial W_0$. Then, $(c + \partial W_0) \cap L^* = \varnothing$. From the surjectivity of β we get the existence of some $P \in X_{P_0}$ with $c_P = c$, hence $\overline{P^*} = c + W_0$. By the choice of c, P^* must be contained in the interior of $c + W_0$. $\square$

In the following, let us fix a $\mathfrak{G}$-invariant probability measure μ on X_{P_0}; on $\mathfrak{T}$, we have of course a unique $\mathfrak{G}$-invariant probability measure, namely the normalized Haar measure (observe that, because $\mathfrak{G}$-orbits are dense in $\mathfrak{T}$, every $\mathfrak{G}$-invariant measure is automatically also $\mathfrak{T}$-invariant; as a consequence, $\mathfrak{G}$-invariant subsets of $\mathfrak{T}$, such as $\pi(\mathfrak{G} + \partial W_0)$, have Haar measure 0 or 1). The $\mathfrak{G}$-mapping β from X_{P_0} onto $\mathfrak{T}$ induces a $\mathfrak{G}$-mapping

$$(4.5) \qquad \beta_\sharp : L^2(\mathfrak{T}) \to L^2(\mu) \quad \xi \mapsto \xi \circ \beta$$

According to the above remark, the image measure $\beta\mu$ must coincide with the Haar probability measure on $\mathfrak{T}$. Therefore, $\beta_\sharp$ is an isometry of $L^2(\mathfrak{T})$ into $L^2(\mu)$; we may identify $L^2(\mathfrak{T})$ with the corresponding subspace of $L^2(\mu)$.

As a consequence, in the dynamical spectrum of $(X_{P_0}, \mathfrak{G})$, there is always a component which is isomorphic to the dynamical spectrum of $(\mathfrak{T}, \mathfrak{G})$. The latter is easy enough to determine: the dual $\widehat{\mathfrak{T}}$ forms a total set of eigenfunctions in $L^2(\mathfrak{T})$. As remarked at the end of the last section, this does not say much about the diffraction spectrum of X_{P_0}. However, in the special case of regular model sets, the mapping $\beta_\sharp$ will turn out to be surjective, and, as a consequence, in this case the diffraction spectrum of $(X_{P_0}, \mathfrak{G})$ must be pure point.

A *regular model set* is, by definition, a repetitive FLC set P_0 which (after a translation) has a regular star map (4.1) such that

$$(4.6) \qquad \{t \in L \mid t^* \in W_0^\circ\} \subseteq P_0.$$

We assume now that P_0 is a regular model set and that a regular star map (4.1) is chosen in this fashion (in particular, we assume S5, i.e. that ∂W has Haar measure 0). Then, if $P \in X_{P_0}$ and $P \subseteq L$, $W := \overline{P^*}$,

$$(4.7) \qquad \{t \in L \mid t^* \in W^\circ\} \subseteq P.$$

In order to see this, take an approximative net $(t_{K_i} + P_0)_{i \in I}$ as in the proof of Lemma 4.1, i.e., $(t_{K_i} + P_0) \cap K_i = P \cap K_i$ for all $i \in I$, the K_i cover $\mathfrak{G}$, and $t^*_{K_i}$ converge to some $c \in \mathfrak{G}_{\text{int}}$. If $t \in L$ and $t^* \in c + W_0^\circ$, i.e., $c \in t^* - W_0^\circ$, then there is some $i \in I$ with $t \in K_i$ and $t^*_{K_i} \in t^* - W_0^\circ$, hence $t \in (t_{K_i} + P_0) \cap K_i$. By construction, $t \in P$.

The main thing to note is that now the torus parametrization β becomes 1–1 on a huge portion of X_{P_0}. Namely, if c is chosen such that $(c + \partial W_0) \cap L^* = \varnothing$, then, by (4.7), if $P \subseteq L$, $W := \overline{P^*} = c + \partial W_0$,

(4.8) $$P = \{t \in L \mid t^* \in W^\circ\}.$$

As a consequence, P is uniquely determined by c. We call the elements of X_{P_0} which are translates of such P *generic*; let $X^g_{P_0}$ be the set of all generic elements of X_{P_0}. By construction, β is injective on $X^g_{P_0}$, and $\beta(X^g_{P_0})$ and $\beta(X_{P_0} \setminus X^g_{P_0})$ are disjoint. The image of $X^g_{P_0}$ under β is the set $\pi(\mathfrak{G} + \partial W_0)$, which is of Haar measure 1. (This shows that the dynamical system $(X_{P_0}, \mathfrak{G})$ is an "almost 1–1 extension" of a minimal rotation action on a "torus", a result that was also obtained, in the Euclidean case, in [**FHK**].) This has an important consequence:

THEOREM 4.5. *Let P_0 be a regular model set. Then*

(1) *The dynamical system of P_0 is uniquely ergodic (informally: frequencies of finite patterns exist uniformly, cf. Section 3).*
(2) *The dynamical spectrum is isomorphic to the dynamical spectrum of $(\mathfrak{T}, \mathfrak{G})$; in particular, it is pure point (as a consequence, the diffraction measure of P_0 is pure point).*

PROOF. (1) We show that our $\mathfrak{G}$-invariant probability measure μ is uniquely determined. As observed above, its image under β must be the Haar probability measure which we will temporarily denote by h. We note that $X^g_{P_0}$ is measurable as pre-image of a measurable set under β. For a measurable $A \subseteq X^g_{P_0}$, the image $\beta(A)$ is measurable, too (to see this, look at relatively closed subsets which generate the σ-algebra of measurable subsets of $X^g_{P_0}$). Furthermore, $\beta^{-1}\beta(A) = A$. We obtain $\mu(X^g_{P_0}) = \mu(\beta^{-1}\beta(X^g_{P_0})) = \beta\mu(\beta(X^g_{P_0})) = h(\beta(X^g_{P_0})) = 1$. By the same calculation, for arbitrary measurable $A \subseteq X_{P_0}$, $\mu(A) = \mu(A \cap X^g_{P_0}) = h(\beta(A \cap X^g_{P_0}))$. Now, the right hand side is independent of the choice of the invariant measure μ.

(2) As noted above, $X^g_{P_0}$ has measure 1; it is, of course, $\mathfrak{G}$-invariant. Therefore, $L^2(\mu \downarrow_{X^g_{P_0}})$ is $\mathfrak{G}$-isomorphic to $L^2(\mu)$. The former is, via $\beta \downarrow_{X^g_{P_0}}$, $\mathfrak{G}$-isomorphic to $L^2(h \downarrow_{X^g_{P_0}})$, which in turn is $\mathfrak{G}$-isomorphic to $L^2(\mathfrak{T})$. The assertion follows. $\square$

Let us put the carrier of the spectrum in a more familiar form. We can identify $\widehat{\mathfrak{T}}$ with the annihilator $\tilde{L}^\perp = \{\chi \in (\mathfrak{G} \times \mathfrak{G}_{\text{int}})\widehat{} \mid \chi(g) = 1 (g \in \tilde{L})\}$ of $\tilde{L}$ in the dual of $\mathfrak{G} \times \mathfrak{G}_{\text{int}}$. The latter can be identified with $\widehat{\mathfrak{G}} \times \widehat{\mathfrak{G}}_{\text{int}}$. Sometimes, $\tilde{L}^\perp$ is called the *reciprocal* or *dual* lattice of $\tilde{L}$. For $\chi \in \widehat{\mathfrak{G}} \times \widehat{\mathfrak{G}}_{\text{int}}$, let $\chi_{\|}$ be the restriction of χ to $\mathfrak{G}$; $\chi_{\|}$ can be identified with the projection image of χ under the canonical projection of $\widehat{\mathfrak{G}} \times \widehat{\mathfrak{G}}_{\text{int}}$ onto the first coordinate. Hence, the dynamical spectrum of $(\mathfrak{T}, \mathfrak{G})$ and, under the conditions of the above theorem, of $(X_{P_0}, \mathfrak{G})$ is concentrated on the projection image of the reciprocal lattice of $\tilde{L}$ in $\widehat{\mathfrak{G}}$.

Conclusion

The work presented here settles the Fourier analysis and diffraction properties of general regular model set. However, it seems possible to explore the techniques further in order to investigate an even more general class of point sets: namely those sets which are repetitive and uniquely ergodic and admit a regular star map in the fashion of (4.1) without necessarily being model sets themselves. In this case, the analysis in the preceding section carries as far as (4.5) which injects the L^2 of the "torus" which carries a discrete spectrum into the L^2 of the dynamical system. Therefore, the *dynamical* spectrum will contain a pure point component, and one might be tempted to conclude that the diffraction spectrum must contain a pure point component as well. Unfortunately, the argumentation in Section 4 does not allow to transfer this result to the diffraction spectrum; although it seems very likely to be true, I was not able to rigorously establish that the part of the L^2 of the dynamical system containing this pure point component is not orthogonal to the part generated by the continuous functions contributing to (3.9).

Another question which, although perhaps not of much practical relevance, keeps puzzling me is the question what happens if the boundary of the window has nonzero Haar measure. In this case, the techniques presented above break down completely; for example, the "torus" parametrization remains injective on a null set only which renders it next to useless. It would be of interest to know whether such a pattern can still be uniquely ergodic.

References

[A]   T. M. Apostol, *Introduction to analytic number theory*, Undergrad. Texts Math., Springer, New York, 1976.

[AG]   L. Argabright and J. Gil de Lamadrid, *Fourier analysis of unbounded measures on locally compact Abelian groups*, Mem. Amer. Math. Soc., vol. 145, Amer. Math. Soc., Providence, RI, 1974.

[BHP]   M. Baake, J. Hermisson, and P. A. B. Pleasants, *The torus parametrization of quasiperiodic LI-classes*, J. Phys. A **30** (1997), 3029–3056.

[BMS]   M. Baake, R. V. Moody, and M. Schlottmann, *Limit-(quasi)periodic point sets as quasicrystals with p-adic internal spaces*, J. Phys. A **31** (1998), 5755–5765; math-ph/9901008.

[C]   J. M. Cowley, *Diffraction physics*, North-Holland Personal Library, 3rd ed., North-Holland, Amsterdam, 1995.

[DK]   M. Duneau and A. Katz, *Quasiperiodic patterns*, Phys. Rev. Lett. **54** (1985), 2688–2691.

[D]   S. Dworkin, *Spectral theory and X-ray diffraction*, J. Math. Phys. **34** (1993), 2965–2967.

[E]   R. Ellis, *Lectures on topological dynamics*, Benjamin, New York, 1969.

[FHK]   A. Forrest, J. Hunton, and J. Kellendonk, *Projection quasicrystals. I. Toral rotations*, preprint no. 340, SFB 288, Berlin, 1998.

[F]   H. Furstenberg, *Recurrence in ergodic theory and combinatorial number theory*, Princeton Univ. Press, Princeton, 1981.

[HR]   E. Hewitt and K. A. Ross, *Abstract harmonic analysis. I. Structure of topological groups, integration theory, group representations*, 2nd ed., Grundlehren Math. Wiss., vol. 115, Springer, Berlin, 1979.

[H]   A. Hof, *Uniform distribution and the projection method*, Quasicrystals and Discrete Geometry (Toronto, ON, 1995) (J. Patera, ed.), Fields Inst. Monogr., vol. 10, Amer. Math. Soc., Providence, RI, 1998, pp. 201–206.

[KKL]   P. A. Kalugin, A. Yu. Kitayev, and L. S. Levitov, *6-dimensional properties of $Al_{0.86}Mn_{0.14}$ alloy*, J. Physique Lett. **46** (1985), L601–L607.

[K]   J. L. Kelley, *General topology*, Van Nostrand, Princeton, NJ, 1955; reprint, Grad. Texts in Math., vol. 27, Springer, New York, 1975.

[KN]   P. Kramer and R. Neri, *On periodic and nonperiodic space fillings of E^m obtained by projection*, Acta Cryst. A **40** (1984), 580–587; Erratum, **41** (1985), 619.

[La]   S. Lang, *Real and functional analysis*, 3rd ed., Grad. Texts in Math., vol. 142, Springer, New York, 1993.

[LP]   W. F. Lunnon and P. A. B. Pleasants, *Quasicrystallographic tilings*, J. Math. Pures Appl. (9) **66** (1987), 217–263.

[Me]   Y. Meyer, *Algebraic numbers and harmonic analysis*, North-Holland Math. Library, vol. 2, North-Holland, Amsterdam, 1972.

[M] R. V. Moody, *Meyer sets and their duals*, The Mathematics of Long-Range Aperiodic Order (Waterloo, ON, 1995) (R. V. Moody, ed.), NATO ASI Ser. C: Math. Phys. Sci., vol. 489, Kluwer, Dordrecht, 1997, pp. 403–41.

[RW] C. Radin and M. Wolff, *Space tilings and local isomorphism*, Geom. Dedicata **42** (1992), 355–560.

[RN] F. Riesz and B. Sz.-Nagy, *Functional analysis*, Ungar, New York, 1955; reprint, Dover, New York, 1990.

[Sch] M. Schlottmann, *Cut-and-project sets in locally compact Abelian groups*, Quasicrystals and Discrete Geometry (Toronto, ON, 1995) (J. Patera, ed.), Fields Inst. Monogr., vol. 10, Amer. Math. Soc., Providence, RI, 1998, pp. 247–264.

[So1] B. Solomyak, *Dynamics of self-similar tilings*, Ergodic Theory Dynam. Systems **17** (1997), 695–738.

[So2] _____, *Spectrum of dynamical systems arising from Delone sets*, Quasicrystals and Discrete Geometry (Toronto, ON, 1995) (J. Patera, ed.), Fields Inst. Monogr., vol. 10, Amer. Math. Soc., Providence, RI, 1998, pp. 265–275.

[T] F. Trèves, *Topological vector spaces, distributions and kernels*, Academic Press, New York, 1967.

DEPARTMENT OF MATHEMATICAL SCIENCES, UNIVERSITY OF ALBERTA, EDMONTON, ALBERTA, CANADA T6G 2G1

E-mail address: `martin@miles.math.ualberta.ca`

Centre de Recherches Mathématiques
CRM Monograph Series
Volume **13**, 2000

On Shelling Icosahedral Quasicrystals

Alfred Weiss

ABSTRACT. A quaternionic construction of three-dimensional model sets with icosahedral symmetry induces a decomposition of them into concentric shells whose structure can be determined. The methods originate from the theory of orders in the quaternion algebra and their connection to the arithmetic of quadratic extensions of $\mathbb{Q}(\sqrt{5}\,)$.

Introduction

Perhaps the simplest quasicrystals with icosahedral symmetry in Euclidean space $\mathbb{R}^3$ are those associated [**CMP**] to a root system of type H_3. These can be obtained by cut and project methods from the root lattice of type D_6 which gives them a shelling in the manner of [**SM**]. The purpose of this paper is to exhibit some of the structure of these shells by using the arithmetic of quaternion algebras. For applications and related shelling problems we refer to [**BGJR**] and references given there.

The usual quaternion algebra $\mathbb{H}$, with $\mathbb{R}$-basis 1, i, j, k and defining relations $i^2 = j^2 = k^2 = ijk = -1$, plays its usual role. Namely the 'pure' quaternions $\mathbb{H}_0$, with $\mathbb{R}$-basis i, j, k, are identified with $\mathbb{R}^3$ and the conjugation action of the group $\mathbb{H}^\times$ of units of $\mathbb{H}$ on $\mathbb{H}_0$ defines a group homomorphism $\mathbb{H}^\times \to \mathrm{SO}(3,\mathbb{R})$ which 'carries' the Euclidean structure.

There is, for any field F, a quaternion algebra $\mathbb{H}(F)$ with F-basis 1, i, j, k and the same defining relations as above. When F is a subfield of $\mathbb{R}$ then structures definable on $\mathbb{H}(F)$ induce structures on $\mathbb{H} = \mathbb{R} \otimes_F \mathbb{H}(F)$. When F is an algebraic number field then $\mathbb{H}(F)$ has a rich 'extra' structure which may be exploited to construct the model sets Σ_E^r of interest here. This is carried out at some length in the first three sections, based on the theory of orders in $\mathbb{H}(F)$ (definitions in [**R**, Chapter 2]). We first give computational proofs, but also sketch a more conceptual approach which shows that the situation is very canonical.

2000 *Mathematics Subject Classification*. Primary 11R52, 52C23; Secondary 11E25, 16H05.

The author is grateful to the Natural Sciences and Engineering Research Council (NSERC) of Canada for its support.

©2000 American Mathematical Society

In Section 4, the structure of the shells $\Sigma_E^r(n)$ is discussed in terms of a pair of counting formulae

$$|\Sigma_E^r(n)| = \sum_{e \in E} \sum_{m \in S(4n,2r)} |X(4n + m\tau, e)|,$$

with

$$|X(b,e)| = 120 \frac{h(F(\sqrt{-b}))}{w(F(\sqrt{-b}))} N(\mathfrak{c}) \prod_{\mathfrak{p}} \left(1 + (1 - \chi(\mathfrak{p})) \frac{1 - N(\mathfrak{p})^{-c_\mathfrak{p}}}{N(\mathfrak{p}) - 1}\right).$$

in Lemma 4.1, Theorem 4.2, respectively, with $F = \mathbb{Q}(\sqrt{5})$. The sets $S(n,r)$ have a simple structure but $X(b,e)$ depends on the arithmetic of the field $F(\sqrt{-b})$.

The proof of Theorem 4.2 takes the next three sections. The first of these decomposes $X(b,e)$ into subsets parametrized by suborders T in $F(\sqrt{-b})$ in Corollary 5.4; these subsets count certain imbeddings of T in $\mathbb{H}(F)$. In Section 6 the subsets are analyzed via the idea of the Skolem-Noether theorem ([R, 7.21]), and then Section 7 combines the results.

Finally Section 8 ties up some loose ends, both number theoretic and historical. In particular, an analogue (with F replaced by $\mathbb{Q}$) of our formula for $|X(b,e)|$ was already known to Gauss, so that there are, by now, many approaches to it: we end with some discussion of these.

1. Icosians

Every quaternion algebra $\mathbb{H}(F)$ has a reduced trace and a reduced norm defined by $\operatorname{tr}(a) = \bar{a} + a$, $\operatorname{nr}(a) = a\bar{a}$, where $^-$ is the standard conjugation on $\mathbb{H}(F)$,

$$a_0 1 + a_1 i + a_2 j + a_3 k \mapsto a_0 1 - a_1 i - a_2 j - a_3 k.$$

A more intrinsic definition [R, Section 9] is that tr, nr correspond to the usual matrix trace, determinant under any isomorphism of $K \otimes_F \mathbb{H}(F) = \mathbb{H}(K)$ with the ring $\operatorname{M}_2(K)$ of 2×2 matrices over any extension field K of F, e.g., $\mathbb{H}(\mathbb{C}) \xrightarrow{\sim} \operatorname{M}_2(\mathbb{C})$ sending $1, i, j, k$ to

$$\begin{bmatrix} 1 & 0 \\ 0 & 1 \end{bmatrix}, \begin{bmatrix} \sqrt{-1} & 0 \\ 0 & -\sqrt{-1} \end{bmatrix}, \begin{bmatrix} 0 & 1 \\ -1 & 0 \end{bmatrix}, \begin{bmatrix} 0 & \sqrt{-1} \\ \sqrt{-1} & 0 \end{bmatrix}.$$

Then $^-$ can be defined by $\bar{a} = \operatorname{tr}(a)\mathbf{1} - a$ and all properties we need are simple matrix calculations; in particular, $a^2 - \operatorname{tr}(a)a + \operatorname{nr}(a)\mathbf{1} = 0$ is the Cayley-Hamilton theorem.

To see icosahedral symmetry, we wish to choose $F \subset \mathbb{R}$ so that $\mathbb{H}(F)^\times$ has a finite subgroup I whose image in $\operatorname{SO}(3,\mathbb{R})$ is the icosahedral group and then form the 'icosian' ring $\mathbb{I}$ consisting of $\mathbb{Z}$-linear combinations of I. Since the quaternion group $Q_8 = \langle i, j \rangle$ is present for every F, the issue is to find an element α of order 5. But $\operatorname{nr}: \mathbb{H}^\times \to \mathbb{R}^\times$ is a homomorphism with image consisting of positive reals so $\operatorname{nr}(I) = 1$ and $\alpha + \alpha^{-1} = \alpha + \bar{\alpha} = \operatorname{tr}(\alpha)$ is in F. Since $\alpha^4 + \alpha^3 + \alpha^2 + \alpha + 1 = 0$ implies $-(\alpha + \alpha^{-1})$ is a root of $X^2 - X - 1$ it follows that F contains $\frac{1 \pm \sqrt{5}}{2}$.

From now on, $F = \mathbb{Q}(\tau)$ with $\tau = \frac{1}{2}(1 + \sqrt{5})$ and $'$ is the nontrivial $\mathbb{Q}$-automorphism of F, i.e., $\tau' = 1 - \tau = \frac{1}{2}(1 - \sqrt{5})$. Then $R = \mathbb{Z}[\tau]$ is the ring of algebraic integers of F and is a Dedekind ring (even a principal ideal domain) with group of units $R^\times$ generated by -1 and τ, i.e., $R^\times = \{\pm \tau^m \mid m \in \mathbb{Z}\}$.

Now the R-span Λ of Q_8 is an R-order, i.e. a subring which is finitely generated as R-module [**R**, Section 8], and our discussion above makes $\mathbb{I}$ an R-order and even forces us to take $\mathbb{I} \supseteq \Lambda$ (up to conjugacy in $\mathbb{H}(F)$: for Q_8 is unique up to conjugacy in $\mathbb{H}(F)^\times$, by ([**R**, 7.21]), because Q_8 spans $\mathbb{H}(F)$). Since Λ is an R-order its complementary module

$$\tilde{\Lambda} = \{a \in \mathbb{H}(F) \mid \operatorname{tr}(a\Lambda) \subseteq R\}$$

contains it [**R**, Section 25], and indeed we have $\tilde{\Lambda} = \frac{1}{2}\Lambda$ since 1, i, j, k has dual basis $\frac{1}{2}$, $-\frac{i}{2}$, $-\frac{j}{2}$, $-\frac{k}{2}$ (relative to tr). Thus $\Lambda \subseteq \mathbb{I} \subseteq \tilde{\mathbb{I}} \subseteq \tilde{\Lambda} = \frac{1}{2}\Lambda$ bounds $\mathbb{I}$ above.

We still have to find α. But knowing α is in $\frac{1}{2}\Lambda$ (if it exists at all in $\mathbb{H}(F)$) and that $\operatorname{nr}(\alpha) = 1$, $\operatorname{tr}(\alpha) = \tau$ or τ' narrows the search considerably, especially as we may choose to make $\operatorname{tr}(\alpha) = -\tau'$ (as then $\operatorname{tr}(\alpha') = -\tau$, where $'$ is defined to fix 1, i, j, k): this choice has $-\tau' = 2\cos(2\pi/5)$.

PROPOSITION 1.1. *Put $\alpha = -\frac{\tau'}{2} - \frac{\tau}{2}i - \frac{1}{2}k$. Then $Q_8 \cup \alpha Q_8$ has R-span a maximal R-order $\mathbb{I}$ of $\mathbb{H}(F)$. We have $\tilde{\mathbb{I}} = \mathbb{I}$.*

PROOF. The R-span is an R-order because it is closed under multiplication, as follows from $i\alpha = j + \alpha i$, $j\alpha = -\tau' j - \alpha j$. To show that $\tilde{\mathbb{I}} = \mathbb{I}$ amounts to showing that the discriminant $d(\mathbb{I}/R)$ is the unit ideal of R ([**R**, 25.2] and its proof). But $d(\mathbb{I}/R) \supseteq d(\Lambda/R) = 2^4 R$ by $\tilde{\Lambda} = \frac{1}{2}\Lambda$, so it suffices to show that $d(\Gamma/R) = 5R$ where $\Gamma \subseteq \mathbb{I}$ is the R-order generated by α and j. This is actually a natural observation: $j\alpha j^{-1} = \bar{\alpha}$ since α has j-coefficient 0, hence conjugation by j is the nontrivial automorphism of $F(\alpha)/F$ and, as $j^2 = -1$, Γ is a crossed product order which makes the discriminant calculation easy (cf. $(*)$ on p. 592 of [**CR**]). Actually Λ is also a crossed product order, as conjugation by j is the nontrivial automorphism of $F(i)/F$, so $\tilde{\Lambda} = \frac{1}{2}\Lambda$ could have been seen this way. Thus $\tilde{\mathbb{I}} = \mathbb{I}$ and we may forget Λ and Γ.

If now Ω is any R-order containing $\mathbb{I}$ then $\mathbb{I} \subseteq \Omega \subseteq \tilde{\Omega} \subseteq \tilde{\mathbb{I}} = \mathbb{I}$ implies $\Omega = \mathbb{I}$, verifying maximality of $\mathbb{I}$, and completing the proof. □

COROLLARY 1.2. *The mapping $\operatorname{tr}: \mathbb{I} \to R$ is surjective.*

PROOF. If this were false, then $\operatorname{tr}(\mathbb{I})$ would be contained in a maximal ideal $\mathfrak{p} \neq R$. Then $\operatorname{tr}(\mathbb{I} \cdot \mathfrak{p}^{-1}\mathbb{I}) \subseteq R$ would imply $\mathfrak{p}^{-1}\mathbb{I} \subseteq \tilde{\mathbb{I}} = \mathbb{I}$. Thus $\mathbb{I} \subseteq \mathfrak{p}\mathbb{I}$, a contradiction by the above proposition. □

The fact that the F-algebra $\mathbb{H}(F)$ has an R-order $\mathbb{I}$ with $\tilde{\mathbb{I}} = \mathbb{I}$, and that $\mathbb{H}(F)$ is contained in the division algebra $\mathbb{H}$ under both imbeddings of F in $\mathbb{R}$ (namely the inclusion and $'$ followed by the inclusion) actually classifies $\mathbb{H}(F)$ up to F-isomorphism (by [**R**, p. 272–274] as all indices at finite primes are 1 and at all infinite primes are 2). Conversely, we could deduce the existence of a maximal order $\mathbb{I}$ in $\mathbb{H}(F)$ with $\tilde{\mathbb{I}} = \mathbb{I}$ by first computing all local indices. This can be done via $\mathbb{H}(F) = F \otimes_\mathbb{Q} \mathbb{H}(\mathbb{Q})$ because of [**R**, 31.90] and the fact that $\mathbb{H}(\mathbb{Q})$ has index 2 at the primes 2, ∞ and index 1 at all others ([**R**, p. 271], and its analogue with $\mathbb{R}$ replaced by $\mathbb{Q}_2$): the point here is that the prime 2 of $\mathbb{Q}$ does not split in F, while ∞ does.

Since $X^2 - X - 1$ is irreducible over $\mathbb{F}_2 = \mathbb{Z}/2\mathbb{Z}$ the ring $R/2R$ is a field and $2R$ is a prime ideal of R. The special role of 2, not surprising in view of the above, will continue. We prepare with

LEMMA 1.3. *Put* $\mathbb{F}_4 = R/2R$. *Then* $\mathbb{I}/2\mathbb{I} \simeq M_2(\mathbb{F}_4)$. *Indeed, there is an isomorphism sending* (*the images of*) α, i, j *in* $\mathbb{I}/2\mathbb{I}$ *to* (*the images of*)
$$\begin{bmatrix} \tau' & 1 \\ 1 & 0 \end{bmatrix}, \begin{bmatrix} \tau' & \tau \\ \tau & \tau' \end{bmatrix}, \begin{bmatrix} 0 & 1 \\ 1 & 0 \end{bmatrix}$$
in $M_2(\mathbb{F}_4)$.

PROOF. Since $d(\mathbb{I}/R) = R$, by Proposition 1.1, $\mathbb{I}/2\mathbb{I}$ is a central simple $\mathbb{F}_4$-algebra (by [**R**, 25.4]), and therefore isomorphic to $M_2(\mathbb{F}_4)$ (by [**R**, 7.24]).

This will also follow from the construction of a homomorphism $\mathbb{I} \twoheadrightarrow M_2(\mathbb{F}_4)$ (which it predicts) since $2\mathbb{I}$ is necessarily in the kernel. We first send α to a matrix of trace $-\tau'$, determinant 1: this must work because of the base change properties of reduced trace, norm. Then we choose an image for j so that the crossed product relations $j\alpha j^{-1} = \alpha^{-1}$, $j^2 = -1$ (in the proof of Proposition 1.1) are respected. Since this crossed product order has odd index in $\mathbb{I}$ (as we saw *loc. cit.*) the ring homomorphism extends uniquely from the crossed product order to $\mathbb{I}$. Then, as $(2+\tau)i = (\bar{\alpha} - \alpha)(\tau - j) = (\text{tr}(\alpha) - 2\alpha)(\tau - j)$ in $\mathbb{I}$, we are forced to map i to the claimed matrix over $\mathbb{F}_4$. Is is clear that these matrices generate $M_2(\mathbb{F}_4)$ as $\mathbb{F}_4$-algebra. □

We have, however, neglected the icosahedral symmetry, i.e., exhibiting a finite subgroup I of $\mathbb{I}^\times$ which spans $\mathbb{I}$ as $\mathbb{Z}$-module and has icosahedral image in $SO(3, \mathbb{R})$. We could exhibit I now, but it's easier to return to this point later (cf. Lemma 2.3).

2. Pure Icosians and the Quasicrystals

Let $\mathbb{H}(F)_0$ be the kernel of $\text{tr}: \mathbb{H}(F) \to F$. Then $a \mapsto (\frac{1}{2}\text{tr}(a), a - \frac{1}{2}\text{tr}(a))$ gives the decomposition $\mathbb{H}(F) \xrightarrow{\simeq} F \oplus \mathbb{H}(F)_0$ of vector spaces over F; this is the same as the decomposition into eigenspaces under the standard conjugation $\bar{}$. Because of the $\frac{1}{2}$ this map does not find an R-module complement to $\mathbb{I}_0 = \mathbb{I} \cap \mathbb{H}(F)_0$ in our maximal order $\mathbb{I}$; the deviation is measured by $\tilde{\mathbb{I}}_0 = \{a \in \mathbb{H}(F)_0 \mid \text{tr}(a\mathbb{I}_0) \subseteq R\}$.

LEMMA 2.1. $\tilde{\mathbb{I}}_0 = \{\frac{1}{2}x \mid x \in \mathbb{I}_0 \text{ with } x + 2\mathbb{I} \in \mathbb{F}_4\}$.

PROOF. Here $\mathbb{F}_4$ is identified with the image of $R \to \mathbb{I}/2\mathbb{I}$, $r \mapsto r + 2\mathbb{I}$. Writing $E(a) = a - \frac{1}{2}\text{tr}(a)$, the critical point is

CLAIM. $\tilde{\mathbb{I}}_0 = E(\tilde{\mathbb{I}})$.

PROOF OF CLAIM. We have
$$ab = \left(\tfrac{1}{2}\text{tr}(a) + E(a)\right)\left(\tfrac{1}{2}\text{tr}(b) + E(b)\right)$$
$$= \tfrac{1}{4}\text{tr}(a)\text{tr}(b) + \tfrac{1}{2}\text{tr}(a)E(b) + \tfrac{1}{2}E(a)\text{tr}(b) + E(a)E(b)$$
with first term in F and the next two terms in $\mathbb{H}(F)_0$; thus $\text{tr}(ab) = \frac{1}{2}\text{tr}(a)\text{tr}(b) + \text{tr}(E(a)E(b))$ and, in particular, $\text{tr}(ab) = \text{tr}(aE(b))$ for $a \in \mathbb{H}(F)_0$. Therefore $a \in \mathbb{H}(F)_0$ has $\text{tr}(a\mathbb{I}) \subseteq R$ if and only if $\text{tr}(aE(\mathbb{I})) \subseteq R$. Since $\tilde{\mathbb{I}} = \mathbb{I}$ this implies that $\tilde{\mathbb{I}}_0 = \tilde{\mathbb{I}} \cap \mathbb{H}(F)_0 = \widetilde{E(\mathbb{I})}$ hence $\tilde{\mathbb{I}}_0 = \widetilde{E(\mathbb{I})} = E(\mathbb{I})$, proving the claim. □

Now if $a \in \tilde{\mathbb{I}}_0$ is written $E(b)$ with b in $\mathbb{I}$ then $a = \frac{1}{2}x$ with $x = 2E(b) = 2b - \text{tr}(b)$ in $\mathbb{I}_0$ having $x + 2\mathbb{I} = \text{tr}(b) + 2\mathbb{I} \in \mathbb{F}_4$. Conversely, if x in $\mathbb{I}_0$ has $x + 2\mathbb{I}$ in $\mathbb{F}_4$ then $x + r = 2b$ for some $r \in R$, $b \in \mathbb{I}$. Taking the reduced trace shows $r = \text{tr}(b)$ hence $\frac{1}{2}x = \frac{1}{2}(2b - \text{tr}(b)) = E(b)$ is in $\tilde{\mathbb{I}}_0$, and the lemma is proved. □

Turning now to the construction, via the order $\mathbb{I}$, of the quasicrystals with which we are concerned, we first should recall that the reduced norm of $a_0+a_1i+a_2j+a_3k$ is $a_0^2+a_1^2+a_2^2+a_3^2$; hence the standard inner product on $\mathbb{H} \simeq \mathbb{R}^4$ is given by polarization of the reduced norm:

$$(2.1) \qquad a \cdot b = \tfrac{1}{2}\bigl(\mathrm{nr}(a+b) - \mathrm{nr}(a) - \mathrm{nr}(b)\bigr) = \tfrac{1}{2}\,\mathrm{tr}(a\bar{b}).$$

This is, of course, valid also for $\mathbb{H}(F) \subseteq \mathbb{H}$ and, indeed, also for the 'other' imbedding of $\mathbb{H}(F)$ in $\mathbb{H}$ given by applying $'$ to the coefficients of $1, i, j, k$ and then inclusion into $\mathbb{H}$. Each of these imbeddings $\mathbb{H}(F) \to \mathbb{H}$ induces an $\mathbb{R}$-algebra homomorphism $\mathbb{R} \otimes_\mathbb{Q} \mathbb{H} \to \mathbb{H}$ which, taken together, gives an $\mathbb{R}$-algebra isomorphism

$$(2.2) \qquad \mathbb{R} \otimes_\mathbb{Q} \mathbb{H}(F) \to \mathbb{H} \times \mathbb{H}$$

(because this is just $\_ \otimes_\mathbb{Q} \mathbb{H}(\mathbb{Q})$ applied to $\mathbb{R} \otimes_\mathbb{Q} F \xrightarrow{\sim} \mathbb{R} \times \mathbb{R}$). We fix notation so that the map is just $r \otimes a \mapsto (ra, ra')$, and note that $\mathbb{R} \otimes_\mathbb{Q} \mathbb{H}(F)$ is thus Euclidean space $\mathbb{R}^8$.

PROPOSITION 2.2. *Under $a \mapsto 1 \otimes a$ the maximal R-order $\mathbb{I}$ is imbedded as a lattice in $\mathbb{R} \otimes_\mathbb{Q} \mathbb{H}(F)$.*

PROOF. Since $\mathbb{I}$ contains an F-basis of $\mathbb{H}(F)$, which gives a $\mathbb{Q}$-basis of $\mathbb{H}(F)$ by multiplying by $1, \tau \in R$, we see that $\mathbb{I}$ spans $\mathbb{H}(F)$ over $\mathbb{Q}$ and thus $\mathbb{R} \otimes_\mathbb{Q} \mathbb{H}(F)$ over $\mathbb{R}$.

The discreteness of $\mathbb{I}$ in $\mathbb{R} \otimes_\mathbb{Q} \mathbb{H}(F)$ will follow if we can show that the product of unit spheres in $\mathbb{H} \times \mathbb{H}$ contains only the origin in $\mathbb{I}$. But if $a \in \mathbb{I}$ has (a, a') in this product of spheres then, writing $\mathrm{nr}(a) = n + m\tau$ with n, m in $\mathbb{Z}$, we have $0 \le n + m\tau < 1$ and $0 \le n + m\tau' < 1$. Adding the negative of the second of these to the first gives $-1 < m\sqrt{5} < 1$ hence $m = 0$. But then $n = 0$, $\mathrm{nr}(a) = 0$ and $a = 0$ follow, completing the proof. $\square$

It is precisely the $\mathbb{Q}$-automorphism $'$ of $\mathbb{H}(F)$ which is the point in our construction:

$$(2.3) \qquad \begin{array}{c} \mathbb{H} \longleftarrow \mathbb{R} \otimes_\mathbb{Q} \mathbb{H}(F) \xrightarrow{\;\;'\;\;} \mathbb{H} \\ \uparrow \\ \mathbb{I} \end{array}$$

As both of our projections $\mathbb{R} \otimes_\mathbb{Q} \mathbb{H}(F) \to \mathbb{H}$ are injective on the imbedded $\mathbb{I}$ (because they are on the imbedded $\mathbb{H}(F)$) and the image of $\mathbb{I}$ in $\mathbb{H}$ is dense (because $\mathbb{I} \supseteq R1 + Ri + Rj + Rk$ with the image of R in $\mathbb{R}$ dense) we have a cut and project scheme which defines a model set, in the sense of [**M**, Section 2], for each choice of window in $\mathbb{H}$. We use the sphere of radius r for this and so get the quasicrystal

$$(2.4) \qquad \Sigma^r = \{a \in \mathbb{I} \mid \mathrm{nr}(a') < r^2\}$$

in $\mathbb{H} \simeq \mathbb{R}^4$ (by inclusion), which is closely related to that described in [**ES**]. Of course, this Σ^r is 4-dimensional. But simply replacing $\mathbb{H}(F), \mathbb{I}$ by $\mathbb{H}(F)_0, \mathbb{I}_0$ immediately gives a 3-dimensional one:

$$(2.5) \qquad \begin{array}{c} \mathbb{H}_0 \longleftarrow \mathbb{R} \otimes_\mathbb{Q} \mathbb{H}(F)_0 \xrightarrow{\;\;'\;\;} \mathbb{H}_0 \\ \uparrow \\ \mathbb{I}_0 \end{array}$$

i.e. $a \mapsto 1 \otimes a$ imbeds $\mathbb{I}_0$ as a lattice in $\mathbb{R} \otimes_{\mathbb{Q}} \mathbb{H}(F)_0$, and we get a cut and project scheme with model set

$$(2.6) \qquad \Sigma_0^r = \{a \in \mathbb{I}_0 \mid \text{nr}(a') < r^2\}$$

in $\mathbb{H}_0 \simeq \mathbb{R}^3$.

Generalizing slightly, each subgroup E of $\mathbb{F}_4$ gives a subgroup

$$(2.7) \qquad \mathbb{I}_E = \{\tfrac{1}{2}x \mid x \in \mathbb{I}_0 \text{ with } x + 2\mathbb{I} \in E\}$$

with $\mathbb{I}_0 \subseteq \mathbb{I}_E \subseteq \tilde{\mathbb{I}}_0$ because of Lemma 2.1. Since $\tilde{\mathbb{I}}_0 \subseteq \tfrac{1}{2}\mathbb{I}_0$, each of these plays the same role as $\mathbb{I}_0$, so we get model sets

$$(2.8) \qquad \Sigma_E^r = \{a \in \mathbb{I}_E \mid \text{nr}(a') < r^2\}$$

in $\mathbb{H}_0 \simeq \mathbb{R}^3$.

Consider the subgroup

$$(2.9) \qquad I = \{a \in \mathbb{I} \mid \text{nr}(a) = 1\}$$

of $\mathbb{I}^\times$. Since nr is multiplicative and each E is central in $M_2(\mathbb{F}_4)$ it is immediate that conjugation by I maps every Σ_E^r into itself. On the other hand, I is finite, being a compact and discrete subset of $\mathbb{R} \otimes_{\mathbb{Q}} \mathbb{H}(F)$ under the imbedding of Proposition 2.2, and spans $\mathbb{I}$ as a $\mathbb{Z}$-module (since α and Q_8 are in I and $\tau' = -\alpha - \alpha^{-1}$, and thus $\tau = 1 - \tau'$, are in its $\mathbb{Z}$-span). So the icosahedral symmetry of Σ_E^r and the gap at the end of Section 1 are both settled (since $\mathbb{H}^\times \to \text{SO}(3, \mathbb{R})$ has kernel $\mathbb{R}^\times$, with only ± 1 of finite order) by

LEMMA 2.3. $I/\langle \pm 1 \rangle$ *is isomorphic to the alternating group* A_5.

PROOF. We could appeal to knowledge of the finite subgroups of $\text{SO}(3, \mathbb{R})$ but prefer to use our homomorphism $\mathbb{I} \to M_2(\mathbb{F}_4)$ to show that the induced map $I \to M_4(\mathbb{F}_4)^\times = \text{GL}_2(\mathbb{F}_4)$ has kernel $\langle \pm 1 \rangle$ and image $\text{SL}_2(\mathbb{F}_4)$.

If $g \in I$ is in the kernel then $g = 1 + 2b$, $b \in \mathbb{I}$, implies $g^2 = 1 + 4(b + b^2)$. But no unit $\neq 1$ in $1 + 4\mathbb{I}$ can have finite order: for if $1 + 2^\nu b$, with $\nu \geq 2$ and $b \in \mathbb{I} \setminus 2\mathbb{I}$, has prime order p then the binomial expansion of $1 = (1 + 2^\nu b)^p$ implies $p \cdot 2^\nu b \in 2^{2\nu}\mathbb{I}$ hence $b \in 2^{\nu-1}\mathbb{I} \subset 2\mathbb{I}$, a contradiction. Thus $g^2 = 1$ and so $g = \pm 1$, since $\mathbb{H}(F)$ is a division algebra.

The image of I is in $\text{SL}_2(\mathbb{F}_4)$ because of the base change property of reduced norms. Since I contains α and Q_8 its image is a subgroup of $\text{SL}_2(\mathbb{F}_4)$ of order divisible by 20. This forces the image to be all of $\text{SL}_2(\mathbb{F}_4)$ by the structure of this group, or, alternatively, by the structure of A_5: for the action of $\text{SL}_2(\mathbb{F}_4)$ on the projective line over $\mathbb{F}_4$ induces an isomorphism $\text{SL}_2(\mathbb{F}_4) \to A_5$. □

Incidentally, the same argument with $2R$ replaced by the prime ideal $\sqrt{5}R$ shows that $\mathbb{I}/\sqrt{5}\mathbb{I} \simeq M_2(\mathbb{F}_5)$ and $I \simeq \text{SL}_2(\mathbb{F}_5)$.

3. Root Systems

Viewing $\mathbb{H}(F)$ inside $\mathbb{H} \simeq \mathbb{R}^4$, we use the terminology [**H**, 1.2] of (noncrystallographic) root systems. The set I is a root system of rank 4 in $\mathbb{R}^4$ [**H**, 2.13] and the set $I_0 = I \cap \mathbb{H}_0$ is a root system of rank 3 in $\mathbb{H}_0 \simeq \mathbb{R}^3$. That I, I_0 have type H_4, H_3 follows from the classification of root systems since $|I| = 120$ and

LEMMA 3.1. $|I_0| = 30$.

PROOF. Clearly $I_0 = \{g \in I \mid g^2 = -1\}$. Using the map $I \to \mathrm{SL}_2(\mathbb{F}_4)$ of Lemma 2.3, the image of I_0 is $\{X \in \mathrm{SL}_2(\mathbb{F}_4) \mid X \neq 1, X^2 = 1\}$: this is because $g \in I$, $g^2 \equiv 1 \bmod 2\mathbb{I}$ implies $g^2 = \pm 1$ by the proof of Lemma 2.3, with $g^2 = 1$ only when $g = \pm 1$. Since the map has kernel $\{\pm 1\}$ and since $X^2 = 1$ implies $\det X = 1$ in $\mathrm{M}_2(\mathbb{F}_4)$, it remains to show that $\{X \in \mathrm{M}_2(\mathbb{F}_4) \mid X^2 = 1\}$ has 16 elements. But writing $X = \begin{bmatrix} a & b \\ c & d \end{bmatrix}$ the condition $X^2 = 1$ amounts to $d = a$, $a^2 + bc = 1$. The point is that the second of these implies $a = a^4 = (1+bc)^2$, and conversely. So b, c are arbitrary and determine a, d by $d = a = (1+bc)^2$, completing the proof. □

Alternatively we may show I, I_0 have type H_4, H_3 by computing sytems of simple roots for them in $\mathbb{I}$, according to

(3.1) $\quad\underset{\alpha_1}{\circ}\overset{5}{\text{---}}\underset{\alpha_2}{\circ}\text{---}\underset{\alpha_3}{\circ}\text{---}\underset{\alpha_4}{\circ}$

We may compute this by choosing α_4 to be the $\alpha \in \mathbb{I} \setminus \mathbb{I}_0$, of Section 1 (because the Weyl group is transitive on simple roots) and then finding α_3, α_2, α_1 successively, being careful to stay inside $\mathbb{I}$. We get, for H_4,

(3.2) $\quad\begin{aligned}\alpha_1 &= \tfrac{1}{2}(-i - \tau'j + \tau k), & \alpha_2 &= \tfrac{1}{2}(i - \tau'j - \tau k), \\ \alpha_3 &= \tfrac{1}{2}(-\tau'i - \tau j + k), & \alpha_4 &= \tfrac{1}{2}(-\tau' - \tau i - k)\end{aligned}$

in agreement with the standard model of [**CMP**, 4.4], up to order, and a system of simple roots for H_3 by omitting α_4. (That we stay in $\mathbb{I}$ follows from $\alpha_3 j - \tau^2 \alpha_4 = \tau i + \tau k$, $\alpha_2 j + \alpha_4 = 0$, $\alpha_1 j + \alpha_4 = -\tau i - k$ by multiplying on the right by $-j \in \mathbb{I}$.)

LEMMA 3.2. α_1, α_2, α_3, α_4 is an R-basis of $\mathbb{I}$ and α_1, α_2, α_3 is an R-basis of $\mathbb{I}_0$.

PROOF. The discriminant of the R-span M of α_1, α_2, α_3, α_4 is $d(M/R) = \det[\mathrm{tr}(\alpha_i \alpha_j)]R = (-5 + 3\tau)R = (-\tau^{-4})R = R$ hence $M \subseteq \mathbb{I}$ implies $M = \mathbb{I}$. Since an R-linear combination of α_1, α_2, α_3, α_4 is in $\mathbb{I}_0$ precisely when the coefficient of α_4 is 0 the result for $\mathbb{I}_0$ follows. □

The point of this is that it identifies $\mathbb{I}$, $\mathbb{I}_0$ with the 'lattices' M_4, M_3 of [**CMP**]. Then (according to [**CMP**, 5.1]) every H_4-'lattice' is 'equivalent' to $\mathbb{I}$, and every H_3-'lattice' is 'equivalent' to $\mathbb{I}_E$ for some subgroup E of $\mathbb{F}_4$. This means that the most 'symmetric' of the 'lattices' are treated here. But there are other 'canonical' windows ([**CMP**, 7.12], for example).

4. The Shelling Problem

The different of $F/\mathbb{Q}$ has a generator $\delta = 2 + \tau$ which is totally positive (i.e., positive under both imbeddings $F \to \mathbb{R}$). It follows that

(4.1) $\quad (x \mid y) = \mathrm{Tr}_{F/\mathbb{Q}}\left(\delta^{-1}\,\mathrm{tr}(x\bar{y})\right)$

defines an even, unimodular (by Proposition 1.1), positive definite $\mathbb{Z}$-valued bilinear form on $\mathbb{I}$, under which it is isometric to the E_8 lattice. This induces the shelling

(4.2) $\quad \mathbb{I}(n) = \{u \in \mathbb{I} \mid (u \mid u) = 2n\}$

on $\mathbb{I}$ which induces the shelling

(4.3) $\quad \Sigma^r(n) = \Sigma^r \cap \mathbb{I}(n)$

of [**SM**] on Σ^r. The cardinality of these shells $\Sigma^r(n)$ has been determined [**MW**]. Just as in Section 2, restricting the form (|) to $\mathbb{I}_E$ then gives the shelling $\mathbb{I}_E(n) = \mathbb{I}_E \cap \mathbb{I}(n)$ of $\mathbb{I}_E$ and thus we get the shelling, for each subgroup E of $\mathbb{F}_4$,

$$(4.4) \qquad \Sigma_E^r(n) = \Sigma_E^r \cap \mathbb{I}_0(n)$$

of Σ_E^r, which we want to investigate. Here we must allow n to take (positive) values in $\frac{1}{4}\mathbb{Z}$: for $\tilde{\mathbb{I}}_0 \subseteq \frac{1}{2}\mathbb{I}$ and $u \mapsto (u|u)$ even on $\mathbb{I}$ implies $(u \mid u)$ takes values in $\frac{1}{2}\mathbb{Z}$ on $\tilde{\mathbb{I}}_0$. It is known [**CMP**, 6.3] that $\mathbb{I}_0$ with (|) is isometric to the D_6 lattice.

These shells $\Sigma_E^r(n)$, $n \in \frac{1}{4}\mathbb{Z}$, can be decomposed into basic chunks. Define

$$(4.5) \qquad X(b,e) = \{x \in \mathbb{I}_0 \mid \mathrm{nr}(x) = b, x + 2\mathbb{I} = e\}$$

for totally positive $b \in R$ and each $e \in \mathbb{F}_4 \subseteq \mathbb{I}/2\mathbb{I}$. Also put

$$(4.6) \qquad S(n,r) = \{m \in \mathbb{Z} \mid n + m\tau \text{ is totally positive and } n + m\tau' < r^2\}$$

for nonnegative $n \in \mathbb{Z}$ and real $r > 0$. Then we have

LEMMA 4.1. *The sets $S(n,r)$ are finite, and, for $n \in \frac{1}{4}\mathbb{Z}$,*

$$(4.7) \qquad \Sigma_E^r(n) = \overset{\bullet}{\underset{e \in E}{\bigcup}} \overset{\bullet}{\underset{m \in S(4n,2r)}{\bigcup}} \{\tfrac{1}{2}x \mid x \in X(4n + m\tau, e)\}.$$

PROOF. An integer m is in $S(n,r)$ if, and only if, $(n - r^2)\tau < m \leq n\tau$ and $m \geq n\tau'$, so finiteness is trivial.

Now $u \in \mathbb{H}(F)_0$ is in $\Sigma_E^r(n)$ if, and only if, $u \in \mathbb{I}_E$, $(u \mid u) = 2n$, $\mathrm{nr}(u') < r^2$. By definition of $\mathbb{I}_E$ this is equivalent to $u = \frac{1}{2}x$ with $x \in \mathbb{I}_0$, $x + 2\mathbb{I} \in E$, $(x \mid x) = 8n$, $\mathrm{nr}(x') < 4r^2$. Since $(x \mid x) = 2\,\mathrm{Tr}_{F/\mathbb{Q}}\bigl(\delta^{-1}\,\mathrm{nr}(x)\bigr)$ this means that, for some $e \in E$, we have $x \in \mathbb{I}_0$, $x + 2\mathbb{I} = e$, $\mathrm{Tr}_{F/\mathbb{Q}}\bigl(\delta^{-1}\,\mathrm{nr}(x)\bigr) = 4n$, $\mathrm{nr}(x)' < (2r)^2$. Finally, as nr takes totally positive values in R on $\mathbb{I}$, and $\mathrm{Tr}_{F/\mathbb{Q}}(\delta^{-1}\tau) = 0$, this reads, for some $e \in E$, that $x \in \mathbb{I}_0$, $x + 2\mathbb{I} = e$, $\mathrm{nr}(x) = 4n + m\tau$, with $m \in \mathbb{Z}$ so $4n + m\tau$ is totally positive with $(4n + m\tau)' < (2r)^2$, i.e., $x \in X(4n + m\tau, e)$ and $m \in S(4n, 2r)$, completing the proof. □

Since $S(n,r)$ has transparent structure we focus now entirely on the sets $X(b,e)$ which we study in terms of the arithmetic of $F(\sqrt{-b})$. To formulate the results we regard b (totally positive in R) and e (in $\mathbb{F}_4$) as fixed *without explicitly including them in the* (considerable) *notation* built on them.

We first factor $b = b_0 b_1^2$ in R with the ideal $b_0 R$ squarefree (using unique factorization, but actually only for convenience). Then $F(\sqrt{-b}) = F(\sqrt{-b_0})$ depends only on b_0 (up to squares of units in R) and our first invariants w, h, χ depend only on $F(\sqrt{-b_0})$ and will be discussed later. For the remaining invariants we write $e = \epsilon + 2R$ for the unique $\epsilon \in \{0, 1, \tau, \tau^{-1}\}$ for which it is true. If $-b \equiv \epsilon^2 \bmod 4R$ then we define an ideal $\mathfrak{c}$ of R by

$$(4.8) \qquad \mathfrak{c} = \begin{cases} b_1 R, & \text{if } -b_0 \equiv \tau^{2w} \bmod 4R \text{ for some integer } w \\ \frac{b_1}{2}R, & \text{if not.} \end{cases}$$

If $-b \not\equiv \epsilon^2 \bmod 4R$ we regard $\mathfrak{c}$ as undefined. Note that our formulations depend only on b_0, b_1 up to multiplication by squares of units, respectively units, so are independent of which factorization of the above type we choose.

Writing $N(\mathfrak{a})$ for the cardinality of $R/\mathfrak{a}$, when $\mathfrak{a}$ is an ideal of R, our goal is

THEOREM 4.2. *If $-b \not\equiv \epsilon^2 \bmod 4R$ then $X(b,e)$ is empty. If $-b \equiv \epsilon^2 \bmod 4R$ then, factoring $\mathfrak{c} = \prod_\mathfrak{p} \mathfrak{p}^{c_\mathfrak{p}}$ into powers of prime ideals $\mathfrak{p}$ of R, we have*

$$(4.9) \qquad |X(b,e)| = 120 \frac{h}{w} N(\mathfrak{c}) \prod_\mathfrak{p} \left(1 + (1 - \chi(\mathfrak{p})) \frac{1 - N(\mathfrak{p})^{-c_\mathfrak{p}}}{N(\mathfrak{p}) - 1} \right).$$

The unexplained invariants of $F(\sqrt{-b_0})$ are as follows. The function χ of prime ideals $\mathfrak{p}$ of R depends on how $\mathfrak{p}$ decomposes in $F(\sqrt{-b_0})$, i.e.,

$$(4.10) \quad \chi(\mathfrak{p}) = \begin{cases} 1, & \text{if } \mathfrak{p} \text{ splits in } F(\sqrt{-b_0}) \text{ (i.e., } \mathfrak{p}\widetilde{R} = \mathfrak{P}_1 \mathfrak{P}_2 \text{ with } \mathfrak{P}_1 \ne \mathfrak{P}_2) \\ 0, & \text{if } \mathfrak{p} \text{ ramifies in } F(\sqrt{-b_0}) \text{ (i.e., } \mathfrak{p}\widetilde{R} = \mathfrak{P}^2) \\ -1, & \text{if } \mathfrak{p} \text{ is inert in } F(\sqrt{-b_0}) \text{ (i.e., } \mathfrak{p}\widetilde{R} = \mathfrak{P}) \end{cases}$$

where $\mathfrak{P}$'s are prime ideals of the ring of integers $\widetilde{R}$ of $F(\sqrt{-b_0})$. The number $w = w(F(\sqrt{-b_0}))$ is simply the number of roots of unity in $F(\sqrt{-b_0})$. Finally $h = h(F(\sqrt{-b_0}))$ is the class number of $\widetilde{R}$. This, the most interesting invariant we need, is the number of isomorphism classes of $\widetilde{R}$-modules we get from the nonzero ideals of $\widetilde{R}$.

The proof of the theorem will be given in Section 7, after the preparations of Sections 5, 6. The first of these breaks the problem into subproblems for certain R-orders T in $F(\sqrt{-b_0})$. The second solves the subproblems in a way which can be recombined in Section 7.

EXAMPLE. We show that $|\Sigma_E^r(n)| = 480$ when $r = 1/\sqrt{\tau}$ and $n = 55$, for every E. The set $S(4n, 2r)$ consists of $220 + m\tau$ for $m = 352$ and the next 3 integers. It follows that $\mathfrak{c}$ is defined only when $m = 352$ and $e = 0$, in which case $\mathfrak{c} = 2R$ and $b_0 = 11$. By Lemma 4.1 and then Theorem 4.2, we have

$$|\Sigma_E^r(n)| = |X(220 + 352\tau, 0)| = 480 \frac{h}{w},$$

since $N(\mathfrak{c}) = 4$ and $\chi(2R) = 1$, so it remains to show that $h = w = 2$ for the field $K = F(\sqrt{-11})$.

This field K has ring of integers $\widetilde{R} = R[\omega]$, $\omega = \frac{1}{2}(1 + \sqrt{-11})$, hence absolute discriminant $5^2 \cdot 11^2$. It follows that $w = 2$, by comparison with ramification in cyclotomic fields, and that the Minkowski bound ([**L**, V, Section 4]) is < 9. Comparing the decomposition of small primes in the subfields F, $\mathbb{Q}(\sqrt{-11})$, $\mathbb{Q}(\sqrt{-55})$ of K, we deduce that the ideal class group of K is generated by the classes of prime ideals $\mathfrak{p}_2, \mathfrak{q}_2, \mathfrak{p}_5, \mathfrak{q}_5$ with $\mathfrak{p}_2 \mathfrak{q}_2 = (2)$, $\mathfrak{p}_5 \mathfrak{q}_5 = (2 + \tau)$. Since, moreover, $\mathfrak{p}_2 \mathfrak{p}_5 = (\tau + \omega)$ is principal, after changing notation if necessary, the class of $\mathfrak{p}_5 = (2 + \tau, \tau + \omega)$ already generates, and, as $\mathfrak{p}_5^2 = (2 - \omega)$, $h = 2$ will follow once we know $\mathfrak{p}_5$ is not principal.

But if $\mathfrak{p}_5$ is generated by $\gamma \in R[\omega]$ then both $N_{K/F}(\gamma)$ and $2 + \tau$ are totally positive generators of $N_{K/F}(\mathfrak{p}_5)$, hence, after changing the generator γ if necessary, $N_{K/F}(\gamma) = 2 + \tau$. Writing $\gamma = c_0 + c_1 \omega$, with c_0, c_1 in R, this implies $2 + \tau = N_{K/F}(\gamma) = (c_0 + c_1/2)^2 + 11(c_1/2)^2 \ge 11 c_1^2/4$ under *both* real imbeddings of F. Hence taking $N_{F/\mathbb{Q}}$ of both sides implies $c_1 = 0$ and the result.

5. Optimal Imbeddings

We regard b, e as fixed and keep the notation of Section 4. Our analysis of $X(b,e)$ is based on observing that $x \in X(b,e)$ implies $\bar{x} = -x$ hence $x^2 = -\bar{x}x = -\operatorname{nr}(x) = -b$, proving one containment of

LEMMA 5.1. $X(b,e) = \{x \in \mathbb{I} \mid x^2 = -b, x + 2\mathbb{I} = e\}$.

PROOF. Conversely if $x \in \mathbb{I}$ has $x^2 = -b$ then $F(x)$ is a subfield of $\mathbb{H}(F)$ isomorphic to $F(\sqrt{-b})$; $F(\sqrt{-b})$ has degree 2 over F since b totally positive in F implies $-b$ is not a square. But $y \mapsto \bar{y} = \operatorname{tr}(y) - y$ is a nontrivial automorphism of $F(x)/F$, since $\bar{}$ fixes only F, which uniquely determines it. Thus $\bar{x} = -x$ and so $x \in \mathbb{I}_0$ has $\operatorname{nr}(x) = \bar{x}x = -x^2 = b$, proving the lemma. $\square$

This description of $X(b,e)$ exhibits its elements x as images of $\sqrt{-b}$ under R-algebra homomorphisms $\rho\colon R[\sqrt{-b}] \to \mathbb{I}$ which send $\sqrt{-b}$ to e when composed with $\mathbb{I} \to \mathbb{I}/2\mathbb{I}$. Since $R[\sqrt{-b}]$ contains an F-basis of $F(\sqrt{-b})$, each such ρ uniquely induces an F-algebra imbedding $\rho\colon F(\sqrt{-b}) \to \mathbb{H}(F)$ in terms of which $\rho^{-1}(\mathbb{I})$ is an R-order satisfying $R[\sqrt{-b}] \subseteq \rho^{-1}(\mathbb{I}) \subseteq \widetilde{R}$, since $\widetilde{R}$ is the unique maximal R-order in $F(\sqrt{-b})$. We capture this relationship by saying that an R-order imbedding $\rho\colon T \to \mathbb{I}$ is *optimal* if $\rho^{-1}(\mathbb{I}) = T$, when ρ is extended to $F(\sqrt{-b}) \to \mathbb{H}(F)$.

To exploit this notion we must survey the possible T's and then study optimal imbeddings of each one. The first part of this is

LEMMA 5.2. (a) *The maximal R-order $\widetilde{R}$ of $F(\sqrt{-b_0})$ is given by $\widetilde{R} = R[\omega]$ where*

(5.1) $$\omega = \begin{cases} \dfrac{\tau^w + \sqrt{-b_0}}{2}, & \text{if } -b_0 \equiv \tau^{2w} \bmod 4R \text{ for some } w \\ \sqrt{-b_0}, & \text{if not.} \end{cases}$$

The R-orders T, which span $F(\sqrt{-b_0})$, are then uniquely expressible as $T = R + \mathfrak{a}\omega$ for an ideal $\mathfrak{a}$ of R.

(b) *The R-orders T whose optimal imbeddings in $\mathbb{I}$ give rise to points of $X(b,e)$ are precisely those which contain $\underset{\sim}{R} := R[\tfrac{1}{2}(\epsilon + \sqrt{-b})]$. There are no R-orders T such that $\underset{\sim}{R} \subseteq T \subseteq \widetilde{R}$ unless $-b \equiv \epsilon^2 \bmod 4R$, in which case they are given by $T = \widetilde{R} + \mathfrak{a}\omega$ with $\mathfrak{a}$ dividing the ideal $\mathfrak{c}$ of Section 4.*

PROOF. (a) Since we have $\widetilde{R} \supseteq R[\sqrt{-b_0}]$, the discriminant $d(\widetilde{R}/R)$ divides $d(R[\sqrt{-b_0}]/R) = 4b_0 R$. As b_0 is squarefree this implies that $d(\widetilde{R}/R)$ is $b_0 R$ or $4b_0 R$ and, in particular, that $\widetilde{R} \subseteq \tfrac{1}{2}R[\sqrt{-b_0}]$. But if u, v are in R then $\tfrac{1}{2}(u + v\sqrt{-b_0})$ has minimal polynomial $X^2 - uX + \tfrac{1}{4}(u^2 + b_0 v^2)$ hence is in $\widetilde{R}$ if and only if $u^2 + b_0 v^2 \in 4R$. Surveying such u, v computes $\widetilde{R}$.

If T is an R-order then the R-module T/R is contained in $\widetilde{R}/R$ which has R-basis $\omega + R$, hence $T/R = \mathfrak{a} \cdot (\omega + R)$ for a unique ideal $\mathfrak{a}$ of R, and so $T = R + \mathfrak{a}\omega$. Conversely, every $R + \mathfrak{a}\omega$ is an R-order.

(b) If $\rho\colon T \to \mathbb{I}$ is an optimal imbedding with $\rho(\sqrt{-b})$ mapping to e under $\mathbb{I} \to \mathbb{I}/2\mathbb{I}$ then $\rho(\epsilon + \sqrt{-b}) \in 2\mathbb{I}$ hence $\tfrac{\epsilon + \sqrt{-b}}{2} \in \rho^{-1}(\mathbb{I}) = T$ implying $T \supseteq \underset{\sim}{R}$. Conversely, if $\underset{\sim}{R} \subseteq T$ then $\epsilon + \sqrt{-b}$ is in $2T$ hence $\rho(\sqrt{-b}) \equiv \epsilon \bmod 2\mathbb{I}$ for any $\rho\colon T \to \mathbb{I}$.

For T to exist with $\underset{\sim}{R} \subseteq T \subseteq \widetilde{R}$ we must have $\underset{\sim}{R} \subseteq \widetilde{R}$, i.e., $\tfrac{1}{2}(\epsilon + \sqrt{-b}) \in \widetilde{R}$. As in (a), this means $\tfrac{1}{4}(\epsilon^2 + b) \in R$, as required.

Under the parametrization of R-orders T by ideals $\mathfrak{a}$, containment is preserved so it remains only to compute the ideal $\mathfrak{c}$ so that $\underset{\sim}{R} = R + \mathfrak{c}\omega$, assuming, of course, that $-b \equiv \epsilon^2 \bmod 4R$.

If $e \neq 0$ then $-b_0 b_1^2 \equiv \epsilon^2 \bmod 4$, with ϵ a unit, implies $-b_0 \equiv \tau^{2w} \bmod 4$ for some w hence $\tau^w b_1 \equiv \epsilon \bmod 2$ and $\omega = \frac{\tau^w + \sqrt{-b_0}}{2}$. Then $\frac{\epsilon + \sqrt{-b}}{2} + R = b_1 \omega + R$ and so $\mathfrak{c} = b_1 R$.

If $e = 0$ then $\epsilon \equiv 0 \bmod 4$ and so $-b \equiv 0 \bmod 4$ implies $b_1 \equiv 0 \bmod 2$. If $\omega = \frac{1}{2}(\tau^w + \sqrt{-b_0})$ this implies $\frac{1}{2}(\epsilon + \sqrt{-b}) + R = b_1 \omega + R$; if $\omega = \sqrt{-b_0}$ it gives $\frac{1}{2}(\epsilon + \sqrt{b}) + R = \frac{1}{2} b_1 \omega + R$, both as claimed, so completing the proof. □

To begin the study of optimal imbeddings $\rho \colon T \to \mathbb{I}$ we observe that the group $\mathbb{I}^\times$ of units of $\mathbb{I}$ acts on the set of optimal imbeddings $\rho \colon T \to \mathbb{I}$ by $(\rho^u)(t) = u^{-1} \cdot \rho(t) \cdot u$, $t \in T$.

LEMMA 5.3. *Let $w(T)$ be the number of roots of unity in T. Then the $\mathbb{I}^\times$-orbit of $\rho \colon T \to \mathbb{I}$ has $120/w(T)$ elements.*

PROOF. If $u \in \mathbb{I}^\times$ then $\mathrm{nr}(u)$ is a totally positive unit of R hence a power τ^{2k} of τ^2: then $\tau^{-k} u$ has reduced norm 1 hence is in I. As τ^k is central in $\mathbb{H}(F)$, the $\mathbb{I}^\times$-orbit of ρ is the I-orbit of ρ. Since $|I| = 120$ it suffices to show that the stabilizer $\{g \in I \mid \rho^g = \rho\}$ has $w(T)$ elements.

However $\rho^g = \rho$ if and only if g centralizes the image of ρ. But $\rho(F(\sqrt{-b_0}))$ is a maximal subfield of $\mathbb{H}(F)$ hence coincides with its own centralizer [**R**, 7.14]. Thus $\rho^g = \rho$ means $g \in \mathbb{I} \cap \rho(F(\sqrt{-b_0})^\times) = \mathbb{I} \cap \rho(T^\times)$ is a root of unity in the image of ρ; every such root of unity has reduced norm 1 (because it is totally positive) and so is in I. The lemma is proved. □

We summarize our analysis of $X(b,e)$ so far in

COROLLARY 5.4. *In the notation of Lemmas 5.2, 5.3 we have*

$$|X(b,e)| = 120 \sum_{\underset{\sim}{R} \subseteq T \subseteq \tilde{R}} \frac{1}{w(T)} (\textit{number of } \mathbb{I}^\times \textit{-orbits of optimal imbeddings } T \to \mathbb{I}).$$

The 'local' analogue of this formulation has a simple answer. Namely if $\mathfrak{p}$ is a prime ideal of R we let $R_\mathfrak{p} = \varprojlim_n R/\mathfrak{p}^n$ be the completion of R at $\mathfrak{p}$ and let $T_\mathfrak{p} = R_\mathfrak{p} \otimes_R T$, $\mathbb{I}_\mathfrak{p} = R_\mathfrak{p} \otimes_R \mathbb{I}$ be the $R_\mathfrak{p}$-orders obtained by extending scalars to $R_\mathfrak{p}$.

PROPOSITION 5.5. *There is precisely one $\mathbb{I}_\mathfrak{p}^\times$-orbit of optimal ($R_\mathfrak{p}$-algebra) imbeddings $\rho \colon T_\mathfrak{p} \to \mathbb{I}_\mathfrak{p}$.*

PROOF. Since $R_\mathfrak{p}$ is a discrete valuation ring $T_\mathfrak{p} = R_\mathfrak{p} + R_\mathfrak{p} \mathfrak{a} \omega$ is generated, as $R_\mathfrak{p}$-algebra, by one element $\omega_\mathfrak{p}$, namely any generator of the ideal $R_\mathfrak{p} \mathfrak{a} \omega$ (which is principal even if $\mathfrak{a}$ was not). Writing $F_\mathfrak{p}$ for the field of fractions of $R_\mathfrak{p}$ then $F_\mathfrak{p} \otimes_F \mathbb{H}(F) = \mathbb{H}(F_\mathfrak{p}) \simeq \mathrm{M}_2(F_\mathfrak{p})$ by the discussion of the classification of central simple algebras in Section 1. And $\mathbb{I}_\mathfrak{p}$ is a maximal $R_\mathfrak{p}$-order in $\mathbb{H}(F_\mathfrak{p})$ hence [**R**, 18.7] isomorphic to $\mathrm{M}_2(R_\mathfrak{p})$. So we need to show that there is precisely one $\mathrm{M}_2(R_\mathfrak{p})^\times$-orbit of optimal imbeddings $T_\mathfrak{p} \to \mathrm{M}_2(R_\mathfrak{p})$, a matrix problem.

Letting $X^2 + uX + v$ be the polynomial satisfied by $\omega_\mathfrak{p}$ over $R_\mathfrak{p}$ then $\rho \colon T_\mathfrak{p} \to \mathrm{M}_2(R_\mathfrak{p})$, $\omega_\mathfrak{p} \mapsto \begin{bmatrix} 0 & -v \\ 1 & -u \end{bmatrix}$ is an $R_\mathfrak{p}$-algebra homomorphism. Because of the entries in the first column we see that $a \begin{bmatrix} 1 & 0 \\ 0 & 1 \end{bmatrix} + b \begin{bmatrix} 0 & -v \\ 1 & -u \end{bmatrix} \in \mathrm{M}_2(R_\mathfrak{p})$ happens for a, b in $F_\mathfrak{p}$ only when a, b are both in $R_\mathfrak{p}$, i.e., ρ is an optimal imbedding.

To show that this determines the only $\mathrm{M}_2(R)^\times$-orbit, choose $\pi \in R_\mathfrak{p}$ so $\pi R_\mathfrak{p}$ is the unique maximal ideal of $R_\mathfrak{p}$, and suppose that there is a different such orbit. Then the images $\rho(\omega_\mathfrak{p})$, for ρ in this orbit, fall into a single $\mathrm{M}_2(R_\mathfrak{p})^\times$-conjugacy class C of matrices in $\mathrm{M}_2(R_\mathfrak{p})$, and our hypothesis is that C contains no matrix $\left[\begin{smallmatrix} c_{11} & c_{12} \\ c_{21} & c_{22} \end{smallmatrix}\right]$ with $c_{11} = 0$, $c_{21} = 1$: for such a matrix satisfying $X^2 + uX + v$ would have to be $\left[\begin{smallmatrix} 0 & -v \\ 1 & -u \end{smallmatrix}\right]$, a contradiction. Thus

(i) C contains no matrix with $c_{11} = 0, c_{21} \not\equiv 0 \bmod \pi$: else $c_{21} \in R_\mathfrak{p}^\times$ and conjugating by $\left[\begin{smallmatrix} 1 & 0 \\ 0 & c_{21} \end{smallmatrix}\right]$ gives the forbidden first column.

(ii) C contains no matrix with $c_{21} \not\equiv 0 \bmod \pi$: else conjugating by $\left[\begin{smallmatrix} 1 & c_{11}/c_{21} \\ 0 & 1 \end{smallmatrix}\right]$ contradicts (i).

(iii) C contains no matrix with $c_{12} \not\equiv 0 \bmod \pi$: else conjugating by $\left[\begin{smallmatrix} 0 & 1 \\ 1 & 0 \end{smallmatrix}\right]$ contradicts (ii).

(iv) C contains no matrix with $c_{11} - c_{22} \not\equiv 0 \bmod \pi$: else conjugating by $\left[\begin{smallmatrix} 1 & 1 \\ 0 & 1 \end{smallmatrix}\right]$ contradicts (iii), since $c_{11} - c_{22} + c_{12} - c_{21} \equiv c_{11} - c_{22} \bmod \pi$ by (ii), (iii).

But now (ii), (iii), (iv) imply that every matrix in C is congruent mod π to c_{11} times the identity matrix. Consequently, $\frac{1}{\pi}(c_{11} - \omega_\mathfrak{p}) \notin T_\mathfrak{p}$ has image in $\mathrm{M}_2(R_\mathfrak{p})$ under our supposed orbit of ρ's. But this contradicts their optimality and finishes the proof. $\square$

6. Genera

Fixing T as above we now analyze the $\mathbb{I}^\times$-orbits of optimal imbeddings $T \to \mathbb{I}$ by associating modules for the R-order $\Gamma = T \otimes_R \mathbb{I}$ to them. Namely, given ρ, we make the set $\{[y] \mid y \in \mathbb{I}\}$ into a right Γ-module $M(\rho)$ by defining the Γ-action (using multiplication in $\mathbb{I}$ inside the bracket):

(6.1) $$[y](t \otimes x) = [\rho(t)yx].$$

Using the canonical maps $T \to \Gamma$, $t \mapsto t \otimes 1$, and $\mathbb{I} \to \Gamma$, $x \mapsto 1 \otimes x$, we observe that each $M(\rho)$ is a Γ-module M satisfying

(a) T acts faithfully on M via $T \to \Gamma$ and, under the induced action of $F(\sqrt{-b_0}) = F \otimes_R T$ on $F \otimes_R M$, we have $\{a \in F(\sqrt{-b_0}) \mid Ma = M\} = T$.

(b) the $\mathbb{I}$-module $\mathrm{res}_\mathbb{I} M$ induced by $\mathbb{I} \to \Gamma$ is isomorphic to $\mathbb{I}$ as right $\mathbb{I}$-module.

PROPOSITION 6.1. *The $\mathbb{I}^\times$-orbits of optimal imbeddings $T \to \mathbb{I}$ are in bijection with the isomorphism classes of Γ-modules M satisfying* (a) *and* (b).

PROOF. Each $M(\rho)$ satisfies (a) and (b), the first by optimality of ρ and the second by construction. If $u \in \mathbb{I}^\times$ then $[y] \mapsto [uy]$ is a Γ-module isomorphism $M(\rho^u) \to M(\rho)$ so an $\mathbb{I}^\times$-orbit of optimal imbeddings ρ gives rise to an isomorphism class of $M(\rho)$'s.

Conversely if M satisfies (a), (b) use (b) to choose $m_0 \in M$ so that $x \mapsto m_0(1 \otimes x)$ is an $\mathbb{I}$-module isomorphism $\mathbb{I} \to M$ and define $\rho: T \to M$ by $m_0(t \otimes 1) = m_0(1 \otimes \rho(t))$. Then ρ is an optimal embedding by a); and a different choice of m_0 is necessarily $m_0(1 \otimes u)$ with $u \in \mathbb{I}^\times$ giving rise to the $\mathbb{I}^\times$-conjugate ρ^u of ρ. Since any module isomorphic to M is isomorphic to $M(\rho)$ this completes the proof. $\square$

This reformulation of the problem is technically useful because it allows the standard method of analyzing modules of the R-order Γ to be applied. Namely

Γ-modules M, M' are placed in the same *genus* if and only if $M_{\mathfrak{p}} \simeq M'_{\mathfrak{p}}$ as $\Gamma_{\mathfrak{p}}$-modules for every prime ideal $\mathfrak{p}$ ($\neq 0$) of R. Here $M_{\mathfrak{p}} = R_{\mathfrak{p}} \otimes_R M$, $\Gamma_{\mathfrak{p}} = R_{\mathfrak{p}} \otimes_R \Gamma$ with $R_{\mathfrak{p}}$ the completion of R at $\mathfrak{p}$ as in Section 7.

Writing $h(T)$ for the number of isomorphism classes of T-modules in the genus of T we come finally to the heart of the matter

PROPOSITION 6.2. *The number of $\mathbb{I}^{\times}$-orbits of optimal imbeddings $T \to \mathbb{I}$ is given by $h(T)$.*

PROOF. In view of Proposition 6.1 we show that there are $h(T)$ isomorphism classes of Γ-modules M satisfying (a) and (b). Any two such modules M, M' are in the same genus. For $M_{\mathfrak{p}}$, $M'_{\mathfrak{p}}$ give rise, by the $\mathfrak{p}$-analogue of Proposition 6.1, to optimal imbeddings $T_{\mathfrak{p}} \to \mathbb{I}_{\mathfrak{p}}$. These are in the same $\mathbb{I}_{\mathfrak{p}}^{\times}$-orbit by Proposition 5.5 hence $M_{\mathfrak{p}} \simeq M'_{\mathfrak{p}}$ by Proposition 6.1 again.

A module M satisfying (a), (b) exists: we see this by patching together local solutions, which exist by Proposition 5.5. Namely applying the $\mathfrak{p}$-analogue of Proposition 6.1 to the imbedding of Proposition 5.5 gives a $\Gamma_{\mathfrak{p}}$-module $M^{(\mathfrak{p})}$ satisfying the $\mathfrak{p}$-analogue ($a_{\mathfrak{p}}$), ($b_{\mathfrak{p}}$) of (a), (b). The point now is

CLAIM. *There exists a Γ-module M with $R_{\mathfrak{p}} \otimes_R M \simeq M^{(\mathfrak{p})}$ for all $\mathfrak{p}$.*

PROOF OF CLAIM. Let A denote the F-algebra $F \otimes_R \Gamma$. Then we have $A \simeq (F \otimes_R T) \otimes_F (F \otimes_R \mathbb{I}) \simeq F(\sqrt{-b_0}) \otimes_F \mathbb{H}(F) \simeq M_2\big(F(\sqrt{-b_0})\big)$ as F-algebra again using the classification results from Section 1: the point is that $F(\sqrt{-b_0})$ has no real imbedding (since $\mathbb{H}(F)$ was already split at all finite primes). So A has a simple module V (unique up to isomorphism) with $\dim_F V = 4$.

Now $M^{(\mathfrak{p})}$ is a faithful $T_{\mathfrak{p}}$-module by ($a_{\mathfrak{p}}$) and $\dim_{F_{\mathfrak{p}}} F_{\mathfrak{p}} \otimes_{R_{\mathfrak{p}}} M^{(\mathfrak{p})} = 4$ by ($b_{\mathfrak{p}}$). So $F_{\mathfrak{p}} \otimes_F V$ and $F_{\mathfrak{p}} \otimes_{R_{\mathfrak{p}}} M^{(\mathfrak{p})}$ are both faithful $F_{\mathfrak{p}} \otimes_F A$-modules of dimension 4 over $F_{\mathfrak{p}}$. By the structure of modules over $F_{\mathfrak{p}} \otimes_F A$ [**CR**, 3.11, 3.12] it follows that $F_{\mathfrak{p}} \otimes_{R_{\mathfrak{p}}} M^{(\mathfrak{p})} \simeq F_{\mathfrak{p}} \otimes_F V$ for all $\mathfrak{p}$. Using these isomorphisms as identifications it follows that $M = \bigcap_{\mathfrak{p}}(V \cap M_{\mathfrak{p}})$ satisfies [**R**, 5.3] the claim.

Then M satisfies condition (a), since $T = \bigcap_{\mathfrak{p}}(F \cap T_{\mathfrak{p}})$ (by [**R**, 4.21 and 5.2]) but only the weaker conclusion, for (b), that $\operatorname{res}_{\mathbb{I}} M$ is in the genus of $\mathbb{I}$. However now the fact (cf. [**T**, **V**]) that all right ideals of $\mathbb{I}$ are principal implies that this genus has a single isomorphism class so (b) holds after all. This proves the claim. □

It thus remains only to show that the isomorphism classes in the genus of M_{Γ} are in bijection with the isomorphism classes in the genus of T_T. Observe that $t \mapsto$ "$m \mapsto m(t \otimes 1)$" is a homomorphism of R-orders $T \to \operatorname{End}(M_{\Gamma})$ which induces the isomorphism $F(\sqrt{-b}) \to \operatorname{End}(V_A)$ on applying $F \otimes_R \_$: it follows from (a) that $T \to \operatorname{End}(M_{\Gamma})$ is an isomorphism. Viewing this as an identification we view M as a $T - \Gamma$ bimodule $_T M_{\Gamma}$ and apply [**CR**, 6.3]. This says that $\operatorname{Hom}(M, \_)_{\Gamma}$ and $\_ \otimes_T M$ induce mutually inverse bijections between the isomorphism classes in the genus of M_{Γ} and the genus of T_T, and the proposition is proved. □

7. Proof of the Theorem

Combining Corollary 5.4 with Proposition 6.2 we have

$$(7.1) \qquad |X(b,e)| = 120 \sum_{R \subseteq T \subseteq \widetilde{R}} \frac{h(T)}{w(T)}$$

with the T's parametrized by Lemma 5.2(b).

CLAIM. *If* $T = R + \mathfrak{a}\omega$ *then*
$$\frac{h(T)}{w(T)} = \frac{h}{w} N(\mathfrak{a}) \prod_{\mathfrak{p}|\mathfrak{a}} \left(1 - \frac{\chi(\mathfrak{p})}{N(\mathfrak{p})}\right)$$
with $\mathfrak{p}$ running through prime ideals of R containing $\mathfrak{a}$.

PROOF. This follows from the relation
$$(7.2) \qquad |\operatorname{Pic}(T)| = \frac{|\operatorname{Pic}(\widetilde{R})|}{(\widetilde{R}^\times : T^\times)} \frac{|(\widetilde{R}/\mathfrak{f})^\times|}{|(T/\mathfrak{f})^\times|}$$
of Dedekind [**N**, I, 12.12]. Here $\mathfrak{f}$ is the conductor $\{a \in \widetilde{R} \mid a\widetilde{R} \subseteq T\}$ of T in $\widetilde{R}$: since $T = R + \mathfrak{a}\omega$ we have $\mathfrak{f} = \mathfrak{a}\widetilde{R}$. Every unit of $\widetilde{R}$ is a power of τ times a root of unity, by the same argument as in the proof of Lemma 5.3 with nr replaced by $N_{F(\sqrt{-b_0})/F}$, hence $(\widetilde{R}^\times : T^\times) = w(\widetilde{R})/w(T)$. We also have $|\operatorname{Pic}(T)| = h(T)$. First, $|\operatorname{Pic}(T)|$ is the number of modules locally isomorphic to T in the sense of [**N**, I, 12.12] because of [**N**, I, 12.4] and because T-isomorphism between ideals is given by multiplication by elements of $F(\sqrt{-b_0})$. Second, this notion of local isomorphism coincides with our notion of genus of T-modules (i.e., viewing T as an R-order) because of [**R**, 27.1]. Since $w(\widetilde{R}) = w, h(\widetilde{R}) = h$, in the notation of Section 4, the latter because of [**N**, 3.8], we now have
$$\frac{h(T)}{w(T)} = \frac{h}{w} \frac{|(\widetilde{R}/\mathfrak{a}\widetilde{R})^\times|}{|(T/\mathfrak{a}\widetilde{R})^\times|}$$
where $\mathfrak{a}\widetilde{R}$ is an ideal of $\widetilde{R}$ contained in $T = R + \mathfrak{a}\widetilde{R}$ with $\mathfrak{a}\widetilde{R} \cap R = \mathfrak{a}$, hence the inclusion $R \hookrightarrow T$ induces a ring isomorphism $R/\mathfrak{a} \to T/\mathfrak{a}\widetilde{R}$. Thus
$$|(T/\mathfrak{a}\widetilde{R})^\times| = |(R/\mathfrak{a})^\times|.$$
However, $|(R/\mathfrak{a})^\times| = N(\mathfrak{a}) \prod_{\mathfrak{p}|\mathfrak{a}} (1 - 1/N(\mathfrak{p}))$, by the Chinese remainder theorem. The analogous formula holds for $|(\widetilde{R}/\mathfrak{a}\widetilde{R})^\times|$ with the difference that $N(\mathfrak{a}\widetilde{R}) = N(\mathfrak{a})^2$ (since $\widetilde{R}/\mathfrak{a}\widetilde{R}$ has $R/\mathfrak{a}$-basis $\bar{1}, \bar{\omega}$) and that the product is over prime ideals $\mathfrak{P}$ of $\widetilde{R}$. So we get
$$\frac{|(\widetilde{R}/\mathfrak{a}\widetilde{R})^\times|}{|(R/\mathfrak{a})^\times|} = \frac{N(\mathfrak{a})^2}{N(\mathfrak{a})} \prod_{\mathfrak{p}|\mathfrak{a}} \left(\frac{\prod_{\mathfrak{P} \supseteq \mathfrak{p}}(1 - 1/N(\mathfrak{P}))}{1 - 1/N(\mathfrak{p})}\right)$$
and the factor at each $\mathfrak{p}$ depends on how $\mathfrak{p}$ decomposes in $F(\sqrt{-b_0})$. When $\mathfrak{p}$ is inert then $\mathfrak{P} = \mathfrak{p}\widetilde{R}$ is unique and has $N(\mathfrak{P}) = N(\mathfrak{p})^2$; in the other two cases each $\mathfrak{P}$ has $N(\mathfrak{P}) = N(\mathfrak{p})$ but there are two $\mathfrak{P}$'s when $\mathfrak{p}$ splits and only one when $\mathfrak{p}$ ramifies. This proves the claim, by definition of χ. □

Combining the above with the parametrization of Lemma 5.2(b), we find our sum is empty unless $-b \equiv \epsilon^2 \mod 4R$, and then it is
$$|X(b,e)| = 120 \frac{h}{w} \sum_{\mathfrak{a}|\mathfrak{c}} N(\mathfrak{a}) \prod_{\mathfrak{p}|\mathfrak{a}} \left(1 - \frac{\chi(\mathfrak{p})}{N(\mathfrak{p})}\right)$$

where we are summing a multiplicative function of $\mathfrak{a}$ over all divisors of $\mathfrak{c}$. By unique factorization of ideals the sum is

$$\prod_{\mathfrak{p}} \left(1 + \sum_{k=1}^{c_{\mathfrak{p}}} N(\mathfrak{p}^k)\left(1 - \frac{\chi(\mathfrak{p})}{N(\mathfrak{p})}\right)\right)$$

from which the theorem follows immediately. $\square$

8. Etcetera

This section contains remarks of two kinds, first concerning the invariants in the theorem, and then about other approaches to its proof.

The character χ of the Theorem is a basic arithmetic invariant of the extension $F(\sqrt{-b_0})/F$. We have $\chi(\mathfrak{p}) = 0$ only ([**R**, 4.37]) for $\mathfrak{p}$ dividing the discriminant

$$(8.1) \qquad d(\widetilde{R}/R) = \begin{cases} b_0 R, & \text{if } -b_0 \equiv \tau^{2w} \mod 4R \text{ for some } w \\ 4b_0 R, & \text{if not} \end{cases}$$

of $F(\sqrt{-b_0})/F$ (from Lemma 5.2(a)). The function $\mathfrak{p} \mapsto \chi(\mathfrak{p})$ specifies $F(\sqrt{-b_0})$ as an Abelian extension of F in the sense of class field theory: the character it induces on the fractional ideals of F (coprime to $d(\widetilde{R}/R)$) is the Artin map ([**L**, X, Section 1]) to $\text{Gal}(F(\sqrt{-b_0})/F)$ composed with the unique nontrivial character of this Galois group. It may be expressed in terms of quadratic residue symbols over F, i.e., for $\mathfrak{p} \nmid 2d(\widetilde{R}/R)$ we have

$$(8.2) \qquad \chi(\mathfrak{p}) = \left(\frac{-b_0}{\mathfrak{p}}\right) \equiv (-b_0)^{(N(\mathfrak{p})-1)/2} \mod \mathfrak{p}$$

([**L**, I, Section 8] and [**N**, V, 3.5]) and if $\mathfrak{p} = 2R$ does not divide $d(\widetilde{R}/R)$ then

$$(8.3) \qquad \chi(2R) = (-1)^{\text{Tr}_{F/\mathbb{Q}}(1+\tau^{-2w}b_0)/4}.$$

And there is a reciprocity law ([**N**, VI,8.3]).

The number $w = w(F(\sqrt{-b_0})/F)$ is 2 with only three exceptions:

$$w\big(F(\sqrt{-1})\big) = 4; \quad w\big(F(\sqrt{-3})\big) = 6; \quad w\big(F(\sqrt{-(2+\tau)})\big) = 10.$$

The class number $h = h(F(\sqrt{-b_0})/F)$ has order of magnitude $N(b_0 R)^{1/2}$ (compare [**L**, XVI]) and may be computed more efficiently than by the classical method of the example of Section 4: there is a 'rational' formula of Shintani (cf. [**N**, VII, 9.8]) and, computationally better, a 'transcendental' formula [**Lo**] with good convergence.

The subject of other approaches is partly one of context. To address this we first distinguish between two basic approaches.

The first, going back to Gauss, is in the language of positive definite quadratic forms over $\mathbb{Z}$. There is considerable literature, centred around Siegel's mass formula, but unfortunately mostly still over $\mathbb{Z}$. This does, however, generalize to totally real number fields ([**S**]).

The second approach is the quaternion analogue (again over $\mathbb{Q}$) of that above, which goes back to V. Venkov (cf. [**Rh**]), around 1920. Again, this still works over totally real number fields (cf. [**B**]). This approach, via optimal imbeddings, generalizes to arbitrary orders in the same way as in Section 6 (cf. [**W**] for group orders).

A common generalization is to be expected. However, it will be more complicated, because the fact that right ideals of $\mathbb{I}$ are principal does not generalize (cf. [**V**, Chap. V]).

References

[BGJR] M. Baake, U. Grimm, D. Joseph, and P. Repetowicz, *Averaged shelling for quasicrystals*, Material Science and Engineering A (to appear); preprint math.MG/9907156.

[B] J. Brzezinski, *On embedding numbers into quaternion orders*, Comm. Math. Helv. **66** (1991), 302–318.

[CMP] L. Chen, R. V. Moody, and J. Patera, *Non-crystallographic root systems*, Quasicrystals and Discrete Geometry (Toronto, ON, 1995) (J. Patera, ed.) Fields Inst. Monogr., vol. 10, Amer. Math. Soc., Providence, RI, 1998, pp. 135–178.

[CR] C. W. Curtis and I. Reiner, *Methods of representation theory*, vol. I, Pure Appl. Math., John Wiley & Sons, New York, 1981.

[ES] V. Elser and N. J. A. Sloane, *A highly symmetric four-dimensional quasicrystal*, J. Phys. A **20** (1987), 6161–6167.

[H] J. E. Humphreys, *Reflection groups and Coxeter groups*, Cambridge Stud. Adv. Math., vol. 29, Cambridge Univ. Press, Cambridge, 1990; 2nd corr. printing, 1992.

[L] S. Lang, *Algebraic number theory*, 2nd ed., Grad. Texts in Math., vol. 110, Springer, New York, 1994.

[Lo] S. Louboutin, *Computation of relative class numbers of CM-fields*, Math. Comp. **66** (1997), 1185–1194.

[M] R. V. Moody, *Meyer sets and their duals*, The Mathematics of Long-Range Aperiodic Order (Waterloo, ON, 1995) (R. V. Moody, ed.), NATO ASI Ser. C: Math. Phys. Sci., vol. 489, Kluwer, Dordrecht, 1997, pp. 403–41.

[MW] R. V. Moody and A. Weiss, *On shelling E_8 quasicrystals*, J. Number Theory **47** (1994), 405–412.

[N] J. Neukirch, *Algebraische Zahlentheorie*, Springer, Berlin, 1992; English transl., *Algebraic number theory*, Grundlehren Math. Wiss., vol. 322, Springer, Berlin, 1999.

[Rh] H. P. Rehm, *On a theorem of Gauss concerning the number of integral solutions of the equation $x^2 + y^2 + z^2 = m$*, Ternary Quadratic Forms and Norms (O. Taussky, ed.), Lecture Notes in Pure and Appl. Math., vol. 79, Marcel Dekker, New York, 1982, pp. 31–38.

[R] I. Reiner, *Maximal orders*, London Math. Soc. Monogr, vol. 5, Academic Press, London, 1975.

[SM] J.-F. Sadoc and R. Mosseri, *The E_8 lattice and quasicrystals*, J. Non-Crystalline Solids **153&154** (1993), 247–252.

[S] G. Shimura, *An exact mass formula for orthogonal groups*, Duke Math. J. **97** (1999), 1–66.

[T] J. Tits, *Quaternions over $\mathbb{Q}(\sqrt{5})$, Leech's lattice and the sporadic group of Hall-Janko*, J. Algebra **63** (1980), 56–75.

[V] M.-F. Vignéras, *Arithmétique des algèbres de quaternions*, Lecture Notes in Math., vol. 800, Springer, Berlin, 1980.

[W] A. Weiss, *Torsion units in integral group rings*, J. Reine Angew. Math. **415** (1991), 175–187.

DEPARTMENT OF MATHEMATICAL SCIENCES, UNIVERSITY OF ALBERTA, EDMONTON, ALBERTA T6G 2G1, CANADA

E-mail address: aweiss@vega.math.ualberta.ca

Tilings, C^*-algebras, and K-theory

Johannes Kellendonk and Ian F. Putnam

ABSTRACT. We describe the construction of C^*-algebras from tilings. We describe the K-theory of such C^*-algebras and discuss applications of these ideas in physics. We do not assume any familiarity with C^*-algebras or K-theory.

1. Introduction

Our starting point for this article is the development of the mathematical theory of tilings, especially that of aperiodic tilings which began with the work of Wang, Robinson, and Penrose, see [**GS**]. The connections of this field with dynamical systems and ergodic theory is, by now, quite well-established. More specifically, there are various ways of viewing a tiling of d-dimensional Euclidean space, $\mathbb{R}^d$, as giving rise to an action of the group $\mathbb{R}^d$ on a topological space. The elements of this space are themselves tilings and the action is by the natural notion of translation. We will explain a version of this in Section 2.

The connection between ergodic theory and von Neumann algebras begins with the pioneering work of Murray and von Neumann. The analogous connection between C^*-algebras and topological dynamics also has a long history. For a general reference to operator algebras, see [**Da, Fi, Pe**]. Basically, there is a construction which begins with a general topological dynamical system and produces a C^*-algebra. By a "general topological dynamical system", we certainly include the actions of locally compact groups on locally compact Hausdorff spaces as well as some topological equivalence relations and foliations of manifolds. (See the references above for various special cases and [**Ren**] for a very general version.) While this study began somewhat later than that in ergodic theory and von Neumann algebras, in the last twenty years it has blossomed. This is mainly due to the development of the technical tools needed. In particular, K-theory has had a major impact on the general theory of C^*-algebras and especially on the aspects relating to dynamics. Thus, it seems natural to try to investigate the special case of the dynamics obtained from tilings and their associated C^*-algebras. This was already observed by Alain Connes in [**Co2**]. The goal is two-fold. First, to produce interesting examples of C^*-algebras. The second point is to use C^*-algebras and techniques from their study to learn more about the tilings.

2000 *Mathematics Subject Classification.* Primary 52C23, 46L80; Secondary 19K14, 82D25.

©2000 Johannes Kellendonk and Ian F. Putnam

While written mainly from the mathematical point of view, the article also aims at explaining briefly the physical aspects of (topological) tiling theory. The tilings have been used by physicists as models in the study of quasicrystals. (See, for example, [**Ja, StO**].) On the other hand, operator algebras began as mathematical models in quantum mechanics. These C^*-algebras are closely related with the physics of quasicrystals. We will discuss this and especially the rôle of K-theory in physics. K-theory enters in physics through Bellissard's formulation of the gap labeling [**Be1, Be2**]. Also see his and his co-authors' contribution to this volume.

The article is written for the reader having little or no background in the theory of C^*-algebras. This means that we will sacrifice some precision in our discussions. We hope that the main ideas are accessible if we avoid getting bogged down in technicalities (even if they are important ones).

We will begin by describing tilings as dynamical systems. The general theory is presented in the next section and in the following section we discuss tilings possessing self-similarity in the form of a substitution rule. Of course, much of this is fairly standard by now. However, there are certain points where our view anticipates the questions we will look at later when dealing with C^*-algebras.

There are several different constructions of C^*-algebras from a tiling. We present two of these in Sections 4 and 5. The first is to proceed from the continuous dynamics of the natural action of Euclidean space as translations of the tilings. The second takes a more discrete view of the situation. It tends to be more combinatorial and probably more accessible for someone unfamiliar with operator algebras. It is also the important one for physics, if one uses the tight-binding approximation. This is discussed in Section 6. In fact, the two C^*-algebras are not so different. They are equivalent to one another in Rieffel's sense of strong Morita equivalence. We will describe this notion and its consequences briefly in Section 5 also. There is a third approach to constructing C^*-algebras from tilings. It has been developed by J. Bellissard and is strongly motivated by physical considerations. It is more operator theoretic than the constructions we consider, which tend to be more geometric.

Section 7 gives a short (and highly incomplete) introduction to K-theory for C^*-algebras and in the following section we discuss its relevance within physics. In particular, we will give a physical motivation for the study of the K-theory of the C^*-algebras we have constructed from tilings.

The final section gives an outline of the computations made of the K-theory of the two C^*-algebras we have constructed earlier. These computations concentrate on the case of substitution tiling systems. The case of tilings obtained from the projection method has been considered recently by Forrest, Hunton and Kellendonk [**FHK**].

The case of the first C^*-algebra (from the continuous dynamics) was done by the second author, in collaboration with Jared Anderson. The second C^*-algebra was done by the first author. The fact that these two are strongly Morita equivalent implies that they will have isomorphic K-theories.

Unfortunately, our desire to provide an introduction forces us to limit our discussions. Let us quickly mention some items which we do not include. The more intricate computations of the K theory are sometimes omitted. In particular, we do not describe the computation of the kernel of the map from $K_0(AF_T)$ to $K_0(A_T)$ which appears in [**Kel2**]. We use the simplest possible definition of a substitution tiling system. There are many generalizations, which actually occur in certain examples of interest. We do not discuss topological equivalence of tilings.

We present an example, the octagonal tiling. More examples can be found in the references, especially in our own papers [**AP, Kel1, Kel2, Kel3**].

2. Tilings as Dynamics

In this section, we show how a tiling T of $\mathbb{R}^d$ gives rise to a topological dynamical system $(\Omega_T, \mathbb{R}^d)$. That is, Ω_T is a compact metric space with an action of $\mathbb{R}^d$ or, equivalently, a d-dimensional flow. The construction is a fairly standard one in dynamics. We refer the reader to [**GS, RW, Ro1, Rud, So1**].

Let us begin with some notation. $\mathbb{R}^d$ denotes the usual d-dimensional Euclidean space. For x in $\mathbb{R}^d$, $r > 0$, $B(x, r)$ denotes the open ball, centered at x with radius r. If $X \subset \mathbb{R}^d$ and $x \in \mathbb{R}^d$, then $X + x = \{x' + x \mid x' \in X\}$, the translate of X by x.

A tiling, T, of $\mathbb{R}^d$ is a collection of subsets $\{t_1, t_2, \ldots\}$, called tiles, such that their union is $\mathbb{R}^d$ and their interiors are pairwise disjoint. We will also assume, for simplicity, that each is homeomorphic to the closed unit ball, $\overline{B(0, 1)}$. We also allow the possibility that our tiles carry labels. So that if two tiles have the same label, then one is a translate of the other. If we include labels, then when we write $t + x = t'$, for t, t' in T, x in $\mathbb{R}^d$, we mean not only that the sets are the same, but the labels on t and t' are the same. Generally, we say two tiles are the same tile type if one is a translate of the other.

If T is a tiling and x is in $\mathbb{R}^d$, then

$$T + x = \{t + x \mid t \in T\}$$

the translate of T by x is also a tiling. Beginning with a single tiling T, we consider all of its translates $T + \mathbb{R}^d$ and endow this set with a metric d as follows. For $0 < \epsilon < 1$, we say the distance between T_1 and T_2 in $T + \mathbb{R}^d$ is less than ϵ if we may find vectors x_1, x_2 in $B(0, \epsilon)$ such that $T_1 + x_1$ and $T_2 + x_2$ are equal on $B(0, \frac{1}{\epsilon})$. If there are no such x_1, x_2 for any ϵ, then we set the distance to be 1. (See also [**RW, Ro1, Rud, So1**].)

The construction is a standard one. Notice already that it is measuring something interesting about the way patterns in T repeat: for x, y in $\mathbb{R}^d$ $d(T - x, T - y)$ is small when the patterns in T at x and y agree, up to a small translation.

DEFINITION 2.1. Given a tiling T, we let Ω_T denote the completion of the metric space $(T + \mathbb{R}^d, d)$. We refer to this as the *continuous hull* of T.

It is important (but fairly easy) to observe that the elements of Ω_T can be viewed as tilings and that the same definition of our metric d extends to Ω_T.

THEOREM 2.2 ([**RW**]). *Let T be a tiling. Suppose that, for any $R > 0$, there are, up to translation, only finitely many patches in T (i.e., subsets of T) whose union has diameter less than R. Then (Ω_T, d) is compact.*

We will refer to the hypothesis of this theorem as the finite pattern condition although it is also called finite local complexity ([**La**]).

It is clear that $\mathbb{R}^d$ acts by translation on the elements of Ω_T; if T' is a tiling in Ω_T, so is $T' + x$, for any x in $\mathbb{R}^d$. It is clear also that $(\Omega_T, \mathbb{R}^d)$ is *topologically transitive*, i.e., there is a dense orbit (namely that of T). More subtly, we can ask whether every orbit is dense. In this case, we say $(\Omega_T, \mathbb{R}^d)$ is minimal.

THEOREM 2.3. $(\Omega_T, \mathbb{R}^d)$ is minimal if and only if, for every finite patch P in T, there is an $R > 0$, such that for every x in $\mathbb{R}^d$, there is a translate of P contained in T and in $B(x, R)$.

The condition in the theorem is also called repetitivity. We say that a tiling T is *aperiodic* if $T + x \neq T$, for any nonzero vector x. We will mainly be interested in aperiodic tilings T and for such a tiling, it is possible that Ω_T will contain periodic tilings. (Consider tiling the plane with unit squares, fitting edge to edge. Remove four of them meeting at a point and replace with a square of side length two. This tiling is aperiodic, but its hull contains the original tiling by unit squares.)

Throughout the rest of the paper we will say that

(i) T is *minimal*, if the conditions of Theorem 2.3 are satisfied.

(ii) T is *strongly aperiodic*, if Ω_T contains no periodic tilings.

We will note, but not prove, the following.

PROPOSITION 2.4. *If the tiling T is aperiodic and minimal, then it is strongly aperiodic.*

3. The Dynamics of Substitution Tilings

Many tilings of interest possess a self-similarity structure. In fact, one can begin with a finite set of tiles with a substitution rule and produce tilings from iteration of the rule. The self-similarity appears as this substitution applied to the resulting tilings. (See also [**GS, Ken, Ro1, So1**].)

We begin as follows. Suppose we have a finite collection of nonempty, compact sets, each being homeomorphic to the closed unit ball, $\{p_1, \ldots, p_N\}$, in $\mathbb{R}^d$. These we call the proto-tiles. (This is not necessarily the standard use of this term.) We suppose we have a substitution rule ω and a scaling factor $\lambda > 1$. This means that, for each p_i, $\omega(p_i)$ is a finite collection of subsets, each one being a translate of one of the proto-tiles, overlapping only on their boundaries. Moreover the union of these sets is exactly λp_i. Thus, ω allows us to replace proto-tiles by patches.

We can extend the definition of ω to translates of the original proto-tiles, p, by setting $\omega(p + x) = \omega(p) + \lambda x$, for x in $\mathbb{R}^d$. If P is any patch made up of such translates (in a nonoverlapping fashion), then we define

$$\omega(P) = \{\omega(t) \mid t \in P\}$$

If T is a tiling, then so is $\omega(T)$. This also means that we can iterate, forming a sequence of patches, $\omega^k(p_i)$, for $k = 1, 2, 3, \ldots$.

We will assume that our substitution is primitive: for some $k > 1$, a translate of each p_i appears inside the patch $\omega^k(p_j)$, for all i, j.

It is a fairly standard argument, which we now sketch, to show that such a system will actually admit tilings. (For more details, see [**GS**].) One can find a translate of one of the proto-tiles t and a $k > 1$ so that the sequence of patches $\omega^{kn}(t)$, for $n = 1, 2, 3, \ldots$, grows to cover the plane and is consistent in the sense that any two agree where they overlap. We let T denote the union of these patches which is a tiling. With the hypothesis of primitivity, the hull Ω_T is independent of the choice of T constructed in this way. To emphasize this fact, we will drop the subscript T from our notation. (In fact, there is another equivalent definition of the space Ω which avoids making a choice of T. The construction above is needed however, to show that this space is nonempty.) We will also assume that T satisfies the finite pattern condition.

As we noted above, there is an extension of ω to tilings. Its restriction to Ω is continuous and surjective [**Mo, AP**] and we also have $\omega(\Omega) \subseteq \Omega$. From now on we will also assume that $\omega \colon \Omega \to \Omega$ is injective as well. This is quite a subtle point. It amounts to what is usually called "recognizability". As an example, consider having a single tile which is a unit square in the plane. The scaling factor λ is 2 and the map ω simply divides the square into four smaller squares, and then rescales by 2. If we center this square at the origin and iterate to obtain our tiling, T, we obtain the tiling of the plane by unit squares with vertices on the integer lattice points. Consider T and $T + (.5, .5)$, which is the same tiling, but aligned so that the centres are on the integer lattice. Now, we have $\omega(T) = \omega(T + (.5, .5)) = T$, and so ω is not injective. In fact, it is fairly easy to generalize this example to show that our hypothesis that ω is injective implies that Ω contains no periodic tilings. (For the converse, see [**So2**].)

Let us mention that this setup can be generalized considerably. First, the constant λ can be replaced by any expansive linear transformation of $\mathbb{R}^d$. In [**Kel2**], there is an even more general version where a substitution is defined as (roughly) a map from the set of patches in a tiling to itself satisfying certain conditions. The expansiveness ($\lambda > 1$) is replaced by a growth condition.

We may consider (Ω, ω) as a dynamical system of its own. As such, it possesses special features which are common in the field of hyperbolic dynamics. Let us give some background for this material.

In his seminal paper [**Sm**], Smale proposed and initiated the study of Axiom A systems. The idea is to consider a compact Riemannian manifold M with a diffeomorphism, f. We then isolate a closed invariant subset Λ of M, based on the idea of recurrence. More specifically, Λ is the set of chain recurrent points of f. We then assume that f is hyperbolic on Λ in the following sense. The tangent bundle to M, TM, when restricted to Λ, $T_\Lambda M$, may be decomposed as a direct sum

$$T_\Lambda M = E^s \oplus E^u$$

where each summand is invariant under the derivative of f and, at least roughly speaking, the derivative contracts vectors in E^s while the derivative of f^{-1} contracts vectors in E^u. (For a general reference to such systems, see [**KH**].)

Smale made the key observation that, although (M, f) is smooth, Λ need not be a manifold. (For example, see Smale's horseshoe [**Sm**].) Now the system $(\Lambda, f \mid \Lambda)$ exists only in the topological category and motivated by this idea, Ruelle gave a definition of a Smale space [**Rue**]. The name is slightly misleading since the object is both a space and a map.

Basically, a Smale space is a compact metric space with a homeomorphism so that, locally, the space may be written as the product of two subsets. This decomposition (or rather its germs) are invariant under the map and the map contracts the first subset, while its inverse contracts the second.

It is fairly easy to see that our system, (Ω, ω), arising from a substitution tiling with hypotheses as above, has the structure of a Smale space [**Ken, Ro1, AP**]. Let T be any tiling in Ω. We want to produce two subsets of Ω containing T, whose Cartesian product is, in a natural way, homeomorphic to a neighborhood of T. For the first, consider all tilings which agree with T on a ball at the origin of radius one. First, notice that the map ω acts as a contraction on this set since iteration of the map on two such tilings, produces tilings which agree on larger and larger balls and so the distance between them contracts at an exponential rate. For the second

set, take all tilings which are translates of T by a small amount. Notice that the equation
$$\omega^{-n}(T+x) = \omega^{-n}(T) + \lambda^{-n}x$$
immediately implies that any two such tilings get closer together under iteration of the map ω^{-1}. Finally, it follows at once from the definition of our metric on Ω that for any tiling, T' close to T, we may find a unique small vector x so that $T'+x$ agrees with T on a ball of radius one. Then the map sending T' to the pair $(T'+x, T-x)$ is a homeomorphism from a neighborhood of T to the Cartesian product of the two sets mentioned above.

We remark that in this local description of Ω, the local contracting direction is totally disconnected while the other local coordinate is homeomorphic to an open set in $\mathbb{R}^d$ [**Ro1**].

Our next objective is to present the space, Ω, as an inverse limit of more tractable spaces. The substitution rule allows us to be very specific about this. In particular, all the spaces are copies of the same space, which we will denote Γ. Moreover, the maps between these spaces will also be stationary.

Let $\widetilde{\Gamma}$ be the disjoint union of the proto-tiles. We define an equivalence relation on this space as follows. For x in p_i and y in p_j, we set $x \sim y$ if there are tiles p_i+z and p_j+w in T such that $x+z = y+w$. This simply means that if, somewhere in T, we see copies of p_i and p_j overlapping at the points corresponding to x and y, then we set $x \sim y$. At this point $\sim$ may not be transitive. We define $\sim$ to be the equivalence relation generated by this relation. The space Γ is the quotient of $\widetilde{\Gamma}$ by $\sim$. It is a compact Hausdorff space.

In specific examples, all of which having tiles which are polygons, this space has a cellular structure. In the top dimension d, the d-cells are the interiors of the proto-tiles. This idea has not been developed in generality.

There is a natural map $\gamma\colon \Gamma \to \Gamma$ which is induced by ω. If x is in $\widetilde{\Gamma}$, then x is in some p_i. Then the point λx lies in some tile in $\omega(p_i)$, say t. This is a translate of some other proto-tile, say $t = p_j + y$. We then define $\gamma(x)$ to be $\lambda x - y$, which is in p_j and hence in $\widetilde{\Gamma}$. It is possible that λx lies in more than one proto-tile in $\omega(p_i)$, but it this case, it is easy to see that the $\sim$-equivalence class of the resulting point is unique. It is easy to see this induces a well-defined map on Γ. Now the inverse limit
$$\Gamma \xleftarrow{\gamma} \Gamma \xleftarrow{\gamma} \Gamma \xleftarrow{\gamma}$$
is denoted by Ω_0. It can be defined as
$$\{(x_0, x_1, x_2, \ldots) \mid x_i \in \Gamma, \gamma(x_{i+1}) = x_i, \text{ for all } i \in \mathbb{N}\}.$$

THEOREM 3.1 ([**AP**]). *If the substitution system "forces its border", then Ω is homeomorphic to Ω_0.*

We will define the condition of "forcing the border" as we give a sketch of the proof. For the moment, let us make a few remarks.

The homeomorphism will actually be a topological conjugacy between the map ω on Ω and the natural shift map on the inverse limit
$$\omega_0(x_0, x_1, \ldots) = (\omega_0(x_0), x_0, x_1, x_2, \ldots).$$

The class of dynamical systems obtained via this inverse limit construction was introduced and studied intensely by R. F. Williams [**W**] as models for "expanding attractors" within the context of Smale's program for Axiom A systems. It seems appropriate to refer to such systems as "Williams solenoids". It is interesting to see that our "forcing the border" condition appears in Williams' work in the form of a "flattening" axiom.

We will give a short sketch of the proof describing the homeomorphism from Ω to Ω_0. The idea is fairly simple. Begin with any tiling T in Ω. We want to define a sequence $(x_0, x_1, \ldots)$ in the inverse limit. First, locate the tile in T which contains the origin. It is the form $p + x$, for some proto-tile p and vector x in $\mathbb{R}^d$. Since the origin is in $p + x$, we have $-x$ in p. This gives us a point in $\widetilde{\Gamma}$ and its image in Γ is x_0. Of course, the origin may lie on two or more tiles in T. In this case, all of the points obtained in $\widetilde{\Gamma}$ will be $\sim$-equivalent and so the point in Γ is unique.

To obtain x_k, for $k \geq 1$, we repeat this procedure using $\omega^{-k}(T)$ instead of T. It is simple to check that $\gamma(x_{k+1}) = x_k$, for all k and so this sequence defines a point in Ω_0. The important issue here is that this map is injective. To see this, first notice that if the origin is in the interior of a tile in T, then the point x_0 uniquely determines the tile covering the origin in T. In the case when the origin lies in more than one tile, things are a bit more subtle and one must work more carefully. We will not go into the details in this case. Similarly, x_k determines the tile covering the origin in $\omega^{-k}(T)$. Let us call this tile t_k. This means that x_k determines a patch, $\omega^k(t_k)$, in T which contains the origin. The idea is that we can hope these patches grow to cover the plane as k increases. In this case, the sequence $(x_0, x_1, \ldots)$ then determines T. This will not be true in general, but it is enough that the substitution "forces its border", in the following sense [**Kel1**]. There is a $k \geq 1$ such that, if T and T' are two tilings containing a tile t, then the patches in $\omega^k(T)$ and $\omega^k(T')$ consisting of all tiles which meet $\omega^k(t)$ are identical.

Consider the following one-dimensional example given as a substitution on the alphabet a, b, c. Suppose we define

$$\omega(a) = baabc \quad \omega(b) = bbbc, \quad \omega(c) = bbcaaac.$$

Notice each word begins in b and ends in c. So we don't really need to know the symbols to the left or right of an a in an infinite string to know that we will see

$$\ldots c \underbrace{baabc}_{\omega(a)} b \ldots$$

after applying ω to the infinite string. This is an example of a substitution forcing its border. The Penrose substitution also forces its border.

Of course, this seems like a strong hypothesis. However, given any substitution tiling system, we can replace it by one that forces its border and has exactly the same collection of tilings. (More precisely, the tilings from the new system are mutually locally derivable in the sense of [**BSJ**] with those of the old.) We form a new set of proto-tiles as follows. For each of our original proto-tiles, p, look at all patches in all tilings in Ω consisting of a translate of p and all its neighboring tiles. For each such patch (there are only finitely many) create a copy of p and give it a label, which consists of this patch. (The patch only functions as a label; the actual points are the same as in p.) It is easy to see how to define a substitution map on these new labeled proto-tiles. It is also easy to see that the collection of tilings will be "the same" as before and that this new system forces it border.

It is worth noting that that although the result is quite nice from an abstract viewpoint, this situation can be difficult from a practical one. As an example, the "chair", "boot" or "triomino" substitution has four proto-tiles. Unfortunately, it does not force its border. Applying the strategy above yields an equivalent system with fifty-six proto-tiles. The space Γ will be a cell complex with fifty-six 2-cells.

The last result which produced a description of the space Ω as an inverse limit was in the context of substitution tiling systems. In fact, a weaker version of this seems possible in much more generality.

PROBLEM. For an aperiodic minimal tiling T, express Ω_T as an inverse limit of spaces which are fundamentally simpler, such as finite cell complexes.

In Section 8, we will use this description of Ω to compute the K-theory of certain C^*-algebras. Before getting quite so involved in the theory of C^*-algebras, it seems interesting to ask at this point, whether the algebraic topology, especially the K-theory and cohomology of the space Ω contains information about the tiling? In a similar spirit, Geller and Propp [**GP**] introduced the notion of the projective fundamental group of a $\mathbb{Z}^2$-action which will generalize to tilings. Presumably, this presentation of Ω as an inverse limit will make the computation of such invariants more accessible. In particular, it seems to be an interesting question: "To what extent does $H^*(\Omega_T)$ or $K^*(\Omega_T)$ determine the long range order structure of T?"

4. The C^*-Algebra of a Tiling I: The Continuous Case

Let T_0 be a fixed tiling. We construct Ω_{T_0}, which we now denote simply by Ω, as in Section 2. We will assume that T_0 is minimal and aperiodic. We want to construct a C^*-algebra from $(\Omega, \mathbb{R}^d)$.

A C^*-algebra is a $\mathbb{C}$-algebra (not necessarily commutative) with an involution $a \to a^*$ and a norm in which it is complete [**Da, Pe**]. There are further hypotheses which are fairly standard. The most important item is the C^*-condition on the norm; that is, for every element, a, in the algebra, we have $\|a^*a\| = \|a\|^2$. The two canonical examples are the following. First, let X be any compact Hausdorff space. The collection of continuous $\mathbb{C}$-functions on X with supremum norm and pointwise operations of addition, multiplication and complex conjugation is a commutative C^*-algebra, denoted $C(X)$. The second example is to begin with a complex Hilbert space $\mathcal{H}$ and let $\mathcal{B}(\mathcal{H})$ denote the set of bounded linear operators on $\mathcal{H}$. With the operator norm and usual algebraic structure it is a C^*-algebra. This second example is noncommutative provided $\dim \mathcal{H} \geq 2$.

To create our C^*-algebra, $C^*(\Omega, \mathbb{R}^d)$, we begin with $C_c(\Omega \times \mathbb{R}^d)$, the continuous $\mathbb{C}$-functions of compact support on $\Omega \times \mathbb{R}^d$. It is a linear space in the obvious way. We define a product and involution on it by

$$(4.1) \qquad f \cdot g(T, x) = \int_{y \in \mathbb{R}^d} f(T, y)\, g(T - y, x - y)\, dy$$

$$(4.2) \qquad f^*(T, x) = \overline{f(T - x, -x)},$$

for f, g in $C_c(\Omega \times \mathbb{R}^d)$, T in Ω and x in $\mathbb{R}^d$. This makes $C_c(\Omega \times \mathbb{R}^d)$ into a $*$-algebra. There are a number of subtle technical points about equipping it with a norm. We will mention here only that this can be done in a natural way. Unfortunately, it will not be complete. Its completion is $C^*(\Omega, \mathbb{R}^d)$ and is indeed a C^*-algebra. This is an example of the construction of the crossed product C^*-algebra. (For more details of this construction, see [**Pe, Z-M**].)

We want to present several other views of this C^*-algebra. First, for the reader who likes to think of operators on Hilbert space, we proceed as follows. Consider the Hilbert space $L^2(\mathbb{R}^d)$. We define, for each f in $C_c(\Omega \times \mathbb{R}^d)$, an operator $\lambda(f)$ on $L^2(\mathbb{R}^d)$ by

$$(4.3) \qquad (\lambda(f)\xi)(x) = \int_{y \in \mathbb{R}^d} f(T_0 + x, y)\xi(x - y)\, dy$$

for ξ in $L^2(\mathbb{R}^d)$ and x in $\mathbb{R}^d$. The map λ is a $*$-homomorphism of $C_c(\Omega \times \mathbb{R}^d)$ into $\mathcal{B}(L^2(\mathbb{R}^d))$ which extends to an isometric $*$-isomorphism between $C^*(\Omega, \mathbb{R}^d)$ and the closure of the collection of $\lambda(f)$'s in the operator norm. It is worth noting that there are many other representations, not all as natural as this.

With this description of operators, this C^*-algebra is still not the most intuitive of objects. Let us now take yet another view. Suppose we return to formulae (4.1) and (4.2), defining our product and involution, and change $T - y$ in the first and $T - x$ in the second to simply T. This then removes any hint that $\mathbb{R}^d$ is acting on Ω. Performing the Fourier transform in the $\mathbb{R}^d$ variable, one obtains functions on $\Omega \times \mathbb{R}^d$ which are continuous and vanish at infinity. The product and involution of (4.1) and (4.2) become pointwise product and conjugate. This Fourier transform extends to an isometric isomorphism of the resulting C^*-algebra onto $C_0(\Omega \times \mathbb{R}^d)$, the continuous functions vanishing at infinity on $\Omega \times \mathbb{R}^d$, which is a commutative C^*-algebra.

If we return to (4.1) and (4.2) again and replace $T - y$ and $T - x$ with $T - hx$ and $T - hy$, where $0 \leq h < \infty$ is a parameter, then we can actually view $C^*(\Omega, \mathbb{R}^d)$ ($h = 1$) as a deformation of $C_0(\Omega \times \mathbb{R}^d)$ ($h = 0$). (For much more general situations viewed in this way, see [**Ri2**].)

The action of $\mathbb{R}^d$ on Ω makes $C^*(\Omega, \mathbb{R}^d)$ noncommutative. In fact, its centre is trivial. We have the following even stronger result.

THEOREM 4.1 ([**EH, GR**]). *If T_0 is aperiodic and minimal, then $C^*(\Omega_{T_0}, \mathbb{R}^d)$ is simple; i.e., it has no nontrivial closed two-sided ideals.*

We describe another formulation of $C^*(\Omega, \mathbb{R}^d)$. Let

$$R_T = \{(T, T') \in \Omega \times \Omega \mid T' \text{ is a translate of } T\}.$$

This is an equivalence relation on Ω whose classes are simply the $\mathbb{R}^d$-orbits. The map sending (T, x) in $\Omega \times \mathbb{R}^d$ to $(T, T+x)$ is obviously a surjection and if we assume that Ω has no periodic tilings, then it is injective as well. If we simply translate our product and involution on $C_c(\Omega \times \mathbb{R}^d)$ to $C_c(R_T)$, they become

$$(4.4) \qquad (f \cdot g)(T, T') = \int_{T''} f(T, T'')g(T'', T')\, dT''$$
$$(4.5) \qquad f^*(T, T') = \overline{f(T', T)},$$

for T, T' in Ω, f, g in $C_c(R_T)$. The nice thing about this formulation is that it reminds one of matrix multiplication and conjugate transpose. Also this definition can then be extended to study other equivalence relations [**Ren**]. We will do this in Section 5. There are, however, some topological subtleties. We have implicitly transferred the topology of $\Omega \times \mathbb{R}^d$ over to R_T via our bijection. This is not the relative topology of $R_T \subset \Omega \times \Omega$; for large x, T and $T + x$ may be close so $(T, T+x)$ is close to (T, T) in the relative topology, but not in that from $\Omega \times \mathbb{R}^d$. (It is worth noting that, in the relative topology, R_T is not locally compact and hence a bit of a disaster from an analytic view.)

We will close this section with a bit of philosophy. We have a group, $\mathbb{R}^d$, which is acting freely on a space, Ω. In Alain Connes' program of noncommutative geometry [**Co2**], the C^*-algebra $C^*(\Omega, \mathbb{R}^d)$ acts as a replacement for the orbit space, $\Omega/\mathbb{R}^d$. This should be interpreted as follows. If the space $\Omega/\mathbb{R}^d$ is reasonable (i.e. Hausdorff), then $C^*(\Omega, \mathbb{R}^d)$ should be equivalent to the commutative C^*-algebra $C(\Omega/\mathbb{R}^d)$. In our situation of a minimal, aperiodic tiling, the orbit space $\Omega/\mathbb{R}^d$ has the indiscrete topology and is effectively useless as a topological space, while the noncommutative C^*-algebra contains much interesting information. Much more on this point of view may be found in [**Co2**].

5. The C^*-Algebra of a Tiling II: The Discrete Case

In this section, we want to construct another C^*-algebra from a tiling T. The description of this algebra will be simpler than that of $C^*(\Omega, \mathbb{R}^d)$ in the last section. The point is that this new C^*-algebra is equivalent in a certain sense (strong Morita equivalence) to $C^*(\Omega, \mathbb{R}^d)$. First, we will describe the new algebra and then discuss its relation to the old one.

To motivate our discussion, let us examine the simplest noncommutative C^*-algebra: $\mathbf{M}_n$, the $n \times n$ complex matrices, for $n \geq 2$.

For each pair, $1 \leq i, j \leq n$, let $e(i,j)$ denote the matrix which is one in the (i,j) entry and zero elsewhere. Clearly, these elements satisfy the relations:

$$(5.1) \qquad e(i,j)^* = e(j,i)$$

$$(5.2) \qquad e(i,j)e(i',j') = \begin{cases} e(i,j') & \text{if } i' = j, \\ 0 & \text{otherwise.} \end{cases}$$

In fact, it turns out that $\mathbf{M}_n$ is the universal C^*-algebra generated by a collection $\{e(i,j) \mid 1 \leq i, j \leq n\}$ satisfying the relations (5.1) and (5.2). (As an aside, the general construction of C^*-algebras from generators and relations is rather tricky. For instance, there is no free C^*-algebra on one element.)

Now let us turn to our tiling T. We look at all triples (P, t_1, t_2) where $P \subset T$ is finite (i.e., a patch) and $t_1, t_2 \in P$ (allowing $t_1 = t_2$). We say two of these triples are equivalent if one is a translate of the other. We let $[P, t_1, t_2]$ denote the equivalence class under translation of (P, t_1, t_2) and call this a doubly pointed pattern class. Our C^*-algebra, which we will denote A_T [**Kel1, Kel2**], is generated by elements $e[P, t_1, t_2]$, where $[P, t_1, t_2]$ is a doubly pointed pattern class, subject to some relations. The first is that

$$(5.3) \qquad e[P, t_1, t_2]^* = e[P, t_2, t_1].$$

The second is that, if (P, t_1, t_2) and (P', t_1', t_2') are both contained in a larger patch in such a way that $t_2 = t_1'$, then

$$(5.4) \qquad e[P, t_1, t_2]e[P', t_1', t_2'] = e[P \cup P', t_1, t_2']$$

whereas otherwise that product is 0. Finally, we require the following. Suppose P is any patch, t is any tile in P and $P_1, P_2, \ldots, P_k$ are a collection of patches, each containing P and so that any tiling which contains P contains exactly one of the P_i, then we have

$$(5.5) \qquad e[P, t, t] = \sum_{i=1}^{k} e[P_i, t, t]$$

Observe first, that if P is fixed, each element $e[P,t,t]$ is a self-adjoint idempotent and also that

(5.6) $$e[P,t_1,t_2]e[P,t_1,t_2]^* = e[P,t_1,t_1]$$
(5.7) $$e[P,t_1,t_2]^*e[P,t_1,t_2] = e[P,t_2,t_2].$$

Secondly, if we list representatives of all tile types: $t_1, t_2, \ldots, t_n$, then

(5.8) $$\sum_{i=1}^{n} e[\{t_i\}, t_i, t_i]$$

is a unit for our C^*-algebra. Now, we want to give an indication of why this C^*-algebra is related to $C^*(\Omega, \mathbb{R}^d)$. At the same time, we will obtain another description of it.

Recall Ω, our space of tilings. For each tile type (or labeled tile type) in T, we choose a point in the interior which we call a puncture. Now in our tiling each tile, t, is given a puncture $x(t)$, so that if two tiles t_1 and t_2 are translates, say $t_1 = t_2 + x$, then $x(t_1) = x(t_2) + x$. In the same way, each tile in every tiling in Ω also gets a puncture.

DEFINITION 5.1 ([**Kel1**]). We define the discrete hull of T, which we denote by Ω_{punc}, to be all the tilings T' in Ω such that the origin is a puncture of some tile t in T'; that is, $x(t) = 0$. (Note that, since the punctures do not lie on the boundaries of tiles, the choice of t is unique.)

First observe the following simple facts about Ω_{punc}.

(1) If T' is any tiling in Ω, $T' + x$ is in Ω_{punc} for some x in $\mathbb{R}^d$.
(2) If T' is in Ω_{punc}, there is an $\epsilon > 0$ such that $T' + x$ is *not* in Ω_{punc}, for any x with $0 < |x| < \epsilon$.
(3) Ω_{punc} is closed in Ω.

We summarize by saying that Ω_{punc} is a transversal to the $\mathbb{R}^d$-action. (See [**MRW**].) Somewhat less obvious is that, if we assume T satisfies the finite pattern condition, then Ω_{punc} is a Cantor set; that is, it is compact, has no isolated points and its topology is generated by sets which are both closed and open. Let us present such sets. Let P be a finite patch in T and let t be an element of P. Then $P - x(t)$ is a patch having a puncture on the origin. Look at all tilings in Ω_{punc} which contain the patch $P - x(t)$; i.e.,

$$U(P,t) = \{T' \in \Omega \mid P - x(t) \subset T'\}.$$

One can check that $U(P,t)$ is both open and closed in the relative topology of Ω_{punc} and that such sets generate the topology of Ω_{punc}.

We now define an equivalence relation R_{punc} on Ω_{punc} as follows:

$$R_{\text{punc}} = \{(T_1, T_2) \mid T_1, T_2 \in \Omega_{\text{punc}} \text{ and } \exists x \in \mathbb{R}^d : T_1 = T_2 + x\}.$$

This means we are taking the equivalence relation on Ω_T whose classes are simply the $\mathbb{R}^d$-orbits, which we called R_T in Section 4, and we are restricting it to Ω_{punc}: $R_{\text{punc}} = R_T \cap (\Omega_{\text{punc}} \times \Omega_{\text{punc}})$. We provide R_{punc} with a topology as follows. A sequence $(T_n, T_n + x_n)$ in R_{punc} converges to $(T, T + x)$ if and only if $d(T_n, T)$ and $|x_n - x|$ both tend to zero. There are a number of technical subtleties here, but we proceed to define a C^*-algebra $A_T = C^*(R_{\text{punc}})$ as follows [**Kel1, Ren**]. Begin with $C_c(R_{\text{punc}})$ as a linear space and define product and involution by

$$(f \cdot g)(T_1, T_2) = \sum_{T' \text{ s.t. } (T_1, T') \in R_T} f(T_1, T')g(T', T_2) \tag{5.9}$$

$$f^*(T_1, T_2) = \overline{f(T_2, T_1)}. \tag{5.10}$$

These formulas should certainly remind one of matrix multiplication and adjoint and (4.4) and (4.5). As with $C^*(\Omega, \mathbb{R}^d)$, this $*$-algebra must be given a norm and completed to get a C^*-algebra. We will not discuss this here, but the result is $C^*(R_T)$ which we will call simply A_T. We point out that Bellissard et al. make a very similar construction with measures instead of tilings (see their contribution in this book).

We must address two issues: first, to see the elements $e[P, t_1, t_2]$ we discussed earlier and secondly to see how this C^*-algebra is related to $C^*(\Omega, \mathbb{R}^d)$.

For the first part, let (P, t_1, t_2) be a doubly pointed pattern class in T. The map sending T' in $U(P, t_1)$ to $T' + x(t_2) - x(t_1)$ in $U(P, t_2)$ is a homeomorphism. The graph of this map is not only contained in R_{punc}, but is actually a compact and open subset. Let $e[P, t_1, t_2]$ denote its characteristic function. Now one checks easily from the definitions (5.10) and (5.9) that these elements satisfy the relations (5.3), (5.4) and (5.5). Also, the graphs of such functions actually generate the topology of R_{punc}, so the linear span of the $e[P, t_1, t_2]$'s is dense in $C_c(R_{\text{punc}})$ and hence in A_T.

To continue our analogy of A_T with M_n, we observe the following analogue of the subalgebra of diagonal matrices.

PROPOSITION 5.2. *The map sending the characteristic function of $U(P, t)$ to $e[P, t, t]$ is a unital injective $*$-homomorphism of $C(\Omega_{\text{punc}})$ to A_T.*

We now turn to the second problem, relating A_T with $C^*(\Omega, \mathbb{R}^d)$. These algebras are strongly Morita equivalent—a concept introduced by Marc Rieffel [**Ri1, MRW**]. In fact, any time one considers a transversal to an equivalence relation satisfying conditions 1, 2, 3 as above, the associated C^*-algebras will be related in this way.

Rather than describe this in detail, we will give some simple examples and some consequences. If A is any C^*-algebra, it is strongly Morita equivalent to $\text{M}_n(A)$, the C^*-algebra of $n \times n$ matrices over A. Also, if h is any self-adjoint element of A whose spectrum is nonnegative, then A is strongly Morita equivalent to the closure of hAh, provided the closed two-sided ideal in A generated by h is all of A.

If two C^*-algebras are strongly Morita equivalent, then there is a natural bijective correspondence between their ideal structures (ideal means closed two-sided ideal), their representation theories and their K-theories. Although the definition of strong Morita equivalence is complicated enough that we omit it here, it is the most natural notion of equivalence for C^*-algebras—perhaps even more natural than isomorphism.

6. C^*-Algebras in Physics

In this section, we will discuss the role of the C^*-algebra, A_T, in physics. In the quantum mechanical model of the motion of a particle in Euclidean space, an observable is a self-adjoint operator. Ignoring internal degrees of freedom (like spin) and provided there are no external forces (like an external magnetic field) such an operator is constructed from position and momentum operators. So we can work entirely inside the algebra generated by the momentum and position operators. We choose to work within the C^*-algebra which is the closure of this algebra and refer to it as the C^*-algebra of observables. We want to study the impact of the topology of this underlying noncommutative space, in the sense of [**Co2**]. A first difficulty is that many of these operators are unbounded. One approach to dealing with this is to pass to resolvents of the operators. Instead, we want to consider the tight binding approximation for a particle in a solid. In this, the solid can be modeled by a tiling; the tiles representing the locations of the atoms so that congruent arrangements are represented by congruent patches in the tiling. The particle motion becomes discrete. The particle hops from tile to tile. The Hilbert space of wave functions is replaced by the space of square summable functions on the set of tiles. This has two immediate consequences. First, (absolute) position is described by a tile in the tiling and second, momentum—usually thought of as a generator of translation—has to be replaced by finite translation (or strictly speaking, its difference with the identity). However, as a consequence of locality of interaction, observables like the potential, which are independent of momentum, depend only on the position of the particle (i.e., a tile) inside a patch whose position in the tiling doesn't matter. (The larger the interaction radius, the larger the patch.) More technically, let P be a patch and t be a tile in P. Then the momentum-independent observables will be functions of the $e[P,t,t]$, operators which describe whether the particle is at t in a patch which is a translate of P. Now, suppose we consider a patch, P in T, consisting of a pair of adjacent tiles (meaning their intersection is codimension one) t_1, t_2. Our operator $e[\{t_1,t_2\},t_1,t_2]$ represents the operator associated with the transition from a tile of type t_2 to an adjacent one of type t_1 in *all* patches which are translations of P. It is not a unitary, but rather a partial isometry. It can be regarded as a "partial translation"; partial in the sense that its domain corresponds only to those tiles which are translates of t_2 in the translate of P. These partial translations are the operators which replace momentum. The C^*-algebra these generate is exactly our algebra A_T. A similar construction has been given by Bellissard [**Be2, BCL**] for the case of some standard tilings obtained from the projection method.

There is another approach which comes from the ideas of disordered systems. This has been developed principally by Bellissard [**Be2**]. The idea is to begin with a bounded operator A acting on the Hilbert space $\mathcal{H} = L^2(\mathbb{R}^d)$, which should be thought of as a bounded function of the Hamiltonian for our particle. There is a natural action of $\mathbb{R}^d$ on this same Hilbert space via translations. We use V_x to denote the unitary operator

$$(V_x\xi)(y) = \xi(y-x),$$

for any ξ in $\mathcal{H}$, x and y in $\mathbb{R}^d$.

One considers the set of translations of A under conjugation by this unitary representation of $\mathbb{R}^d$ and its closure in the strong operator topology. This we define to be the strong operator hull of A

$$\mathrm{Hull}(A) = \overline{\{V_x^* A V_x \mid x \in \mathbb{R}^d\}}^{-\mathrm{SOT}}$$

(Here, SOT refers to convergence in the strong operator topology on $\mathcal{B}(\mathcal{H})$. We recall that a sequence of operators, A_n, converges to an operator A in the strong operator topology in $\|A_n\xi - A\xi\|$ converges to zero, for every vector ξ in (H).)

The idea is to think of measures on Hull(A) as probabilities of the different translates of A. Now, one can view Hull(A) and the action of $\mathbb{R}^d$ on it as a dynamical system and perform with it the same constructions we did with $(\Omega_T, \mathbb{R}^d)$. In fact, in many cases one can show that these dynamical systems are conjugate, provided the operator A reflects the structure of the tiling (e.g., by being quasiperiodic with respect to the tiling), but to make the last statement precise in general is still partly an open problem. The analogous approach works also in the tight binding approximation if the tiling can be identified with an (amenable) discrete group which plays the role of $\mathbb{R}^d$.

7. K-Theory for C^*-Algebras

The subject of K-theory has revolutionized the subject of C^*-algebras in the past twenty-five years. For longer discussions on the matter, we refer the reader to [**Da, Bl, W-O**].

Let A be a C^*-algebra with unit. (The nonunital case is a minor but annoying adaptation.) There are Abelian groups $K_0(A)$ and $K_1(A)$ associated to A. For separable C^*-algebras, such as all those appearing from tilings, these groups are countable. For physics, $K_0(A)$ seems the more interesting; it is basically a calculus for dealing with the projections in the C^*-algebra. If a self-adjoint operator has a spectrum which may be decomposed into disjoint closed pieces, the spectral projections for the pieces determine elements in $K_0(A)$. In addition, the simple notion that projections in a C^*-algebra may be compared (determined by containment of their ranges, if they are operators acting on a Hilbert space) produces a natural pre-order on $K_0(A)$. In most of our examples here, this seems to be an actual order. This makes $K_0(A)$ a rather rich invariant.

To define $K_0(A)$, we proceed (in a rather heuristic fashion) as follows. We want to look at all projections or self-adjoint idempotents in A. That is,

$$P_1(A) = \{p \in A \mid p^2 = p = p^*\}.$$

Two are equivalent if they are similar, that is

$$p \approx q \quad \text{if } p = uqu^{-1},$$

for some invertible u in A. Two projections p and q are called *orthogonal* if $pq = 0$, which implies $qp = 0$ also. In this case, $p + q$ is again a projection. This definition can be extended to equivalence classes:

$$[p] + [q] = [p + q] \quad \text{if } pq = 0.$$

We face the question, given two equivalence classes, whether we can find an orthogonal pair of representatives? This cannot always be done (suppose one is the class of the identity!), but we can solve the problem as follows. Let

$$P(A) = \{p \mid p \in \mathrm{M}_n(A), \text{ for some } n, p^2 = p = p^*\},$$

where $\mathrm{M}_n(A)$ denotes the $n \times n$ matrices with entries from A. For each n, $\mathrm{M}_n(A)$ is a C^*-algebra. Here, we implicitly assume $\mathrm{M}_n(A) \subset \mathrm{M}_{n+1}(A)$ by identifying (a_{ij}) and

$$\begin{pmatrix} & & 0 \\ a_{ij} & & \vdots \\ & & 0 \\ 0 & \cdots & 0 & 0 \end{pmatrix}.$$

We extend the definition of $\approx$ to $M_n(A)$. Now if p and q are in $M_n(A)$, $p^2 = p = p^*$, $q^2 = q = q^*$, regard both in $M_{2n}(A)$, where $q = \begin{pmatrix} q & 0 \\ 0 & 0 \end{pmatrix}$ is similar to $\begin{pmatrix} 0 & 0 \\ 0 & q \end{pmatrix}$ which is orthogonal to $p = \begin{pmatrix} p & 0 \\ 0 & 0 \end{pmatrix}$. So

$$[p] + [q] = \left[\begin{pmatrix} p & 0 \\ 0 & q \end{pmatrix} \right].$$

We are on our way to turning $P(A)/\approx$ into an Abelian group. At the moment we have a semigroup with identity ($p = 0$). The remainder of the construction (usually referred to as the Grothendieck group) is standard. The first problem is that our semigroup may not have cancellation. We define

$$p \sim q \quad \text{if} \quad \begin{pmatrix} p & 0 \\ 0 & 1_k \end{pmatrix} \approx \begin{pmatrix} q & 0 \\ 0 & 1_k \end{pmatrix},$$

for some k, where 1_k denotes the multiplicative identity in $M_k(A)$. Then $P(A)/\sim$ is a semigroup with an identity and cancellation, but perhaps no inverses. Then

$$K_0(A) = \{[p] - [q] \mid p, q \in P(A)\}$$

is all formal differences. The semi-group $P(A)/\sim$ appears in $K_0(A)$ as

$$K_0(A)^+ = \{[p] - [0] \mid p \in P(A)\}$$

which is a generating cone. If $K_0(A)^+ \cap (-K_0(A)^+) = \{0\}$ then we may define an order by $x \geq y$ if and only if $x - y \in K_0(A)^+$.

Let us compute this for the simplest of all C^*-algebras, the complex numbers. In fact, this will give us some useful insights for later. Let Tr denote the trace on $M_n(\mathbb{C})$: $\text{Tr}(a) = \sum_i a_{ii}$. For a projection p in $M_n(\mathbb{C})$, $\text{Tr}(p)$ is the rank of p, hence an integer. (Observe that, if we view p in $M_n(\mathbb{C})$ or $M_{n+1}(\mathbb{C})$, its trace is the same.) Now trace has several important properties. It is invariant under similarity. Hence it is well-defined on $P(A)/\approx$. It is additive and hence well-defined on $P(A)/\sim$ and is, in fact, a morphism of semigroups. Finally, two projections in $M_n(\mathbb{C})$ are similar if and only if they have the same rank or trace. Thus Tr induces an isomorphism

$$\widehat{\text{Tr}} \colon K_0(\mathbb{C}) \to \mathbb{Z}; \ \widehat{\text{Tr}}([p] - [q]) = \text{Tr}(p) - \text{Tr}(q).$$

Another nice feature of Tr is that it is positive $\text{Tr}(a^*a) \geq 0$ for all a in $M_n(\mathbb{C})$. So $\text{Tr}(p) = \text{Tr}(p^*p) \geq 0$. This means that $\widehat{\text{Tr}}$ is a homomorphism of ordered groups (usual order on $\mathbb{Z}$) and, in this case, is actually an order isomorphism.

The idea of using the trace on $M_n(\mathbb{C})$ is something which can be applied to many more general C^*-algebras. If A is any unital C^*-algebra, a *trace*, τ, on A is a linear functional $\tau \colon A \to \mathbb{C}$ such that

(i) $\tau(ab) = \tau(ba)$, for all a, b in A,
(ii) $\tau(1) = 1$,
(iii) $\tau(a^*a) \geq 0$, for all a in A.

Given such a functional, we can define a group homomorphism

$$\widehat{\tau} \colon K_0(A) \to \mathbb{R}$$

by

$$\widehat{\tau}(p) = \sum_{i=1}^n \tau(p_{ii}),$$

for p in $M_n(A)$, $p^2 = p = p^*$. The fact that similar projections have the same trace follows from (i) above. The rest of the argument that $\widehat{\tau}$ is a well-defined group

homomorphism is exactly as for the complex numbers. Also, condition (iii) ensures that $\hat{\tau}$ is positive:
$$\hat{\tau}(K_0(A)^+) \subset [0, \infty).$$
The two important differences from the special case of $\mathrm{M}_n(\mathbb{C})$ are that $\hat{\tau}$ may not be integer valued in general, and that it need not be injective.

We will briefly discuss the existence of such functionals for our C^*-algebras $C^*(\Omega_T, \mathbb{R}^d)$ and A_T associated to a minimal aperiodic tiling, T. For the former algebra, the amenability of $\mathbb{R}^d$ implies the existence of a probability measure μ on Ω_T which is invariant under the action of $\mathbb{R}^d$. We can then define a functional on $C_c(\Omega_T \times \mathbb{R}^d)$ by
$$\tau(f) = \int_\Omega f(T', 0) \, d\mu(T')$$
for f in $C_c(\Omega_T \times \mathbb{R}^d)$.

This has the correct positivity and trace properties, but some subtleties arise because it does not extend to a continuous linear functional on the completion of $C_c(\Omega_T \times \mathbb{R}^d)$, which is $C^*(\Omega_T, \mathbb{R}^d)$. This can still be useful, as it is often finite on the projections in the algebra.

For substitution systems satisfying our earlier hypotheses, there is a natural choice for such a measure, namely the measure which maximizes the entropy of the transformation or the so-called Bowen measure.

If we look at the situation for A_T, we want to do something similar on the equivalence relation R_{punc}. We say a measure ν on Ω_{punc} is R_{punc}-invariant if, for any open set E in R_{punc} such that the two canonical projection maps from E to Ω_{punc},
$$r(T_1, T_2) = T_1, \quad s(T_1, T_2) = T_2$$
are local homeomorphisms to their images, then we have
$$\nu\bigl(r(E)\bigr) = \nu\bigl(s(E)\bigr).$$
Using the notation of Section 4, this is equivalent to saying that
$$\nu\bigl(U(P, t_1)\bigr) = \nu\bigl(U(P, t_2)\bigr),$$
for any patch P and any two tiles t_1, t_2 in P.

From such a measure we may construct a functional on $C_c(R_{\mathrm{punc}})$ by setting
$$\tau(f) = \int_{\Omega_{\mathrm{punc}}} f(T', T') \, d\nu(T')$$
for f in $C_c(R_{\mathrm{punc}})$. In terms of our earlier description of the generators of A_T, this means
$$\tau(e[P, t_1, t_2]) = \begin{cases} \nu\bigl(U(P, t_1)\bigr) & \text{if } t_1 = t_2 \\ 0 & \text{if } t_1 \neq t_2 \end{cases}$$

In this case, the functional will extend continuously to A_T and have all the desired properties. Again for substitution tilings, the situation is quite good. Again one takes the Bowen measure and uses the fact that it will decompose into a product measure in the local coordinates. The local contracting coordinate will contain Ω_{punc} and the measure on this will have the desired properties.

For more information on the existence and uniqueness of such invariant measures, see [**Kel1, Kel2**].

Let us turn briefly to the group $K_1(A)$. In this case, our interest is in the invertible elements of A modulo homotopy. We let

$$U_n(A) = \{u \in M_n(A) \mid u \text{ is unitary}\}$$

and set

$$u \sim v \text{ in } U_n(A)$$

if there is a continuous path u_t, $0 \leq t \leq 1$ in $U_n(A)$ with $u_0 = u$, $u_1 = v$. We also regard $U_n(A) \subset U_{n+1}(A)$ by equating a and

$$\begin{pmatrix} & & & 0 \\ & a & & \vdots \\ & & & 0 \\ 0 & \cdots & 0 & 1 \end{pmatrix}.$$

The group structure is multiplication. It is a standard calculation to show that, for u, v in $U_n(A)$,

$$\begin{pmatrix} uv & 0 \\ 0 & 1 \end{pmatrix}, \begin{pmatrix} u & 0 \\ 0 & v \end{pmatrix}, \begin{pmatrix} vu & 0 \\ 0 & 1 \end{pmatrix}$$

are all homotopic in $U_{2n}(A)$. This shows that we could have also defined the product by direct sum and that our group operation is commutative. $K_1(A)$ is defined as the union of the $U_n(A)/\sim$.

For any complex, unitary matrix a it is quite easy to construct a path of unitary matrices from a to 1. It follows that $K_1(\mathbb{C})$ is the trivial group.

8. Gap-Labeling

We want to discuss the relevance to physics of the K-theory of the C^*-algebras which we are discussing, particularly A_T. This is summarized by the term "gap-labeling", which we will explain. A more thorough treatment can be found in [**Be2, Kel1**]. The one-dimensional case is treated in [**BBG**].

Suppose that H is the Hamiltonian of a particle moving in our solid which we have modeled by a tiling. More accurately, suppose H is the Hamiltonian in our tight-binding approximation. This means it will be a bounded self-adjoint operator and will lie in our C^*-algebra A_T. Its spectrum is a bounded subset of $\mathbb{R}$. A maximal connected subset of its complement in $\mathbb{R}$ is an open interval which is called a gap. We let $\text{Gap}(H)$ denote the set of all gaps of H. Notice that the gaps are naturally ordered like energy on the real line. As the spectrum is bounded, there is an unbounded gap of the form $(-\infty, a)$ and an unbounded gap of the form (b, ∞). These we denote by $-\infty$ and ∞, respectively. For any two gaps, $g_1 < g_2$, the spectral projection of the operator H associated with the spectrum between g_1 and g_2 is an element of the C^*-algebra, since it is the result of an application of a function which is continuous of the spectrum of H. For a single gap g, we let P_g denote the spectral projection of the interval from $-\infty$ to g. The gap-labeling is based on the map

$$g \in \text{Gap}(H) \to [P_g] \in K_0(A_T)$$

and we call $[P_g]$ a label for the gap g. Notice the following properties:
- $[P_{-\infty}] = [0]$
- $[P_\infty] = [1]$
- $g_1 < g_2$ implies $[P_1] < [P_2]$

- The label of a gap is a topological invariant in that, if we perturb H along a norm continuous path in such a way that the gap does not disappear, then the label does not change.

The third property ensures that the labeling is injective. If we can compute the K-theory of A_T, we have a list of labels for the gaps in the spectrum of H, even if we cannot exactly determine the spectrum.

Suppose further that we have a normalized trace τ on our C^*-algebra of observables A_T. As we noted before, this induces a map, $\hat{\tau}$ from $K_0(A_T)$ to the real numbers and we may apply this to our gap-labeling scheme. So to any gap $g \in \text{Gap}(H)$, we associate the real number

$$\hat{\tau}([P_g]) = \tau(P_g).$$

This labeling is the most interesting for physics. If we use the trace which is the trace per unit volume, then this is the density of eigenstates integrated up to the gap. This density of states is accessible to physical experiments. With this choice of trace, we call $\hat{\tau}(K_0(A_T))$ the gap-labeling group. It is often smaller and easier to compute than the K_0 group itself. It is an open problem to find a physical interpretation for the elements of the kernel of $\hat{\tau}$.

There is no general reason why, for a specific operator H, $g \to \hat{\tau}([P_g])$ should map onto the elements in $\hat{\tau}(K_0(A_T))$ between 0 and 1. In one dimension one has found many examples for which this is the case but in higher dimensions one does not expect this.

Let us remark why the choice of C^*-algebra A_T is important. We have already argued that it is the C^*-algebra which contains the Hamiltonian, H, we are interested in. We could, in fact, just use the C^*-algebra generated by H. This is a commutative algebra isomorphic to the continuous functions on the spectrum of H. This is something which we were not very optimistic about computing in the first place. With A_T as our choice of C^*-algebra, the computation of the range of our labeling scheme, that is the gap-labeling group, is actually independent of the operator H. It is an invariant which characterizes the influence of the structure of the space on the spectrum of an operator and its density of states.

Baake et al. have observed phenomena in other areas of physics which resemble the gap-labeling of Schrödinger operators [**BGJ, GrBa**]. Although it is not clear how these are related to the K-theory of some C^*-algebra we mention one of these, namely the distribution of zeros for the partition function of a classical quasiperiodic Ising chain in a constant magnetic field. To explain this a little bit consider a one dimensional tiling, i.e., an infinite sequence of intervals, and put spins at the points where the intervals touch. The spins interact with the external magnetic field in the usual way but the constants of interaction between nearest neighbors depend on the interval (e.g. are proportional to its length) which lies between the two spins. One then is interested in the zeros of the partition function as a function of the strength h of the external magnetic field, or more precisely as a function of $z = e^{\beta h}$ with β proportional to the inverse of the temperature. It is known that these zeros lie all on the unit circle. Consider now the integrated density of zeros, that is the integral of the density of zeros over a segment of the circle starting at the real line. The surprising observation which has been made for the simplest 1-dimensional substitution tilings is that the dependence of this integrated density of zeros on the length of the segment looks like the devil's staircase given by the integrated density of states (as a function of energy) of a typical tight binding operator on the tiling. The values of the integrated density of zeros at points where it is constant seem to belong to the gap-labeling group of the tiling.

9. K-Theory of the C^*-Algebras of Tilings

We begin with a minimal aperiodic tiling T. We first construct two C^*-algebras: $C^*(\Omega_T, \mathbb{R}^d)$ and A_T and we then want to compute their K-theories. Since these C^*-algebras are strongly Morita equivalent, the answers are the same. But the rather different views we have of the two will provide different information.

Let us begin with $C^*(\Omega_T, \mathbb{R}^d)$. The careful reader will have noticed little to recommend this algebra so far. We have done no computations involving its elements. (As we are about to ask "what are the projections in this algebra?", one might be worried by the fact that we haven't yet written one down. Actually, we won't.) The biggest single advantage of $C^*(\Omega_T, \mathbb{R}^d)$ is the following theorem of Connes.

THEOREM 9.1 ([**Bl**, **Co1**]).
$$K_i(C^*(\Omega_T, \mathbb{R}^d)) \cong K_{i-d}(C(\Omega_T)) \cong K^{i-d}(\Omega_T),$$
where K^{i-d} is topological K-theory, and $i - d$ is interpreted mod 2.

The result has nothing particular to do with tilings; it is simply a statement about $\mathbb{R}^d$-actions on spaces (or even on C^*-algebras). It requires no hypotheses of aperiodicity or minimality. Connes originally referred to this as an analogue of the Thom isomorphism which states that if E is a vector bundle over a space X, then $K^*(E) \cong K^{*-d}(X)$, where d is the dimension of E. That is, the K-theory of E is independent of how E twists over X and it is the same for all vector bundles of a fixed dimension. In our situation of $\mathbb{R}^d$ acting on Ω_T, Connes' theorem says that the K-theory is independent of the action. Recall from Section 4 that if $\mathbb{R}^d$ acts trivially, then the C^*-algebra is isomorphic to $C_0(\Omega_T \times \mathbb{R}^d)$. Finally,
$$K_i(C_0(\Omega_T \times \mathbb{R}^d)) \cong K_{i-d}(C(\Omega_T))$$
is the famous Bott periodicity result. (See both 1.6.4 and 9.1 of [**Bl**].) So Connes' result leaves us with the problem of understanding the topology of Ω. One small word of warning; Connes' isomorphism, like much of the machinery of K-theory, does *not* respect the order structure on K_0.

We are left with the problem of computing the K-theory of the space Ω_T. For this, we restrict our attention to substitution tiling systems. In this case, we will make use of Theorem 3.1 which expresses the space Ω as an inverse limit. We also use the fact that K-theory is continuous in the sense that the K-theory of an inverse limit of a sequence of topological spaces is isomorphic to the direct limit of their K-theory groups. Putting this together, we obtain the following.

THEOREM 9.2 ([**AP**]). *For a substitution tiling system satisfying our earlier hypotheses, we have*
$$K_i(C^*(\Omega, \mathbb{R}^d)) = \varinjlim K^{i-d}(\Gamma) \to K^{i-d}(\Gamma) \to K^{i-d}(\Gamma) \to \cdots.$$

It is important to remember that the space Γ is fundamentally simpler than Ω. In many examples arising from polygonal tiling schemes, it has the structure of a finite cell complex. In practical terms, this allows for the computation of the K-theory of the C^*-algebras $C^*(\Omega, \mathbb{R}^d)$. This is carried out completely for several examples, including the Penrose tilings, in [**AP**].

As a final remark, we repeat that the method we have provided for computing $K_i(C^*(\Omega_T, \mathbb{R}^d))$ says nothing about the order on the K_0-group. This is unfortunate, since this is a valuable part of the data.

We turn to our other C^*-algebra, A_T, where T is generated as above from a substitution system. Again we make the hypotheses that the substitution is primitive, the map ω is injective and the tiling satisfies the finite pattern condition.

The first step in constructing the discrete hull is to select some punctures for our tiles. We will make the assumption (and we lose no generality in doing so) that each of our proto-tiles contains the origin in its interior. We then select the origin as our puncture.

Recall that if P is a patch in T and t is in P, then $U(P, t)$ is a clopen set in Ω_{punc}, consisting of all tilings with a copy of $t \in P$ at the origin. Recall also, our description of A_T in Section 4 as being generated by elements $e[P, t_1, t_2]$, where (P, t_1, t_2) is a doubly pointed pattern class.

We will discuss a method for the computation of $K_0(A_T)$ in [**Kel1, Kel2**]. (We will have nothing to say about K_1.)

Recall that the subalgebra generated by the elements $e[P, t, t]$ is isomorphic to $C(\Omega_{\text{punc}})$. Our first observation is that the K-theory of this C^*-algebra is computable. If (f_{ij}) is any matrix with elements from $C(\Omega_{\text{punc}})$ which is a projection, then its trace, $\sum f_{ii}$ is a continuous integer-valued function on Ω_{punc}. Just as the case for a matrix algebra, this trace map extends to an isomorphism from

$$K_0(C(\Omega_{\text{punc}})) \xrightarrow{\cong} C(\Omega_{\text{punc}}, \mathbb{Z}),$$

where the range is the continuous integer-valued functions with pointwise addition. The fact that this is an isomorphism depends on the space Ω_{punc} being totally disconnected. Such a result is certainly not true in higher dimensions. Observe that the map sends the projection $e[P, t, t]$ to the characteristic function of the set $U(P, t)$.

The inclusion of $C(\Omega_{\text{punc}})$ in A_T induces a map on K_0-groups. Unlike the map on algebras it is far from injective. To see this, suppose $[P, t_1, t_2]$ is a doubly pointed pattern class. Let $v = e[P, t_1, t_2]$ and

$$u = \begin{pmatrix} v & 1 - vv^* \\ 1 - v^*v & v^* \end{pmatrix}$$

be in $M_2(A_T)$. It is easy to check that $u^{-1} = u^*$ and that

$$u \begin{pmatrix} e[P, t_1, t_1] & 0 \\ 0 & 0 \end{pmatrix} u^{-1} = \begin{pmatrix} e[P, t_2, t_2] & 0 \\ 0 & 0 \end{pmatrix}$$

so that $e[P, t_1, t_1]$ and $e[P, t_2, t_2]$ now determine the same element in $K_0(A_T)$, while the characteristic functions of $U(P, t_1)$ and $U(P, t_2)$ represent distinct elements in $K_0(C(\Omega_{\text{punc}}))$.

Motivated by this observation, we let E_T be the subgroup of $C(\Omega_{\text{punc}}, \mathbb{Z})$ generated by all elements of the form

$$\chi_{U(P,t_1)} - \chi_{U(P,t_2)},$$

where $[P, t_1, t_2]$ is a doubly pointed pattern class. We call

$$H(T) = C(\Omega_{\text{punc}}, \mathbb{Z})/E_T$$

the integer group of coinvariants of T. It is an invariant of the tiling T and R_{punc} (but not of A_T).

It is a rather interesting question to ask how close this is to $K_0(A_T)$. We have seen that there is a homomorphism of $C(\Omega_{\text{punc}}, \mathbb{Z})$ to $K_0(A_T)$ whose kernel contains E_T. Could the kernel be larger? In fact, in nice situations it is not. Is the map onto? That is, can one find projections in A_T other than the $e[P, t, t]$'s? (In the case of a matrix algebra, this amounts to the observation that every projection is similar to a diagonal one.) In fact, there are other, less obvious, projections in many cases.

Suppose for a moment that we are dealing with a situation where all of our tiles are unit squares, but carry labels so that T is minimal and aperiodic. We can put our punctures in the center of each tile. Then there is an obvious action of $\mathbb{Z}^2$ on Ω_{punc} so that the map

$$\Omega_{\text{punc}} \times \mathbb{Z}^2 \to R_{\text{punc}}$$

defined by $(T, \nu) \to (T, T + \nu)$ is a homeomorphism. In this case,

$$A_T = C^*(R_{\text{punc}}) = C^*(\Omega_{\text{punc}}, \mathbb{Z}^d)$$

can be constructed very much like the case for $C^*(\Omega_T, \mathbb{R}^d)$ of Section 4. See [**Pe, Da**] for more details. Moreover, there is machinery which will compute the K-theory (again without order) of such a C^*-algebra. The case $d = 1$ was a great breakthrough of Pimsner and Voiculescu [**Bl, Da, W-O**]. It can be extended to higher values of d, although spectral sequences become involved in the calculations. The case of 2-dimensional square tilings has been carefully analyzed by van Elst [**El**] and later Forrest and Hunton [**FH**] have investigated the K-theory of C^*-algebras associated with actions of $\mathbb{Z}^d$ on Cantor sets for general d.

As a sample, in the case $d = 2$, the result shows

$$K_0(A_T) \cong C(\Omega_{\text{punc}}, \mathbb{Z})/E_T \oplus \mathbb{Z}.$$

Let us take a moment to explain the $\mathbb{Z}$-term. We let u and v denote the characteristic functions of the graphs of the two maps

$$T' \to T' + (1, 0)$$
$$T' \to T' + (0, 1).$$

These graphs are compact open sets in R_{punc}, so u and v are in A_T. They also commute and the C^*-algebra they generate, denoted by $C^*(u, v)$, is isomorphic to $C(\mathbb{T}^2)$, the continuous functions on the 2-torus. This C^*-algebra contains a projection in $M_2(C(\mathbb{T}^2))$ which is rank one at each point of $\mathbb{T}^2$, but not similar to $\begin{pmatrix} 1 & 0 \\ 0 & 0 \end{pmatrix}$. (See [**W-O**] for further details.) The similarity exists pointwise, but cannot be made continuous. So this projection (or rather its formal difference with $\begin{pmatrix} 1 & 0 \\ 0 & 0 \end{pmatrix}$) generates the $\mathbb{Z}$ in $K_0(A_T)$. It is really present because of the fact the tiling is in $\mathbb{R}^2$. It has little to do with the tiling itself. (To make this statement a little more precise, the intersection of $C(\Omega_{\text{punc}})$ and $C^*(u, v)$ in A_T is just the scalar multiples of the identity.)

Of course, much of the interest in aperiodic tilings comes from the fact that the tiles aren't always squares. However, it is shown in [**Kel2**] that the same techniques as above can be applied much more generally. We say that the tiling T is a decoration of $\mathbb{Z}^d$ if we may choose a set of punctures for some (but perhaps not all) tile types such that there is an action (denoted by ·) of $\mathbb{Z}^d$ on $\Omega_{\text{punc}'}$ (new punctures!) by homeomorphism such that the map

$$(T', x) \in \Omega_{\text{punc}'} \times \mathbb{Z}^d \to (T', x \cdot T') \in R_{\text{punc}'}$$

is a homeomorphism. The topology on $\Omega_{\text{punc}'} \times \mathbb{Z}^d$ is the product topology, while that on $R_{\text{punc}'}$ is as defined earlier.

The first question is naturally: how much change do we make in A_T by selecting only a subset of tile types to have punctures? Let

$$P = \sum e(\{t_i\}, t_i, t_i),$$

where the sum is over all tile types, t_i, having a puncture in the new system. Then the C^*-algebra of the new $R_{\text{punc}'}$ is isomorphic to PA_TP and P is a projection. At least if the tiling is minimal, the ideal generated by P is all of A_T and hence the new C^*-algebra is strongly Morita equivalent to our old one (provided T is minimal and aperiodic so A_T is simple).

The second important question is how often does this situation arise? In fact, this is possible for any tiling obtained by the generalized grid method [**SSL**], including the Penrose tilings and the Ammann-Beenker tilings [**Kel2**].

Also, in [**FHK**], this point of view is developed completely for tilings which are obtained by the so-called projection method. It is shown that generically, there is a natural choice of transversal so that the C^*-algebra, $C^*(\Omega, \mathbb{R}^d)$ is strongly Morita equivalent to the crossed product arising from an action of $\mathbb{Z}^d$ on this transversal.

At this point, we have a description of the K_0 group of A_T in terms of our integer group of coinvariants at least for $d = 2$. Unfortunately, we do not yet have a very good grasp of this invariant. In particular, we want to be able to compute it in specific cases. We now restrict our attention to the case of substitution tilings satisfying the condition of Section 3.

To obtain a better description, we will introduce a new C^*-algebra, AF_T, which will be intermediate

$$C(\Omega_{\text{punc}}) \subset AF_T \subset A_T$$

This new algebra will be from a special class of algebras called AF-algebras (for approximately finite dimensional), which are both well-understood and rather rich [**Bl, Da, W-O, Ef**]. In particular, the K-theory of this C^*-algebra will be computable (including the order!) and give us a better approximation to that of A_T.

Our analysis will now proceed as follows. We will define a sequence of C^*-subalgebras of A_T. In fact, these will be nested so that the completion of their union is also a C^*-subalgebra. Moreover, each one of them will be finite dimensional as a vector space and isomorphic to a finite direct sum of matrix algebras. Of course, the dimension and the size of the matrix algebras will grow as we pass out in the sequence. The closure of their union is a so-called "approximately finite dimensional" or AF-algebra, which will be our AF_T.

We begin with a small observation. Recall our notation from Section 5: for a patch, P and tile t in P, we let $U(P, t)$ denote the set of all tilings T in Ω_{punc} containing P and with t covering the origin. Suppose that p and p' are two prototiles and x and x' are points in their respective interiors. Then the sets $U(\{p\}, p) - x$ and $U(\{p'\}, p') - x'$ are disjoint in Ω, unless $p = p'$ and $x = x'$. Since ω is injective on Ω, the same is true of the sets $\omega^N(U(\{p\},p)) - \lambda^N x$ and $\omega^N U(\{p'\},p')) - \lambda^N x'$, for any positive integer N.

We are now ready to begin our definition of the sequence of C^*-subalgebras. Let N be any nonnegative integer. For each proto-tile p, let $\mathrm{Punc}(N,p)$ denote the set of all the punctures in $\omega^N(p)$. Now, for each pair x,y in $\mathrm{Punc}(N,p)$, we define the set
$$E_p^N(x,y) = \{(\omega^N(T) - x, \omega^N(T) - y) \mid T \in U(\{p\},p)\}$$
We note the following properties of this set. First, for any T in $U(\{p\},p)$, the tilings $\omega^N(T) - x$ and $\omega^N(T) - y$ are both in Ω_{punc} and the second is the translate of the first by $x - y$. This means our set $E_p^N(x,y)$ is contained in R_T. The fact that it is a clopen subset is easy to verify. So we define $e_p^N(x,y)$ to be the characteristic function of $E_p^N(x,y)$, which is then an element of A_T.

The following relations follow easily from the first of our observation above. For any proto-tiles p and p', punctures x,y in $\mathrm{Punc}(N,p)$ and x',y' in $\mathrm{Punc}(N,p')$, we have
$$e_p^N(x,y) e_{p'}^N(x',y') = 0, \quad \text{if } p \neq p'$$
$$e_p^N(x,y) e_{p'}^N(x',y') = 0, \quad \text{if } p = p' \text{ and } y \neq x'$$
$$e_p^N(x,y) e_{p'}^N(x',y') = e_p^N(x,y') \quad \text{if } p = p' \text{ and } y = x'.$$
The second and third relations mean that if we fix both N and p and define
$$A_{N,p} = \mathrm{span}\{e_p^N(x,y) \mid x,y \in \mathrm{Punc}(N,p)\}$$
then $A_{N,p}$ is isomorphic to the algebra of complex $n \times n$ matrices, where n is the number of punctures in $\omega^N(p)$. Moreover the first relation means that these algebras, for different values of p, are orthogonal. So we define
$$A_N = \mathrm{span}\{e_p^N(x,y) \mid p \text{ a proto-tile and } x,y \in \mathrm{Punc}(N,p)\}$$
and we have
$$A_N = \bigoplus_p A_{N,p}.$$
Observe that the number of matrix summands is the number of proto-tiles and this is independent of N. Of course, the sizes of the matrices, which is the number of punctures in the inflations of the proto-tile, will grow with N.

We want to show that, for all N, we have $A_N \subseteq A_{N+1}$. For any proto-tile p, we let I_p denote the set of all pairs (p', x'), where p' is a proto-tile and x' is in $\mathbb{R}^d$ satisfying $p + x' \in \omega(p')$. The first step is to verify that
$$U(\{p\},p) = \bigcup_{(p',x') \in I_p} \omega\big(U(\{p'\},p')\big) - x'$$
and the sets in the union are pairwise disjoint. From this it follows that
$$e_p^0(0,0) = \sum_{(p',x') \in I_p} e_{p'}^1(x',x').$$
If we apply the map ω^N to the equality of sets above, we also obtain
$$e_p^N(x,y) = \sum_{(p',x') \in I_p} e_{p'}^{N+1}(\lambda^N x' + x, \lambda^N x' + y)$$
for any N and x,y in $\mathrm{Punc}(N,p)$. This equation then shows the inclusion of A_N in A_{N+1}.

We note that the identity of A_T is

$$\sum_p e_p^0(0,0)$$

which is in A_0, hence in all A_N. One verifies that the C^*-algebra generated by the elements $e_p^N(x,x)$, as N, p and x vary is the commutative C^*-algebra $C(\Omega_{\text{punc}})$.

As we mentioned above, we can form the union of this sequence of C^*-algebras, which is a subalgebra of A_T, but will not be closed. Its closure is a C^*-subalgebra, which we denote by AF_T. It is an approximately finite-dimensional C^*-algebra, or AF-algebra. We have seen in the construction that the elements generating this AF-algebra are functions on R_{punc}, just as the elements of A_T are. However, they are nonzero only on a proper sub-equivalence relation of R_{punc}.

Having given a description of the algebra AF_T, we want to show how its K-theory may be computed directly in the following manner [**Ef, Bl, W-O, Da**].

As each A_N is contained in A_{N+1}, the inclusion induces a map

$$K_0(A_N) \to K_0(A_{N+1})$$

which is a (not necessarily injective) positive group homomorphism. Finally, it is a theorem that, since AF_T is the closure of the union of the A_N's, we have

$$K_0(AF_T) = \varinjlim K_0(A_1) \to K_0(A_2) \to \cdots$$

where the limit is taken in the category of ordered Abelian groups.

In our case, this is really quite tractable. First of all, recall our calculation from Section 7 of the K-theory of the complex numbers. The same calculation shows that, for any n,

$$K_0(M_n) \cong \mathbb{Z}, \quad K_1(M_n) \cong 0.$$

and that the K_0-group is generated by the class of any rank one projection in M_n. We apply this to $A_{N,p}$, for any N and proto-tile p to assert that the group $K_0(A_{n,p})$ is generated by the class of $e_p^N(x,x)$, where x is any puncture in $\text{Punc}(N,p)$. Now the K-theory of a direct sum of C^*-algebras is the direct sum of their K-theories and so

$$K_0(A_N) \cong \bigoplus_p K_0(A_{N,p}) \cong \mathbb{Z}^n,$$

where n is the number of proto-tiles. The next step is to understand the map induced on K-theory by the inclusion of A_N into A_{N+1}. Fix a proto-tile p and consider the generator $[e_p^N(x,x)]$ of the pth summand in $K_0(A_N)$. (Here we have chosen some puncture x in $\text{Punc}(N,p)$.) The formula above in the case of $x = y$ becomes

$$e_p^N(x,x) = \sum_{(p',x') \in I_p} e_{p'}^{N+1}(\lambda^N x' + x, \lambda^N x' + x)$$

Each element in the sum on the right is a rank one projection in one of the matrix summands of A_{N+1}. In fact, in the p'-summand, it is the sum of exactly $B(p',p)$ rank one projections, where $B(p',p)$ is the number of occurrences of copies of p in $\omega(p')$.

Putting all of this together, we see that $K_0(AF_T)$ is the inductive limit, in the category of ordered Abelian groups, of the system

(9.1) $$\mathbb{Z}^n \xrightarrow{B} \mathbb{Z}^n \xrightarrow{B} \cdots .$$

Here, each $\mathbb{Z}^n$ is given the standard or simplicial order where an element is positive if and only if each entry is nonnegative. The structure of such groups is well-understood [**Ef, Ha**]. There is a unique trace, τ, on the algebra AF_T given by

$$\tau\bigl(e_p^N(x,y)\bigr) = \begin{cases} \lambda^{-N}\xi_p & \text{if } x = y \\ 0 & \text{if } x \neq y \end{cases}$$

where (ξ_p) is the Perron-Frobenius normalized eigenvector of the matrix B^t. The normalization of ξ is related to the order unit of $K_0(A_T)$, i.e., determined by the requirement that $\widehat{\tau}([\sum_p e_p^0(0,0)]) = 1$. Thus it is $\sum_p \xi_p = 1$. It follows that $\widehat{\tau}(K_0(AF_T))$ is the subgroup of $\mathbb{R}$ generated by $\lambda^{-N}\xi_p$, where N is a nonnegative integer and p is a proto-tile. It is interesting to note that this depends only on the combinatorics of the substitution, not on the geometry.

Recall that AF_T is a subalgebra of A_T and the inclusion induces a map on K_0 groups. The range of the map is exactly the same as the range of the map induced on $C(\Omega_{\text{punc}})$, namely the integer group coinvariants $H(T)$. The kernel of the map can be computed and this provides a method for computing $H(T)$ and hence, if $d \leq 2$, for $K_0(A_T)$. This is a long calculation for which we refer the reader to [**Kel2**]. But what becomes quickly clear is that $\widehat{\tau}$ applied to the above kernel is 0, or stated differently

$$\widehat{\tau}\bigl(H(T)\bigr) = \widehat{\tau}\bigl(K_0(AF_T)\bigr).$$

For $d = 1$ we have $K_0(A_T) = H(T)$. For $d = 2$ the above may be combined with results of [**El**] provided T reduces to a decoration of $\mathbb{Z}^2$. We already mentioned that in that case $K_0(A_T) \cong H(T) \oplus \mathbb{Z}$ and van Elst has shown (by explicit calculation) that $\widehat{\tau}$ evaluated on the second summand $\mathbb{Z}$ already belongs to $\widehat{\tau}(H(T))$. Thus in that case

$$(9.2) \qquad \widehat{\tau}\bigl(K_0(A_T)\bigr) = \widehat{\tau}\bigl(H(T)\bigr) = \widehat{\tau}\bigl(K_0(AF_T)\bigr) = \widehat{\tau}\bigl(C(\Omega_{\text{punc}})\bigr)$$

which effectively solves the problem of computing the gap labeling group in theses cases: it is the group generated by $\lambda^{-N}\xi_p$, N a nonnegative integer and p a proto-tile. For more general substitution systems, the last two equalities still hold and it seems reasonable to ask whether the first does also.

We mention that in [**Kel2**] the same result was stated even for $d = 3$. It was based, however, on the calculations made in [**El**] for $d = 3$ and the latter are not correct.

We remark that in the case where the substitution does not force its border, then we must use the method of decorated proto-tiles mentioned earlier.

10. Example: Octagonal Tilings

The example we provide, the (undecorated) octagonal tilings, gives us also the opportunity to explain a little of what we had to leave out in the general discussion.

In the common version the tiles of an octagonal tiling are squares and rhombi but to make the substitution unique and simpler we divide the squares into triangles and decorate these triangles in a symmetry breaking way. We call the resulting tiling the triangle version and denote it by T_3. This operation yields a mutually locally derivable tiling and thus doesn't change the ordered K-zero group. It does alter, however, the order unit and we have to be careful about this point when it comes to the gap-labeling group. The substitution of the tiling looks as in Figure 1.

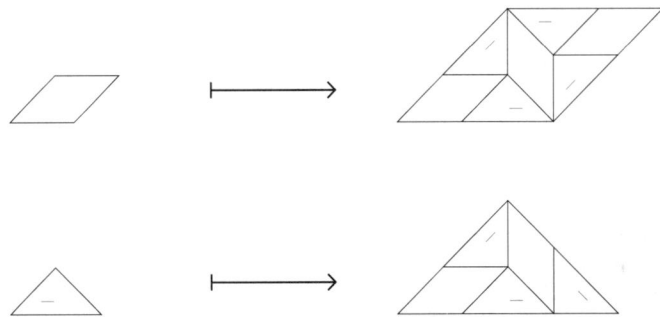

FIGURE 1. Substitution of the octagonal tiling (triangle version).

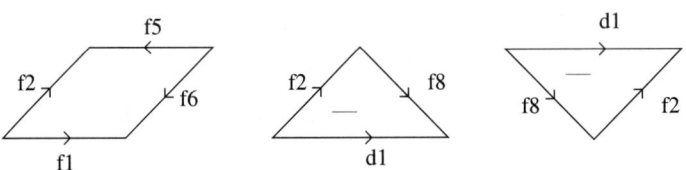

FIGURE 2. CW-complex for the octagonal tiling (triangle version).

Not only the above two tiles on the l.h.s. appear in the tiling but also all its rotates around $n\pi/4$ and reflections along the boundaries of the tiles, the substitution on these is extended in the obvious way: by symmetry. Thus an octagonal tiling has 20 proto-tiles: 4 of them congruent to the rhombus, the remaining 16 congruent to the triangle. The substitution is primitive, recognizable and forces its border.

We apply Theorem 9.2 to calculate the K-groups. We start with the calculation of $K(\Gamma)$. What we haven't explained in the main text is that $K(\Gamma)$ is isomorphic to $H(\Gamma)$, the cohomology of the CW-complex Γ, and that the inductive limit $\varinjlim H(\Gamma) \xrightarrow{\gamma} H(\Gamma)$ yields the cohomology $H(\Omega)$ of Ω and is isomorphic to $K(\Omega)$, see [**AP**]. But let us describe $H(\Gamma)$. Γ gives rise to a chain complex

$$0 \to C^2 \xrightarrow{\partial_2} C^1 \xrightarrow{\partial_1} C^0 \to 0$$

where C^k is the $\mathbb{Z}$-module generated by the k-cells of Γ and ∂_k the boundary operator. ∂_k applied to (a generator corresponding to) a k-cell is a signed sum of (the generators corresponding to) the $k-1$-cells which make up its boundary. There is some freedom to choose the signs, such a choice corresponding to the choice of an orientation for each cell. $H(\Gamma)$ is the cohomology of the dual complex.

Recall that the 2-cells of the CW-complex Γ for a 2-dimensional tiling are the (interior of) the proto-tiles. We get Γ by identifying two edges of two proto-tiles if we can find two tiles in the tiling, one a translate of the first and the other a translate of the second proto-tile, such that the two corresponding edges become identical. To determine this for the octagonal tiling requires a little work but since the tiling is of finite pattern type this is feasible. For the result see Figure 2.

We have only drawn three of the proto-tiles. The remaining 17 can be obtained from the above by rotation around $n\pi/4$. The labels on the edges then change as follows: f_i and d_i have to be replaced by f_{i+n} and d_{i+n} counting the index modulo 8.

Equal label on edges means that they have to be identified. In particular, in contrast to what the above picture suggests the complex is connected (as it should be). Note that the identifications imply that there is only one vertex. This is not a general feature. It implies that ∂_1 is the zero map. We have indicated a choice of orientation on the edges by an arrow. ∂_2 applied to a 2-cell, i.e., a proto-tile, is equal to the signed sum of its edges and we choose its sign to be $+$ if the arrow of the edges points left around the proto-tile (otherwise $-$). Hence the dual ∂_2' of ∂_2 applied to an edge (we may again identify the generators for $Hom(\Gamma, \mathbb{Z})$ with the cells) is the signed sum of the proto-tiles which contain that edge and the sign is $+$ if the proto-tile lies left of the edge (w.r.t. the direction of the latter). Thus the dual complex looks like

$$0 \to \mathbb{Z} \xrightarrow{0} \mathbb{Z}^{16} \xrightarrow{\partial_2'} \mathbb{Z}^{20} \to 0$$

and ∂_2' is the 20×16-matrix

$$\begin{pmatrix} 1 & -1 & 0 & 0 & 1 & -1 & 0 & 0 & 0 & 0 & 0 & 0 & 0 & 0 & 0 & 0 \\ 0 & 1 & -1 & 0 & 0 & 1 & -1 & 0 & 0 & 0 & 0 & 0 & 0 & 0 & 0 & 0 \\ 0 & 0 & 1 & -1 & 0 & 0 & 1 & -1 & 0 & 0 & 0 & 0 & 0 & 0 & 0 & 0 \\ -1 & 0 & 0 & 1 & -1 & 0 & 0 & 1 & 0 & 0 & 0 & 0 & 0 & 0 & 0 & 0 \\ 0 & -1 & 0 & 0 & 0 & 0 & 0 & -1 & 1 & 0 & 0 & 0 & 0 & 0 & 0 & 0 \\ -1 & 0 & -1 & 0 & 0 & 0 & 0 & 0 & 0 & 1 & 0 & 0 & 0 & 0 & 0 & 0 \\ 0 & -1 & 0 & -1 & 0 & 0 & 0 & 0 & 0 & 0 & 1 & 0 & 0 & 0 & 0 & 0 \\ 0 & 0 & -1 & 0 & -1 & 0 & 0 & 0 & 0 & 0 & 0 & 1 & 0 & 0 & 0 & 0 \\ 0 & 0 & 0 & -1 & 0 & -1 & 0 & 0 & 0 & 0 & 0 & 0 & 1 & 0 & 0 & 0 \\ 0 & 0 & 0 & 0 & -1 & 0 & -1 & 0 & 0 & 0 & 0 & 0 & 0 & 1 & 0 & 0 \\ 0 & 0 & 0 & 0 & 0 & -1 & 0 & -1 & 0 & 0 & 0 & 0 & 0 & 0 & 1 & 0 \\ -1 & 0 & 0 & 0 & 0 & 0 & -1 & 0 & 0 & 0 & 0 & 0 & 0 & 0 & 0 & 1 \\ 0 & 1 & 0 & 0 & 0 & 0 & 0 & 1 & -1 & 0 & 0 & 0 & 0 & 0 & 0 & 0 \\ 1 & 0 & 1 & 0 & 0 & 0 & 0 & 0 & 0 & -1 & 0 & 0 & 0 & 0 & 0 & 0 \\ 0 & 1 & 0 & 1 & 0 & 0 & 0 & 0 & 0 & 0 & -1 & 0 & 0 & 0 & 0 & 0 \\ 0 & 0 & 1 & 0 & 1 & 0 & 0 & 0 & 0 & 0 & 0 & -1 & 0 & 0 & 0 & 0 \\ 0 & 0 & 0 & 1 & 0 & 1 & 0 & 0 & 0 & 0 & 0 & 0 & -1 & 0 & 0 & 0 \\ 0 & 0 & 0 & 0 & 1 & 0 & 1 & 0 & 0 & 0 & 0 & 0 & 0 & -1 & 0 & 0 \\ 0 & 0 & 0 & 0 & 0 & 1 & 0 & 1 & 0 & 0 & 0 & 0 & 0 & 0 & -1 & 0 \\ 1 & 0 & 0 & 0 & 0 & 0 & 1 & 0 & 0 & 0 & 0 & 0 & 0 & 0 & 0 & -1 \end{pmatrix}.$$

Its rank is 11. Further computation shows $H^0(\Gamma) \cong \mathbb{Z}$, $H^1(\Gamma) \cong \mathbb{Z}^5$, and $H^2(\Gamma) \cong \mathbb{Z}^9$.

It remains to determine the induced maps $H^k(\Gamma) \xrightarrow{\gamma_k} H^k(\Gamma)$ and to compute the inductive limit. γ_2 is the map induced from the map $\mathbb{Z}^{20} \xrightarrow{B} \mathbb{Z}^{20}$ of (9.1) which already appeared for the construction of AF_T. It turns out that B it is invertible over $\mathbb{Z}$ and preserves the image of ∂_2'. Therefore the inductive limit $H^2(\Omega) = \varinjlim H^2(\Gamma) \xrightarrow{\gamma_2} H^2(\Gamma) \xrightarrow{\gamma_2} \cdots$ is isomorphic to $H^2(\Gamma)$. With the same reason one obtains $K_0(AF_T) \cong \mathbb{Z}^{20}$. The map γ_1 is induced from the substitution map for the edges which can be read off the substitution (Figure 1) if one adds to the tiles on both sides the labels for the edges. It turns out that γ_1 is as well invertible over $\mathbb{Z}$ and it preserves the kernel of ∂_2'. Therefore $H^k(\Omega) \cong H^k(\Gamma)$ in all degrees (again a result which does not hold in general) and the result above was already final:

$$H^0(\Omega) \cong \mathbb{Z}, \quad H^1(\Omega) \cong \mathbb{Z}^5, \quad H^2(\Omega) \cong \mathbb{Z}^9.$$

Thus $K_0(A_T) \cong \mathbb{Z}^9 \oplus \mathbb{Z}$ ($H^2(\Omega)$ are the coinvariants) and $K_1(A_T) \cong \mathbb{Z}^5$.

To determine the gap-labeling group we note that the octagonal tilings may as well be constructed with the grid method so that we can apply (9.2). For that we have to compute the Perron-Frobenius eigenvalue λ and a corresponding eigenvector ξ of B^t. One finds $\lambda = (1 + \sqrt{2})^2$ and

(10.1) $$\xi \propto (2\sqrt{2}, 2\sqrt{2}, 2\sqrt{2}, 2\sqrt{2}, 1, 1, 1, 1, 1, 1, 1, 1, 1, 1, 1, 1, 1, 1, 1, 1)$$

expressed in a basis in which the first four indices are identified with the rhombi and the remaining sixteen with triangles. The proper normalization of ξ for the triangle version T_3 is $\sum_p \xi_p = 1$. One finds that the group generated by $\lambda^{-N}\xi_p$, N nonnegative integer and $1 \leq p \leq 20$ is

(10.2) $$\hat{\tau}(K_0(A_{T_3})) = \frac{\mathbb{Z} + 2\sqrt{2}\mathbb{Z}}{8(2 + \sqrt{2})}.$$

Looking now at the original version T of the octagonal tiling by squares and rhombi we have to take a different normalization for ξ into account. It corresponds roughly speaking to giving the triangles only half the weight, i.e., $\sum_{p=1}^{12} \xi_p = 1$, see [**Kel3**]. With this normalization the gap labeling group becomes

(10.3) $$\hat{\tau}(K_0(A_T)) = \frac{\mathbb{Z} + 2\sqrt{2}\mathbb{Z}}{8(1 + \sqrt{2})}.$$

This is in agreement with [**BCL**] where the authors have computed $\tau(C(\Omega_{\text{punc}}))$ by determining the measures of the clopen sets which generate the topology of Ω_{punc}. (The formulae (2.2) and (2.3) stated in [**Kel3**] were incorrectly derived from (10.1) and should be replaced by (10.2) and (10.3).)

We finish with a remark on the approach to calculate the coinvariants with the method developed in [**Kel2**]. We mentioned that $H^2(\Omega)$ are the coinvariants which according to the main text can be expressed as a quotient $K_0(AF_T)/E$ where E is the kernel of the map on the K_0-groups induced from the embedding $AF_T \hookrightarrow A_T$. But in fact, the similarity with the above calculation for $H^2(\Omega)$ is deeper. When it comes to the calculations the only difference between the two approaches is that instead of looking at edges and counting which tiles they separate one considers the doubly pointed patterns which are formed by two neighboring tiles and counts these tiles. The choice of order for these tiles reflects the choice of orientation for their common edge.

References

[AP] J. E. Anderson and I. F. Putnam, *Topological invariants for substitution tilings and their associated C^*-algebras*, Ergodic Theory Dynam. Systems **18** (1998), 509–537.

[BGJ] M. Baake, U. Grimm, and D. Joseph *Trace maps, invariants, and some of their applications*, Internat. J. Modern Phys. B **7** (1993), 1527–1550; math-ph/9907156.

[BSJ] M. Baake, M. Schlottmann, and P. D. Jarvis, *Quasiperiodic tilings with tenfold symmetry and equivalence with respect to local derivability*, J. Phys. A **24** (1991), 4637–4654.

[Bel] J. Bellissard, *K-theory of C^*-algebras in solid state physics*, Statistical Mechanics and Field Theory: Mathematical Aspects (Groningen, 1985) (T. C. Dorlas, N. M. Hugenholtz, and M. Winnik, eds.), Lecture Notes in Phys., vol. 257, Springer, Berlin, 1986, pp. 99–156.

[Be2] _____, *Gap labelling theorems for Schrödinger's operators*, From Number Theory to Physics (Les Houches, 1989) (M. Waldschmidt, P. Moussa, J. M. Luck, and C. Itzykson, eds.), Springer, Berlin, 1992, pp. 539–630.

[BBG] J. Bellissard, A. Bovier, and J. M. Ghez, *Gap labelling theorems for one-dimensional discrete Schrödinger operators*, Rev. Math. Phys. **4** (1992), 1–38.

[BCL] J. Bellissard, E. Contensou, and A. Legrand, *K-théorie des quasi-cristaux, image par la trace: le cas du réseau octogonal*, C. R. Acad. Sci. Paris Sér. I Math. **326** (1998), 197–200.

[Bl] B. Blackadar, *K-theory for operator algebras*, Math. Sci. Res. Inst. Publ., vol. 5, Springer, New York, 1986; 2nd ed., Cambridge Univ. Press, Cambridge, 1998.

[Co1] A. Connes, *An analogue of the Thom isomorphism for crossed products of a C^*-algebra by an action of $\mathbb{R}$*, Adv. in Math. **39** (1981), 31–55.

[Co2] ———, *Noncommutative geometry*, Academic Press, San Diego, CA, 1994.

[Da] K. R. Davidson, *C^*-algebras by example*, Fields Inst. Monogr., vol. 6, Amer. Math. Soc., Providence, RI, 1996.

[Ef] E. G. Effros, *Dimensions and C^*-algebras*, CBMS Reg. Conf. Ser. Math., vol. 46, Amer. Math. Soc., Providence, RI, 1981.

[EH] E. G. Effros and F. Hahn, *Locally compact transformation groups and C^*-algebras*, Mem. Amer. Math. Soc., vol. 75, Amer. Math. Soc., Providence, RI, 1967.

[El] A. van Elst *Gap-labelling theorems for Schrödinger operators on the square and cubic lattice*, Rev. Math. Phys. **6** (1994), 319–342.

[Fi] P. A. Fillmore, *A user's guide to operator algebras*, Canad. Math. Soc. Ser. Monogr. Adv. Texts, Wiley, New York, 1996.

[FH] A. Forrest and J. Hunton, *Cohomology and K-theory of commuting homeomorphisms of the Cantor set*, Ergodic Theory Dynam. Systems **19** (1999), 611–625.

[FHK] A. Forrest, J. Hunton, and J. Kellendonk, *Projection quasicrystals. I. Toral rotations* (preprint); *II. Versus substitutions* (preprint); *III. Cohomology* (preprint) 1998/1999.

[GP] W. Geller and J. Propp, *The projective fundamental group of a $\mathbb{Z}^2$-shift*, Ergodic Theory Dynam. Systems **15** (1995), 1109–1118.

[GR] E. C. Gootman and J. Rosenberg, *The structure of crossed product C^*-algebras: a proof of the generalized Effros-Hahn conjecture*, Invent. Math. **52** (1979), 283–298.

[GrBa] U. Grimm and M. Baake, *Aperiodic Ising models*, The Mathematics of Long-Range Aperiodic Order (Waterloo, ON, 1995) (R. V. Moody, ed.), NATO ASI Ser. C: Math. Phys. Sci., vol. 489, Kluwer, Dordrecht, 1997, pp. 199–237.

[GS] B. Grünbaum and G. C. Shephard, *Tilings and patterns*, Freeman, New York, 1987.

[Ha] D. Handelman, *Positive matrices and dimension groups affiliated to C^*-algebras and topological Markov chains*, J. Operator Theory **6** (1981), 55–74.

[Ja] C. Janot, *Quasicrystals. A primer*, 2nd ed., Monographs on the Physics and Chemistry of Materials, Clarendon Press, Oxford, 1994.

[KH] A. Katok and B. Hasselblatt, *Introduction to the modern theory of dynamical systems*, Encyclopedia Math. Appl., vol. 54 Cambridge Univ. Press, Cambridge, 1995.

[Kel1] J. Kellendonk, *Noncommutative geometry of tilings and gap labelling*, Rev. Math. Phys. **7** (1995), 1133–1180.

[Kel2] ———, *The local structure of tilings and their integer group of coinvariants*, Commun. Math. Phys. **187** (1997), 115–157.

[Kel3] ———, *Integer groups of coinvariants associated to octagonal tilings*, Operator Algebras and Their Applications (Waterloo, ON, 1994/1995) (P. Fillmore and J. Mingo, eds.) Fields Inst. Commun., vol. 13, Amer. Math. Soc., Providence, RI, 1997, pp. 155–169.

[Ken] R. Kenyon, *The construction of self-similar tilings*, Geom. Funct. Anal. **6** (1996), 471–488.

[La] J. C. Lagarias, *Geometric models for quasicrystals. I. Delone sets of finite type*, Discrete Comput. Geom. **21** (1999), 161–191; *II. Local rules under isometries*, 345–372.

[Mi] J. A. Mingo, *C^*-algebras associated with one-dimensional almost periodic tilings*, Commun. Math. Phys. **183** (1997), 307–337.

[Mo] S. Mozes, *Tilings, substitution systems and dynamical systems generated by them*, J. Analyse Math. **53** (1989), 139–186.

[MRW] P. S. Muhly, J. N. Renault, and D. P. Williams, *Equivalence and isomorphism for groupoid C^*-algebras*, J. Operator Theory **17** (1987), 3–22.

[Pe] G. K. Pedersen, *C^*-algebras and their automorphism groups*, London Math. Soc. Monogr., vol. 14, Academic Press, London, 1979.

[RW] C. Radin and M. Wolff, *Space tilings and local isomorphism*, Geom. Dedicata **42** (1992), 355–360.

[Ren] J. N. Renault, *A groupoid approach to C^*-algebras*, Lecture Notes in Math., vol. 793, Springer, Berlin, 1980.

[Ri1]   M. A. Rieffel, *Morita equivalence for operator algebras*, Operator Algebras and Applications. Part 1 (Kingston, 1980) (R. V. Kadison, ed.), Proc. Symp. Pure Math., vol. 38, Amer. Math. Soc., Providence, RI, 1982, pp. 285–298.

[Ri2]   _____, *Lie group convolution algebras as deformation quantizations of linear Poisson structures*, Amer. J. Math. **112** (1990), 657–686.

[Ro1]   E. A. Robinson, Jr., *The dynamical theory of tilings and quasicrystallography*, Ergodic Theory of $\mathbb{Z}^d$-Actions (Warwick, 1993–1994) (M. Pollicott and K. Schmidt, eds.) London Math. Soc. Lecture Note Series, Cambridge Univ. Press, Cambridge (1996), pp. 451–473.

[Ro2]   _____, *The dynamical properties of Penrose tilings*, Trans. Amer. Math. Soc. **384** (1996), 4447–4464.

[Rud]   D. J. Rudolph, *Markov tilings of $\mathbb{R}^d$ and representations of $\mathbb{R}^d$ actions*, Measure and Measurable Dynamics (Rochester, NY, 1987) (R. D. Mauldin, R. M. Shortt, and C. E. Silva, eds.), Contemp. Math., vol. 94, Amer. Math. Soc., 1989, pp. 271–289.

[Rue]   D. Ruelle, *Thermodynamic formalism. The mathematical structures of classical equilibrium statistical mechanics*, Encyclopedia Math. Appl., vol. 5, Addison-Wesley, Reading, MA, 1978.

[Sm]   S. Smale, *Differentiable dynamical systems*, Bull. Amer. Math. Soc. **73** (1967), 747–817.

[SSL]   J. E. S. Socolar, P. J. Steinhardt, and D. Levine. *Quasicrystals with arbitrary orientational symmetry*, Phys. Rev. B **32** (1985), 5547–5550.

[So1]   B. Solomyak, *Dynamics of self-similar tilings*, Ergodic Theory Dynam. Systems **17** (1997), 695–738.

[So2]   _____, *Nonperiodicity implies unique composition for self-similar translationally finite tilings*, Discrete Comput. Geom. **20** (1998), 265–279.

[StO]   P. J. Steinhardt and S. Ostlund (eds.), *The physics of quasicrystals*, World Scientific, Singapore, 1987.

[W-O]   N. E. Wegge-Olsen, *K-theory and C^*-algebras*, Oxford Sci. Publ., Clarendon Press, Oxford, 1993.

[W]   R. F. Williams, *Expanding attractors*, Inst. Hautes Études Sci. Publ. Math. **43** (1974), 169–203.

[Z-M]   G. Zeller-Meier, *Produits croisés d'une C^*-algèbre par un groupe d'automorphismes*, J. Math. Pures et Appl. (9) **47** (1968), 101–239.

FACHBEREICH MATHEMATIK, TECHNISCHE UNIVERSITÄT BERLIN, 10623 BERLIN, GERMANY
E-mail address: `kellen@math.tu-berlin.de`

DEPARTMENT OF MATHEMATICS AND STATISTICS, UNIVERSITY OF VICTORIA, VICTORIA, BC V8W 3P4, CANADA
E-mail address: `putnam@math.uvic.ca`

Hulls of Aperiodic Solids and Gap Labeling Theorems

J. Bellissard, D. J. L. Herrmann, and M. Zarrouati

ABSTRACT. We review the basic constructions liable to replace Bloch theory for aperiodic solids. Point sets describing atomic positions lead to the notion of the hull, a topological dynamical system with an action of the translation group. We establish that quantities like the hull, the diffraction measure or the electronic density of states are uniquely determined by the Gibbs state describing thermal equilibrium of the solid. We recall the construction of the corresponding noncommutative Brillouin zone for electrons or phonons. We describe its topology through its algebraic K-theory. The gap labeling theory is reviewed and completed by a general conjecture for the case of a transversally totally discontinuous hull and by results obtained for two-dimensional media.

Introduction[1]

In 1981, J. Moser [**Mos**], exhibited the first example of a Schrödinger operator with nowhere dense spectrum. Soon after his preprint appeared, several other examples were found (see [**Sim2, Bel1, Bel2, Bel4, Bel6**] for reviews). It was realized at the same time that there was a need for a labeling of the gaps which was robust (namely stable under suitable perturbations of the Hamiltonian) and natural enough from both a mathematical and a physical point of view. Two results were announced in 1981 and published in 1982 in this direction: Johnson & Moser [**JM**] showed that for a Schrödinger equation with almost periodic potential, the set of gap labels was the so-called *frequency module*, whereas one of us realized the connection with the K-theory of C^*-algebras [**Bel1**]. In order to show that it was really the K-group that mattered instead of the frequency module, a counter-example was designed [**BS**]. For a long time, all examples of Hamiltonians with nowhere dense spectra were given by ordinary differential equations (ODE) with almost periodic coefficients or discretized version of them. It was not until 1988 that another class of potentials leading to nowhere dense spectra was found, namely potentials generated by *automatic sequences* [**Bel5**] (see also [**Sut2**]). In 1989, an example in higher dimensions was produced in connection with spectral properties of quasicrystals (QC) [**Sire**]. However, due to the *Bethe-Sommerfeld conjecture* [**DT, Skri, HM, Kar**], Schrödinger operators in higher dimension should have no gaps at high energy. For compounds with metallic behavior, no gaps should

---

2000 *Mathematics Subject Classification.* Primary 52C23, 19K14; Secondary 82D25, 47C15.
[1]This work will be part of the Ph.D. theses of Daniel Herrmann and Marc Zarrouati.

occur near the Fermi level, so that the gap labeling theorem should be of limited use in such cases. However, the K-theory still applies in any dimension raising the question of its physical interpretation.

The main idea behind this construction goes as follows [**Bel6**]. Given a Schrödinger operator H on $\mathbb{R}^d$, or a discretized version of it on $\mathbb{Z}^d$, one constructs a canonical C^*-algebra $\mathcal{A}$ attached to it. In particular, all bounded functions of H and of its translates belong to $\mathcal{A}$. If $\mathfrak{g} = (E_-, E_+)$ is a spectral gap, the characteristic function $\chi_{\mathfrak{g}}$ of the interval $(-\infty, E_-]$ is continuous on the spectrum of H so that $P(\mathfrak{g}) = \chi_{\mathfrak{g}}(H)$ is a projection in $\mathcal{A}$. $\mathcal{A}$ is separable in most cases of interest, so that the number of projections in $\mathcal{A}$, modulo "unitary equivalence" is at most countable [**Ped**]. Since the spectrum is an algebraic invariant, one can associate to each gap $\mathfrak{g}$ the equivalence class $n(\mathfrak{g}) = [P(\mathfrak{g})]$ of its gap projection (see §5.2). It turns out that, modulo a slight extension of the notion of equivalence, this set of equivalence classes is endowed with the structure of a discrete Abelian group, with the direct sum of projections, namely $[P] + [Q] = [P \oplus Q]$. This group is called $K_0(\mathcal{A})$. Since the equivalence class of a projection is a homotopy invariant, the gap labels are invariant by norm resolvent perturbations of H, as long as the corresponding gap does not close. This implies, in particular, sum rules and conservation rules as the dynamics is perturbed. Such rules had been observed for a long time without being conceptually understood. The K-theory gives the explanation.

The previous construction is rather abstract so that one can wonder whether there is a more concrete aspect of it that is used by physicists. The link is given by the so-called *Shubin formula* [**Shu, Bel2, Bel6**]. The *trace per unit volume* defines a *trace* $\mathcal{T}$ on $\mathcal{A}$. Since a trace is invariant under unitary transformations, the trace of a projection depends only on the equivalence class of that projection. Moreover a trace is additive on a direct sum. Consequently, $\mathcal{T}$ induces a group homomorphism $\mathcal{T}_*$ from $K_0(\mathcal{A})$ into $\mathbb{R}$. On the other hand, Shubin showed (see also [**Bel6**] for a generalization of Shubin's formula) that

$$\mathcal{N}(E) = \mathcal{T}\{\chi_{(-\infty, E]}(H)\}$$

where $\mathcal{N}(E)$ is the so-called *integrated density of states* (IDOS), defined as the number of eigenstates of H per unit volume with eigenvalues lower than or equal to E. It turns out that this function is nondecreasing and nonnegative, vanishing below the infimum of the spectrum. Therefore its derivative $d\mathcal{N}$ defines a Stieltjes-Lebesgue measure called the *density of states* (DOS). Note that physicists have access to the DOS, especially near the Fermi level, through various experimental techniques. Numerical calculations of band spectra in solids give also a computation of the DOS. In 1D, thanks to the Sturm-Liouville theory for ODE's, the IDOS is also given by the *rotation number* of the wave function (i.e., the real solution of the equation $-\psi''(x) + V(x)\psi(x) = E\psi(x)$ vanishing at $-\infty$), obtained by counting the number of sign changes of ψ per unit length. As a consequence of all these remarks

$$E \in \mathfrak{g} \implies \mathcal{N}(E) = \mathcal{T}_*([P(\mathfrak{g})]) \in \mathcal{T}_*(K_0(\mathcal{A})).$$

This is a remarkable result implying that the IDOS on gaps must take values in some specific countable subgroup of $\mathbb{R}$ that depends on H only through the C^*-algebra $\mathcal{A}$. Since this set is model independent it can, in principle, be computed without knowing much about the spectrum of H, giving an *a priori* constraint on it.

Since 1981, the set of gap labels has mainly been computed for 1D systems (see [**Bel6**] for a review). In some smooth cases, using the Connes index theorem [**Co3**], one also get the gap labels in higher dimension. However, for a long time, the case of QC's was an open problem. It is only since the middle of the nineties that we start to have some results. A deep theorem by Forrest and Hunton [**FH**] gives a rational isomorphism (the Chern character) between the K-group and some group homology in some class of systems. It is valid in any dimension and applies to QC's. It has recently been supplemented by explicit calculations [**FHK1, GK**] of the group K_0 for QC's with codimension 1 and 2. The computation of its image under the trace is still open to a large extent, even though results for 2D QC's are now available [**Elst, Kel1, Kel2, BCL, AP**]. The 3-dimensional or 3-codimensional cases are still under scrutiny. Recent progress led to the computation of the cohomology groups [**FK**] in this case. However, the gap labeling is not computed yet, mainly because the structure of the K-groups is qualitatively more involved. The aim of this paper is to review the general construction explained above and to give an account of the calculations of gap labels for 2D systems. We will adopt a point of view slightly different from the one given in [**Bel2, Bel6**]. We will take advantage of the recent work of Lagarias and Pleasants on point sets [**LP**]. To each point set, liable to represent the set of atomic positions, we associate the point measure supported by this set and give weight n to each point on which n atoms are lying. Such measures will be called *quasi uniformly discrete* and we shall denote the space of such measures by $\mathrm{QD}(\mathbb{R}^d)$. Thanks to this representation, the weak-$*$ topology on the space of measures induces a natural topology on the set $\mathrm{QD}(\mathbb{R}^d)$ of atomic configurations. This makes it quite an easy task to use compactness arguments, leading to the notion of the *hull* of an atomic configuration, which is the closure of the orbit of one atomic configuration under the translation group $\mathbb{G}$. This closure is, in most cases of interest, a compact space endowed with an action of $\mathbb{G}$. This dynamical system gives rise to a C^*-algebra in a canonical way which describes electronic observables that are considered for solving the Schrödinger equation. The spectral gaps of the electronic Hamiltonian are then labeled by means of the K-theory of that C^*-algebra.

We also introduce a complementary point of view taking the thermodynamics of the atomic motion into account. By an axiomatic approach, we propose a framework in which the Gibbs state describing the thermal equilibrium of atomic configurations becomes a probability measure on $\mathrm{QD}(\mathbb{R}^d)$. Since the translation group acts on $\mathrm{QD}(\mathbb{R}^d)$ by homeomorphisms, it acts on such probabilities as well. If the Gibbs state is unique it should be translation invariant and ergodic. Using such properties, we show that the Gibbs state uniquely determines:

(1) The hull of the point set: *the thermodynamic equilibrium defines entirely the relevant family of atomic configurations.*
(2) The associated diffractive measure, which corresponds to the diffraction pattern of the point set.
(3) The electronic DOS of the Hamiltonian: the Gibbs state defines in a unique way a *trace per unit volume* on the C^*-algebra of observables. Hence, thermodynamic principles give a natural choice of a trace and this trace is exactly what is needed to obtain the set of gap labels.
(4) The density of phonon modes (this aspect will not be developed here): the matrix defining phonon modes belongs to a similar C^*-algebra.

Examples of solid compounds for which this construction applies will be described in more detail. Three cases will be emphasized due to their importance in modern technology and as paradigm for aperiodic systems: the case of perfect crystals, like most metals, the case of crystals with impurities, like doped semiconductors, and the case of quasicrystals.

The construction of the C^*-algebra of observables will follow together with a description of the construction of the K-group for beginners (the reader is invited to see [**Bla**] for more details). Various standard techniques used to compute the K-group and the set of gap labels will be described. These results will be applied to QC's especially in dimension two. The calculation of the K-groups and gap labels will be given for the known cases. A conjecture for the set of gap labels will be formulated.

Acknowledgments. This work has benefited from many contacts and exchanges. J. B. would like to thank M. Baake and R. V. Moody for inviting him to the workshop held at Oberwolfach in 1998. He is indebted to J. C. Lagarias for his major work on point sets and discrete geometry. M. Z. and J. B. thank A. Legrand for his constant help in dealing with K-theory and J. Kellendonk for numerous discussions and his permanent interest in the gap labeling theorem. D. H. would like to thank M. Baake and T. Janssen for constant support.

1. Atomic Sites and Their Hull

A typical equation of motion for conduction electrons in a solid is the Schrödinger equation $H\psi = E\psi$, where H is the following operator acting on $\mathcal{H} = L^2(\mathbb{R}^d)$

$$(1.1) \qquad H = -\frac{\hbar^2}{2m}\Delta + \sum_{j=1}^{c}\sum_{y \in L_j} v_j(. - y).$$

Here d is the physical space dimension, m is the mass of the charge carrier, $\hbar$ is Planck's constant, $j = 1, \ldots, c$ labels the chemical atomic species, L_j is the point set of equilibrium positions of atoms of type j and v_j the effective potential for valence electrons near an atom of type j.

1.1. Bloch Theory Versus Aperiodicity. If the solid is a perfect crystal, all point sets L_j are invariant under a *translation group* $\mathcal{R}$, i.e., a discrete subgroup of $\mathbb{R}^d$ generating $\mathbb{R}^d$, namely what mathematicians call a *lattice*. In such a case, $\mathcal{R}$ is represented in $\mathcal{H}$ by unitary operators $U(a)$, $a \in \mathcal{R}$ and

$$(1.2) \qquad U(a)HU(a)^{-1} = H \quad \forall a \in \mathcal{R}.$$

Therefore one can simultaneously diagonalize H and the $U(a)$'s. Since $\mathcal{R}$ is Abelian, diagonalization of the $U(a)$'s is performed through its character group $\mathcal{R}^*$. Standard results in Pontryagin duality theory imply that $\mathcal{R}^*$ is isomorphic to the quotient $\mathbb{B} = \mathbb{R}^{d*}/\mathcal{R}^\perp$ of the dual group of $\mathbb{R}^d$ (isomorphic to $\mathbb{R}^d$) by the orthogonal $\mathcal{R}^\perp$ of $\mathcal{R}$ in this group. It is a well known fact that $\mathcal{R}^\perp$ is a lattice in $\mathbb{R}^d$ (called *the reciprocal lattice* in solid state physics [**Jon**]) so that $\mathbb{B} = \mathbb{R}^{d*}/\mathcal{R}^\perp$ is a compact group homeomorphic to a d-torus, even if analytically, point symmetries of the crystal may provide it with extra structures. Throughout this paper $\mathbb{B}$ will be called the *Brillouin zone* (strictly speaking this is slightly different from what crystallographers call Brillouin zone).

The concrete calculation of $\mathbb{B}$ goes as follows: any character of $\mathbb{R}^d$ is represented by an element $k \in \mathbb{R}^{d*}$. Since $\mathbb{R}^{d*}$ and $\mathbb{R}^d$ can be identified canonically, by using the usual Euclidean structure, one can see k as a vector $k = (k_1, \ldots, k_d) \in \mathbb{R}^d$. The corresponding character is given by the map

$$\eta_k \colon x \in \mathbb{R}^d \mapsto e^{i\langle k | x \rangle} \in U(1).$$

In particular η_k restricts to a character of $\mathcal{R}$, with the condition that $\eta_k \upharpoonright_\mathcal{R} = \eta_{k'} \upharpoonright_\mathcal{R}$ if and only if $k - k' \in \mathcal{R}^\perp$, where

$$\mathcal{R}^\perp = \{b \in \mathbb{R}^d; \langle b | a \rangle \in 2\pi \mathbb{Z}, \forall a \in \mathcal{R}\}.$$

Since $\mathbb{B}$ is a compact group, the diagonalization of the $U(a)$'s requires the use of a direct integral decomposition of $\mathcal{H}$ over $\mathbb{B}$, so that

$$\mathcal{H} = \int_{k \in \mathbb{B}}^\oplus d^d k \, \mathcal{H}_k, \quad H = \int_{k \in \mathbb{B}}^\oplus d^d k \, H_k,$$

where H_k is an operator acting on the Hilbert space $\mathcal{H}_k$. In the specific case given by (1.1), $\mathcal{H}_k$ is the space of functions ψ on $\mathbb{R}^d$ such that $\psi(x+a) = e^{i\langle k|x\rangle}\psi(x)$ for all $a \in \mathcal{R}$ and that $\int_\mathbb{V} d^d x \, |\psi(x)|^2 = \|\psi\|_{\mathcal{H}_k}^2 < \infty$, where $\mathbb{V} = \mathbb{R}^d/\mathcal{R}$. H_k is then the partial differential operator formally given by the same expression as H, but with domain $\mathcal{D}_k$ given by the space of elements $\psi \in \mathcal{H}_k$ such that $\partial_i \psi/\partial x_i \in \mathcal{H}_k$, for $1 \leq i \leq d$, and $\Delta_x \psi \in \mathcal{H}_k$. Then H_k is unitarily equivalent to an elliptic operator on the torus $\mathbb{R}^d/\mathcal{R} = \mathbb{V}$. In solid state physics, $\mathbb{V}$ is called the *Wigner-Seitz cell*, whereas it is called the *Voronoï cell* in tiling theory.

As a consequence, it follows that, for each $k \in \mathbb{B}$, the spectrum of H_k is discrete and unbounded. If $E_i(k)$ denotes the eigenvalues, with a convenient labeling i, the maps $k \in \mathbb{B} \mapsto E_i(k) \in \mathbb{R}$ are called the *band functions*. The spectrum of H is recovered as $\mathrm{spec}(H) = \bigcup_{i, k \in \mathbb{B}} E_i(k)$ and is called a *band spectrum*. A discrete spectrum is usually liable to be computable by suitable algorithms, since it restricts to diagonalizing large matrices.

This is a short summary of *Bloch theory*. Strutt first realized the existence of band functions [**Str1, Str2**], but soon after Bloch wrote his important paper [**Blo**]. In 1930, Peierls gave a perturbative treatment of the band calculations [**Pei**] and Brillouin discussed the $2D$ and $3D$ cases [**Bri**]. The reader is invited to look at [**Jon, AM**] to understand why this theory has been so successful in solid state physics. Let us simply mention that the first explicit calculations of bands in $3D$ were performed in 1933 by Wigner & Seitz [**WS**] on sodium using the cellular method that holds their names. The symmetry properties of the wave function were explicitly used in an important paper by Bouckaert, Smoluchowski & Wigner [**BSW**] leading to noticeable simplifications of the band calculation.

If the solid is no longer a perfect crystal, (1.2) is violated so that there may not be any space symmetry anymore, even though some point symmetry may survive. How can one deal with such situations in general? In specific cases, where the breaking of translation symmetry can be considered as a small perturbation, one can find ways around. For instance, if there are isolated impurities in the crystal, a good approximation was developed by Slater in 1949 [**Sla**] to compute the change produced in the energy spectrum. But many examples of compounds require another treatment taking into account the aperiodicity: semiconductors at low temperature for impurity band electrons [**SE**], quasicrystals [**HG**], amorphous materials, even metallic liquids. To deal with this situation, one of us has proposed, a long time ago [**Bel2, Bel6**], to replace Bloch theory by the formalism of noncommutative geometry [**Co6**]. It consists in replacing the set of continuous functions over the Brillouin zone $\mathcal{C}(\mathbb{B})$, by a noncommutative algebra over a virtual space that will be called the *noncommutative Brillouin zone*. This algebra is built from the family $\{U(a)HU(a)^{-1}; a \in \mathbb{R}^d\}$ of the translates of H. The construction leads to $\mathcal{C}(\mathbb{B})$ whenever the crystal is perfect. This is explained in the next sections.

1.2. The Hull of a Point Set. For $x = (x_1, \ldots, x_d) \in \mathbb{R}^d$, we set $|x|_\infty = \max_{1 \leq i \leq d} |x_i|$. If $x \in \mathbb{R}^d$ and $r > 0$ we set $B(x,r) = \{y \in \mathbb{R}^d; |y-x|_\infty < r\}$: it is the open ball of radius r centered at x for the metric $|\cdot|_\infty$ namely the open hypercube with faces perpendicular to the vectors of the canonical basis, centered at x and with side $2r$.

Atomic sites of type j are located on the discrete point set $L = L_j$ contained in $\mathbb{R}^d$. Following Lagarias and Pleasants [**LP**], we define the following hierarchy of properties:

- L is *discrete* if the intersection of any compact set of $\mathbb{R}$ with L is finite.
- L is *uniformly discrete* if there is $r > 0$, such that any ball of radius r contains at most one point of L. This means that there is a nonzero minimum distance between points of L.
- L is *relatively dense* if there is $R > 0$ such that any ball of radius R contains at least one point of L.
- L is a *Delone* (or Delaunay) *set* if it is both uniformly discrete and relatively dense.
- A Delone set L is *finitely generated* if the $\mathbb{Z}$-module generated by $L - L$ in $\mathbb{R}^d$ is finitely generated.
- A Delone set L has *finite type* if $L - L$ is discrete and closed.
- L is a *Meyer set*, whenever both L and $L - L$ are Delone sets.

If L is a Delone set, we set $r_0 = \sup\{r > 0; |B(x,r) \cap L| \leq 1, \forall x \in \mathbb{R}^d\}$ and $r_1 = \inf\{R > 0; |B(x,R) \cap L| \geq 1, \forall x \in \mathbb{R}^d\}$. Then the minimal distance r_L between two distinct points of L is $r_L = 2r_0$. We will say that L is (r,R)-Delone whenever $r \leq r_0$ and $R > r_1$.

EXAMPLE 1.1. A point set in $\mathbb{R}^d$, randomly distributed with respect to a Poisson process, is discrete but not uniformly discrete with probability one.

EXAMPLE 1.2. In practice, due to quantum mechanics, the equilibrium positions of atoms in any solid medium form a uniformly discrete set. Impurities in a semiconductor, distributed randomly, are located on a uniformly discrete set which is not a Delone set in general. This is also the case for zeolithes which may have empty holes of arbitrary size.

EXAMPLE 1.3. Most solids, e.g., amorphous systems, glasses, crystals, quasicrystals, have their atoms on a Delone set. Random tilings built from a quasilattice [**Moss**], have their vertices on finitely generated Delone sets.

EXAMPLE 1.4. Point sets constructed by cut-and-project method [**HG**], namely model sets [**Moo**], and more generally point sets obtained by inflation, matching rules or by covering clusters [**JS, Gum, Kra2**], are Meyer sets [**Mey1, Mey2**].

Now, we denote by $\mathcal{M}(\mathbb{R}^d)$ the space of measures on $\mathbb{R}^d$. By construction it is the set of linear continuous maps from $\mathcal{C}_c(\mathbb{R}^d)$, the space of continuous functions with compact support on $\mathbb{R}^d$, into $\mathbb{C}$.

Let $\mathrm{QD}(\mathbb{R}^d) = \{\nu \in \mathcal{M}(\mathbb{R}^d); \nu \text{ is pure point and } \nu(\{x\}) \in \mathbb{N}, \forall x \in \mathbb{R}^d\}$. For each $\nu \in \mathrm{QD}(\mathbb{R}^d)$, $L^{(\nu)} = \mathrm{supp}(\nu) = \{x; \nu(\{x\}) \geq 1\}$ is a discrete set. Given $x \in L^{(\nu)}$, the integer $N_x = \nu(\{x\})$, called the *multiplicity of x*, can be interpreted as the number of atoms located at x. Conversely, given a discrete set L and for each $x \in L$ an integer $n(x) \in \mathbb{N}_*$, one can define the measure

$$\nu^{(L,\underline{n})}(dx) = \sum_{y \in L} n(y)\delta(x-y)$$

This gives a one-to-one correspondence between discrete sets with multiplicity and measures in $\mathrm{QD}(\mathbb{R}^d)$.

One can extend the correspondence between points sets in $\mathbb{R}^d$ and measures by defining $\mathrm{UD}_r(\mathbb{R}^d) = \{\nu \in \mathrm{QD}(\mathbb{R}^d); \forall x \in \mathbb{R}^d, \nu(B(x,r)) \leq 1\}$. This set of measures corresponds to r-uniformly discrete point sets with no degeneracy. We shall set $\mathrm{UD}(\mathbb{R}^d) = \bigcup_{r>0} \mathrm{UD}_r(\mathbb{R}^d)$, the space of uniformly discrete point sets. In much the same way, (r,R)-Delone sets will be represented by elements $\nu \in \mathrm{UD}_r(\mathbb{R}^d)$ such that, for any open ball of radius R, $\nu(B(x,R)) \geq 1$. This subspace will be denoted by $\mathrm{Del}_{(r,R)}(\mathbb{R}^d)$. The space of all Delone sets will then be $\mathrm{Del}(\mathbb{R}^d) = \bigcup_{0<r<R} \mathrm{Del}_{(r,R)}(\mathbb{R}^d)$.

The main properties of these spaces are the following (see Section 2):

THEOREM 1.5. *Let $\mathcal{M}(\mathbb{R}^d)$ be endowed with the weak-$*$ topology with respect to $\mathcal{C}_c(\mathbb{R}^d)$, then:*

(i) *The subspace $\mathrm{QD}(\mathbb{R}^d)$ is a closed set. It is therefore a Polish space (i.e., a complete metrizable space).*
(ii) *For all $r > 0$ the subspace $\mathrm{UD}_r(\mathbb{R}^d)$ of $\mathrm{QD}(\mathbb{R}^d)$ is compact.*
(iii) *For all $0 < r \leq R$ the subspace $\mathrm{Del}_{(r,R)}(\mathbb{R}^d)$ of $\mathrm{UD}_r(\mathbb{R}^d)$ is compact.*
(iv) *The subspaces $\mathrm{UD}(\mathbb{R}^d)$ and $\mathrm{Del}(\mathbb{R}^d)$ are dense in $\mathrm{QD}(\mathbb{R}^d)$.*

Therefore the correspondence between point sets and point measures becomes quite powerful in that it gives immediately a good topological structure. The last Claim (iv) in Theorem 1.5 indicates that it is not possible to get away from point sets with multiplicity unless we accept to work on the nonclosed F_σ-sets $\mathrm{UD}(\mathbb{R}^d)$ or $\mathrm{Del}(\mathbb{R}^d)$.

Let us give an intuitive description of what the weak-$*$ topology means (see Section 2.7 for a precise statement). Given $\nu \in \mathrm{QD}(\mathbb{R}^d)$, one associates $L^{(\nu)}$ as its support, but we consider each point $x \in L^{(\nu)}$ as a finite set with N_x points in it (its multiplicity). Then a sequence $(\nu_n)_{n\in\mathbb{N}}$ of elements of $\mathrm{QD}(\mathbb{R}^d)$ converges to ν if and only if, for any open ball $B(x,r)$, the sets $L^{(\nu_n)}$ counted with their multiplicities, converge to $L^{(\nu)}$ in the sense of the Hausdorff distance [**Bar**]. In other words, convergence of $L^{(\nu_n)}$ towards $L^{(\nu)}$ means convergence of points of $L^{(\nu_n)}$ in any bounded window.

More generally, given the atomic sites with several species, one associates the vector-valued point measure $\vec{\nu} = (\nu_1, \ldots, \nu_c)$ with $\nu_j = \nu^{(L_j)}$. The vector-valued measure $\vec{\nu}$ belongs to $\mathcal{M}(\mathbb{R}^d) \otimes \mathbb{C}^c$ that can be seen as the dual space of $\mathcal{C}_c(\mathbb{R}^d) \otimes \mathbb{C}^c$. The duality between them is given by $\langle \vec{\nu} \mid \vec{f} \rangle = \int_{\mathbb{R}^d} d\vec{\nu}(x) \vec{f}(x)$.

We endow these spaces with the weak-$*$ topology, namely a sequence $\vec{\nu}_n$ of vector-valued measures converges to $\vec{\nu}$ if and only if $\lim_{n\to\infty} \langle \vec{\nu}_n \mid \vec{f} \rangle = \langle \vec{\nu} \mid \vec{f} \rangle$ for every $\vec{f} \in \mathcal{C}_c(\mathbb{R}^d) \otimes \mathbb{C}^c$.

The main result of this subsection is the following theorem, which is a direct consequence of Theorem 1.5.

THEOREM 1.6. *Let $\vec{\nu} = (\nu_1, \ldots, \nu_c)$, with $\nu_j = \nu^{(L_j)}$, the vector-valued measure attached to the atomic positions. Assume the L_j's are all uniformly discrete. Then* the family $\{\tau^a \vec{\nu}; a \in \mathbb{R}^d\}$ of translates of $\vec{\nu}$ by elements of $\mathbb{R}^d$ has a compact weak closure in $\mathcal{M}(\mathbb{R}^d) \otimes \mathbb{C}^c$.

Recall that a *topological dynamical system* is a pair $(X, \mathbb{G})$ where X is a topological space, $\mathbb{G}$ is a locally compact group acting on X by homeomorphisms [**GH**]. In all cases considered here, X will be compact metrizable. $(X, \mathbb{G})$ is called *topologically transitive* if X admits one dense $\mathbb{G}$-orbit. It is called *minimal* if all $\mathbb{G}$-orbits are dense. If X is compact, the set $\mathcal{M}_1(X, \mathbb{G})$ of $\mathbb{G}$-invariant probability measures is a nonempty compact convex set if endowed with the weak-$*$ topology. The extremal elements of this set are exactly the $\mathbb{G}$-invariant ergodic probability measures on x. $(X, \mathbb{G})$ is *uniquely ergodic* if $\mathcal{M}_1(X, \mathbb{G})$ reduces to one point only. Two dynamical systems $(X_1, \mathbb{G}_1)$ and $(X_2, \mathbb{G}_2)$ are *semi-conjugate* if it exists a surjective continuous map ϕ from X_1 to X_2 such that $\phi \circ T_1 = T_2 \circ \phi$, where T_i is the action of $\mathbb{G}_i$ on X_i. When ϕ is one-to-one, $(X_1, \mathbb{G}_1)$ and $(X_2, \mathbb{G}_2)$ are *conjugate*. A family $(X_i, \mathbb{G})$ of topological dynamical systems over the group $\mathbb{G}$ is *structurally stable* whenever for any $i \neq j$ there is a homeomorphism $\phi_{i,j} \colon X_i \mapsto X_j$ that conjugates the actions.

We are now ready to rigorously define the hull:

DEFINITION 1.7. Given a finite family $L = (L_1, \ldots, L_c)$ of uniformly discrete point sets in $\mathbb{R}^d$, the closure $\Omega_{\vec{\nu}}$ of the family of translates of the vector-valued measure $\vec{\nu}^{(L)} = (\nu_1, \ldots, \nu_c)$ (with $\nu_j = \nu^{(L_j)}$), is called the *hull* of L. Endowed with the canonical action T of $\mathbb{R}^d$ by translation, it becomes a topological dynamical system $(\Omega, \mathbb{R}^d, T)$ which is topologically transitive.

The notion of the hull was first introduced by one of us in earlier papers [**Bel2, Bel6**] as the compact strong closure Ω_H, when it exists, of the family $\{U(a)HU(a)^{-1}, a \in \mathbb{R}^d\}$ of the translates of the Schrödinger operator $H = \Delta + V$. These two definitions are not strictly equivalent in general. However we can prove that $(\Omega_\nu, \mathbb{R}^d)$ and $(\Omega_H, \mathbb{R}^d)$ are semi-conjugate (Section 2.7).

EXAMPLE 1.8. The point set of a perfect crystal in $\mathbb{R}^d$ is $\mathcal{R}$-periodic, $\mathcal{R}$ being a lattice in $\mathbb{R}^d$ (cf. Section 1.1). Its hull is then the quotient $\mathbb{R}^d/\mathcal{R}$, namely it is homeomorphic to $\mathbb{T}^d$. Others examples are quoted in Section 3.

In many cases, it is quite convenient to work with a tight-binding representation (see Section 4) instead of using the Schrödinger equation (1.1). This means that (1.1) is replaced by a finite difference equation where the wave functions live in $\ell^2(L)$ instead of belonging to $L^2(\mathbb{R}^d)$ [**Bel2**]. The breaking of translation invariance can be described through the notion of *the canonical transversal* and its *groupoid* (see Section 2.5).

DEFINITION 1.9. Let Ω be a closed $\mathbb{R}^d$-invariant subset of $\mathrm{QD}(\mathbb{R}^d)$. Its *canonical transversal* Υ is the closed subset

$$\Upsilon = \{\omega \in \Omega; \omega(\{0\}) \geq 1\}.$$

The set of pairs $(\omega, a) \in \Omega \times \mathbb{R}^d$, such that both ω and $\tau^{-a}\omega$ belong to Υ, is a locally compact groupoid G_Υ [**Ren**] called the *groupoid of the transversal*.

1.3. Gibbs Measures on the Hull. Up to now we have supposed that the atomic positions in a solid are fixed once and for all. As we know however, even solids are subject to phase transitions. The shape of the point set on which atoms are located is usually a consequence of thermodynamics. The description of the Gibbs state in such a situation is still today an open problem to a large extent.

Nevertheless let us try to give a (nonrigorous) description of what it should be, so that we can extract which axioms a Gibbs measure is likely to satisfy.

Due to the large mass difference between electrons and atomic nuclei, one usually treats atomic motion in the Born-Oppenheimer approximation. Namely, one diagonalizes the electronic Hamiltonian considering the atomic positions as adiabatic parameters. Then the electronic energy depends upon atomic positions and acts as an extra attractive potential between nuclei. For this reason, it is enough to consider the Hamiltonian for nuclei alone.

In principle, even the atomic motion is quantized. But there are two kinds of simplifications that should be taken into account. First of all, since the system is solid, atoms only vibrate around their equilibrium positions. The atomic vibrations built up acoustic waves called *phonons*. With a very good accuracy, one can treat phonons as harmonic waves, at least at low enough energies, namely at low enough temperatures. Nevertheless, one may take into account a small amount of anharmonicity if one wishes to.

The second simplification consists in using the Feynman path integral [**Gin1, Gin2**] to represent the Gibbs state describing atomic equilibrium. This gives the atomic partition function only in terms of the atomic potential energy, the contribution of the kinetic energy being represented by random fluctuations around the equilibrium positions.

For all these reasons, the Gibbs state can be described with a rather good accuracy through the potential energy alone. The construction of such a state can be found in [**Rue, Lan, Sin**]. The main properties of the potential energy are the following.

U1. In any finite ball Λ, the potential energy is a function $U_\Lambda(x_1, \ldots, x_N)$ of the positions $x_1, \ldots, x_N$ of atoms located in Λ. Hence one can view it as a function over the space $\mathrm{QD}(\mathbb{R}^d)$.

U2. The potential energy is *translation invariant* that is to say
$$U_{\Lambda+a}(x_1 + a, \ldots, x_N + a) = U_\Lambda(x_1, \ldots, x_N)$$
for all $a \in \mathbb{R}^d$.

U3. The potential energy is *repulsive at short distances*, i.e., it diverges as two atoms become too close to each other. In other words, the potential energy is finite if and only if the atomic sites are located on a uniformly discrete point set, namely an element of $\mathrm{UD}(\mathbb{R}^d)$.

U4. The potential energy is a continuous function of the atomic positions away from coincident points. That means the U_Λ's are continuous on $\mathrm{UD}(\mathbb{R}^d)$.

U5. The potential energy is *asymptotically extensive*, namely:
$$U_{\Lambda_1 \cup \Lambda_2 + a}(x_1, \ldots, x_M, y_1 + a, \ldots, y_N + a)$$
$$- U_{\Lambda_1}(x_1, \ldots, x_M) - U_{\Lambda_2 + a}(y_1 + a, \ldots, y_N + a) \to 0$$
as $a \to \infty$ and $x_i \in \Lambda_1$ and $y_j \in \Lambda_2$.

U6. The potential energy is *attractive at large distances*. The exact description of this property is technical. Let us simply say that one decreases the potential energy in a large box by restricting it to the space $\mathrm{Del}(\mathbb{R}^d)$ of Delone sets.

In the limit for which quantum effects on the atomic motion become negligible, the Gibbs state $\mathbb{P}$ is then described as a limit point, whenever it exists, of the following family of probability measures on $\mathrm{QD}(\mathbb{R}^d)$

$$\mathbb{P}_\Lambda(F) = \frac{1}{\Xi_\Lambda(\beta,\mu)} \sum_{N=0}^\infty \frac{e^{\beta\mu N}}{N!} \int_{\Lambda^{\times N}} d^d x_1 \ldots d^d x_N \, e^{-\beta U_\Lambda(x_1,\ldots,x_N)} F(x_1,\ldots,x_N)$$

where $\beta = 1/kT$, μ is the chemical potential, $\Xi_\Lambda(\beta,\mu)$ (the normalization factor) is the *grand partition function* and F is a uniformly continuous bounded function on $\mathrm{QD}(\mathbb{R}^d)$. Using U1–U6, one expects that $\mathbb{P}$ will be supported by the space $\mathrm{Del}(\mathbb{R}^d)$. As usual, *pure phases*, in the sense given to that word by physicists, are described by extremal points of the set of Gibbs states and are usually ergodic with respect to the translation group. Moreover, we may expect $\mathbb{P}$ to be translation invariant. Such a property is a consequence of the uniqueness of the Gibbs state. However, this last requirement is not always satisfied even in practice, since inhomogeneous boundary conditions may lead to mixed phases with phase boundaries.

Whatever the hypothesis made to built the Gibbs measure $\mathbb{P}$, we expect that it should satisfy the following axioms:

G1. $\mathbb{P}$ is *uniformly discrete* namely, it gives probability one to $\mathrm{UD}(\mathbb{R}^d)$.
G1'. $\mathbb{P}$ is *Delone*, namely it gives probability one to $\mathrm{Del}(\mathbb{R}^d)$.
G2. $\mathbb{P}$ is translation invariant.
G3. $\mathbb{P}$ is ergodic with respect to the translation group.

Remarkably enough, the mathematical framework developed previously gives useful information about the nature of typical atomic configurations. The first ones are summarized in the following (see Section 3.1):

THEOREM 1.10. (i) *Let $\mathbb{P}$ be a probability measure on $\mathrm{QD}(\mathbb{R}^d)$ such that G1, G2 and G3 hold. Then there is $r_0 > 0$ such that for $\mathbb{P}$-almost all ν, its support $L^{(\nu)}$ is r_0-uniformly discrete and not r-uniformly discrete for $r > r_0$.*
(ii) *If in addition $\mathbb{P}$ satisfies G1', there is $R_0 > 0$ such that for $\mathbb{P}$-almost all ν, its support $L^{(\nu)}$ is R_0-relatively dense and not R-relatively dense for $R < R_0$.*

The next result concerns the hull of a typical atomic configuration. If $\mathbb{P}$ satisfies G1', then it is supported by the compact space $\mathrm{Del}_{(r_0,R_0)}(\mathbb{R}^d)$. The *topological support* of $\mathbb{P}$ is the smallest closed subset of $\mathrm{Del}_{(r_0,R_0)}(\mathbb{R}^d)$ of probability one. Then (see Section 3.1):

THEOREM 1.11. *Let $\mathbb{P}$ be a probability measure on $\mathrm{QD}(\mathbb{R}^d)$ obeying G1', G2 and G3. Then, for $\mathbb{P}$-almost all $\nu \in \mathrm{QD}(\mathbb{R}^d)$, its hull Ω_ν coincides with the topological support Ω of $\mathbb{P}$.*

This last theorem shows that the hull is entirely defined by the thermodynamic properties of the solid. We may also wonder whether the hull is structurally stable as the Gibbs state $\mathbb{P}$ varies in a region of uniqueness of the phase diagram. If this is correct, it means that the hull cannot bifurcate unless a phase boundary is crossed. This conjecture would give an explanation of why the lattice symmetry of a perfect crystal is fixed within a region of uniqueness.

Now, if we identify an element $\nu \in \mathrm{QD}(\mathbb{R}^d)$ with a point set in $\mathbb{R}^d$ representing the position of atoms in a solid, the diffraction pattern seen on a screen in an X-ray diffraction experiment or in a transmission electronic microscope (T.E.M.), can be computed from the Fourier transform of ν restricted to the domain Λ occupied by the sample in $\mathbb{R}^d$. Namely the intensity seen on the screen is proportional to

$$(1.3) \qquad I_\Lambda(k) = \frac{1}{|\Lambda|} \left| \sum_{x \in \Lambda} \nu(\{x\}) e^{i\langle k | x \rangle} \right|^2$$

where $k \in \mathbb{R}^d$ represents the wave vector of the diffraction beam, the direction of which gives the position on the screen. In practice, however, the intensity seen is $f(k)I_\Lambda(k)$ instead, where f is a *form factor* which takes into account that the incident beam sees atoms as composite objects rather than as points. The main problem is whether such quantity converges as $\Lambda \uparrow \mathbb{R}^d$. Before answering that question, let us remark that the Fourier transform of $I_\Lambda(k)$ is given by the following expression: if $f \in \mathcal{C}_c(\mathbb{R}^d)$, its Fourier transform is denoted by $\tilde{f}$ and

$$(1.4) \qquad \int_{k \in \mathbb{R}^d} dk\, \tilde{f}(k) I_\Lambda(k) = \frac{1}{|\Lambda|} \sum_{x,y \in \Lambda} \nu(\{x\})\nu(\{y\}) f(x-y) = \rho_\nu^{(\Lambda)}(f)$$

where $\rho_\nu^{(\Lambda)}$ will be called the *finite volume diffraction measure*. From (1.4), it follows that $\rho_\nu^{(\Lambda)} \in \mathcal{M}(\mathbb{R}^d)$ is a positive measure with a Fourier transform being also a positive measure.

The next theorem gives conditions under which convergence holds as $\Lambda \uparrow \mathbb{R}^d$ (see Section 3.2).

THEOREM 1.12. *Let $\mathbb{P}$ be a uniformly discrete translation invariant probability measure on $\mathrm{QD}(\mathbb{R}^d)$ supported by $\mathrm{UD}_r(\mathbb{R}^d)$ for some $r > 0$. Then:*

(i) *The averaged diffraction measure $\rho_\mathbb{P}^{(\Lambda)} = \int_{\nu \in \mathrm{QD}(\mathbb{R}^d)} d\mathbb{P}(\nu) \rho_\nu^{(\Lambda)}$ converges as $\Lambda \uparrow \mathbb{R}^d$ (Λ varying within the set of hypercubes in $\mathbb{R}^d$), to a positive measure $\rho_\mathbb{P} \in \mathcal{M}(\mathbb{R}^d)$.*
(ii) *The distributional Fourier transform of $\rho_\mathbb{P}$ is also a positive measure on $\mathbb{R}^d$.*
(iii) *If in addition $\mathbb{P}$ is ergodic, then for $\mathbb{P}$-almost every $\nu \in \mathrm{QD}(\mathbb{R}^d)$ the family $\rho_\nu^{(\Lambda)}$ of measures on $\mathbb{R}^d$ converges to $\rho_\mathbb{P}$.*

In other words, each invariant ergodic uniformly discrete probability measure on $\mathrm{QD}(\mathbb{R}^d)$ determines the diffraction pattern in a unique way. In particular, if we look into the phase diagram in a region of uniqueness of the Gibbs state, the diffraction pattern is entirely defined.

1.4. C^*-Algebra of Observables. The previous subsections have shown that the set of atomic sites gives rise to a canonical dynamical system $(\Omega, \mathbb{R}^d)$, its hull, over the translation group $\mathbb{R}^d$. There is always a C^*-algebra associated to this dynamical system, namely the crossed product $\mathcal{C}(\Omega) \rtimes \mathbb{R}^d$ [**Ped**]. We will show that this C^*-algebra, suitably modified if a uniform magnetic field is applied to the system, is nothing but the smallest C^*-algebra $\mathcal{A}$ generated by bounded continuous functions of the Schrödinger operator and its translates by $\mathbb{R}^d$, at least for an atomic potential satisfying sufficient regularity conditions (see Section 2.7). Moreover, when applied to a perfect crystal, we will see that $\mathcal{A}$ is isomorphic to $\mathcal{C}(\mathbb{B}) \otimes \mathcal{K}$, namely the space of matrix-valued continuous functions on the Brillouin zone. If the solid is not a perfect crystal, $\mathcal{A}$ is no longer of type I, namely it becomes noncommutative in a nontrivial way. It then replaces the C^*-algebra of continuous functions on the Brillouin zone. That is why we will associate to it a *noncommutative manifold* that we will call the *noncommutative Brillouin zone* (NCBZ) [**Bel2, Bel6**].

Given a uniform magnetic field $B = (B_{\nu\mu})$, namely a real-valued antisymmetric $d \times d$-matrix, we associate to it the C^*-algebra $C^*(\Omega \times \mathbb{R}^d, B)$ defined as follows. We first consider the topological vector space $\mathcal{C}_c(\Omega \times \mathbb{R}^d)$ of continuous functions with compact support in $\Omega \times \mathbb{R}^d$. It is endowed with the following structure of *-algebra

$$(1.5) \qquad fg(\omega, x) = \int_{\mathbb{R}^d} dy\, f(\omega, y) g(T^{-y}\omega, x-y) e^{i\pi(e/h) B.x \wedge y},$$

$$(1.6) \qquad f^*(\omega, x) = \overline{f(T^{-x}\omega, -x)},$$

where $f, g \in C_c(\Omega \times \mathbb{R}^d)$, $B.x \wedge y = \sum B_{\nu\mu} x_\nu y_\mu$ and $\omega \in \Omega, x \in \mathbb{R}^d$. Here e is the electric charge of the particle and $h = 2\pi\hbar$. This *-algebra is represented on $L^2(\mathbb{R}^d)$ by the family of representations $\{\Pi_\omega; \omega \in \Omega\}$ given by

$$(1.7) \qquad \Pi_\omega(f)\psi(x) = \int_{\mathbb{R}^d} dy\, f(T^{-x}\omega, y-x) e^{i\pi(e/h) B.x \wedge y} \psi(y), \quad \psi \in L^2(\mathbb{R}^d),$$

where Π_ω is linear, $\Pi_\omega(fg) = \Pi_\omega(f)\Pi_\omega(g)$ and $\Pi_\omega(f)^* = \Pi_\omega(f^*)$. In addition $\Pi_\omega(f)$ is a bounded operator and the representations $(\Pi_\omega)_{\omega \in \Omega}$ are related by the covariance condition:

$$(1.8) \qquad U(a)\Pi_\omega(f)U(a)^{-1} = \Pi_{T^a\omega}(f).$$

Here the U's are the so-called *magnetic translations* [**Zak**] defined by:

$$(1.9) \qquad U(a)\psi(x) = \exp\left\{\frac{ie}{\hbar} \int_{[x-a,x]} dy^\mu\, A_\mu(y)\right\} \psi(x-a),$$

where $\vec{A} = (A_1, \ldots, A_d)$ is the vector potential defined by $B_{\mu\nu} = \partial_\mu A_\nu - \partial_\nu A_\mu$, $a \in \mathbb{R}^d$, $\psi \in L^2(\mathbb{R}^d)$, and $[x-a, x]$ is the line segment joining $x-a$ to a in $\mathbb{R}^d$. Now, we set

$$(1.10) \qquad \|f\| = \sup_{\omega \in \Omega} \|\Pi_\omega(f)\|,$$

which defines a C^*-norm.

DEFINITION 1.13. *The noncommutative Brillouin zone is the topological manifold associated to the C^*-algebra $\mathcal{A} = C^*(\Omega \rtimes \mathbb{R}^d, B)$ obtained by completion of $\mathcal{C}_c(\Omega \times \mathbb{R}^d)$ under the norm $\|\cdot\|$ defined in (1.10).*

For $B = 0$ we recover the construction of the C^*-crossed product $\mathcal{C}(\Omega) \rtimes \mathbb{R}^d$ [**Ped, Bla**]. In the case of a perfect crystal (see Section 1.1), with lattice translation group $\mathcal{R}$, the hull $\Omega = \mathbb{R}^d/\mathcal{R}$ is homeomorphic to $\mathbb{T}^d$ (see Example 1.8). We get (see [**Rie2**]):

THEOREM 1.14 ([**Bel6**]). *The C^*-algebra $C^*(\mathbb{R}^d/\mathcal{R} \rtimes \mathbb{R}^d, B = 0)$, associated to a perfect crystal with lattice translation group $\mathcal{R}$, is isomorphic to $\mathcal{C}(\mathbb{B}) \otimes \mathcal{K}$, where $\mathcal{C}(\mathbb{B})$ is the space of continuous functions over the Brillouin zone and $\mathcal{K}$ the algebra of compact operators.*

Even though the algebra $\mathcal{C}(\mathbb{B}) \otimes \mathcal{K}$ is already noncommutative, its noncommutative parts come from $\mathcal{K}$, the smallest C^*-algebra generated by finite rank matrices. It describes the possible vector bundles over $\mathbb{B}$. Theorem 1.14 is the reason to claim that $\mathcal{A}$ generalizes the Brillouin zone to aperiodic systems.

At last, a very similar construction can be performed within the tight-binding representation. One starts from the space $\mathcal{C}_c(G_\Upsilon)$ of continuous functions with compact support on the groupoid of the transversal (see Definition 1.9 and Section 2.5). We proceed as before replacing the Hilbert space $L^2(\mathbb{R}^d)$ by $\ell^2(G_\Upsilon^\omega)$ where G_Υ^ω is the fiber of range ω in G_Υ. In this way we construct $C^*(G_\Upsilon, B = 0)$ (see Section 4).

Given any translation invariant probability measure $\mathbb{P}$ on Ω, one can define a *trace* on $\mathcal{A} = C^*(\Omega \rtimes \mathbb{R}^d, B)$ as follows. If $f \in \mathcal{C}_c(\Omega \times \mathbb{R}^d)$ we set [**Bel2, Bel6**]:

$$(1.11) \qquad \mathcal{T}_\mathbb{P}(f) = \int_\Omega d\mathbb{P}(\omega) \, f(\omega, x = 0).$$

This is a densely-defined linear form on $\mathcal{A}$ such that $\mathcal{T}_\mathbb{P}(f^* \cdot f) = \mathcal{T}_\mathbb{P}(f \cdot f^*) \geq 0$, namely it is an unbounded trace on $\mathcal{A}$. It has been shown [**Bel6**] that, whenever $\mathbb{P}$ is $\mathbb{R}^d$-ergodic, this trace satisfies

$$(1.12) \qquad \mathcal{T}_\mathbb{P}(f) = \lim_{\Lambda \uparrow \mathbb{R}^d} \frac{1}{|\Lambda|} \mathrm{Tr}_\Lambda\big(\Pi_\omega(f)\big) \quad \text{for } \mathbb{P}\text{-almost all } \omega\text{'s},$$

if $f \in \mathcal{C}_c(\Omega \times \mathbb{R}^d)$ and Λ varies in the family of open balls centered at one point in $\mathbb{R}^d$. It is therefore the *trace per unit volume*. For a perfect crystal, it is easy to show that this trace equals the integral over the Brillouin zone. Using a Gibbs state to build the equilibrium atomic configurations, we see that it not only uniquely defines the hull, but also endows the C^*-algebra of the hull with a physically natural trace.

1.5. K-Theory: Main Results. We recalled from the Introduction the main ideas leading to the definition and the construction of the K_0-group of a C^*-algebra. Let us give more details and summarize the main results here.

Together with $K_0(\mathcal{A})$ there is another K-group, denoted by $K_1(\mathcal{A})$, necessary in dealing with exact sequences and spectral sequences in K-theory. $K_1(\mathcal{A})$ classifies the homotopy classes of invertible elements of $\mathcal{A} \otimes \mathcal{K}$ (or $(\mathcal{A} \otimes \mathcal{K})^+$, the algebra obtained by adding a unit).

By construction of the K-groups, $\mathcal{A}$ and $\mathcal{A} \otimes \mathcal{K}$ always have the same K-theory (Morita equivalence). As a matter of fact, this leads to group isomorphisms between $K_i\big(C^*(\Omega \rtimes \mathbb{R}^d, B)\big)$ and $K_i\big(C^*(G_\Upsilon, B)\big)$.

For $B = 0$ the situation becomes a bit simpler. Through the Thom-Connes theorem [**Co2**], one gets an isomorphism between $K_i(\mathcal{C}(\Omega) \rtimes \mathbb{R}^d)$ and $K_{i+d}(\mathcal{C}(\Omega))$, where $i + d$ is defined modulo 2. On $\mathcal{C}(\Omega)$, the K-theory is isomorphic to the topological K-group $K^*(\Omega)$ over Ω, that is the classification modulo stable isomorphisms of vector bundles over Ω. Whenever Ω is a smooth manifold, the Chern character gives a rational isomorphism between $K^*(\Omega)$ and the cohomology of Ω with integer coefficients $H^*(\Omega, \mathbb{Z})$. Unfortunately, in most cases of interest in solid state physics, the topological space Ω is totally disconnected transversally to the $\mathbb{R}^d$-action. This means that its canonical transversal is totally disconnected. Therefore, we need to develop other techniques to compute the K-theory. This is exactly the purpose of this work.

As indicated in the Introduction, the trace per unit volume $\mathcal{T}_\mathbb{P}$, attached to any invariant probability measure $\mathbb{P}$ on the hull, gives rise to a group homomorphism $\mathcal{T}_{\mathbb{P}*}: K_0\big(C^*(\Omega \rtimes \mathbb{R}^d, B)\big) \mapsto \mathbb{R}$. The *gap labels* are the elements of the image of this map. Our main conjecture is given as follows

PROBLEM 1.15. *Prove or disprove that, if the hull Ω of a point set has a totally disconnected canonical transversal, the set of gap labels is given by*

$$\mathcal{T}_{\mathbb{P}*}\big(K_0\big(C^*(\Omega \rtimes \mathbb{R}^d, B=0))\big)\big) = \int_\Omega d\mathbb{P}\,\mathcal{C}(\Omega, \mathbb{Z}),$$

where $\mathcal{C}(\Omega, \mathbb{Z})$ is the space of continuous functions on Ω with values in $\mathbb{Z}$.

Evidence for this conjecture is provided by the following results:

THEOREM 1.16. *Let Ξ be a totally disconnected compact space with a $\mathbb{Z}^d$-action. For any $\mathbb{Z}^d$-invariant ergodic probability measure $\mathbb{P}$ on Ξ one has*

$$\mathcal{T}_{\mathbb{P}*}\big(K_0(\mathcal{C}(\Xi) \rtimes \mathbb{Z}^d)\big) = \int_\Xi d\mathbb{P}\,\mathcal{C}(\Xi, \mathbb{Z}) \quad \text{for } d = 1, 2.$$

This theorem was proved in [**Bel6**] for $d = 1$. Note that it solves Problem 1.15 in this case. For indeed every $\mathbb{R}$-action on a compact space reduces to a $\mathbb{Z}$-action on any transversal (the Poincaré first return map). Moreover, integrating integer-valued continuous functions on Ω is equivalent to integrating integer-valued continuous functions on a transversal.

For $d \geq 2$, an $\mathbb{R}^d$-action on a compact space does not reduce, in general, to a $\mathbb{Z}^d$-action on a transversal, so that Problem 1.15 remains open in general. However, in many cases, including quasicrystals, the K-theory can be computed through a $\mathbb{Z}^d$-action. The proof of Problem 1.15 for $d = 2$ for some class of $\mathbb{Z}^2$-actions was given by A. van Elst [**Elst**] (see Section 6.3). Unfortunately, van Elst's proof for $d = 3$ is not correct. The explicit computations provided by physicists for 3D quasicrystals (see for instance [**KaGr**]), give strong indications that Problem 1.15 should have a positive answer in any dimension. In the last chapter a list of useful cases will be given together with the explicit values of the set of gap labels.

2. Construction and Properties of the Hull

2.1. Points Sets with Multiplicity. Let $\mathrm{QD}(\mathbb{R}^d) = \{\nu \in \mathcal{M}(\mathbb{R}^d); \forall x \in \mathbb{R}^d, \nu$ is pure point and $\nu(\{x\}) \in \mathbb{N}\}$ (cf. Section 1.2). It is straightforward to check that for each $\nu \in \mathrm{QD}(\mathbb{R}^d)$, $L^{(\nu)} = \mathrm{supp}(\nu) = \{x; \nu(x) \geq 1\}$ is a discrete set. Let us call N_x the *multiplicity* of x, and define $N_\nu = (N_x)_{x \in L^{(\nu)}}$.

LEMMA 2.1. $\mathrm{QD}(\mathbb{R}^d)$ *is closed in $\mathcal{M}(\mathbb{R}^d)$.*

PROOF. Let $\nu \in \mathcal{M}(\mathbb{R}^d)$ be such that $\nu = \lim_{n\to\infty} \nu_n$ where $(\nu_n)_{n\in\mathbb{N}} \in \mathrm{QD}(\mathbb{R}^d)$. Let $x \in \mathbb{R}^d$ and $N \in \mathbb{N}$ be such that $N \leq \nu(\{x\}) < N+1$. Continuity properties of ν imply the existence of a real r such that $N \leq \nu(B(x,s)) \leq \nu(\overline{B(x,s)}) < N+1$ for $0 \leq s \leq r$. A sequence $(\nu_n)_{n\in\mathbb{N}}$ of positive Radon measures in $\mathcal{M}(\mathbb{R}^d)$ converges weakly to ν if and only if for every compact set K and for every relatively compact open set U,

(2.1) $\qquad \limsup_{n\to\infty} \nu_n(K) \leq \nu(K), \quad \text{and} \quad \liminf_{n\to\infty} \nu_n(U) \geq \nu(U).$

Using these inequalities, for $0 < s < s' < r$ we get

(2.2) $\quad N \leq \limsup_{n\to\infty} \nu_n\big(B(x,s)\big) \leq \nu\big(\overline{B(x,s)}\big) \leq \cdots \leq \nu\big(B(x,s')\big)$
$$\leq \liminf_{n\to\infty} \nu_n\big(B(x,s')\big) < N+1$$

Since $\nu_n \in \mathrm{QD}(\mathbb{R}^d)$, it follows that $\{\nu_n(B(x,s)); n \in \mathbb{N}\} \subset \mathbb{N}$ so that the liminf and the limsup must be integers. Thus $\nu(B(x,s)) = N$ for all $0 < s < r$ implying that for all $x \in \mathbb{R}^d$, $\nu(\{x\}) \in \mathbb{N}$ and that the support $L^{(\nu)}$ of ν is a discrete set. □

Now, we are interested in defining natural compact subsets of $\mathrm{QD}(\mathbb{R}^d)$. By "natural" we mean compact subsets which we expect to be the actual support of some Gibbs measure of the system (cf. Section 3.1). Let us first mention a general result on compactness which we will extensively use afterwards.

THEOREM 2.2 ([**Bau**]). *Let E be a locally compact space and $\mathcal{M}(E)$ be the set of all Radon measures endowed with the weak-$*$ topology. A set $\mathcal{F} \subset \mathcal{M}(E)$ has a weakly compact closure if and only if*

$$\sup_{\nu \in \mathcal{F}} |\nu(f)| < \infty$$

holds for all $f \in \mathcal{C}_c(E)$.

Following Section 1.2, let us introduce

$$\mathrm{UD}_r(\mathbb{R}^d) = \{\nu \in \mathrm{QD}(\mathbb{R}^d); \forall x \in \mathbb{R}^d,\ \nu(B(x,r)) \leq 1\}.$$

As previously noticed, for such measures, N_ν is trivial and $L^{(\nu)}$ is r-discrete. Elements of $\mathrm{UD}_r(\mathbb{R}^d)$ can then be seen as point sets in $\mathbb{R}^d$. The choice of these spaces comes from the fundamental result

LEMMA 2.3. *$\mathrm{UD}_r(\mathbb{R}^d)$ is a compact subset of $\mathrm{QD}(\mathbb{R}^d)$ which is invariant under the action of $\mathbb{R}^d$.*

PROOF. Let $f \in \mathcal{C}_c(\mathbb{R}^d)$. Let us define

$$\delta(f) = \inf\{R > 0; \exists x \in \mathbb{R}^d, \mathrm{supp}(f) \subset \overline{B(x,R)}\}.$$

If $\nu \in \mathrm{UD}_r(\mathbb{R}^d)$, we have $\nu(f) = \sum_{x \in L^{(\nu)}} f(x)$, then $|\nu(f)| \leq \|f\|\, |L^{(\nu)} \cap B(0,R)|$. Uniform discreteness gives

$$|L^{(\nu)} \cap B(0,R)|r^d \leq \left| \bigsqcup_{x \in L_\nu \cap B(0,R)} B(x, r/2) \right| < |B(0, R+r)|$$

where $\bigsqcup$ is a disjoint union, due to r-discreteness of ν. It implies the following bound, uniformly in ν: $|\nu(f)| \leq \|f\|.(\delta(f)/r + 1)^d$. Thanks to Theorem 2.2, $\mathrm{UD}_r(\mathbb{R}^d)$ is relatively compact. As in the proof of Lemma 2.1, a similar argument shows that $\mathrm{UD}_r(\mathbb{R}^d)$ is closed, then compact. The translation invariance and metrizability of $\mathrm{UD}_r(\mathbb{R}^d)$ are left to the reader. □

Let $\mathrm{UD}(\mathbb{R}^d) := \bigcup_{r>0} \mathrm{UD}_r(\mathbb{R}^d)$. It is the set of all uniformly discrete subsets of $\mathbb{R}^d$.

LEMMA 2.4. $\overline{\mathrm{UD}(\mathbb{R}^d)} = \mathrm{QD}(\mathbb{R}^d)$.

PROOF. Since $\mathrm{QD}(\mathbb{R}^d)$ is closed, we only have to prove $\mathrm{QD}(\mathbb{R}^d) \subset \overline{\mathrm{UD}(\mathbb{R}^d)}$. Let ν belong to $\mathrm{QD}(\mathbb{R}^d)$. For $n \in \mathbb{N}$ we set $L_n = L^{(\nu)} \cap B(0,n)$. For each $x \in \mathbb{R}^d$, let $e_x \in \mathbb{R}^d$ be chosen such that $|e_x|_\infty = 1$ and let $r_n > 0$ be the minimal distance between points of L_n. For $x \in L_n$ we set $y_i(x) = x + ([(i-1)r_n]/2nN_x)e_x$ for $1 \leq i \leq N_x$ and $L^{(n)} = \{y_i(x); x \in L_n, 1 \leq i \leq N_x\} + 2(n+r_n)\mathbb{Z}$. Remark that $(e_x)_{x \in L_n}$ is chosen such that $\{y_i(x), x \in L_n\} \subset B(0,n)$. Clearly, $L^{(n)}$ is uniformly discrete. We set $\nu_n = \nu^{(L^{(n)})} \in \mathrm{UD}(\mathbb{R}^d)$. Now, let $f \in \mathcal{C}_c(\mathbb{R}^d)$ and $n_f \in \mathbb{N}$ be such that $\mathrm{supp}(f) \subset B(0, n_f)$. For $n \geq n_f$ we get

$$|\nu(f) - \nu_n(f)| \leq \sum_{x \in L_n} \sum_{i=1}^{N_x} \left| f\left(x + \frac{r_n(i-1)}{2nN_x} e_x\right) - f(x) \right| \leq \eta_f\left(\frac{r_n}{2n}\right)\nu\big(B(0, n_f)\big)$$

where η_f is the modulus of continuity of f. Thus $\nu = \lim_{n \to \infty} \nu_n$. □

2.2. Delone Measures. Following the Lagarias classification of point sets, we introduce $\mathrm{Del}_{(r,R)}(\mathbb{R}^d) = \{\nu \in \mathrm{UD}_r(\mathbb{R}^d); \nu\big(\overline{B(x,R)}\big) \geq 1, \forall x \in \mathbb{R}^d\}$.

PROPOSITION 2.1. *ν belongs to $\mathrm{Del}_{(r,R)}(\mathbb{R}^d)$ if and only if $L^{(\nu)}$ is a Delone set.*

PROOF. Obvious. □

LEMMA 2.5. *$\mathrm{Del}_{(r,R)}(\mathbb{R}^d)$ is a compact and metrizable set, invariant under the action of $\mathbb{R}^d$.*

PROOF. Since it is included in the compact set $\mathrm{UD}_r(\mathbb{R}^d)$, it is enough to show that $\mathrm{Del}_{(r,R)}(\mathbb{R}^d)$ is a closed set. The same argument as in Lemma 2.1 gives this result. □

We set $\mathrm{Del}_r(\mathbb{R}^d) = \bigcup_{R>0} \mathrm{Del}_{(r,R)}(\mathbb{R}^d)$.

LEMMA 2.6. $\overline{\mathrm{Del}_r(\mathbb{R}^d)} = \mathrm{UD}_r(\mathbb{R}^d)$

PROOF. Let $\nu \in \mathrm{UD}_r(\mathbb{R}^d)$ and for $n \in \mathbb{N}$, $L_n = L^{(\nu)} \cap B(0,n)$, $L^{(n)} = L_n + 2(n+r)\mathbb{Z}$. $2R_n = \max\{2n, \sup\{d(x,y); (x,y) \in L_n \times L_n\}\}$. Then, if $L_n \neq \emptyset$, $\nu_n = \nu^{(L^{(n)})} \in \mathrm{Del}_{(r,R_n)}(\mathbb{R}^d)$. Now, let $f \in \mathcal{C}_c(\mathbb{R}^d)$ and let $n_f \in \mathbb{N}$ be such that $\mathrm{supp}(f) \subset B(0, n_f)$. $\nu(f)$ coincides with $\nu_n(f)$ for each n greater than n_f. So, $\nu = \lim_{n \to \infty} \nu_n$. □

REMARK 2.7. $\mathrm{Del}_r(\mathbb{R}^d)$ is not closed. However it is an F_σ, namely a countable union of closed sets.

PROOF OF THEOREM 1.5. It follows from Lemmas 2.1, 2.3, 2.4, 2.5, and 2.6. □

2.3. General Properties of the Hull. Let ν be a measure in $\mathrm{UD}_r(\mathbb{R}^d)$. Since $\mathrm{UD}_r(\mathbb{R}^d)$ is compact, the closure Ω_ν of the $\mathbb{R}^d$-orbit of any $\nu \in \mathrm{UD}_r(\mathbb{R}^d)$ is compact, too. Ω_ν has been called the hull of ν (see Definition 1.7).

In what follows, we will consider the hull as an abstract compact metrizable set Ω and we will denote the vector-valued measures corresponding to $\omega \in \Omega$ by $\vec{\nu}_\omega$. The first important result about the hull is the following consequence of Theorem 1.5:

THEOREM 2.8.
(i) *Let $L = (L_1, \ldots, L_c)$ be a finite family of uniformly discrete point sets in $\mathbb{R}^d$ and let Ω be its hull. Then for every $\omega \in \Omega$, there is a finite family $L_\omega = (L_{\omega,1}, \ldots, L_{\omega,c})$ of uniformly discrete sets such that the vector-valued measure corresponding to ω is $\vec{\nu}_\omega = (\nu_{\omega,1}, \ldots, \nu_{\omega,c})$, with $\nu_{\omega,j} = \nu^{(L_{\omega,j})}$.*
(ii) *If, in addition, L_j is a (r,R)-Delone set, so is $L_{\omega,j}$ for any $\omega \in \Omega$.*

REMARK 2.9. Let L be a Delone set with hull Ω, and let r_0, r_1 be defined as in Section 1.2. For $\omega \in \Omega$, L_ω may have $r_0(L_\omega) > r_0$ and $r_1(L_\omega) < r_1$. An example of such a set is given as follows: let L_0 be the lattice $\mathbb{Z}^d$ in $\mathbb{R}^d$. Then L is obtained by removing 0 from L_0 and by adding $x_0 = (\frac{3}{2}, \frac{1}{2}, \ldots, \frac{1}{2})$. Then $r_0 = \frac{1}{4}$, $r_1 = 2$, whereas L_0 is clearly an element of the hull with $r_0 = r_1 = 1$.

REMARK 2.10. If L is a finitely-generated Delone set with hull Ω, for $\omega \in \Omega$, L_ω need not be finitely generated anymore. This can be seen on the following counterexample: let $\mathcal{Z}$ be a finitely-generated dense subgroup of $\mathbb{R}^d$. Let $0 < \delta < \frac{1}{4}$ and let $D = B(0, \delta)$. Let now $\underline{u} = (u_m)_{m \in \mathbb{Z}^d}$ be a sequence such that $u_m \in \mathcal{Z} \cap D$, $\forall m \in \mathbb{Z}^d$. Our example is given by $L_{\underline{u}} = \{m + u_m ; m \in \mathbb{Z}^d\}$. By construction $L_{\underline{u}}$ is a Delone set and it is finitely generated since the differences $m + u_m - m' - u_{m'}$ all belong to $\mathcal{Z}$. Now choose $\underline{u}$ as follows: let $\underline{v} = (v_m)_{m \in \mathbb{Z}^d}$ be a sequence chosen randomly where the v_m's are considered as independent random variables, uniformly distributed on D. Then we choose $u_m \in \mathcal{Z} \cap D$ so that $|u_m - v_m|_\infty \leq 2^{-|m|_\infty} \delta$. Using the same argument as the one proved in Section 3.3, it is possible to show that the hull is homeomorphic to the *mapping torus* of $D^{\mathbb{Z}^d}$ and that, if $\underline{v} \in D^{\mathbb{Z}^d}$, the corresponding point set is $L_{\underline{v}}$. Then it is clear that $\underline{v}$ can be chosen such that $L_{\underline{v}}$ is not finitely generated.

2.4. Dynamical Properties of the Hull. One can wonder whether global properties of the dynamical system defined by the hull Ω of a uniformly discrete point set L can be read off from its local properties. Among these properties, let us examine the structure of orbits. By construction the orbit of L is dense. Which conditions are necessary and sufficient for the hull to define a minimal system? The following result give a necessary condition.

THEOREM 2.11. *Let L be uniformly discrete but not Delone. Then the hull admits one fixed point corresponding to the empty set.*

PROOF. If L is not Delone, there is a sequence $(x_n)_{n \in \mathbb{N}}$ in $\mathbb{R}^d$ such that $B(x_n, n) \cap L = \emptyset$. Therefore given any $f \in \mathcal{C}_c(\mathbb{R}^d)$ one gets $\lim_{n \to \infty} T^{-x_n} \nu^{(L)}(f) = 0$. Hence the measure $\nu_\infty = 0$ belongs to the hull and the corresponding point set is empty. It is clearly a fixed point of the translation group. $\square$

This results shows that if L is not Delone, the hull cannot be minimal unless $L = \emptyset$. The following definition will characterize minimal hulls.

DEFINITION 2.12. Let L be a Delone set in $\mathbb{R}^d$. It is called *uniformly distributed* if and only if for every $f \in \mathcal{C}_c(\mathbb{R}^d), f \geq 0$ and every $s \in \mathbb{R}_+$ the set $L^{f,s} = \{x \in L; T^{-x}\nu^{(L)}(f) > s\}$ is either empty or a Delone set.

THEOREM 2.13. *Let L be a Delone set in $\mathbb{R}^d$. Then the hull Ω of L is minimal if and only if L is uniformly distributed.*

PROOF. (I) Let L be a uniformly distributed Delone set.

(i) Let ω be a point in the hull of L. Then there is a sequence $(a_l)_{l \in \mathbb{N}}$ in $\mathbb{R}^d$ such that $\nu_\omega = \nu^{(L_\omega)} = \lim_{l \to \infty} T^{a_l} \nu^{(L)}$. If $f \in \mathcal{C}_c(\mathbb{R}^d)$, $f \geq 0$, let $t \geq 0$ so that $L^{f,t} \neq \emptyset$. Then one can find $s > t$ such that $L^{f,s} \neq \emptyset$ so that it is a Delone set. Moreover, by definition, $L^{f,s} + a = (L+a)^{f,s}$ for any $a \in \mathbb{R}^d$, so that this is also a Delone set.

(ii) We claim that $L_\omega^{f,t}$ is a Delone set for all $\omega \in \Omega$. Let us consider $L_l = L^{f,s} + a_l$, which is a Delone set, and denote by $\nu_l = \nu^{(L_l)}$ the corresponding measure.

Using the Corollary 1.6 and the Theorem 2.8, it follows that ν_l has a limit point ν in $\mathcal{M}(\mathbb{R}^d)$ corresponding to a Delone set denoted by L_ν. Using the properties of the weak limit, if $x \in L_\nu$, one can find $x_l \in L_l$ such that $x = \lim_{l \to \infty} x_l$. Since $x_l \in L_l \subset L + a_l$ it follows that $x \in L_\omega$ so that $L_\nu \subset L_\omega$. By definition of the x_l's, $T^{a_l - x_l} \nu^{(L)}(f) > s$ for all $l \in \mathbb{N}$ leading to:

$$T^{a_l - x_l} \nu^{(L)}(f) = \nu^{(L+a_l)}(f \circ T^{-x_l}).$$

Since f is continuous with compact support, it is uniformly continuous so that, since x_l converges to x, there is $R_1 > 0$ with $\operatorname{supp}(f \circ T^{x_l}) \subset B(0, R_1)$ for all l. In addition, for every $\varepsilon > 0$ there is $l_0 \in \mathbb{N}$ with $\|f \circ T^{-x} - f \circ T^{-x_l}\| \leq (R_1/r_0+1)^{-d} \varepsilon/2$ for $l \geq l_0$. In particular, using the inequality (2.1)

$$|\nu^{(L+a_l)}(f \circ T^{-x}) - \nu^{(L+a_l)}(f \circ T^{-x_l})| \leq \varepsilon/2.$$

Moreover, since ν_ω is the limit of the $\nu^{(L+a_l)}$'s there is $l_1 \geq l_0$ so that if $l \geq l_1$, $|\nu_\omega(f \circ T^{-x}) - \nu^{(L+a_l)}(f \circ T^{-x})| \leq \varepsilon/2$. It follows that $\nu_\omega(f \circ T^{-x}) = T^{-x} \nu_\omega(f) \geq s > t$, thus $x \in L_\omega^{f,t}$. As a consequence

$$L_\nu \subset L_\omega^{f,t} \subset L_\omega.$$

Since L_ν is relatively dense, so is $L_\omega^{f,t}$ and since L_ω is uniformly discrete, so is $L_\omega^{f,t}$. Hence $L_\omega^{f,t}$ is a Delone set.

(iii) To prove that the hull is minimal, it is enough to show that the orbit of L_ω is dense for all $\omega \in \Omega$. Since the orbit of L is dense by definition of the hull, it is sufficient to show that there is a sequence $(x_l)_{l \in \mathbb{N}}$ in $\mathbb{R}^d$ such that L is the limit of $L_\omega + x_l$. Without loss of generality we can assume $0 \in L$, otherwise we choose $x \in L$ and we replace L by $L - x$.

We build a sequence g_l of continuous functions with compact support as follows. Let r_0 be such that any open ball of radius r_0 contains at most one point of L. For $l \geq 1$ let $f_l(x) = \phi(l|x|_\infty/r_0)$ with

$$\phi(\xi) = \begin{cases} 1 - \xi & \text{if } 0 \leq \xi \leq 1 \\ 0 & \text{otherwise} \end{cases}$$

and

$$g_l(x) = \sum_{y \in L \cap B(0, lr_0)} f_l(y - x).$$

The properties of g_l are the following:

(a) Its support is $\operatorname{supp}(g_l) = \bigcup_{y \in L \cap B(0, lr_0)} \overline{B(y, r_0/l)}$. For $l \geq 2$, these balls are disjoint.
(b) $0 \leq g_l(x) \leq 1$ for all $x \in \mathbb{R}^d$, and $g_l(x) = 1$ if and only if $x \in L \cap B(0, lr_0)$.
(c) Let $n_l = |L \cap B(0, lr_0)|$ and let L' be a point set in $\mathbb{R}^d$. Let us define

$$\hat{g}_l(L; L') = \sum_{x \in L'} g_l(x).$$

Then $0 \leq \hat{g}_l(L; L') \leq n_l$ and if $\operatorname{dist}(L, L') \geq \rho$ then $\hat{g}_l(L; L') \leq n_l - l\rho/r_0$.

Thanks to (c) above, we get $\nu^{(L)}(g_l) = n_l > s_l = n_l - 1/l$, namely, since $0 \in L$, $0 \in L^{g_l, s_l}$. Since it is not empty, it must be a Delone set, because L is uniformly distributed. From part (ii) it follows that $L_\omega^{g_l,s_l}$ is a Delone set. Let then $x_l \in L_\omega^{g_l,s_l}$. This means

$$n_l - \frac{1}{l} < T^{-x_l}\nu_\omega(g_l) = \sum_{y \in L_\omega - x_l} g_l(y) = \hat{g}_l(L; L_\omega - x_l)$$

which, by the property (c) above, implies that the distance between points of L and of $L_\omega - x_l$ contained in $B(0, lr_0)$ is smaller than or equal to r_0/l^2. Hence $L_\omega - x_l$ converges (in the sense of the weak topology for measures), to L. Consequently, the hull is indeed minimal.

(II) Let L be a nonempty Delone set with minimal hull. If L is not uniformly distributed, using (Ii), there is $\omega \in \Omega$, $f \in \mathcal{C}_c(\mathbb{R}^d)$, $f \geq 0$ and $s > 0$ such that $L_\omega^{f,s} \neq \emptyset$ and $L_\omega^{f,s}$ is not a Delone set. Since $L_\omega^{f,s} \subset L_\omega$, it is uniformly discrete. Let $r > 0$ be such that every ball of radius r meets L_ω at one point at most. One can find a sequence $(x_l)_{l \in \mathbb{N}}$ in $\mathbb{R}^d$ such that the open balls $B(x_l, lr)$ never meet $L_\omega^{f,s}$. Extracting a subsequence if necessary, there is $\sigma \in \Omega$ such that $\lim_{l \to \infty} T^{-x_l}\nu_\omega = \nu_\sigma$.

Let $x \in L_\omega^{f,s}$, then $T^{-x}\nu_\omega(f) > s$. Moreover, for $a \in \mathbb{R}^d$, $T^{-a-x_l}\nu_\omega(f)$ converges to $T^{-a}\nu_\sigma(f)$. But we remark that

$$T^{-a-x_l}\nu_\omega(f) = \sum_{y \in L_\omega - x_l} f(y - a).$$

If this sum were not vanishing, there would be $y \in L_\omega - x_l$ such that $y - a \in \mathrm{supp}(f)$, namely $|a - y|_\infty \leq R$ for some $R > 0$. Since the ball $B(0, lr_0)$ does not intersect $L_\omega - x_l$, it would follow that $|a|_\infty \geq lr_0 - R$. This condition cannot be satisfied for all l's, so that the sum vanishes eventually, leading to $T^{-a}\nu_\sigma(f) = 0$ for all $a \in \mathbb{R}^d$. Thus

$$\inf_{a \in \mathbb{R}^d} |T^{-x}\nu_\omega(f) - T^a\nu_\sigma(f)| \geq s > 0$$

showing that the orbit of σ is not dense, a contradiction, since we assumed the hull to be minimal. □

2.5. The Canonical Transversal. For usual dynamical systems where the group $\mathbb{G} = \mathbb{R}$ represents the time evolution, a transversal, also called a *Poincaré section*, is used to replace the continuous time evolution by a discrete time, through the so-called *first return map*. For other groups, this construction can be generalized by using the notion of a *groupoid* [**Co1, Co6, Ren**].

A *groupoid* G is the data of two sets $G^{(0)}$, the set of *objects* or the *basis* of G, and $G^{(1)}$, the set of *arrows*, together with the following structure:

GR1. There are two maps r, $s\colon G^{(1)} \mapsto G^{(0)}$, called the *range* and the *source*, respectively. If $\gamma \in G^{(1)}$ is such that $x = s(\gamma)$, $y = r(\gamma)$ we set $\gamma\colon x \mapsto y$.

GR2. The subset $G^{(2)}$ of $G^{(1)} \times G^{(1)}$ of pairs (γ_1, γ_2) of arrows such that $s(\gamma_1) = r(\gamma_2)$ is the set of *composable arrows*.

GR3. There is a *composition* $(\gamma_1, \gamma_2) \in G^{(2)} \mapsto \gamma_1 \circ \gamma_2 \in G^{(1)}$, which is associative and satisfies: $r(\gamma_1 \circ \gamma_2) = r(\gamma_1)$, $s(\gamma_1 \circ \gamma_2) = s(\gamma_2)$.

GR4. To each object $x \in G^{(0)}$ there is an arrow e_x, *the unit at x*, such that $r(e_x) = s(e_x) = x$ and $\forall \gamma\colon x \mapsto y$, $\gamma = e_y\gamma = \gamma e_x$.

GR5. There is a map called the *inverse* $\gamma \in G^{(1)} \mapsto \gamma^{-1} \in G^{(1)}$ such that $r(\gamma^{-1}) = s(\gamma)$, $s(\gamma^{-1}) = r(\gamma)$, with $\gamma \circ \gamma^{-1} = e_{r(\gamma)}$, $\gamma^{-1} \circ \gamma = e_{s(\gamma)}$ and also $(\gamma_1 \circ \gamma_2)^{-1} = \gamma_2^{-1} \circ \gamma_1^{-1}$.

Thanks to GR4, $G^{(0)}$ can be identified with the set of units and becomes a subset of $G^{(1)}$. Thus G can be identified with the set $G = G^{(1)}$. The groupoid G is *topological* if both $G^{(0)}$ and $G^{(1)}$ are topological spaces and if all maps defined in GR1, ..., GR5 are continuous. If $x \in G^{(0)}$ we denote by G^x the *fiber of range* x, namely the set of arrows with range x. If $X \subset G^{(0)}$, the groupoid G_X induced by X is defined by $G_X^{(0)} = X$ and $G_X^{(1)} = \{\gamma \in G^{(1)} ; r(\gamma), s(\gamma) \in X\}$.

Given a topological dynamical system $(X, \mathbb{G})$, there is a canonical groupoid attached to it, denoted by $G = X \rtimes \mathbb{G}$ and called the *crossed product of X by the action of* $\mathbb{G}$. Its set of objects is X and an arrow is a pair $\gamma = (x, g) \in X \times \mathbb{G}$ with $r(x, g) = x$, $s(x, g) = g^{-1}x$ and $(x, g) \circ (g^{-1}x, g') = (x, gg')$.

A *transversal* of this (topological) dynamical system $(X, \mathbb{G})$ is a closed subset Y of X which meets every $\mathbb{G}$-orbit and such that $\{g \in \mathbb{G} ; g^{-1}x \in Y\}$ is nonempty, discrete for all $x \in X$ and continuous with respect to x (let us remark that this definition is compatible with Definition 1.9).

Let $L \subset \mathbb{R}^d$ be a uniformly discrete point set with hull Ω and let $\Omega \rtimes \mathbb{R}^d$ be the corresponding groupoid. Then we get

PROPOSITION 2.2. *Let $L \subset \mathbb{R}^d$ be a uniformly discrete point set with hull Ω. The set $\Upsilon = \{\omega \in \Omega; L_\omega \ni 0\}$ is a compact transversal such that for all $\omega \in \Omega$, $\{a \in \mathbb{R}^d; T^a\omega \in \Upsilon\} = L_\omega$.*

PROOF. The proof is straightforward and left to the reader. □

DEFINITION 2.14. The transversal Υ defined in Proposition 2.2 is the canonical transversal attached to L. The corresponding groupoid will be denoted by G_Υ.

REMARK 2.15. Υ is compact in Ω, thus the transversal can be endowed with the topology induced by $\Omega \subset \mathrm{QD}(\mathbb{R}^d)$.

REMARK 2.16. This groupoid plays an important rôle in our approach since it is at the origin of the so-called *tight-binding representation* (see Section 4) [**Bel2**] in which the electronic wave function is restricted to atomic sites and the Schrödinger operator is discretized to become a matrix indexed by atomic sites.

There are special cases for which the groupoid of the transversal is itself a dynamical system. An example is given as follows: let $\mathcal{R}$ be a lattice in $\mathbb{R}^d$, namely a discrete subgroup generating $\mathbb{R}^d$ as a vector space. The point set L is built as the image of a map $a \in \mathcal{R} \mapsto x_a \in \mathbb{R}^d$ such that $\sup_{a \in \mathcal{R}} |x_a - a|_\infty \leq r_2$ for some $r_2 > 0$. If $D = B(0, r_2)$ one has $u_a = x_a - a \in D$. Let then Σ be the closure in $D^\mathcal{R}$ of the family $\{T^a\underline{u}; a \in \mathcal{R}\}$ where $T^a\underline{u} = (u_{b-a})_{b \in \mathcal{R}}$. This is a compact space on which $\mathcal{R}$ acts through the family of homeomorphisms $(T^a)_{a \in \mathcal{R}}$. It is a simple fact that the canonical transversal of L is homeomorphic to Σ and that the corresponding groupoid is isomorphic to $\Sigma \rtimes \mathcal{R}$. One can then reconstruct the hull of L by means of the *mapping torus*, also called *suspension*. Namely, on the space $\Sigma \times \mathbb{R}^d$ there are two types of actions: (i) $\mathbb{R}^d$ itself acts through $\phi_s \colon (\underline{u}, x) \mapsto (\underline{u}, x + s)$; (ii) $\mathcal{R}$ acts through $\eta_a \colon (\underline{u}, x) \mapsto (T^a\underline{u}, x - a)$. These actions commute with each other. The quotient space $M\Sigma = \Sigma \times \mathbb{R}^d / \mathcal{R}$ inherits an action of $\mathbb{R}^d$ through the quotient map associated to ϕ. It then follows that the hull of L is homeomorphic to $M\Sigma$ and the corresponding actions of $\mathbb{R}^d$ are conjugate through this homeomorphism.

2.6. Hulls with Totally Disconnected Transversal. If $\nu \in \mathrm{UD}(\mathbb{R}^d)$, let $(\Omega_\nu, \mathbb{R}^d)$ be its hull. Very often, the hull is totally disconnected transversally. Which conditions on ν are sufficient to provide such a property?

Let us endow Υ with the metric δ_H defined as

$$(2.3) \qquad \delta_H(\omega_1,\omega_2) = \inf\left\{\frac{1}{R+1}; \delta_{R,H}(\omega_1,\omega_2) \le \frac{1}{R}, R \in \mathbb{R}_+\right\}$$

where $\delta_{R,H}(\omega_1,\omega_2) = d_H\left((L_{\omega_1} \cap B(0,R)) \cup \partial B(0,R), (L_{\omega_2} \cap B(0,R)) \cup \partial B(0,R)\right)$ and d_H is the Hausdorff distance on point set spaces ([**RW, FHK2**]). It is elementary to show that the topology induced by δ_H is the weak-$*$ topology on $\mathrm{UD}(\mathbb{R}^d)$. This confirms the intuitive view of the weak-$*$ topology outlined in Section 1.2, and actually makes the link with the hulls introduced in [**KP**] more precise. As a direct consequence, δ_H endows Υ with the weak-$*$ topology (Remark 2.15).

LEMMA 2.17. *If a metric d on Υ takes values in a discrete set away from zero, Υ is totally disconnected for the topology induced by d.*

PROOF. If $B(\omega_0, r)$ is an open ball it is closed. For indeed, let $\omega \ne \omega_0$ belong to the closed ball $\overline{B(\omega_0,r)}$. ω is a limit of points $(\omega_n)_n$ of $B(\omega_0, r)$ so that $d(\omega_0, \omega) = \lim_n d(\omega_0, \omega_n) \le \sup_n d(\omega_0, \omega_n)$. Discreteness of values taken by d away from zero implies that $\sup\{d(\omega_0, \omega_n), n \in \mathbb{N}\} \subset \{d(\omega_0, \omega_n), n \in \mathbb{N}\}$. Thus ω belongs to $B(\omega_0, r)$, too, and the ball is closed. $\square$

PROPOSITION 2.3. *Let $\Omega_R(\nu) = \overline{\bigcup_{x \in \mathrm{supp}(\nu)} \mathrm{supp}(T^{-x}\nu) \cap B(0,R)}$. If for every $R > 0$, $\Omega_R(\nu)$ has no accumulation point, then the transversal Υ of Ω_ν is totally disconnected for the topology induced by the metric δ_H.*

PROOF. Let $\rho(R)$ be the number of points in $\Omega_R(\nu)$. Since $\Omega_R(\nu)$ is finite for all $R > 0$, ρ is finite on $\mathbb{R}_+$, integer-valued, and nondecreasing. Thus the set $D(\nu) = \{R; |\Omega_{R+\epsilon}(\nu)| > |\Omega_{R-\epsilon}(\nu)|, \epsilon \in \mathbb{R}_*^+, R \in \mathbb{R}_+\}$ is discrete. Let $d_{u.m}$ be the distance defined by $d_{u.m}(\omega_1,\omega_2) = \inf\{1/(R+1); d_H(L_{\omega_1} \cap B(0,R), L_{\omega_2} \cap B(0,R)) = 0, R \in \mathbb{R}_+\}$. The set of values of $d_{u.m}$ on Υ is included in $\{1/(R+1); R \in D(\nu), R \in \mathbb{R}_+\}$ and is discrete away from zero. By Lemma 2.17, the topology induced by $d_{u.m}$ on Υ is totally disconnected. It is elementary to check that this topology coincides with the weak-$*$ topology on Υ. $\square$

REMARK 2.18. Kellendonk proves a similar result in [**Kel1**, p. 122].

COROLLARY 2.1. *Let $\nu \in \mathrm{UD}(\mathbb{R}^d)$ with L_ν a Delone set of finite type. Then the transversal Υ of Ω_ν is totally disconnected.*

PROOF. For every $R > 0$ and every Delone set of finite type L the set $(L - L) \cap B(0,R)$ is finite. Therefore $\Omega_R(\nu)$ in Proposition 2.3 has no accumulation point. $\square$

2.7. The Hull of a Schrödinger Operator.
Let us consider a Schrödinger operator on $L^2(\mathbb{R}^d)$

$$(2.4) \qquad H = (\vec{P} - e\vec{A})^2/2m + V = H_0 + V,$$

where V is the effective potential seen by an electron and $\vec{A}$ is the vector potential corresponding to a uniform magnetic field. In general H is not translation invariant. However, the physical properties of a homogeneous medium do not depend upon the choice of an origin. In particular, H can be replaced by any of its translates $H_x = U(x)HU(x)^{-1}$, $x \in \mathbb{R}^d$, and the physics will be the same. This choice being arbitrary, the smallest possible algebra of observables should contain all bounded functions of the $\{H_x, x \in \mathbb{R}^d\}$'s. Following [**Bel6**], the homogeneity of the Hamiltonian H will be defined by:

DEFINITION 2.19. Let $\mathcal{H}$ be a Hilbert space with a countable basis. Let $\mathbb{G}$ be a locally compact group (for instance $\mathbb{R}^d$ or $\mathbb{Z}^d$). Let U be a unitary projective representation of $\mathbb{G}$, namely for each $a \in \mathbb{G}$ there is a unitary operator $U(a)$ acting on $\mathcal{H}$ such that the family $U = \{U(a); a \in \mathbb{G}\}$ satisfies the following properties:

 (i) $U(a)U(b) = U(a+b)e^{i\phi(a,b)}$, where $\phi(a,b)$ is some phase factor.
 (ii) For each $\psi \in \mathcal{H}$, the map $a \in \mathbb{G} \to U(a)\psi \in \mathcal{H}$ is continuous.

Then a self-adjoint operator H on $\mathcal{H}$ is *homogeneous* with respect to $\mathbb{G}$ if the family $S = \{R_a(z) = U(a)(z\mathbf{1} - H)^{-1}U(a)^{-1}; a \in \mathbb{G}\}$ admits a compact closure in the strong operator topology (for some $z \in \mathbb{C}$).

Let us now define the "hull" of a homogeneous operator H. For z in the resolvent set $\rho(H)$ of H, let $\Omega_H(z)$ be the strong closure of the family $\{R_a(z) = U(a)(z\mathbf{1} - H)^{-1}U(a)^{-1}; a \in \mathbb{G}\}$. By definition of homogeneity, it is a metrizable compact space. Moreover, $\Omega_H(z)$ is endowed with a $\mathbb{G}$-action by means of the representation U of $\mathbb{G}$. Actually, $\Omega_H(z)$ does not depend on the choice of z [**Bel6**]. Identifying $\Omega_H(z)$ and $\Omega_H(z')$ for $z, z' \in \mathbb{C}$ gives rise to an abstract compact space Ω_H endowed with an action of $\mathbb{G}$. If $\omega \in \Omega_H$ and $a \in \mathbb{G}$ we will denote by $T^a\omega$ the result of the action of a on ω, and $R(z, \omega)$ the representative of ω in $\Omega_H(z)$. Then one gets

$$U(a)R(z,\omega)U(a)^{-1} = R(z, T^a\omega),$$
$$R(z',\omega) - R(z,\omega) = (z-z')R(z,\omega)R(z',\omega) = (z-z')R(z',\omega)R(z,\omega).$$

In addition, $z \to R_\omega(z)$ is norm-holomorphic in $\rho(H)$ for every $\omega \in \Omega_H$, and $\omega \to R_\omega(z)$ is strongly continuous.

DEFINITION 2.20. Let H be a homogeneous operator on the Hilbert space $\mathcal{H}$ with respect to the representation U of the locally compact group $\mathbb{G}$. Then the *hull* of H is the dynamical system $(\Omega_H, \mathbb{G}, T)$, where Ω_H is the (abstract) compact space given by the strong closure of the family $\{R_a(z) = U(a)(z\mathbf{1} - H)^{-1}U(a)^{-1}; a \in \mathbb{G}\}$ for some $z \in \rho(H)$, and the $\mathbb{G}$ action T on Ω_H is induced by U.

In the case of a Schrödinger operator (2.4) the situation has been clarified in [**Bel6**]. The vector potential $\vec{A}$ satisfies

(2.5) $$\partial_\mu A_\nu - \partial_\nu A_\mu = B_{\mu\nu} = \text{const.}$$

and therefore the kinetic part H_0 is actually translation invariant provided one uses the magnetic translations (1.9).

THEOREM 2.21 ([**Bel6, NB**]). *Let $H = H_0 + V$ be given by (2.4) with $V \in L^\infty_\mathbb{R}(\mathbb{R}^d)$, i.e., a real, measurable, essentially bounded function over $\mathbb{R}^d$. Then H is homogeneous with respect to the representation of $\mathbb{R}^d$ given by (1.9). Moreover, the hull of H is homeomorphic to the weak closure Ω_V of the family $\{U(a)VU(a)^{-1}; a \in \mathbb{R}^d\}$ in the space $L^\infty_\mathbb{R}(\mathbb{R}^d)$ endowed with the weak topology w.r.t. $L^1(\mathbb{R}^d)$. Furthermore, there exists a Borelian function v on Ω_V such that $V_\omega(x) = v(T^{-x}\omega)$ for almost every $x \in \mathbb{R}^d$ and every $\omega \in \Omega_V$. If in addition V is uniformly continuous, then v is continuous.*

The Schrödinger operator (1.1) acting on $\mathcal{H} = L^2(\mathbb{R}^d)$ was given by

$$H = -\frac{\hbar^2}{2m}\Delta + \sum_{j=1}^{c}\sum_{y\in L_j} v_j(.-y) = -\frac{\hbar^2}{2m}\Delta + \sum_{j=1}^{c}\nu_j * v_j. \tag{2.6}$$

where $v_j(.-y)$ is the effective potential for valence electrons due to an atom of species j at position y. Here $\nu_j * v_j$ denotes the convolution of the measure $\nu_j = \nu^{(L_j)}$ and the potential v_j. For simplicity, we assume in the following that we have only one species of atoms and that the point set of atomic positions is r-discrete. The second sum in (2.6) being infinite, the following lemma provides a sufficient condition for its convergence. Let

$$L^1_{K,r}(\mathbb{R}^d) = \left\{ f \in L^1(\mathbb{R}^d); |f(x)| \leq \frac{K}{r^d}\int_{|x-y|<r/2} d^dy\, |f(y)|, \text{ for a.e. } x \right\}$$

be the set of integrable K-subharmonic functions on $\mathbb{R}^d$. It is elementary to check that $L^1_{K,r}(\mathbb{R}^d)$ is closed in $L^1(\mathbb{R}^d)$ and contained in $L^\infty(\mathbb{R}^d)$.

LEMMA 2.22. *Let $v \in L^1_{K,r}(\mathbb{R}^d)$. Then $\nu * v \in L^\infty_\mathbb{R}(\mathbb{R}^d)$ and the map $v \mapsto \nu * v$ is continuous.*

PROOF. For almost all $x \in \mathbb{R}^d$,

$$|\nu * v(x)| = \left|\sum_{y\in L^{(\nu)}} v(x-y)\right| \leq \sum_{y\in L^{(\nu)}} |v(x-y)| \leq \sum_{y\in L^{(\nu)}} \left\{ \frac{K}{r^d}\int_{|y'-(x-y)|<r/2} d^dy'\, |v(y')| \right\}.$$

Since $\nu \in \mathrm{UD}_r(\mathbb{R}^d)$, $L^{(\nu)}$ is r-discrete and we get

$$\sum_{y\in L^{(\nu)}} \left\{ \frac{K}{r^d}\int_{|y'-(x-y)|<r/2} d^dy'\, |v(y')| \right\} \leq \frac{K}{r^d}\int_{\mathbb{R}^d} d^dy'|v(y')| \leq \frac{K}{r^d}\|v\|_1. \qquad \square$$

According to Theorem 2.21 and Lemma 2.22, the Hamiltonian (2.6) is homogeneous if $\nu \in \mathrm{UD}_r(\mathbb{R}^d)$ and its hull is well defined. As outlined in Section 1.2, the definition of the hull of an operator came earlier than the definition of the hull of a point set. The following gives a link between these two notions:

THEOREM 2.23. *Let $v \in L^1_{K,r}(\mathbb{R}^d)$ be a real-valued atomic potential. For $\nu \in \mathrm{UD}_r(\mathbb{R}^d)$ the map $\varphi_v \colon T_1^a(\nu) \in \Omega_\nu \to U(a)R_z(H_\nu)U(a)^{-1}$ can be continued in a unique way as a surjective, strongly continuous function from Ω_ν onto $\Omega_H(z)$ fulfilling*

$$\varphi_v \circ T_1^x(\omega) = U(x)R_z(H_\omega)U(x)^{-1}. \tag{2.7}$$

In particular, φ_v semi-conjugates the hull of ν, $(\Omega_\nu, \mathbb{R}^d, T_1)$, and the hull $(\Omega_H, \mathbb{R}^d, T_2)$ of the Schrödinger operator H.

PROOF. By Lemma 2.22, we have $\nu_\omega * v \in L^\infty_\mathbb{R}(\mathbb{R}^d)$. We will prove that $\omega \in \Omega_\nu \to \nu_\omega * v \in L^\infty_\mathbb{R}(\mathbb{R}^d)$ is continuous in the sense of Theorem 2.21 implying the continuity of φ_v.

For $f \in L^1(\mathbb{R}^d)$, let $f_n \in C_c(\mathbb{R}^d)$, $n \in \mathbb{N}$ be such that $\lim_{n\to\infty}\|f_n - f\|_1 = 0$ and let $v_k \in C_c(\mathbb{R}^d) \cap L^1_{K,r}(\mathbb{R}^d)$, $k \in \mathbb{N}$, be such that $\lim_{k\to\infty}\|v_k - v\|_1 = 0$. Let

$\omega, \omega_l \in \Omega_\nu$, $l \in \mathbb{N}$ be such that $\lim_{l \to \infty} \omega_l = \omega$ and let us define $L_l = \text{supp}(\nu_{\omega_l})$, $L = \text{supp}(\nu_\omega)$. Then

$$|\langle f|(\nu_{\omega_l} * v - \nu_\omega * v)\rangle|$$
$$\leq |\langle f - f_n|(\nu_{\omega_l} * v - \nu_\omega * v)\rangle| + |\langle f_n|(\nu_{\omega_l} * v - \nu_{\omega_l} * v_k) - (\nu_\omega * v - \nu_\omega * v_k)\rangle|$$
$$+ |\langle f_n|\nu_{\omega_l} * v_k - \nu_\omega * v_k)\rangle|.$$

The first term is bounded by:

$$\|(\nu_{\omega_l} * v - \nu_\omega * v)\|_\infty \int |f(y) - f_n(y)|\, dy \leq 2\frac{K}{r^d}\|v\|_1\|f - f_n\|_1,$$

the second term by:

$$\|(\nu_{\omega_l} - \nu_\omega) * (v - v_k)\|_\infty \|f_n\|_1 \leq 2\frac{K}{r^d}\|f_n\|_1\|v - v_k\|_1$$

(we use Lemma 2.22), and the third by:

$$|\langle f_n * \check{v}_k|\nu_\omega - \nu_{\omega_l}\rangle|$$

where $\check{v}(x) = v(-x)$. Since $f_n * \check{v}_k \in \mathcal{C}_c(\mathbb{R}^d)$, this third term converges to zero for $l \to \infty$ by definition of the weak-$*$ topology on the space of measures. The continuity follows from a 3ε argument. The covariance property (2.7) is obvious and is left to the reader. The surjectivity follows from the continuity of φ_v: the image of Ω_ν is compact and the orbit of $R_z(H_\nu)$ is dense in it. Therefore this image coincides with the operator hull. Hence, φ_v semi-conjugates the two dynamical systems. □

COROLLARY 2.2. *Let $\nu \in \text{UD}_r(\mathbb{R}^d)$ and $v \in L^1_{K,r}(\mathbb{R}^d)$ be a real-valued atomic potential with $\text{supp}(v) \subset B(0, r_v)$ for some $r_v \leq r$. Then the map φ_v of Theorem 2.23 is a homeomorphosm.*

PROOF. According to Theorem 2.23 it is enough to prove that the map φ_v is one-to-one. Let $\nu_1, \nu_2 \in \Omega_\nu$ and let x be such that $\nu_1 * v(x) = \nu_2 * v(x) \neq 0$. Then, there exists $(y_1, y_2) \in L^{(\nu_1)} \times L^{(\nu_2)}$ such that $x \in \Delta = B(y_1, r_v) \cap B(y_2, r_v)$. If $y_1 \neq y_2$ then $B(y_2, r_v) \setminus \Delta \neq \emptyset$. Since $r_v \leq r$, $I = B(y_1, r) \cap [B(y_2, r_v) \setminus \Delta] \neq \emptyset$. Let $x' \in I$, then $\forall y \in L^{(\nu_1)}$ $v(x' - y) = 0$, but $v(x' - y_2) \neq 0$, a contradiction. Thus $\nu_1 = \nu_2$. □

Let us conclude this section by the following result.

THEOREM 2.24 ([**Bel6**]). *Let H be the Schrödinger operator $H = \Delta + V$ with $V \in L^\infty(\mathbb{R}^d)$ and let $(\Omega_H, \mathbb{R}^d, T)$ be the hull of H. Then for each z in the resolvent set of H and for every $x \in \mathbb{R}^d$ there is an element $r(z; x) \in C^*(\Omega_H \times \mathbb{R}^d, B)$, such that for each $\omega \in \Omega_H$, we have $\Pi_\omega(r(z; x)) = (z - H_{T^{-x}\omega})^{-1}$.*

COROLLARY 2.3. *Let $\nu \in \text{UD}_r(\mathbb{R}^d)$ and let $(\Omega_\nu, \mathbb{R}^d)$ be the hull of ν. Let H be the Schrödinger operator (2.6). Then for each z in the resolvent set of H, and for every $x \in \mathbb{R}^d$, there is an element $r(z, x) \in C^*(\Omega_\nu \times \mathbb{R}^d, B)$, such that for each $\omega \in \Omega_\nu$, we have $\Pi_\omega(r(z, x)) = (z - H_{T^{-x}\omega})^{-1}$.*

PROOF. By Theorem 2.23 the two hulls are semi-conjugate. Hence $C^*(\Omega_H \times \mathbb{R}^d, B)$ is $*$-isomorphic to a subalgebra of $C^*(\Omega_\nu \times \mathbb{R}^d, B)$. □

Though Theorem 2.24 shows that the C^*-algebra generated by H and its translates lies in $C^*(\Omega_H \times \mathbb{R}^d)$, we do not know under which conditions $C^*(\Omega_H \times \mathbb{R}^d)$ is strictly larger, except whenever Corollary 2.2 holds..

3. Hulls and Thermodynamics

In this section, we will explicitly compute the hull of the few examples of realistic solids we quote in Section 1.2. We develop here a new point of view in comparison with [**Bel2, Bel6**], by emphasizing the rôle of the Gibbs measure in describing the thermal equilibrium of the solid under consideration, as a natural choice of an invariant ergodic measure for taking space averages. This point of view is developed in Section 3.1 below. We follow an approach already suggested by Radin [**Rad**] on the basis of work done in rigorous statistical mechanics [**Rue, Sin**]. Then in Section 3.3, we build the hull for a random distribution of impurities in a crystal. In Section 3.4 we consider the case of a quasicrystal for which the hull is explicitly constructed supplementing various results obtained previously [**BCL, Kel1, Kel2, AP**].

3.1. Thermal Equilibrium for Atoms.

It follows from the definition of $\mathrm{QD}(\mathbb{R}^d)$ (see Section 2) that it is a *Polish space* [**Par**], namely a metrizable space with a metric making it complete. In $\mathrm{QD}(\mathbb{R}^d)$ there are two special families of compact spaces given by the $\mathrm{UD}_r(\mathbb{R}^d)$'s and the $\mathrm{Del}_{(r,R)}(\mathbb{R}^d)$'s with $r > 0$, $R > 0$ and $\overline{\mathrm{UD}(\mathbb{R}^d)} = \overline{\mathrm{Del}(\mathbb{R}^d)} = \mathrm{QD}(\mathbb{R}^d)$ (see Theorem 1.5). This structure motivates the following definitions (for a discussion see Section 1.3). Let $\mathbb{P}$ be a probability measure on $\mathrm{QD}(\mathbb{R}^d)$:

G1. $\mathbb{P}$ is *uniformly discrete*, i.e., it gives probability one to $\mathrm{UD}(\mathbb{R}^d)$.

G1'. $\mathbb{P}$ is *Delone*, i.e., it gives probability one to $\mathrm{Del}(\mathbb{R}^d)$.

G2. $\mathbb{P}$ is translation invariant.

G3. $\mathbb{P}$ is ergodic with respect to the translation group.

PROOF OF THEOREM 1.10. Let $\mathbb{P}$ satisfy G1. One remembers that

$$r_1 > r_2 \implies \mathrm{UD}_{r_2}(\mathbb{R}^d) \subset \mathrm{UD}_{r_1}(\mathbb{R}^d), \forall r_1, r_2 > 0.$$

Let then $(r_n)_{n \in \mathbb{N}}$ be a decreasing sequence of positive numbers, converging to zero. We get $\lim_n \mathbb{P}(\mathrm{UD}_{r_n}(\mathbb{R}^d)) = \mathbb{P}(\lim_n \mathrm{UD}_{r_n}(\mathbb{R}^d)) = \mathbb{P}(\mathrm{UD}(\mathbb{R}^d)) = 1$. Thus, $\forall \epsilon > 0$ there exists $r_0 > 0$ such that $\forall r \leq r_0$, $\mathbb{P}(\mathrm{UD}_r(\mathbb{R}^d)) \geq 1 - \epsilon$. Since $\mathbb{P}$ is $\mathbb{R}^d$-invariant and ergodic, and since $\mathrm{UD}_r(\mathbb{R}^d)$ is an $\mathbb{R}^d$-invariant compact set, for all $r > 0$, $\mathbb{P}(\mathrm{UD}_r(\mathbb{R}^d)) = 0$ or 1. Let $r_0 = \sup\{r > 0; \mathbb{P}(\mathrm{UD}_r(\mathbb{R}^d)) = 1\}$. Since $\bigcap_{r<r_0} \mathrm{UD}_r(\mathbb{R}^d) = \mathrm{UD}_{r_0}(\mathbb{R}^d)$, by σ-additivity, we get $\mathbb{P}(\mathrm{UD}_{r_0}(\mathbb{R}^d)) = 1$. Moreover, by definition of r_0 we also get $\mathbb{P}(\mathrm{UD}_r(\mathbb{R}^d)) = 0$ for $r > r_0$. Hence $\mathbb{P}$ is concentrated on $\mathrm{UD}_{r_0}(\mathbb{R}^d) \setminus \bigcup_{r>r_0} \mathrm{UD}_r(\mathbb{R}^d)$. If $\mathbb{P}$ satisfies condition G1' instead of G1, a similar argument proves that $\mathbb{P}$ is concentrated on $\mathrm{Del}_{(r_0, R_0)}(\mathbb{R}^d) \setminus \bigcup_{r>r_0, R<R_0} \mathrm{Del}_{(r,R)}(\mathbb{R}^d)$ in $\mathrm{QD}(\mathbb{R}^d)$. □

PROOF OF THEOREM 1.11. The proof of Theorem 1.11 is a simple corollary of the following lemma, that is a general property of topological dynamical systems. □

LEMMA 3.1. *Let X be a compact metrizable set, $\mathbb{G}$ a group acting on X, and $\mathbb{P}$ an ergodic and $\mathbb{G}$-invariant probability measure on X. Then* $\mathrm{supp}(\mathbb{P}) = \overline{\mathrm{Orb}(x)}$ *for $\mathbb{P}$-almost every $x \in X$.*

PROOF. In the following the support of $\mathbb{P}$ will be denoted by Ω and the closure of the orbit of $x \in X$ by Ω_x.

(I) Let f be a continuous function on Ω. By Birkhoff's theorem there is a $\mathbb{P}$-measurable set $\Sigma_f \subset \Omega$ such that $\mathbb{P}(\Sigma_f) = 1$ and $\forall y \in \Sigma_f$:

$$\lim_{\Lambda \to \mathbb{R}^d} \frac{1}{|\Lambda|} \int_{a \in \Lambda} f(T^a y) = \int_\Omega d\mathbb{P}(y)\, f(y)$$

Since X is metrizable, so is Ω, so that there is a countable dense subset $(f_n)_{n \in \mathbb{N}}$ in $\mathcal{C}(\Omega)$. If $\Sigma_\infty = \cap_n \Sigma_{f_n}$, the σ-additivity of $\mathbb{P}$ gives $\mathbb{P}(\Sigma_\infty) = 1$.

(II) Now, for $f \in \mathcal{C}(\Omega)$ and $y \in \Sigma_\infty$ we set:

$$I_{(\Lambda, y)}(f) = \int_\Omega d\mathbb{P}(y')\, f(y') - \frac{1}{|\Lambda|} \int_{a \in \Lambda} f(T^a y)$$

If $\epsilon > 0$, let $N \in \mathbb{N}$ be such that $\|f - f_n\| < \epsilon/2$, $\forall n \geq N$. Using $|I_{(\Lambda, y)}| \leq 2\|f\|$, we get $|I_{(\Lambda, y)}(f) - I_{(\Lambda, y)}(f_n)| = |I_{(\Lambda, y)}(f - f_n)| \leq \epsilon$. This result holds for every $\epsilon > 0$. Since $\lim_\Lambda |I_{(\Lambda, y)}(f_n)| = 0$, one has $\limsup_\Lambda |I_{(\Lambda, y)}(f)| = 0$ for all $y \in \Sigma_\infty$.

(III) Ω being the support of $\mathbb{P}$, it is the smallest closed subset of X such that any open subset O of X with $O \cap \Omega \neq \emptyset$ satisfies $\mathbb{P}(O) > 0$. Let $x \in \Omega$ be such that $\Omega_x \subset \Omega$ and $\Omega_x \neq \Omega$. Let then $y \in \Omega \setminus \Omega_x$. Since Ω_x is closed, there is an open set $O \ni y$ such that $O \cap \Omega_x = \emptyset$. Since X is metrizable, there is an open set $O' \subset O$, such that $\overline{O'} \subset O$ and $\mathbb{P}(O') = \epsilon > 0$. By Urysohn's lemma [**RS**], there is a continuous function f, vanishing on $\overline{O'}$, taking the value one on Ω_x and such that $0 \leq f \leq 1$. In particular $0 \leq \int_\Omega d\mathbb{P} f \leq 1 - \epsilon$. Then $\forall x' \in \Omega_x$,

$$(3.1) \qquad 1 = \lim_{\Lambda \to \mathbb{R}^d} \frac{1}{|\Lambda|} \int_{a \in \Lambda} f(T^a x') > \int_\Omega d\mathbb{P}(y)\, f(y).$$

It follows from (I) and (II) that $x \notin \Sigma_\infty$, so that the set of x's for which $\Omega_x \neq \Omega$ has $\mathbb{P}$-measure zero. $\square$

3.2. Diffraction Measure.

PROOF OF THEOREM 1.12. The existence of the limit will be proved by using subadditivity and Birkhoff's ergodic theorem.

(I) Since $\mathbb{P}$ is translation invariant, one has $\rho_\mathbb{P}^{(\Lambda)} = \rho_\mathbb{P}^{(\Lambda + x)}$ for all $x \in \mathbb{R}^d$. Therefore one can always choose Λ of the form $\Lambda_R = (0, R)^{\times d}$. We also remark that if $R' > R$ then

$$\frac{|\Lambda_{R'} \setminus \Lambda_R|}{|\Lambda_R|} \leq d \left(\frac{R'}{R} \right)^{d-1} \frac{R' - R}{R}.$$

Let now f be a continuous function with compact support contained in the ball $B(0, r(f))$. For $\nu \in \mathrm{QD}(\mathbb{R}^d)$, we set $f_\nu(x) = \sum_{y \in \mathbb{R}^d} f(x - y) \nu(\{y\})$. Since $\mathbb{P}$ is supported on $\mathrm{UD}_r(\mathbb{R}^d)$, for $\mathbb{P}$-almost every ν we get (if $L^{(\nu)}$ is the support of ν)

$$\|f_\nu\| = \sup_x \left| \sum_{y \in L^{(\nu)}} f(x - y) \right| \leq \|f\| \left(\frac{r(f)}{r} + 1 \right)^d = C_0(f)$$

where $\|f\| = \sup_{x \in \mathbb{R}^d} |f(x)|$. Hence $|\rho_\nu^{(\Lambda)}(f)| \leq C_0(f)$ uniformly with respect to $R > 0$ and to $\nu \in \mathrm{UD}_r(\mathbb{R}^d)$. This shows in particular that the family $\rho_\nu^{(\Lambda)}$ is compact in $\mathcal{M}(\mathbb{R}^d)$, so that limit points do exist. We also conclude that if

$$S_\nu^{\Lambda,\Lambda'}(f) = \sum_{x\in\Lambda\cap L^{(\nu)}}\left(\sum_{y\in\Lambda'\cap L^{(\nu)}} f(x-y)\right)$$

then, if both Λ and Λ' are hypercubes,

(3.2) $$|S_\nu^{\Lambda,\Lambda'}(f)| \leq C(f)\min(|\Lambda|,|\Lambda'|)$$

with $C(f) = C_0(f)\max((2/r)^d, 1)$. Let R_1, R be real numbers such that $2r(f) < R_1 < R$. Let $a \in \mathbb{N}$ be the integer such that $aR_1 < R \leq (a+1)R_1$. Let $\Lambda, \Lambda_a, \delta\Lambda, \Lambda_1$ denote respectively Λ_R, $(0, aR_1]^d$, $\Lambda_R \setminus \Lambda_a$ and $(0, R_1]^d$. Then

$$S_\nu^{\Lambda,\Lambda}(f) = S_\nu^{\Lambda_a,\Lambda_a}(f) + S_\nu^{\Lambda_a,\delta\Lambda}(f) + S_\nu^{\delta\Lambda,\Lambda_a}(f) + S_\nu^{\delta\Lambda,\delta\Lambda}(f).$$

Using the previous bound (3.2), we get

$$|\rho_\nu^{(\Lambda)}(f) - \rho_\nu^{(\Lambda_a)}(f)| \leq \frac{2^{d+1}dC(f)}{a}.$$

One now decomposes Λ_a into the (disjoint) union of the smaller hypercubes $\Lambda(k) = \Lambda_1 + kR_1$ where $k \in I_a = \{0,\ldots,a-1\}^d$

$$S_\nu^{\Lambda_a,\Lambda_a}(f) = \sum_{k,k'\in I_a} S_\nu^{\Lambda(k),\Lambda(k')}(f).$$

Since $f(x-y) = 0$ whenever $|x-y|_\infty \geq r(f)$, we conclude that $\sum_{k'\in I_a} S_\nu^{\Lambda(k),\Lambda(k')}(f)$ coincides with $S_\nu^{\Lambda(k),\Lambda(k)'}(f)$ if $\Lambda(k)'$ is the set of points in $\mathbb{R}^d$ within a distance at most $r(f)$ from $\Lambda(k)$. Thus if $\delta\Lambda(k) = \Lambda(k)' \setminus \Lambda(k)$, we obtain

$$S_\nu^{\Lambda_a,\Lambda_a}(f) = \sum_{k\in I_a}\left(S_\nu^{\Lambda(k),\Lambda(k)}(f) + S_\nu^{\Lambda(k),\delta\Lambda(k)}(f)\right).$$

This implies, using again the same estimates,

(3.3) $$\left|\rho_\nu^{(\Lambda_a)}(f) - \frac{1}{a^d}\sum_{k\in I_a}\rho_{T^{-kR_1}\nu}^{(\Lambda_1)}(f)\right| \leq \frac{2^{d-1}dr(f)C(f)}{R_1}.$$

In particular, integrating over ν with respect to $\mathbb{P}$ gives

$$|\rho_\mathbb{P}^{(\Lambda_a)}(f) - \rho_\mathbb{P}^{(\Lambda_1)}(f)| \leq \frac{2^{d-1}dr(f)C(f)}{R_1}.$$

There is no loss of generality in assuming that $f \geq 0$. Then, gluing the previous estimates together leads to

$$\limsup_{R\uparrow\infty}\rho_\mathbb{P}^{(\Lambda_R)}(f) \leq \inf_{R_1}\rho_\mathbb{P}^{(\Lambda_1)}(f)$$

showing that both sides converge to some measure $\rho_\mathbb{P}(f)$.

(II) Since the Fourier transform of $\rho_\mathbb{P}^{(\Lambda)}$ is a positive measure for any Λ, one has equivalently

$$\int_{\mathbb{R}^d}d\xi\,\overline{f(\xi)}\int_{\mathbb{R}^d}d\rho_\mathbb{P}^{(\Lambda)}(\eta)\,f(\xi-\eta) \geq 0$$

for all Λ. Therefore this is also true in the limit, showing that the Fourier transform of $\rho_\mathbb{P}$ is a positive measure (Bochner's theorem).

(III) If $\mathbb{P}$ is ergodic the sum $a^{-d} \sum_{k \in I_a} \rho_{T^{-k R_1}\nu}^{(\Lambda_1)}(f)$ converges $\mathbb{P}$-almost surely to $\rho_{\mathbb{P}}^{(\Lambda_1)}(f)$ using Birkhoff's ergodic theorem. It follows from (3.3) that $\mathbb{P}$-almost surely

$$\lim_{R \uparrow \infty} \rho_{\nu}^{(\Lambda_R)}(f) = \rho_{\mathbb{P}}(f).$$
□

3.3. Impurities in a Crystal. Our first example of point sets is given by impurities in a crystal. A typical example in solid state physics is given by doped semiconductors. Let $\mathcal{R}$ be a lattice in $\mathbb{R}^d$. We denote by $\mathfrak{C}$ a discrete $\mathcal{R}$-invariant subset in $\mathbb{R}^d$. Any crystal is represented by such a point set. To each type of atomic species that can be found in the crystal, one associates a *letter* $\mathfrak{a}, \mathfrak{b}, \ldots$. The set $\mathfrak{A}$ of such letters will be called an *alphabet* and will be assumed to be finite. An *atomic configuration* is a sequence $\omega = (\omega_x)_{x \in \mathfrak{C}}$ with $\omega_x \in \mathfrak{A}, \forall x \in \mathfrak{C}$. We will set $\Omega_\infty = \mathfrak{A}^\mathfrak{C}$ for the space of atomic configurations. It is compact if endowed with the product topology. If $a \in \mathcal{R}$ we set $T^a \omega = (\omega_{x-a})_{x \in \mathfrak{C}}$. The maps $T^a (a \in \mathcal{R})$ define an action of $\mathcal{R}$ by homeomorphisms. To each $\omega \in \Omega_\infty$ we associate the following subsets of $\mathfrak{C}$

$$\mathfrak{a} \in \mathfrak{A} \implies L_\omega(\mathfrak{a}) = \{x \in \mathfrak{C}; \omega_x = \mathfrak{a}\}.$$

By construction, $L_\omega(\mathfrak{a})$ is uniformly discrete and $L_\omega(\mathfrak{a}) + a = L_{T^a \omega}(\mathfrak{a})$. Conversely, any partition of $\mathfrak{C}$ by subsets $L_\mathfrak{a}, \mathfrak{a} \in \mathfrak{A}$ defines a unique atomic configuration by setting $\omega^L = (\omega_x)_{x \in \mathfrak{C}}$ with $\omega_x = \mathfrak{a}$ if and only if $x \in L(\mathfrak{a})$. Therefore the *hull* of the family $\{L(\mathfrak{a}); \mathfrak{a} \in \mathfrak{A}\}$ under translations by $\mathcal{R}$ can be identified with the closure of the family of translates of ω^L. Then we set $\forall \omega \in \Omega_\infty, \Omega_\omega = \overline{\{T^a \omega; a \in \mathcal{R}\}}$. Ω_ω is called the $\mathcal{R}$-hull of ω.

Let now $\mathbb{P}$ be an $\mathcal{R}$-invariant ergodic probability measure on Ω_∞. We say that $\mathbb{P}$ is *doping* whenever for any finite subset Λ of $\mathfrak{C}$ and any configuration $\sigma_\Lambda \in \mathfrak{A}^\Lambda$,

$$\mathbb{P}(\{\omega \in \Omega_\infty; \omega \restriction_\Lambda = \sigma_\Lambda\}) > 0.$$

Good examples of such probability measures are provided by *Gibbs states* describing *pure phases* at nonzero temperature for the thermal distribution of impurities [**Rue**]. More precisely, Gibbs measures always have exponential factors which forbid finite volume configurations of zero probability (doping property).

THEOREM 3.2. *Let $\mathbb{P}$ be an $\mathcal{R}$-invariant ergodic and doping probability measure on Ω_∞. Then, for $\mathbb{P}$-almost every $\omega \in \Omega_\infty$, the $\mathcal{R}$-hull Ω_ω coincides with $\Omega_\infty = \mathfrak{A}^\mathfrak{C}$.*

PROOF. Doping property implies that each cylinder of $\mathfrak{A}^\mathfrak{C}$ is contained in $\mathrm{supp}(\mathbb{P})$. Since the set of cylinders is a basis of open sets for the topology of $\mathfrak{A}^\mathfrak{C}$, $\mathrm{supp}(\mathbb{P}) = \Omega_\infty$. Lemma 3.1 then gives the result. □

3.4. The Hull of Quasicrystals. In 1984, Shechtman, Blech, Gratias and Cahn [**SBGC**] identified, in a quenched melt of AlMn alloys, an apparently new object in condensed matter, different from crystals or amorphous bodies in terms of order and symmetry, for which the electron diffraction pattern was point like but five-fold symmetric.

The direct evidence by high resolution electron microscopy imaging [**Boi**] of the existence of Bragg peaks is the signature of a long-range order of the atomic structure. However, this corresponds to a noncrystalline order because the Bragg peaks are not periodically distributed in the reciprocal space, or equivalently because the diffraction pattern exhibits rotation axes of order forbidden by translational symmetry (only rotation axes of order 2, 3, 4 or 6 are consistent with 3-dimensional periodicity). In the case of AlMn, Shechtman et al. showed that the alloy admits an icosahedral point group symmetry, with six 5-fold axis, ten 3-fold axis, fifteen 2-fold axis and the inversion operation. Thus they called this AlMn phase icosahedral (i-AlMn). Other types of quasicrystalline symmetries were soon reported: octagonal, decagonal and dodecagonal phases, with 8-fold, 10-fold and 12-fold order, respectively, with periodicity along the corresponding axis. Very small at the beginning, single quasicrystal grains reach now a size of the order of a centimeter, with very good quality in the i-AlPdMn system (see [**Tsa**] for a review), allowing for precise measurements of physical properties.

The simplest example of a nonperiodic tiling was provided by R. Penrose [**Pen**]. It is built from two types of tiles in the 2D plane, through inflation rules, and exhibits a fivefold symmetry. It was extensively studied by de Bruijn [**Bru1, Bru2**]. But it was recognized only later on by physicists that it is quasiperiodic. However, de Bruijn and also Kramer & Neri [**Kra1, KN**] built examples of quasiperiodic tilings. Most models describing the quasiperiodic order in quasicrystals are based upon the so-called *cut-and-project method*, independently proposed by Duneau & Katz [**DK, KD**], Kalugin, Kitaev & Levitov [**KKL1, KKL2**], Elser [**Els1, Els2**] and Levine & Steinhardt [**LS**]. It was not until 1995 that this method was recognized as equivalent [**Moo**] to the notion of *model sets* provided by Meyer in his thesis work [**Mey1, Mey2**].

The cut-and-project method is defined with spaces and maps as follows

(3.4)
$$\begin{array}{ccccc} \mathbb{R}^d & \xleftarrow{\pi_1} & \mathbb{R}^d \times \mathbb{R}^n & \xrightarrow{\pi_2} & \mathbb{R}^n \\ & & \cup & & \\ \Lambda(M) & \xleftarrow{\pi_1} & \mathcal{R} & \xrightarrow{\pi_2} & M \end{array}$$

where $\mathcal{R} \subset \mathbb{R}^d \times \mathbb{R}^n$ is a lattice and π_1 and π_2 are the projections onto $\mathbb{R}^d$ and $\mathbb{R}^n$, respectively. Furthermore π_1 restricted on $\mathcal{R}$ is *injective* and $\pi_2(\mathcal{R})$ is *dense* in $\mathbb{R}^n$. We call $\mathbb{R}^d$ the *physical space* and $\mathbb{R}^n$ the *internal space*. We assume that π_1 and π_2 are the restriction maps on the corresponding coordinate of $\mathbb{R}^d \times \mathbb{R}^n$. Therefore the setting of a cut-and-project scheme is given by the triple $(\pi_1, \pi_2, \mathcal{R})$. For a subset M in the internal space $\mathbb{R}^n$ we define the corresponding point set in the physical space $\mathbb{R}^d$ as

(3.5) $$\Lambda(M) = \{\pi_1(a); a \in \mathcal{R}, \pi_2(a) \in M\}.$$

M is called the *acceptance domain* (or *window*) of the point set $\Lambda(M)$. For a lattice vector $a \in \mathcal{R}$ we have

(3.6) $$\Lambda(M + \pi_2(a)) = \Lambda(M) + \pi_1(a).$$

DEFINITION 3.3. *A point set L in $\mathbb{R}^d$ is called a model set if there exists a bounded set M with nonempty interior such that $L = \Lambda(M)$.*

PROPOSITION 3.1. *A model set is a Meyer set.*

PROOF. If M is bounded, $\Lambda(M)$ and $\Lambda(M) - \Lambda(M)$ are uniformly discrete. For indeed, let $\Sigma_M = \pi_2^{-1}(M)$ and let $r_1, r_2 > 0$ be such that $M \subset B(0, r_2)$. We set $B = \overline{B(0, r_1)} \times \overline{B(0, 2r_2)}$. If x belongs to $\Lambda(M)$ there is $\xi \in \mathcal{R} \cap \Sigma_M$ such that $\pi_1(\xi) = x$. Then $\overline{B(x, r_1)} \times \overline{B(0, r_2)} \subset \overline{B(\pi_1(\xi), r_1)} \times \overline{B(\pi_2(\xi), 2r_2)} = \xi + B$. If now $y \in \Lambda(M)$, $y \neq x$, either $|y - x|_\infty > r_1$ or there is $b \in \mathcal{R} \cap B$ such that $y - x = \pi_1(b)$. Thus, $|y - x|_\infty \geq \min\{|\pi_1(b)|_\infty; b \in \mathcal{R} \cap B, \pi_1(b) \neq 0\} = r_0 > 0$, because $\mathcal{R} \cap B$ is finite. The same argument applies to $\Lambda(M) - \Lambda(M)$ using the fact that $\mathcal{R} - \mathcal{R} = \mathcal{R}$.

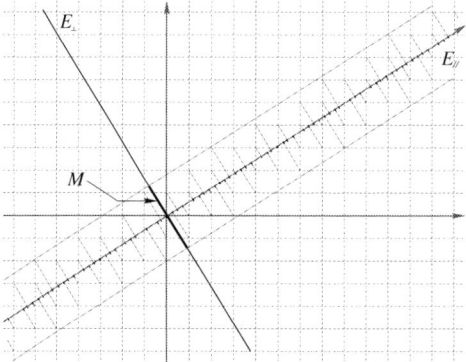

FIGURE 1. Cut-and-project construction

If M has a nonempty interior, $\Lambda(M)$ and $\Lambda(M) - \Lambda(M)$ are relatively dense in $\mathbb{R}^d$. For indeed, one remarks that a point set L in $\mathbb{R}^N$ is relatively dense if and only if there is a closed ball B such that $L + B$ covers $\mathbb{R}^N$. Since $\mathcal{R}$ is a lattice, it is relatively dense and there is such a ball B in $\mathbb{R}^d \times \mathbb{R}^n$. If B_1, B_2 denote its projections, they are compact subsets of $\mathbb{R}^d$ and $\mathbb{R}^n$, respectively, and $B \subset B_1 \times B_2$. By hypothesis, $\pi_2(\mathcal{R})$ is dense in $\mathbb{R}^n$ and since M has a nonempty interior, the subsets $\pi_2(a) - M$ cover $\mathbb{R}^n$, whenever a varies in $\mathcal{R}$. By compactness, there is a finite subset J of $\mathcal{R}$ such that $\pi_2(J) - M$ covers B_2. Let B_3 be the smallest closed ball containing $B_1 - \pi_1(J)$. We claim that $\Lambda(M) + B_3$ covers $\mathbb{R}^d$. For if $x \in \mathbb{R}^d$, there is $a \in \mathcal{R}$ such that $(x, 0) - a \in B$. Thus $\pi_2((x,0) - a) = -\pi_2(a) \in B_2$, so that there is $j \in J$ such that $\pi_2(a + j) \in M$, hence $\pi_1(a + j) \in \Lambda(M)$. On the other hand, $x - \pi_1(a) \in B_1$, implying that $x - \pi_1(a + j) \in B_3$ and therefore that $x \in \Lambda(M) + B_3$. Since $M - M$ has also a nonempty interior, the same argument shows that $\Lambda(M) - \Lambda(M)$ is relatively dense in $\mathbb{R}^d$. □

REMARK 3.4. The cut-and-project construction used by physicists for describing the atomic positions in a quasicrystal is obtained by choosing M as the π_2-projection of a suitable unit cell of $\mathcal{R}$ [**KaGr**] (see Figure 1).

DEFINITION 3.5. Let M be a bounded subset of $\mathbb{R}^n$ with nonempty interior.
  (1) M is called *admissible* if, for every ball $B(x, \epsilon)$ with $\epsilon > 0$ and x in M, there exists a finite family $\{a_1, \ldots, a_p\}$ in $\pi_2(\mathcal{R})$ such that $M \cap (M + a_1) \cap \cdots \cap (M + a_p)$ is a subset of $B(x, \epsilon)$ with nonempty interior.
  (2) M is called *strongly admissible* if it is admissible and if, for every finite family $\{a_1, \ldots, a_p\}$ in $\pi_2(\mathcal{R})$, the set $M \cap (M + a_1) \cap \cdots \cap (M + a_p)$ is either empty or contains a nonempty open ball.
  (3) A model set L is called *admissible* (resp. *strongly admissible*) if there exists an admissible (resp. strongly admissible) set M such that $L = \Lambda(M)$.

EXAMPLE 3.6.   (1) A convex polytope is an admissible set.
  (2) In Figure 1, let $\omega \in \mathbb{R}_+ \setminus \mathbb{Q}$ be the slope of $E_{//}$. Let then $a = -\omega/\sqrt{1+\omega^2}$ and $b = 1/\sqrt{1+\omega^2}$. Then the intervals $M = (a,b), [a,b], (a,b], [a,b]$, comntained in $E_\perp$ are all admissible. However whereas the three first ones are strongly admissible, $[a,b]$ is not.

Let us now consider an admissible model set $L = \Lambda(M)$ and let us denote by $\mathcal{A}_M$ the C^*-algebra generated by the set of functions $\{f \otimes (\chi_M \circ T_n^{\pi_2(a)}); a \in \mathcal{R}, f \in \mathcal{C}_c(\mathbb{R}^d)\}$, where $T_n^{\pi_2(a)}$ denotes the translation in $\mathbb{R}^n$ by $\pi_2(a)$. Here χ_M denotes the characteristic function of M. Let $\mathbb{R}_M^{n+d}$ be the set of characters of $\mathcal{A}_M$ so that, by Gelfand's theorem, $\mathcal{A}_M$ is isomorphic to $\mathcal{C}_0(\mathbb{R}_M^{n+d})$. Since M is admissible, $\mathcal{C}_0(\mathbb{R}^{n+d})$ is a closed subalgebra of $\mathcal{A}_M$. By duality, there is a surjective continuous map $\pi_M \colon \mathbb{R}_M^{n+d} \to \mathbb{R}^{n+d}$. Therefore $\mathbb{R}_M^{n+d}$ can be seen as the completion of $\mathbb{R}^{n+d}$ for a finer topology than the usual one, that will be called the M-topology, in which the sets $\mathbb{R}^d \times M + a$, for $a \in \mathcal{R}$, are open and closed. By construction, for $a \in \mathcal{R}$, the map $x \in \mathbb{R}^d \times \mathbb{R}^n \to x + a \in \mathbb{R}^d \times \mathbb{R}^n$ extends to $\mathbb{R}_M^{n+d}$ by continuity. One sets $\mathbb{T}_M^{n+d} = \mathbb{R}_M^{n+d}/\mathcal{R}$. By construction, for $y \in \mathbb{R}^d \times \{0\}$, the map $x \in \mathbb{R}^d \times \mathbb{R}^n \to x + y \in \mathbb{R}^d \times \mathbb{R}^n$ extends also by continuity to $\mathbb{R}_M^{n+d}$ and commutes with the action of $\mathcal{R}$. Thus it defines an $\mathbb{R}^d$-action $\widehat{T}$ on $\mathbb{T}_M^{n+d}$. Similarly one can define $\mathbb{R}_M^n$ as the set of characters of the C^*-algebra generated by the set $\{\chi_M \circ T_n^{\pi_2(a)}; a \in \mathcal{R}\}$. Since it is a set of idempotents, $\mathbb{R}_M^n$ is totally disconnected and $\mathbb{T}_M^{n+d}$ is transversally totally disconnected. If M is strongly admissible, the the orbit of $(0,0) \in \mathbb{R}^{n+d}$ under the action of $R^d \times \mathcal{R}$ is dense in $\mathbb{R}^{n+d}$. Consequently, the orbit of $(0,0)$ under the action of $\mathbb{R}^d$ ins dense in $\mathbb{T}_M^{n+d}$.

DEFINITION 3.7. *The dynamical system $(\mathbb{T}_M^{n+d}, \mathbb{R}^d)$ is called the pseudo-torus associated to the window M.*

THEOREM 3.8. *Let $L = \Lambda(M)$ be a strongly admissible model set in $\mathbb{R}^d$, and $\nu = \nu^{(L)}$. Then $(\Omega_\nu, \mathbb{R}^d, T)$ is topologically conjugated to $(\mathbb{T}_M^{n+d}, \mathbb{R}^d, \widehat{T})$. This dynamical system is minimal. It is uniquely ergodic, providing M is a Borel set in $\mathbb{R}^n$.*

PROOF. By definition,
$$\nu = \nu^{(\mathcal{R},M)} = \sum_{a \in \mathcal{R}} \chi_M(\pi_2(a))\delta(x - \pi_1(a)).$$

Let $\omega \in \Omega_\nu$. Then there is a sequence $(x_n)_{n \in \mathbb{N}}$ in $\mathbb{R}^d$ such that $\omega = \lim_{n \to \infty} T^{-x_n}\nu$. Then $\forall f \in \mathcal{C}_c(\mathbb{R}^d)$,
$$\omega(f) = \lim_{n \to \infty} \nu(f \circ T^{x_n}) = \lim_{n \to \infty} \sum_{a \in \mathcal{R}} (f \otimes \chi_M) \circ \widehat{T}^a((x_n, 0))$$

Let $f_M := f \otimes \chi_M$ and let $\hat{f}_M := \sum_{a \in \mathcal{R}} f_M \circ \widehat{T}^a$. Since f_M has compact support, the sum defining $\hat{f}_M$ converges uniformly. Clearly, $\hat{f}_M \in \mathcal{A}_M = \mathcal{C}_0(\mathbb{R}_M^{n+d})$ and since $\hat{f}_M$ is invariant under the action of $\mathcal{R}$, $\hat{f}_M \in \mathcal{C}(\mathbb{T}_M^{n+d})$. Consequently $\hat{f}_M(x_n, 0)$ converges to $\omega(f)$. Since f is arbitrary in $\mathcal{C}_c(\mathbb{R}^d)$, $\widehat{T}^{(x_n,0)}(0)$ converges in $\mathbb{T}_M^{n+d}$. Let Φ_ω denote this limit point. Hence $\omega(f) = \hat{f}_M(\Phi_\omega)$ for all $f \in \mathcal{C}_c(\mathbb{R}^d)$.

From this formula it follows immediately that the map Φ is well defined because the family $\{\hat{f}_M; f \in \mathcal{C}_c(\mathbb{R}^d)\}$ separates the points of $\mathbb{T}_M^{n+d}$. Moreover it also shows that Φ is continuous and one-to-one. Since $\overline{\text{Orb}(0)} = \mathbb{T}_M^{n+d}$, the map is also surjective. At last the same formula shows that Φ conjugates the two actions of $\mathbb{R}^d$.

Since $\pi_2(\mathcal{R})$ is dense in $\mathbb{R}^n$ and since M is strongly admissible, actually every orbit in $\mathbb{T}_M^{n+d}$ is dense, implying minimality [**KH**]. Moreover, since any function of the form $\hat{f}_M$ with $f \in \mathcal{C}_c(\mathbb{R}^d)$ is a Borel function over $\mathbb{T}^{n+d}$, any probability on

Ω_ν is uniquely defined by a probability on $\mathbb{T}^{n+d}$. Since $\mathbb{T}^{n+d}$ is a compact group, its Haar measure is the unique ergodic and $\mathbb{R}^d$-invariant measure, showing that the hull is uniquely ergodic. □

PROPOSITION 3.2. *Let $\mathcal{O}$ be a bounded subset of $\mathbb{R}^n$. Then $\pi_M^{-1}(\mathcal{O})$ is a relatively compact subset of $\mathbb{R}_M^n$.*

The proof requires the following two lemmas.

LEMMA 3.9. *Let X be a locally compact space and let $\chi \in \mathcal{C}_0(X)$ satisfy $\chi^2 = \chi$. Then the support of χ is open and compact.*

PROOF. Since $\chi^2 = \chi$, for any $x \in X$, $\chi(x) \in \{0,1\}$. Hence, its support is both open and closed. Moreover, since χ vanishes at infinity, there is a compact subset $K \subset X$ such that, whenever $x \notin K$, then $\chi(x) = 0$. □

LEMMA 3.10. *Let $x \in \mathbb{R}^n$ and $\epsilon > 0$. Then there is a bounded set $\mathcal{U}_x$ of the form $\mathcal{U}_x = \bigcap_{i=1}^p (M + a_i)$, containing x in its interior and contained in the ball $B(x, \epsilon)$.*

PROOF. M being admissible, there are $b_1, \ldots, b_p$ in $\pi_2(\mathcal{R})$ such that $\mathcal{V}_x = \bigcap_{i=1}^p (M + b_i)$ has a nonempty interior and is contained in the ball $B(x, \epsilon/2)$. Therefore there is $y \in B(x, \epsilon/2)$ and $\epsilon_1 < \epsilon/2$ with $B(y, \epsilon_1) \subset \mathcal{V}_x$. Since $\pi_2(\mathcal{R})$ is dense in $\mathbb{R}^n$, there is $a \in \pi_2(\mathcal{R})$ such that $x - a \in B(y, \epsilon_1)$. Thus $|a|_\infty < \epsilon$ so that $\mathcal{U}_x = \mathcal{V}_x + a$ gives the required set. □

PROOF OF THE PROPOSITION 3.2. Since $\mathcal{O}$ is bounded, its closure is compact. The family $(\mathcal{U}_x)_{x \in \overline{\mathcal{O}}}$ is such that the interiors of its elements cover $\overline{\mathcal{O}}$. There is therefore a finite family $x_1, \ldots, x_l$ in $\overline{\mathcal{O}}$ such that $\overline{\mathcal{O}} \subset \bigcup_{i=1}^l \mathcal{U}_{x_i}$. In particular

$$\chi_{\overline{\mathcal{O}}} \leq \sum_{i=1}^l \chi_{\mathcal{U}_{x_i}} = F \in \mathcal{A}_M.$$

Clearly, F takes values in $\mathbb{N}$. So that if g is the continuous bounded function on $\mathbb{R}$ defined by $g(s) = 0$ for $s \leq 0$, $g(s) = s$ for $0 \leq s \leq 1$ and $g(s) = 1$ for $s \geq 1$, the function $g \circ F = \chi$ belongs to $\mathcal{A}_M$ and satisfies $\chi^2 = \chi$. Thanks to Lemma 3.9 the support K of χ is open and compact. Since $\overline{\mathcal{O}}$ is closed, it is closed for the M-topology too, its characteristic function $\chi_{\overline{\mathcal{O}}}$ is Borelian in this topology so that it extends as the characteristic function of $\pi_M^{-1}(\overline{\mathcal{O}})$ showing that $\pi_M^{-1}(\mathcal{O}) \subset K$. □

Since $\pi_2(\mathcal{R})$ is dense in $\mathbb{R}^n$, the π_2-image of the generators of $\mathcal{R}$ on $\mathbb{R}^n$ give a generating set $\{e_1, \ldots, e_{n+d}\}$ of vectors. After relabeling if necessary, they can be chosen such that $\{e_1, \ldots, e_n\}$ are linearly independent in $\mathbb{R}^n$ leading to a splitting of the $\mathcal{R}$-action on $\mathbb{R}^n$ into a $(\mathbb{Z}^n \times \mathbb{Z}^d)$-action. By construction, one gets such a splitting of the $\mathcal{R}$-action on $\mathbb{R}_M^n$, too. One sets $\mathbb{T}^n = \mathbb{R}^n/\mathbb{Z}^n$ and $\mathbb{T}_M^n = \mathbb{R}_M^n/\mathbb{Z}^n$. Since $\{e_1, \ldots, e_n\}$ are linearly independent vectors, $\mathbb{T}^n$ is a compact set. Since the $\mathbb{Z}^n$-action and the $\mathbb{Z}^d$-action commute on $\mathbb{R}_M^n$, $\mathbb{T}_M^n$ is canonically endowed with a $\mathbb{Z}^d$-action.

PROPOSITION 3.3. *$\mathbb{T}_M^n$ is a compact, totally disconnected, metrizable space, endowed with a $\mathbb{Z}^d$-action. Moreover $\mathcal{C}(\mathbb{T}_M^n) \rtimes \mathbb{Z}^d$ and $\mathcal{C}_0(\mathbb{R}_M^n) \rtimes \mathbb{Z}^{n+d}$ are Morita equivalent.*

PROOF. Let $\mathbb{D}^n$ be the open fundamental domain for the $\mathbb{Z}^n$-action defined by $\mathbb{D}^n = \{\sum_{i=1}^n s_i e_i; 0 < s_i < 1\}$. Then the closure $\mathbb{D}_M^n$ of $\pi_M^{-1}(\mathbb{D}^n)$ in $\mathbb{R}_M^n$ is a fundamental domain for the $\mathbb{Z}^n$-action. Since $\mathbb{D}^n$ is relatively compact in $\mathbb{R}^n$, by Proposition 3.2, $\mathbb{D}_M^n$ is a compact set in $\mathbb{R}_M^n$. Let $f_0 \in \mathcal{C}_c(\mathbb{R}_M^n)$ be such that $f_0(x) = 1$, $\forall x \in \mathbb{D}_M^n$, and $0 \leq f_0 \leq 1$ elsewhere. Let $\hat{f}_0 := \sum_{a \in \mathbb{Z}^n} f_0 \circ T_n^a$. $\hat{f}_0 \in \mathcal{C}(\mathbb{T}_M^n)$. One gets $\hat{f}_0(x) \neq 0$, $\forall x \in \mathbb{T}_M^n$. Thus $\hat{f}_0$ is invertible in $\mathcal{C}(\mathbb{T}_M^n)$ and $\mathbb{T}_M^n$ is a compact set.

Since the vectors $\{e_1, \ldots, e_n\}$ are linearly independent, the $\mathbb{Z}^n$-action on $\mathbb{R}^n$ is free and wandering, namely there is no orbit reduced to a single point, and for each compact set K in $\mathbb{R}^n$ the set $\{g \in \mathbb{Z}^n, g.K \cap K \neq \emptyset\}$ is finite. Since π_M is a continuous surjection, the $\mathbb{Z}^n$-action on $\mathbb{R}_M^n$ is free and wandering, too. Then $\mathcal{C}(\mathbb{T}_M^n)$ and $\mathcal{C}_0(\mathbb{R}_M^n) \rtimes \mathbb{Z}^n$ are Morita equivalent [**Rie2**] and such an equivalence can be extended to the $\mathbb{Z}^d$-crossed products. □

Up to now, M has only been assumed to be admissible. The following property actually holds in quasicrystals:

DEFINITION 3.11. A polytope in $\mathbb{R}^n$ is said $\mathcal{R}$-*compatible* if its vertices belong to $\pi_2(\mathcal{R})$.

If the acceptance domain M is an $\mathcal{R}$-compatible polytope, let $F_1, \ldots, F_p$ be the hyperplanes of $\mathbb{R}^{n+d}$ parallel to the maximal faces of $\mathbb{R}^n \times M$. For each $j \in \{1, \ldots, p\}$, let $u_j \in \mathbb{R}^{n+d}$ be a unit vector perpendicular to F_j so as to define $F_j^+ = \{x \in \mathbb{R}^{n+d}; \langle u_j | x \rangle \geq 0\}$ and $F_j^- = \mathbb{R}^{n+d} \setminus F_j^+$. Let then $\mathcal{F}$ be the family of affine hyperplanes $F_j + a$ with $j \in \{1, \ldots, p\}$ and $a \in \pi_2(\mathcal{R})$. $\mathbb{R}^d \times \mathbb{R}^n$ is then endowed with the coarsest topology for which, given any $F \in \mathcal{F}$, the closed half-space F^+ is both closed and open. It will be called the $\mathcal{F}$-topology. The same construction can be performed in $\mathbb{R}^n$. Let $\mathbb{R}_\mathcal{F}^{n+d}$ and $\mathbb{R}_\mathcal{F}^n$ be the completions of $\mathbb{R}^d \times \mathbb{R}^n$ and $\mathbb{R}^n$ with this topology, respectively. In much the same way, $\mathbb{T}_\mathcal{F}^{n+d} = \mathbb{R}_\mathcal{F}^{n+d}/\mathcal{R}$ is well defined and can be endowed with an $\mathbb{R}^d$-action. Let us assume now that M equals the intersection of all the half-spaces $F_j^\sigma + a$, $j \in \{1, \ldots, p\}$, $\sigma \in \{\pm\}$ and $a \in \mathcal{R}$, such that $F_j^\sigma + a$ contains the interior of M. The following proposition is easy to prove:

PROPOSITION 3.4. *If M is $\mathcal{R}$-compatible, the M-topology and the $\mathcal{F}$-topology are equivalent on $\mathbb{R}^{n+d}$. In particular $\mathbb{T}_\mathcal{F}^{n+d} = \mathbb{T}_M^{n+d}$.*

REMARK 3.12. An alternative description of this pseudo-torus is proposed in [**Le, FHK1**] under very general hypotheses on M.

REMARK 3.13. In a Meyer set obtained through the cut-and-project method, any bounded pattern repeats itself infinitely often. More generally, each Meyer set is a Delone set of finite type. Thanks to Corollary 2.1, this implies that the canonical transversal is totally disconnected. This can also be seen from the particular topology described just before.

4. Tight-Binding Representation

The Schrödinger operator H describing the electronic motion (see Section 2.7), is an unbounded self-adjoint operator on $L^2(\mathbb{R}^d)$. The unboundedness introduces useless technical difficulties in most cases. For indeed, only those electrons with an energy close to the Fermi level really matter for the description of the electronic property of the material in which they evolve. Restricting H to such an energy window leads to the so-called *tight-binding representation*, at least if the atomic orbitals contributing to such energy range are sufficiently well localized on the atom. In the physics literature, such a construction leads to the *tight-binding approximation* [**AM**], obtained by truncating the effective Hamiltonian to a finite set of hopping terms, such as the nearest neighbors.

Let H be defined by (2.6), and $v \in L^1(\mathbb{R}^d)$ be the atomic potential describing the electron-ion interactions. Then v is attractive at large distance in order to bind the electrons, and it vanishes at infinity. Consequently, the atomic Hamiltonian $H_{\mathrm{at}} = \Delta + v$ has bound states. Let E be the energy of such a bound state and let ψ be the corresponding wave function. ΔE will denote the energy gap between E and the energy level closest to E. Since the atomic potential vanishes at infinity, the wave function ψ decreases exponentially fast with a rate of decay given by $|E|$ [**Agm**]. Therefore, if $\nu \in \mathrm{UD}_r(\mathbb{R}^d)$ with r sufficiently large, the functions $(\psi_y(x)) = (\psi(x-y)_{y \in L^{(\nu)}})$ are approximate eigenfunctions of H, and E is the corresponding approximate eigenvalue. More precisely, let M^ν be the matrix indexed by $L^{(\nu)}$ and defined by:

$$(4.1) \qquad M^\nu_{y,y'} = \langle \psi_y | \psi_{y'} \rangle = \delta_{yy'} + O(e^{-E|y-y'|_\infty}) \quad \text{as } r \to \infty.$$

This matrix, seen as an operator acting on $\ell^2(L^{(\nu)})$, is nonnegative and there is an r_0 such that if $r > r_0$ then $\|M^\nu - 1\| < 1$. Thus, $(M^\nu)^{-1/2}$ is well defined leading to the following lemma, the proof of which being left to the reader:

LEMMA 4.1. *Let $\mathcal{H}$ be the the subspace of $L^2(\mathbb{R}^d)$ spanned by the ψ_y's. Let $(\phi_y)_{y \in L^{(\nu)}}$ be the family of functions in $\mathcal{H}$ defined by:*

$$(4.2) \qquad \phi_y = \sum_{y' \in L^{(\nu)}} ((M^\nu)^{-1/2})_{y',y} \psi_{y'}$$

Then, $(\phi_y)_{y \in L^{(\nu)}}$ is an orthogonal basis of $\mathcal{H}$. In addition, $\|\phi_y - \psi_y\| = O(e^{-|E|r})$ as $r \to \infty$.

Let now P be the orthogonal projection onto the subspace $\mathcal{H}$ and $Q = 1 - P$. For r large enough the spectrum of PHP is contained in the open interval $I_E = (E - \frac{1}{2}\Delta E, E + \frac{1}{2}\Delta E)$ which does not meet the spectrum of $(1-P)H(1-P)$. The part of the spectrum of H contained in I_E can be investigated through the projection method, known as the Schur complement formula or Feshback method [**Fes1, Fes2, Bel3**] which consists in defining the following energy dependent operator:

$$(4.3) \qquad H_{\mathrm{eff}}(z) = PHP + PHQ \frac{1}{z - QHQ} QHP,$$

$H_{\mathrm{eff}}(z)$ is the effective Hamiltonian associated to H. The following theorem describes the relation between H and $H_{\mathrm{eff}}(z)$ corresponding to the energy window I_E.

THEOREM 4.2 ([**Bel2**]). *Let $z \in \mathbb{C} \backslash \mathrm{spec}(QHQ)$. Then z belongs to the spectrum of $H_{\mathrm{eff}}(z)$ if and only if z belongs to the spectrum of H. λ is an eigenvalue of H with eigenvector φ, if and only if λ is an eigenvalue of $H_{\mathrm{eff}}(\lambda)$ with eigenvector $P\varphi$. φ can be recovered from the formula:*

$$(4.4) \qquad Q\varphi = \frac{1}{\lambda - QHQ} QHP\varphi.$$

Since the two parts of the spectrum of H are separated by a gap, $H_{\text{eff}}(z)$ is norm-analytic in z near E. It follows from (4.3) that if P is close enough to an eigenprojection of H, the second term is a second-order perturbative term. If λ_0 is the first order approximation for an eigenvalue, (4.3) can be systematically expanded in powers of $z - \lambda_0$, providing a method for calculating the eigenvalues and eigenfunctions.

The exponential decay of ψ leads to the boundedness of QHP, and $PHQ \times (z - QHQ)^{-1}QHP$ is bounded by $O(e^{-2|E|r})$ as $r \to \infty$, uniformly for z in any compact subset of I_E. Consequently:

LEMMA 4.3 ([**Bel2**]). *The matrix of $H_{\text{eff}}(z)$ in the basis of the $(\phi_y)_{y \in L^{(\nu)}}$ is diagonally dominant, namely*:

$$(4.5) \qquad \langle \phi_y | H_{\text{eff}}(z) \phi_{y'} \rangle = O(e^{-2|E||y-y'|}) \quad \text{as } r \to \infty.$$

REMARK 4.4. $(M^\nu)_{y',y}$ is a function of $y - y'$ only; thus there is a function $\phi \in L^2(\mathbb{R}^d)$ such that $\forall y \in L^{(\nu)}$, $\phi_y = U(y)\phi$ (where $U(y)$ is the operator of translation by y), so that $\langle \phi_y | H_\omega(z) \phi_{y'} \rangle = \langle \phi | H_{T^{-y}\omega}(z) U(y - y') \phi \rangle$.

The unbounded operator H, acting on $L^2(\mathbb{R}^d)$, has been replaced by a bounded operator acting on $\ell^2(L^{(\nu)})$, the Hilbert space attached to the atomic sites. The tight-binding approximation is obtained by replacing all matrix elements between sites farther than a given distance ℓ from zero. Choosing ℓ large enough, this gives a good approximation since the error is of order $O(e^{-|E|\ell})$. The discrete hull for $H_{\text{eff}}(z)$ can be directly defined only whenever $L^{(\nu)}$ is invariant by some lattice $\mathcal{R}$ of translations. In the general case of a tight-binding Hamiltonian on a point set, the hull can be defined through using the notion of groupoid G_Υ of the transversal Υ (see Proposition 2.2 and Theorem 2.23).

4.1. The C^*-Algebra of a Groupoid. The construction of the C^*-algebra of a groupoid can be found in [**Ren**]. In the special case given here, it goes as follows. Thanks to Proposition 2.2, the groupoid of the transversal can be defined as $G_\Upsilon = \{(\omega, a) \in \Upsilon \times \mathbb{R}^d, a \in L^{(\omega)}\}$. Since $L^{(\omega)}$ is uniformly discrete and depends continuously on ω, G_Υ is a closed subspace of $\Omega \times \mathbb{R}^d$ (Remark 2.15). Let $\mathcal{C}_c(G_\Upsilon)$ be the vector space of continuous functions on G_Υ with compact support. It is endowed with the structure of a *-algebra in the following way:

$$(4.6) \qquad fg(\omega, a) = \sum_{t \in L^{(\omega)}} f(\omega, t) g(T^{-t}\omega, a - t),$$

$$(4.7) \qquad f^*(\omega, a) = \overline{f(T^{-a}\omega, -a)}.$$

For $\omega \in \Upsilon$, we denote by Π_ω the *-representation of $\mathcal{C}_c(G_\Upsilon)$ on $\ell^2(L^{(\omega)})$ given by [**Co1, Co3**]:

$$(4.8) \qquad \Pi_\omega(f)\psi(a) = \sum_{t \in L^{(\omega)}} f(T^{-a}\omega, t - a)\psi(t), \quad a \in L^{(\omega)}.$$

These representations satisfy also a covariance condition. Namely, for $\gamma = (\omega, t) \in G_\Upsilon$, let $U_\omega(t)$ be the unitary operator from $\ell^2(L^{(T^{-t}\omega)})$ into $\ell^2(L^{(\omega)})$ given by

$$U_\omega(t)\psi(y) = \psi(y - t).$$

It satisfies:

$$(4.9) \qquad U_\omega(t)\Pi_{T^{-t}\omega}(f) = \Pi_\omega(f)U_\omega(t).$$

Then a C^*-norm is given by

(4.10) $$\|f\| = \sup_{\omega \in \Upsilon} \|\Pi_\omega(f)\|.$$

$C^*(G_\Upsilon)$ will denote the completion of $\mathcal{C}_c(G_\Upsilon)$ under this norm. For $\omega \in \Upsilon$ let H_ω be given by

(4.11) $$H_\omega = H_0 + \sum_{y \in L^{(\omega)}} v(.-y) = H_0 + \nu_\omega * v,$$

(where v is an atomic potential, see the precise definition in Section 2.7). Then, according to Lemma 4.3 and Remark 4.4, the effective Hamiltonian for H_ω acts on $\mathcal{H}_\omega = \overline{\mathrm{Span}\{U(y)\phi; y \in L^{(\omega)}\}} \cong \ell^2(L^{(\omega)})$. Thanks to its covariance and continuity properties, it satisfies:

PROPOSITION 4.1 ([**Bel2**]). *Let $h(z)$ be the function on G_Υ defined by:*

$$h(z)(\omega, y) = \langle \phi | H_\omega(z) U(y) \phi \rangle,$$

where ϕ is defined in Remark 4.4. Then $h(z)$ belongs to $C^*(G_\Upsilon)$ for any $z \in I_E$. Furthermore, the matrices of $H_\omega(z)$ and of $\Pi_\omega(h(z))$, with respect to the basis $(\phi_y)_{y \in L^{(\omega)}}$, coincide.

Therefore, $C^*(G_\Upsilon)$ can be used to describe the spectrum of H locally. The following connection between the discretized and the continuum algebras should be noted:

THEOREM 4.5 ([**Rie1**]). *The C^*-algebras $\mathcal{B} = C^*(\Omega \times \mathbb{R}^d)$ and $\mathcal{C} = C^*(G_\Upsilon)$ are Morita equivalent, namely $\mathcal{B} \otimes \mathcal{K}$ and $\mathcal{C} \otimes \mathcal{K}$ are isomorphic (not in a canonical way).*

The proof consists in constructing a Hilbert $\mathcal{B}$-$\mathcal{C}$-bimodule implying an isomorphism of the stabilized algebras of $\mathcal{B}$ and $\mathcal{C}$. Another strong consequence of the existence of this bimodule will be seen in Theorem 6.1.

4.2. Physical Models on Quasilattices. This section is devoted to the description of the tight-binding representations for quasicrystals. As described in Section 3.4, the set of atomic positions is an admissible model set L that will be called a *quasilattice*. Let $\mathcal{R}$ be a lattice in $\mathbb{R}^{n+d}$, let $M \in \mathbb{R}^n$ be an *admissible* acceptance domain for a cut-and-project scheme $(\pi_1, \pi_2, \mathcal{R})$ and $L = \Lambda(M)$ be the corresponding model set.

Then π_1 induces an isomorphism between the physical Hilbert space $\mathcal{H} = \ell^2(L)$ and $\ell^2(\mathcal{S}_M)$, where $\mathcal{S}_M$ is the strip $\mathcal{S}_M = \{a \in \mathcal{R}; \pi_2(a) \in M\}$. By definition π_1 is injective on $\mathcal{R}$. Therefore $\mathcal{H}$ can be seen as a subspace of the large Hilbert space $\mathcal{E} = \ell^2(\mathcal{R})$. If χ_M denotes the characteristic function of $\mathcal{S}_M$ in $\mathcal{R}$, the orthogonal projection onto $\mathcal{H}$ in $\mathcal{E}$ can be identified with the multiplication operator by χ_M also denoted by χ_M. In general for $K \subset \mathbb{R}^n$ and $b \in \mathcal{R}$, χ_{K+b} will denote the projection onto $\ell^2(\{a \in \mathcal{R}, \pi_2(a - b) \in K\})$.

The C^*-algebra generated by the translation operators $T_a (a \in \mathcal{R})$ contains all $\mathcal{R}$-invariant observables. It is isomorphic to the algebra $\mathcal{C}(\mathbb{T}^{d+n})$ of continuous functions on the $(n + d)$-dimensional torus. Physical models on L will be rather obtained by restricting the action of $\mathcal{R}$ to $\mathcal{H}$. More precisely, the quasitranslation operators are defined by $S_a = \chi_M T_a \chi_M$ for $a \in \mathcal{R}$ and the C^*-algebra they generate will be denoted by $\mathcal{Q}(n, d, M)$. The operators S_a are partial isometries which do not commute. Thus, since $\mathcal{Q}(n, d, M)$ is unital, it can be seen as the set of continuous functions on a noncommutative compact space. The commutation rules for the partial isometries are given as follows:

$$S_{a_1} S_{a_2} \cdots S_{a_k} = \chi_{M(\underline{a})} T_{a_1 + \cdots + a_n} \chi_{M(\bar{a})} \tag{4.12}$$

where $\underline{a} = (a_1, \ldots, a_n)$, $M(\underline{a}) = M \cap (M + a_1) \cap \cdots \cap (M + a_1 + \cdots + a_n)$, and $M(\bar{a}) = M \cap (M - a_n) \cap \cdots \cap (M - a_1 - \cdots - a_n)$. In particular, we then obtain $S_a = \chi_{M \cap (M+a)} T_a \chi_{M \cap (M-a)}$ and

$$S_a S_a^* = \chi_{M \cap (M+a)}, \quad S_a^* S_a = \chi_{M \cap (M-a)}. \tag{4.13}$$

An integral and differential calculus can be defined on $\mathcal{Q}(n, d, M)$ [**Bel6**].

Let $(\Omega, \mathbb{R}^d)$ be the hull of the corresponding quasilattice $L = \Lambda(M)$. The topology of Ω is given by Theorem 3.8. Let $\mathbb{R}^n$ be the internal space.

THEOREM 4.6 ([**BCL**]). *The algebras $\mathcal{Q}(n, d, M)$ and $\mathcal{C}(\mathbb{R}_M^n) \rtimes \mathbb{Z}^{n+d}$ are Morita equivalent.*

THEOREM 4.7. *Let Υ be the canonical transversal of $(\Omega, \mathbb{R}^d)$ defined in Proposition 2.2. Then $\mathcal{Q}(n, d, M)$ is isomorphic to $C^*(G_\Upsilon)$, where $C^*(G_\Upsilon)$ is the groupoid C^*-algebra constructed in Section 4.1.*

PROOF. Let be $L = L^{(\nu)} = \Lambda(M)$. We consider $\mathcal{Q}(n, d, M)$ as acting on $\ell^2(L)$. Since L has finite type we know by Corollary 2.1 that Υ is totally disconnected. Furthermore, since $\Omega_R(\nu)$, defined in Proposition 2.3, is finite for every $R > 0$, we conclude that G_Υ is also totally disconnected and therefore $C_c(G_\Upsilon)$ is already generated by the locally constant functions, i.e., by the characteristic functions of clopen (closed and open) sets. Consider $T \subset \Omega_R(\nu)$ for some $R > 0$ and $x \in T$, then we define the set $A(T, x) = \{(\omega, x) \in G_\Upsilon, T \subset L(\omega)\}$ which is clopen (cf. Section 2.6). This type of clopen sets form a basis for the topology of G_Υ. Actually, every clopen set in G_Υ is given by a finite, pairwise disjoint union of such clopen sets. Let $e_T \otimes e_x(\omega, t)$ be the characteristic function of the set $A(T, x)$.

For simplicity we assume $0 \in \Lambda(M) = L$ (we have $\ell^2(L) \cong \ell^2(L(T^x\nu))$ for every $x \in \mathbb{R}^d$). Let Π_ν be the representation given by (4.8). By construction, the orbit of ν is dense in Ω_ν and therefore this representation is faithful. Let $a \in \mathcal{R}$ such that $y_a = \pi_1(a) \in L - L$. Then

$$S_a = \Pi_\nu(e_{\{0, y_a\}} \otimes e_{y_a}), \quad \chi_{M \cap (M+a)} = \Pi_\nu(e_{\{0, -y_a\}} \otimes e_0), \tag{4.14}$$

as operators in $B(\ell^2(L))$. For $a \in \mathcal{R}$ with $y_a \notin L - L$ we have $S_a = 0$. On the other hand, let $T = \{y_0, \ldots, y_n\} \subset \Omega_R(\nu)$ for some $R > 0$. Then there exist $a_0, \ldots, a_n \in \mathcal{R}$ with $y_0 = \pi_1(a_0), \ldots, y_n = \pi_1(a_n)$ and we have

$$\Pi_\nu(e_T \otimes e_{y_0}) = S_{a_n}^* S_{a_n} \cdots S_{a_2}^* S_{a_2} S_{a_1}^* S_{a_1} S_{a_0}. \tag{4.15}$$

Therefore we have $\Pi_\nu(C_c(G_\Upsilon)) = \mathcal{Q}(n, d, M)$. □

EXAMPLE 4.8. Here is the situation for $d = 1$. Then the lattice L is contained in a line in $\mathbb{R}^{1+n}$. Choosing an orientation on it and an origin in L, the points of L can uniquely be labeled in increasing order by integers. Hence $\mathcal{H} = \ell^2(L)$ is isomorphic to $\ell^2(\mathbb{Z})$. In the specific case where $n = 1$, let ω be the slope of the line containing L and let $\alpha = \omega/(\omega + 1)$ (see Fig. 1). The acceptance domain M is then given by the projection of the unit square, namely the interval $[-\omega/\sqrt{1 + \omega^2}, 1/\sqrt{1 + \omega^2})$. The labeling of points of L can be explicitly computed from the points in the strip $\mathcal{S}$ through the map $a = (m, n) \in \mathcal{S} \to l(a) = m + n - n_0 \in \mathbb{Z}$. The inverse map is then given by $n = n_0 + 1 + [l\alpha - \theta]$ and $m = l - 1 - [l\alpha - \theta]$, provided $\theta = \alpha - (1-\alpha)\eta \in [0,1]$ (where $[l\alpha - \theta]$ denotes the integer part of $l\alpha - \theta$).

Then, setting $T = S_1 + S_2$, T is unitary and is represented by the translation by one in $\mathcal{H}$. On the other hand, it can be checked that $\chi_{M \cap (M-e(2))} = S_2^* S_2$ is represented by the operator $\chi_{\alpha,\theta}$ of multiplication by $\chi_{[1-\alpha,1)}(l\alpha - \theta)$ in $\mathcal{H}$ (where $\chi_{[1-\alpha,1)}$ is the characteristic function of $[1-\alpha,1)$) on the unit circle. Moreover the algebra $\mathcal{Q}(2,1,M) = \mathcal{Q}_\alpha$ is generated by T and $\chi_{\alpha,\theta}$. The Abelian C^*-algebra generated by the family $\chi_{\alpha,n} = T^n \chi_{\alpha,\theta} T^{-n}$ is thus isomorphic to the algebra $\mathcal{C}_\alpha$ generated by the characteristic functions $\chi_{[1+(n-1)\alpha,1+n\alpha)}$ of the interval $[1+(n-1)\alpha, 1+n\alpha)$ of the unit circle. $\mathcal{Q}_\alpha$ appears therefore as the crossed product of $\mathcal{C}_\alpha$ by the rotation on the unit circle. In particular, if α is an irrational number, $\mathcal{C}_\alpha$ contains all continuous functions on the unit circle, and consequently $\mathcal{Q}_\alpha$ contains the irrational rotation algebra $\mathcal{A}_\alpha$.

The Kohmoto model is the archetype of a discrete Hamiltonian which belongs to this algebra:

$$(4.16) \quad H_x \psi(n) = \psi(n+1) + \psi(n-1) + V\chi_{[1-\alpha,1)}(n\alpha - x)\psi(n), x \in \mathbb{T}, \psi \in \ell^2(\mathbb{Z}).$$

The algebra is generated by the characteristic function $\chi_{[1-\alpha,1)}$ of the interval $[1-\alpha,1)$ on the torus $\Omega = \mathbb{T}$, with $T = R_\alpha$ the rotation by $\alpha \in [0,1] \setminus \mathbb{Q}$, and $\mathbb{P}$ is the normalized Lebesgue measure. As an admissible model set, $(\mathbb{T}, R_\alpha)$ is uniquely ergodic.

5. General Gap Labeling Theorem

5.1. Integrated Density of States and Shubin's Formula. Let H be the Schrödinger operator, acting on the Hilbert space $L^2(\mathbb{R}^d)$, given by (2.4), where $V \in L^\infty(\mathbb{R}^d)$ is real. For any rectangular box Λ, let H_Λ denote the restriction of H to Λ with some boundary conditions (e.g., Dirichlet or periodic boundary conditions). Since H is elliptic and Λ is compact, the spectrum of H_Λ is discrete and bounded from below. Let $N_\Lambda(E)$ be the number of eigenvalues of H_Λ smaller than or equal to E, counted with their multiplicities. Thanks to the homogeneity of H, $N_\Lambda(E)$ should not vary much as Λ is translated. Moreover, since H connects only nearby regions, $N_\Lambda(E)$ should be additive, namely, if Λ and Λ' are two large nonintersecting boxes, $N_{\Lambda \cup \Lambda'}(E) = N_\Lambda(E) + N_{\Lambda'}(E) + o(|\Lambda \cup \Lambda'|)$. In particular, $N_\Lambda(E)$ should grow proportionally to $|\Lambda|$ as $\Lambda \uparrow \mathbb{R}^d$. This is why the *integrated density of states* (IDOS) is defined as follows:

$$(5.1) \quad \mathcal{N}(E) = \lim_{\Lambda \uparrow \mathbb{R}^d} \frac{N_\Lambda(E)}{|\Lambda|} \in \mathbb{R}_+,$$

where the limit, whenever it exists, is understood in the Følner sense [**Gre**]. The first rigorous work on the IDOS goes back to Benderskiĭ and Pastur [**BP**], who proved the existence of the limit for a one-dimensional Schrödinger operator on a lattice with random potential, with probability one. Then the existence and smoothness properties of the derivative $d\mathcal{N}$ as a Stieltjes-Lebesgue measure were proved by different methods with an increasing degree of generality for the Schrödinger operator with random potential [**Pas, Nak, KM**]. The algebraic approach goes back to the work of Shubin [**Shu**] inspired by the index theory [**CMS**]. The extension to more general coefficients is elementary and has been given by one of us in the discrete case [**Bel2**] and in the continuum case [**Bel6**].

Actually, the value $N_\Lambda(E)$ is nothing but the trace of the projection $\chi(H_\Lambda \leq E)$ onto eigenstates of H_Λ with energy less than or equal to E. So the IDOS appears as:

$$(5.2) \qquad \mathcal{N}_\mathbb{P}(E) = \lim_{\Lambda \uparrow \mathbb{R}^d} \frac{1}{|\Lambda|} \operatorname{Tr}_\Lambda(\chi(H_\Lambda \leq E)).$$

The IDOS depends on the choice of the translation invariant ergodic probability measure $\mathbb{P}$ on the hull of the Hamiltonian.

This formula is very reminiscent of the formula defining the trace per unit volume $\mathcal{T}_\mathbb{P}$ (see (1.12)) of the eigenprojector $\chi(H \leq E)$ in the infinite volume limit. Notice that this projector does not belong, in general, to the C^*-algebra $\mathcal{A}$ generated by H but to the von Neumann algebra $L^\infty(\mathcal{A}, \mathcal{T}_\mathbb{P})$ associated to the trace per unit volume by means of the GNS construction.

DEFINITION 5.1 (Shubin's formula). The Schrödinger operator (2.4) obeys *Shubin's formula*, whenever the IDOS satisfies:

$$(5.3) \qquad \mathcal{N}_\mathbb{P}(E) = \mathcal{T}_\mathbb{P}(\chi(H \leq E)),$$

where $\mathcal{T}_\mathbb{P}$ is the trace per unit volume defined in (1.11).

Shubin's formula will follow from showing that $\operatorname{Tr}_\Lambda(\chi(H_\Lambda \leq E))/|\Lambda|$ and $\operatorname{Tr}_\Lambda(\chi(H \leq E))/|\Lambda|$ as $\Lambda \uparrow \mathbb{R}^d$ have the same limit. Here is a short review of the cases in which Shubin's formula has been established [**Bel6**].

Let $(\Omega, \mathbb{R}^d, T)$ be a dynamical system with Ω a compact space. We suppose $(\Omega, \mathbb{R}^d, T)$ is topologically transitive, namely there is $\omega_0 \in \Omega$ the orbit of which is dense in Ω. Let $U^\infty(\Omega, \mathbb{R}^d)$ be the set of smooth functions f on $\mathbb{R}^d$, such that there is $F \in C(\Omega)$ for which $f(x) = F(T^{-x}\omega_0)$ and that, for all $\omega \in \Omega$, the function $f_\omega(x) = F(T^{-x}\omega)$ has bounded derivatives at all orders.

THEOREM 5.2. *Let H_ω be a uniformly elliptic self-adjoint operator of the form*

$$(5.4) \qquad H_\omega = \sum_{|\alpha| \leq m} h_\omega^{(\alpha)}(x) D^\alpha$$

with coefficients $h^{(\alpha)} \in U^\infty(\Omega, \mathbb{R}^d)$, $\alpha = (\alpha_1, \ldots, \alpha_d) \in \mathbb{N}^d$ *is a multi-index,* $|\alpha| = \alpha_1 + \cdots + \alpha_d$ *and* $D^\alpha = \prod_{1 \leq \mu \leq d} (-i\partial_\mu)^{\alpha_\mu}$.

Then there is an $r(z) \in C^*(\Omega \times \mathbb{R}^d)$ *such that* $\Pi_\omega(r(z)) = \{z - H_\omega\}^{-1}$, *for every complex z in the resolvent set of H_ω, and Shubin's formula holds.*

In the discrete case, the situation is technically much simpler. Let $\mathbb{G}$ be a countable discrete amenable group [**Gre**] (not necessarily Abelian). By amenable we mean that there exists a Følner sequence in $\mathbb{G}$.

THEOREM 5.3 ([**Bel2**]). *Let $(\Omega, \mathbb{G}, T)$ be a topological dynamical system with Ω compact and metrizable, and $\mathbb{G}$ discrete countable and amenable. Given any $\mathbb{G}$-invariant ergodic probability measure on Ω, any self-adjoint element of the crossed product $C(\Omega) \rtimes \mathbb{G}$ satisfies Shubin's formula.*

From Shubin's formula it is easy to obtain [**Bel6**]:

PROPOSITION 5.1. *Let H be a homogeneous operator, and let $\mathcal{A}$ be the C^*-algebra generated by H and its translates. Let $\mathcal{T}_\mathbb{P}$ be a translation invariant trace, for which H obeys Shubin's formula. Then its IDOS is a nonnegative, nondecreasing function of E, which is constant on each gap of* $\operatorname{spec}(H)$.

PROPOSITION 5.2. *Let H be as in Proposition 5.1. If $\mathcal{T}_\mathbb{P}$ is faithful, the spectrum of H coincides with the set of points in $E \in \mathbb{R}^d$ in the vicinity of which the IDOS is not constant.*

REMARK 5.4. Let H be the Schrödinger operator $H = \Delta + V$ on $L^2(\mathbb{R}^d)$ where V converges to zero at infinity. Then the hull of H is the one-point compactification of $\mathbb{R}^d$. The only translation-invariant ergodic probability measure on the hull $\Omega \cong \mathbb{R}^d \cup \{\infty\}$ is the Dirac measure at ∞. Therefore, the trace per unit volume cannot be faithful. In particular, the IDOS does not take into account the discrete spectrum of H, since the density of such eigenvalues is zero.

PROPOSITION 5.3. *With the assumptions of Proposition 5.1, any discontinuity point of the IDOS is an eigenvalue of H with infinite multiplicity.*

5.2. General Gap Labeling and the K_0-Group. This section is devoted to a short review of K-theory [**Bla, Weg**] and to the statement of the gap labeling theorem for homogeneous media (all the proofs can be found in [**BBG, Bel6**]).

Let H be a homogeneous self-adjoint operator affiliated to the C^*-algebra $\mathcal{A}$. Let $\mathfrak{g}$ be a spectral gap of H and let $P(\mathfrak{g})$ be the eigenprojection on the spectral interval $(-\infty, E]$ for any $E \in \mathfrak{g}$. Being a smooth function of H, $P(\mathfrak{g}) \in \mathcal{A}$. Now, changing H by a unitary transformation does not modify the spectrum as a set. Therefore, the gap $\mathfrak{g}$ can be labeled by the equivalence class of $P(\mathfrak{g})$ under unitary transformations. In order to take into account the case of nonunital algebras, the following definition of equivalence must be used:

DEFINITION 5.5. Two projections P and Q of a C^*-algebra $\mathcal{A}$ are equivalent if there is $U \in \mathcal{A}$ such that $UU^* = P$ and $U^*U = Q$. The equivalence will be denoted by $P \approx Q$.

The set of equivalence classes of projections in $\mathcal{A}$ will be denoted by $\mathcal{P}(\mathcal{A})$, and the equivalence class of P by $[P]$. Two projections P and Q are orthogonal whenever $PQ = QP = 0$. Then $P + Q$ is a new projection, called the direct sum of P and Q, denoted by $P \oplus Q$.

PROPOSITION 5.4 ([**Ped**]). *Let $\mathcal{A}$ be a separable C^*-algebra.*
  (i) *The set $\mathcal{P}(\mathcal{A})$ of equivalence classes of projections in $\mathcal{A}$ is countable.*
  (ii) *Let P and Q be two projections in $\mathcal{A}$. Then the equivalence class of their direct sum, if it exists, depends only on the equivalence classes of P and of Q. In particular, a sum is defined on the set Θ of pairs $([P], [Q])$ in $\mathcal{P}(\mathcal{A})$, such that there are $P' \approx P$ and $Q' \approx Q$ with $P'Q' = Q'P' = 0$, by $[P] + [Q] = [P' \oplus Q']$. This composition law is commutative and associative.*

The main problem is that the direct sum may not be defined everywhere. To overcome this difficulty, $\mathcal{A}$ is replaced by its *stabilization* $\mathcal{A} \otimes \mathcal{K}$, where $\mathcal{K}$ is the algebra of compact operators. A C^*-algebra $\mathcal{A}$ is *stable* if $\mathcal{A}$ and $\mathcal{A} \otimes \mathcal{K}$ are isomorphic. For any C^*-algebra $\mathcal{A}$, $\mathcal{A} \otimes \mathcal{K}$ is always stable, because $\mathcal{K} \otimes \mathcal{K} \cong \mathcal{K}$. Two C^*-algebras $\mathcal{A}$ and $\mathcal{B}$ are *Morita equivalent* whenever $\mathcal{A} \otimes \mathcal{K}$ is isomorphic to $\mathcal{B} \otimes \mathcal{K}$.

PROPOSITION 5.5. *Given any pair P and Q of projections in $\mathcal{A} \otimes \mathcal{K}$, there is always a pair P', Q' of mutually orthogonal projections in $\mathcal{A} \otimes \mathcal{K}$ such that $P' \approx P$ and $Q' \approx Q$. Therefore the sum $[P] + [Q] = [P' \oplus Q']$ is always defined.*

In this way, if $\mathcal{A}$ is a stable algebra, the set $\mathcal{P}(\mathcal{A})$ of equivalence classes of projections is an Abelian monoid with neutral element given by the class of the zero projection. If $\mathcal{A}$ is not stable, $\mathcal{P}(\mathcal{A})$ will be replaced by $\mathcal{P}(\mathcal{A} \otimes \mathcal{K})$. The

Grothendieck construction gives a canonical way to construct a group from such a monoid. This is a direct generalization of the construction of the group of integers $\mathbb{Z}$ from $\mathbb{N}$. The formal difference $[P] - [Q]$ is defined as the equivalence class of pairs $([P], [Q]) \in \mathcal{P}(\mathcal{A} \otimes \mathcal{K}) \times \mathcal{P}(\mathcal{A} \otimes \mathcal{K})$ under the relation

$$([P], [Q])\mathfrak{R}([P'], [Q']) \iff \exists [S] \in \mathcal{P}(\mathcal{A} \otimes \mathcal{K}); [P] + [Q'] + [S] = [P'] + [Q] + [S].$$

The corresponding quotient is the Abelian group $K_{00}(\mathcal{A}) = \mathcal{P}(\mathcal{A} \otimes \mathcal{K}) \times \mathcal{P}(\mathcal{A} \otimes \mathcal{K})/\mathfrak{R}$. Whenever $\mathcal{A}$ is unital, $K_0(\mathcal{A}) := K_{00}(\mathcal{A})$. Otherwise, $\mathcal{A}$ must be enlarged to $\mathcal{A}^+$ obtained by adjoining a unit, so that $\mathcal{A}$ becomes a two-sided closed ideal of $\mathcal{A}^+$. The quotient map $\pi\colon \mathcal{A}^+ \to \mathcal{A}^+/\mathcal{A}$ induces a group homomorphism $\pi_*: K_{00}(\mathcal{A}^+) \to K_{00}(\mathcal{A}^+/\mathcal{A})$, the kernel of which is the group $K_0(\mathcal{A})$ (see [**Bla**] for details). It leads to:

PROPOSITION 5.6. *Let $\mathcal{A}$ be a separable C^*-algebra.*
  (i) *The set $K_0(\mathcal{A})$ is countable and has a canonical Abelian group structure.*
  (ii) *Any *-isomorphism $\varphi\colon \mathcal{A} \mapsto \mathcal{B}$ between C^*-algebras induces a group homomorphism $\varphi_*\colon K_0(\mathcal{A}) \mapsto K_0(\mathcal{B})$, so that K becomes a functor from the category of C^*-algebras into the category of Abelian discrete groups.*
  (iii) *Any trace $\mathcal{T}$ on $\mathcal{A}$ defines, in a unique way, a group homomorphism $\mathcal{T}_*$ from $K_0(\mathcal{A})$ to $\mathbb{R}$ such that if P is a projection on $\mathcal{A}$, $\mathcal{T}(P) = \mathcal{T}_*([P])$ where $[P]$ is the class of P in $K_0(\mathcal{A})$.*
  (iv) *If $\mathcal{A}$ and $\mathcal{B}$ are two Morita equivalent C^*-algebras then $K_0(\mathcal{A})$ and $K_0(\mathcal{B})$ are isomorphic.*

Putting together Shubin's formula with the last proposition leads to:

THEOREM 5.6 (Gap labeling theorem for homogeneous media). *Let H be a homogeneous self-adjoint operator satisfying Shubin's formula (5.1), let $\mathcal{A}$ be the C^*-algebra generated by H and by its translates and let $\mathrm{spec}(H)$ be its spectrum in $\mathcal{A}$. Then*:
  (i) *For any gap $\mathfrak{g}$ in the spectrum of H, the value of the IDOS of H on $\mathfrak{g}$ belongs to the countable set of real numbers $\mathcal{T}_*(K_0(\mathcal{A})) \cap [0, \mathcal{T}(\mathbf{1})]$.*
  (ii) *The equivalence class $n(\mathfrak{g}) = [P(\mathfrak{g})] \in K_0(\mathcal{A})$ gives a labeling which is invariant under norm perturbations of the Hamiltonian H within $\mathcal{A}$.*
  (iii) *If $S \subset \mathbb{R}$ is a closed and open subset in $\mathrm{spec}(H)$, then $n_S = [P_S] \in K_0(\mathcal{A})$, where $[P_S]$ is the eigenprojection of H corresponding to S, is a labeling for each such part of the spectrum.*

Let $t \in \mathbb{R} \to H(t)$ be a continuous family of self-adjoint operators (in the norm of the resolvent) with resolvent in $\mathcal{A}$.
  (iv) *(homotopy invariance) The gap edges of $H(t)$ are continuous, and the labeling of a gap $\{\mathfrak{g}(t)\}$ is independent of t as long as the gap does not close.*
  (v) *(additivity) If, for $t \in [t_0, t_1]$, the spectrum of H contains a clopen subset $S(t)$ such that $S(0) = S_+ \cup S_-$ and $S(1) = S'_+ \cup S'_-$, where $S_\pm$ and $S'_\pm$ are clopen sets in $\mathrm{spec}(H(t_0))$ and $\mathrm{spec}(H(t_1))$, respectively, then $n_{S_+} + n_{S_-} = n_{S'_+} + n_{S'_-}$.*

5.3. Higher K-Groups and Exact Sequences.

The explicit computation of K-groups can be performed using the methods developed in homological algebra. The main tools are exact sequences and spectral sequences. However, these methods require introducing higher order K-groups. Let $\mathcal{A}$ be a C^*-algebra and let $\mathrm{GL}_n(\mathcal{A})$

be the group of invertible elements of the algebra $M_n(\mathcal{A})$ (when $\mathcal{A}$ is nonunital, $\mathrm{GL}_n(\mathcal{A}) = \{u \in \mathrm{GL}_n(\mathcal{A}^+); u \equiv \mathbf{1}_n \bmod M_n(\mathcal{A})\}$). $\mathrm{GL}_n(\mathcal{A})$ is embedded as a subgroup of $\mathrm{GL}_{n+1}(\mathcal{A})$ using

$$\begin{pmatrix} \mathrm{GL}_n(\mathcal{A}) & 0_n \\ 0_n^* & 1 \end{pmatrix}$$

(with $0_n = (0, \ldots, 0)$). Let $\mathrm{GL}_\infty(\mathcal{A})$ be the inductive limit of $\mathrm{GL}_n(\mathcal{A})$, namely the norm closure of their union, and let $[\mathrm{GL}_\infty(\mathcal{A})]_0$ be the connected component of the identity in $\mathrm{GL}_\infty(\mathcal{A})$. K_1 is defined as follows:

(5.5) $\qquad K_1(\mathcal{A}) = \mathrm{GL}_\infty(\mathcal{A})/\mathrm{GL}_\infty(\mathcal{A})_0 = \varinjlim\{\mathrm{GL}_n(\mathcal{A})/[\mathrm{GL}_n(\mathcal{A})]_0\}.$

If $\mathcal{A}$ is separable, then $K_1(\mathcal{A})$ is countable, since nearby invertible elements are in the same component. For $u \in \mathrm{GL}_n(\mathcal{A})$, let $[u]$ be its class in $K_1(\mathcal{A})$. The relation $[u][v] = [\mathrm{diag}(u,v)]$ defines a product in $K_1(\mathcal{A})$.

PROPOSITION 5.7 ([**Bla**]). $K_1(\mathcal{A})$ *is an Abelian group.*

The *suspension* of $\mathcal{A}$ is the C^*-algebra $S\mathcal{A}$ of continuous functions $f: \mathbb{R} \to \mathcal{A}$ vanishing at $\pm\infty$, endowed with pointwise addition, multiplication and adjoint, and the sup-norm. Hence $S\mathcal{A} \cong C_0(\mathbb{R}) \otimes \mathcal{A}$. Then

THEOREM 5.7 ([**Bla**]). $K_1(\mathcal{A})$ *is canonically isomorphic to* $K_0(S\mathcal{A})$.

Therefore we can also define higher K-groups by

$$K_2(\mathcal{A}) = K_1(S\mathcal{A}) = K_0(S^2\mathcal{A}), \ldots, K_n(\mathcal{A}) = \cdots = K_0(S^n\mathcal{A}).$$

THEOREM 5.8 (Bott periodicity). $K_0(\mathcal{A}) \cong K_2(\mathcal{A})$.

More precisely $K_0(\mathcal{A})$ *is isomorphic to the group* $\pi_1(\mathrm{GL}_\infty(\mathcal{A}))$ *of homotopy classes of closed paths in* $\mathrm{GL}_\infty(\mathcal{A})$. *Furthermore, if* $\mathcal{T}$ *is a trace on* $\mathcal{A}$ *and if* $t \in [0,1] \to U(t)$ *is a closed path in* $\mathrm{GL}_\infty(\mathcal{A})$ [**Co2**]:

(5.6) $\qquad \mathcal{T}_*([U]) = \dfrac{1}{2\pi i} \displaystyle\int_{[0,1]} dt\, \mathcal{T}(U(t)^{-1}U'(t)),$

where $\mathcal{T}_*$ *is the map induced by* $\mathcal{T}$ *on* $K_0(\mathcal{A})$.

K_i is a covariant functor with the following properties:

THEOREM 5.9. *Let* $\mathcal{J}, \mathcal{A}, \mathcal{A}_n, \mathcal{B}$ *be* C^*-*algebras and* n, i *nonnegative integers*:

(i) *If* $f: \mathcal{A} \to \mathcal{B}$ *is a* *-*homomorphism, then* f *induces a group homomorphism* $f_*: K_i(\mathcal{A}) \to K_i(\mathcal{B})$. *Then* $\mathrm{id}_* = \mathrm{id}$, *and* $(f \circ g)_* = f_* \circ g_*$.
(ii) $K_i(\bigoplus_n \mathcal{A}_n) \cong \bigoplus_n K_i(\mathcal{A})$
(iii) *If* $\mathcal{A}$ *is the inductive limit of the sequence* $(\mathcal{A}_n)_{n>0}$ *of* C^*-*algebras then* $K_i(\mathcal{A})$ *is the inductive limit of the groups* $K_i(\mathcal{A}_n)$.
(iv) *If* $\phi: \mathcal{J} \to \mathcal{A}$, *and* $\psi: \mathcal{A} \to \mathcal{B}$ *are* *-*homomorphisms such that the sequence*

$$0 \to \mathcal{J} \to \mathcal{A} \to \mathcal{B} \to 0$$

be exact, there is a six-term exact sequence of the form:

(5.7)
$$\begin{array}{ccccc} K_0(\mathcal{J}) & \xrightarrow{\phi_*} & K_0(\mathcal{A}) & \xrightarrow{\psi_*} & K_0(\mathcal{B}) \\ {\scriptstyle\mathrm{Ind}}\uparrow & & & & \downarrow{\scriptstyle\mathrm{Exp}} \\ K_1(\mathcal{B}) & \xleftarrow{\psi_*} & K_1(\mathcal{A}) & \xleftarrow{\phi_*} & K_1(\mathcal{J}) \end{array}$$

In the previous theorem, Ind et Exp are the connection automorphisms defined as follows (whenever $\mathcal{A}$ is unital): let P be a projection in $\mathcal{B} \otimes \mathcal{K}$, and let A be a self-adjoint element of $\mathcal{A} \otimes \mathcal{K}$ such that $\psi \otimes \mathrm{id}(A) = P$. Then $\psi \otimes \mathrm{id}(e^{2i\pi A}) = e^{2i\pi P} = \mathbf{1}$, so that $B = e^{2i\pi A} \in (\mathcal{J} \otimes \mathcal{K})^+$ and is unitary in $(\mathcal{J} \otimes \mathcal{K})^+$. The class of B gives an element of $K_1(\mathcal{J})$ which is, by definition, $\mathrm{Exp}([P])$. In much the same way, let now U be an unitary element of $\mathbf{1} + (\mathcal{B} \otimes \mathcal{K})$. Without loss of generality it is the image under $\psi \otimes \mathrm{id}$ of a partial isometry W in $(\mathcal{A} \otimes \mathcal{K})$. Then $\mathrm{Ind}([U])$ is the class of $[WW^*] - [W^*W]$ in $K_0(\mathcal{J})$. These definitions actually make sense.

The Connes-Thom Isomorphism. The C^*-algebra of a dynamical system introduced in Section 1.4 is a special case of the C^*-crossed product construction. Let $\mathcal{A}$ be a C^*-algebra, $\mathbb{G}$ be a locally compact group, and α be a continuous homomorphism from $\mathbb{G}$ into $\mathrm{Aut}(\mathcal{A})$ (namely the group of *-automorphisms of $\mathcal{A}$ endowed with the topology of pointwise norm convergence). A *covariant representation* of the triple $(\mathcal{A}, \mathbb{G}, \alpha)$ is a pair of representations (Π, ρ) of $\mathcal{A}$ and $\mathbb{G}$ on the same Hilbert space such that $\rho(g)\Pi(a)\rho(g)^* = \Pi(\alpha_g(a))$ for all $a \in \mathcal{A}$ and $g \in \mathbb{G}$. Each covariant representation of $(\mathcal{A}, \mathbb{G}, \alpha)$ gives a representation of the twisted convolution algebra $C_c(\mathbb{G}, A)$ by integration (compare with Section 1.4), and hence a pre-C^*-norm on this *-algebra. The supremum of all these norms is a C^*-norm, and the completion of $\mathcal{C}_c(\mathbb{G}, A)$ with respect to this norm is called the *crossed product* of $\mathcal{A}$ by $\mathbb{G}$ under the action α, denoted by $A \rtimes_\alpha \mathbb{G}$. The *-representations of $A \rtimes_\alpha \mathbb{G}$ are in natural one-to-one correspondence with the covariant representations of the dynamical system $(\mathcal{A}, \mathbb{G}, \alpha)$.

THEOREM 5.10 ([**Co2**]). $K_i(A \rtimes_\alpha \mathbb{R}) \cong K_{1-i}(A)$, *for* $i = 0, 1$.

The Pimsner & Voiculescu exact sequence.

THEOREM 5.11 ([**PV**]). *Let $\mathcal{A}$ be a separable C^*-algebra, and let α be a *-automorphism of $\mathcal{A}$. There exists a six-term exact sequence:*

(5.8)
$$\begin{array}{ccccc} K_0(\mathcal{A}) & \xrightarrow{\mathrm{id} - \alpha_*} & K_0(\mathcal{A})) & \xrightarrow{j_*} & K_0(\mathcal{A} \rtimes_\alpha \mathbb{Z}) \\ \uparrow{\scriptstyle \mathrm{Ind}} & & & & \downarrow{\scriptstyle \mathrm{Exp}} \\ K_1(\mathcal{A} \rtimes_\alpha \mathbb{Z})) & \xleftarrow{j_*} & K_1(\mathcal{A}) & \xleftarrow{\mathrm{id} - \alpha_*} & K_1(\mathcal{A}) \end{array}$$

where j is the canonical injection of $\mathcal{A}$ into the crossed product.

6. Results for 1D and 2D Quasicrystals

Let $L = \Lambda(M)$ be an admissible model set (see Section 3.4), and $(\Omega, \mathbb{R}^d) = (\mathbb{T}_M^{n+d}, \mathbb{R}^d)$ its hull. Let Υ be its (totally disconnected) transversal, and $C^*(G_\Upsilon)$ be the algebra of the corresponding groupoid. Let $\mathbb{T}_M^n$ be the quotient of $\mathbb{R}_M^n$ by the $\mathbb{Z}^n$-action defined in Section 3.4. In the following M will be a Borel set. According to Theorem 3.8, the Lebesgue measure, denoted here by $\mathbb{P}$, is the unique translation invariant, ergodic measure on $\Omega = \mathbb{T}_M^{n+d}$. It induces a transverse measure, also denoted by $\mathbb{P}$, on Υ.

THEOREM 6.1. $\mathcal{T}_{\mathbb{P}*}(K_0(\mathcal{C}(\mathbb{T}_M^n) \rtimes \mathbb{Z}^d)) = \mathcal{T}_{\mathbb{P}*}(K_0(\mathcal{C}(\Omega) \rtimes \mathbb{R}^d))$

PROOF. Proposition 3.3, Theorems 4.6 and 4.7 ensure that the algebras are Morita equivalent. But, for separable algebras $\mathcal{A}$ and $\mathcal{B}$, Morita equivalence implies the existence of a $\mathcal{A}$-$\mathcal{B}$-bimodule [**BGR, Co6**]. Then, not only are the K-groups of these two algebras isomorphic but also their images under the traces coincide [**Bel6**]. □

REMARK 6.2. Kellendonk obtained a similar result in [**Kel1, Kel2**] by proving that model sets can actually be seen as "decorations" of $\mathbb{Z}^d$.

REMARK 6.3. In the one-dimensional case, the Poincaré first return map ensures that the transversal Υ can be endowed with a $\mathbb{Z}$-action. The Pimsner and Voiculescu six-term exact sequence permits to compute the K-groups and the set of gap labels. For higher dimensions however, the Poincaré map is replaced by the groupoid of the transversal, which is not necessarily identical to a group action. Nevertheless Theorem 6.1 permits to reduce the K-theory problem to a group action.

6.1. Completely Disconnected Hulls. The following results have been known for a long time (see [**BBG**] for a review).

LEMMA 6.4. *Let Ξ be a totally disconnected compact metrizable space. Then, $K_1(\mathcal{C}(\Xi)) = 0$.*

PROOF. Let d be a compatible metric on Ξ, and let r be a positive real number. Since Ξ is totally disconnected, each $\xi \in \Xi$ admits a clopen neighborhood $\mathcal{O}_\xi$ of diameter less than r. Ξ being compact, a finite covering of Ξ by clopen sets of diameter less than r can be extracted. Since the complement of a clopen set is a clopen set, this covering can be chosen to be a finite partition denoted by $\mathfrak{P}$. Let now f be an $M_n(\mathbb{C})$-valued continuous invertible function on Ξ. To prove the lemma, it is enough to show that f is homotopic to the constant function equal to one. Since Ξ is compact, f is uniformly continuous, so that given any $\epsilon > 0$, there is $r > 0$ such that $d(\xi, \xi') \leq r \Rightarrow \|f(\xi) - f(\xi')\| \leq \epsilon$. For each $\mathcal{O}_k \in \mathfrak{P}$, let $\xi_k \in \mathcal{O}_k$ and χ_k be the characteristic function of $\mathcal{O}_k$. Hence $\chi_k \in \mathcal{C}(\Xi)$. Then if $f_\epsilon = \sum_k f(\xi_k)\chi_k$, $\|f - f_\epsilon\| \leq \epsilon$. Thus f and f_ϵ are homotopic for ϵ small enough. By hypothesis, all the $f(\xi_k)$'s are invertible $n \times n$ matrices, so that there are matrices H_k such that $f(\xi_k) = \exp(H_k)$. Setting $f_\epsilon(\xi, t) = \sum_k \exp(tH_k)\chi_k(\xi)$ gives a homotopy between f_ϵ and 1 in the set of invertible elements of $M_n(\mathcal{C}(\Xi))$. □

Let P be a projection in $\mathcal{C}(\Xi) \otimes \mathcal{K}$. P may be viewed as a continuous mapping from Ξ into $\mathcal{K}$ leading to the map $f_P(\omega) = \text{Tr}(P(\omega))$ (where Tr is the usual trace on $M_n(\mathbb{C})$). Let then $f_*: P \mapsto f_P$, so that:

LEMMA 6.5. *The map f_* is an isomorphism from $K_0(\mathcal{C}(\Xi))$ to the group $\mathcal{C}(\Xi, \mathbb{Z})$ of integer-valued continuous functions on Ξ.*

PROOF. f_P depends only on the equivalence class of P. For indeed, if $P \approx Q$, there is $U \in \mathcal{C}(\Xi) \otimes \mathcal{K}$ such that $UU^* = P$ and $U^*U = Q$. In particular, $f_P(\xi) = \text{Tr}(U(\xi)U^*(\xi)) = \text{Tr}(U^*(\xi)U(\xi)) = f_Q(\xi)$. Moreover, if $PQ = QP = 0$, then $f_{P \oplus Q}(\xi) = \text{Tr}(P(\xi) \oplus Q(\xi)) = \text{Tr}(P(\xi)) + \text{Tr}(Q(\xi)) = f_P(\xi) + f_Q(\xi)$, and so $f_{P \oplus Q} = f_P + f_Q$. Hence f defines a group homorphism f_* between $K_0(\mathcal{C}(\Xi))$ and $\mathcal{C}(\Xi, \mathbb{Z})$.

f_* IS SURJECTIVE. For if $h \in \mathcal{C}(\Xi, \mathbb{Z})$, let Ξ_n denote the set of $\xi \in \Xi$ such that $h(\xi) = n$. Since h is continuous and integer-valued, the family $\{\Xi_n, n \in \mathbb{N}\}$ gives a finite partition of Ξ into clopen sets. Let χ_n denote the characteristic function of Ξ_n, and π_n be the n-dimensional projection $1_n \oplus 0$ in $\mathcal{K}$. Then if $P' = \sum_{n>0} \chi_n \otimes \pi_n$ and $Q' = \sum_{n>0} \chi_{-n} \otimes \pi_n$, h can be written as $h = f_{P'} - f_{Q'}$.

f_* IS INJECTIVE. It is sufficient to show that any projection in $\mathcal{C}(\Xi) \otimes \mathcal{K}$ is equivalent to a projection of the same type as P' above. For indeed, $f_P - f_Q = 0$ implies that P and Q admit the same partition $\{\Xi_n, n \in \mathbb{N}\}$ into clopen subsets of Ξ on which their dimension is a fixed integer. So if $P \approx P'$, then $Q \approx P'$ namely $[P] - [Q] = 0$. Let then P be a projection in $\mathcal{C}(\Xi) \otimes \mathcal{K}$. Let $\{\Xi_n; n \in \mathbb{N}\}$ be the partition into clopen subsets on which the dimension of P is n. Proceeding as in the proof of Lemma 6.4, given $0 < \epsilon < 1$, for each $n \in \mathbb{N}$, there is a finite partition $(\Xi_{n,k})_k$ of Ξ_n into clopen subsets, such that, if $\xi_{n,k} \in \Xi_{n,k}$, the projection $P_\epsilon = \sum_{n,k} P(\xi_{n,k})\chi_{n,k}$ satisfies $\|P - P_\epsilon\| \leq \epsilon < 1$. Thus $P \approx P_\epsilon$. Now $P(\xi_{n,k})$ is a fixed projection of dimension n so that there is $S_{n,k}$ in $\mathcal{K}$ such that $P(\xi_{n,k}) = S_{n,k}S_{n,k}^*$ and $\pi_n = S_{n,k}^* S_{n,k}$. Therefore, setting $S = \sum_{n,k} S_{n,k}\chi_{n,k}$, S is an element of $\mathcal{C}(\Xi) \otimes \mathcal{K}$ such that $P_\epsilon = SS^*$ and $P' = S^*S$, establishing the equivalence between P and P'. □

6.2. Gap Labeling for 1D Quasicrystals.

DEFINITION 6.6. Let $\mathbb{G}$ be an Abelian group and let $T: \mathbb{G} \mapsto \mathbb{G}$ be a group isomorphism. The set $\mathcal{E} = \{g \in \mathbb{G}; \exists h \in \mathbb{G}, g = h - T(h)\}$ is a subgroup. The set $\mathbb{G}^T = \{g \in \mathbb{G}; T(g) = g\}$ is called the group of *invariants* whereas $\mathbb{G}_T = \mathbb{G}/\mathcal{E}$ is called the groups of *co-invariants*.

THEOREM 6.7 ([**Bel6**]). *Let Ξ be a totally disconnected compact metrizable space, endowed with a $\mathbb{Z}$-action T.*

(i) *If T is topologically transitive, $K_1(\mathcal{C}(\Xi) \rtimes_T \mathbb{Z})$ is isomorphic to $\mathbb{Z}$.*

(ii) *$K_0(\mathcal{C}(\Xi) \rtimes_T \mathbb{Z})$ is isomorphic to $\mathcal{C}(\Xi, \mathbb{Z})_T$.*

(iii) *Let $\mathbb{P}$ be a T-invariant ergodic probability measure on Ξ, and $\mathcal{T}_\mathbb{P}$ be the corresponding trace on $\mathcal{C}(\Xi) \rtimes_T \mathbb{Z}$. Then, the image of $K_0(\mathcal{C}(\Xi) \rtimes_T \mathbb{Z})$ by $\mathcal{T}_{\mathbb{P}*}$ is equal to the countable subgroup $\mathbb{P}(\mathcal{C}(\Xi, \mathbb{Z}))$ of $\mathbb{R}$ (see Theorem 1.16).*

REMARK 6.8. In the one-dimensional case, any transversal Υ of $(\Omega, \mathbb{R}, T)$ is canonically endowed with a $\mathbb{Z}$-action. Thus this theorem can be applied to Υ.

PROOF. Due to Lemmas 6.4 and 6.5, the Pimsner and Voiculescu six-term exact sequence (5.8) for $\mathcal{A} = \mathcal{C}(\Xi)$ leads to the following exact sequence:

$$(6.1) \quad 0 \to K_1(\mathcal{C}(\Xi) \rtimes_T \mathbb{Z}) \xrightarrow{f_* \circ \text{Ind}} \mathcal{C}(\Xi, \mathbb{Z}) \xrightarrow{\text{id} - T_*} \mathcal{C}(\Xi, \mathbb{Z}) \xrightarrow{j_* \circ f_*^{-1}} K_0(\mathcal{C}(\Xi) \rtimes \mathbb{Z}) \to 0$$

Since the sequence is exact, it follows that $K_1(\mathcal{C}(\Xi) \rtimes_T \mathbb{Z})$ is isomorphic to the kernel of $\text{id} - T_*$. If T is topologically transitive, this kernel is the set of constant functions in $\mathcal{C}(\Xi, \mathbb{Z})$, that is $\mathbb{Z}$ and (i) is proved.
For the same reason, since the sequence is exact, j_* is surjective. Thus $K_0(\mathcal{C}(\Xi) \rtimes \mathbb{Z})$ $\approx \mathcal{C}(\Xi, \mathbb{Z})/\ker(j_*) = \mathcal{C}(\Xi, \mathbb{Z})_T$. Since j is the canonical injection from $\mathcal{C}(\Xi)$ into $\mathcal{C}(\Xi) \rtimes_T \mathbb{Z}$, it follows from the definition of the trace given by $\mathbb{P}$ that $\mathcal{T}_\mathbb{P} \circ j = \mathbb{P}$. By functoriality, $\mathcal{T}_{\mathbb{P}*} \circ j_* = \mathbb{P}_*$ on the corresponding K_0-groups. Since $\mathbb{P}_*$ is nothing but $\mathbb{P}$ acting on $\mathcal{C}(\Xi, \mathbb{Z})$, and j_* is surjective, (iii) holds. (iii) gives a proof of Theorem 1.16 for the case $d = 1$. □

COROLLARY 6.1. *Let L be a one-dimensional admissible model set with acceptance domain M, M being a Borel set. Let $(\mathbb{T}_M^{n+1}, \mathbb{R})$ be its hull, $\mathbb{T}_M^n$ be the quotient of $\mathbb{R}_M^n$ by the $\mathbb{Z}^n$-action, and $\mathbb{P}$ be the Lebesgue measure on $\mathbb{T}_M^n$ normalized such that the measure of the acceptance domain M is one. Then*

$$\mathcal{T}_{\mathbb{P}*}(K_0(\mathcal{C}(\mathbb{T}_M^{n+1}) \rtimes \mathbb{R})) = \mathcal{T}_{\mathbb{P}*}(K_0(\mathcal{C}(\mathbb{T}_M^n))) = \mathbb{P}(\mathcal{C}(\mathbb{T}_M^n, \mathbb{Z})),$$

and the values on gaps of the IDOS of any Hamiltonian on this one-dimensional quasicrystal belong to $\mathbb{P}(\mathcal{C}(\mathbb{T}_M^n, \mathbb{Z})) \cap [0, 1]$.

PROOF. Theorem 5.6, Theorem 6.1, Lemma 6.5 and Theorem 6.7. □

In Example 4.8 we introduced one-dimensional quasicrystals obtained with $n = 1$ and with M being the projection of a unit cell of $\mathcal{R}$. In this case, M is $\mathcal{R}$-compatible and we get an explicit label for the gaps.

THEOREM 6.9. *The values on gaps of the IDOS of the Kohmoto model (4.16) or of any Hamiltonian belonging to the same algebra belong to* $(\mathbb{Z} + \mathbb{Z}\alpha) \cap [0, 1]$.

REMARK 6.10. Such a result could alternatively be obtained through the use of the irrational rotation algebra (see Example 4.8 and [**Bel6**]).

PROOF. We use Corollary 6.1 and the fact that $\mathcal{C}(\mathbb{T}_M^1, \mathbb{Z})$ is generated by a set of characteristic functions associated with the partition of $\mathbb{T}_M^1$ in polytopes (see the description of the topology induced by M in Section 3.4). Each such one-dimensional model set can be related to an irrational number ω, namely the slope of the physical space relative to the lattice (see Figure 1 and Example 4.8). Integrating the generic characteristic functions can then be performed by computing the lengths of the segments of the partition of $\mathbb{T}_M^1$. It is easy to convince oneself that such lengths belong to $\mathbb{Z} + \alpha \mathbb{Z}$, where $\alpha = \omega/(\omega + 1)$, once the Lebesgue measure is normalized such that the measure of the acceptance domain is one. □

6.3. Gap Labeling for 2D Quasicrystals. Let Ξ be a totally disconnected compact metrizable space, endowed with a $\mathbb{Z}^2$-action, given by two commuting *-automorphisms α_1 and α_2. Let $\mathbb{P}$ be an (α_1, α_2)-invariant ergodic probability measure on Ξ, and let $\mathcal{T}_{\mathbb{P}}$ be the corresponding trace on $\mathcal{C}(\Xi) \rtimes_{\alpha_1 \alpha_2} \mathbb{Z}^2$. Following [**Elst**], an iteration of the Pimsner and Voiculescu six-term exact sequence can be performed, leading to a short exact sequence for $K_0(\mathcal{C}(\Xi) \rtimes_{\alpha_1 \alpha_2} \mathbb{Z}^2)$. Let $\mathcal{A} = \mathcal{C}(\Xi)$ and let $\mathcal{B} = \mathcal{A} \rtimes_{\alpha_1} \mathbb{Z}$. Then α_2 induces a *-automorphism on $\mathcal{B}$ so that $\mathcal{B} \rtimes_{\alpha_2} \mathbb{Z}$ is isomorphic to $\mathcal{C}(\Xi) \rtimes_{\alpha_1 \alpha_2} \mathbb{Z}^2$. Theorem 5.11 leads to:

(6.2)
$$\begin{array}{ccccc} K_0(\mathcal{B}) & \xrightarrow{\mathrm{id} - \alpha_{2*}} & K_0(\mathcal{B}) & \xrightarrow{j_*} & K_0(\mathcal{B} \rtimes_{\alpha_2} \mathbb{Z}) \\ {\scriptstyle \mathrm{Ind}_2} \uparrow & & & & \downarrow {\scriptstyle \mathrm{Exp}_2} \\ K_1(\mathcal{B} \rtimes_{\alpha_2} \mathbb{Z}) & \xleftarrow{j_*} & K_1(\mathcal{B}) & \xleftarrow{\mathrm{id} - \alpha_{2*}} & K_1(\mathcal{B}) \end{array}$$

Moreover, the exact sequence (6.1) gives:

(6.3) $\quad 0 \to K_1(\mathcal{B}) \xrightarrow{f_* \circ \mathrm{Ind}_1} \mathcal{C}(\Xi, \mathbb{Z}) \xrightarrow{\mathrm{id} - \alpha_{1*}} \mathcal{C}(\Xi, \mathbb{Z}) \xrightarrow{j_{1*} \circ f_*^{-1}} K_0(\mathcal{B}) \to 0.$

That is, $K_1(\mathcal{B}) \approx \mathcal{C}(\Xi, \mathbb{Z})^{\alpha_1}$ and $K_0(\mathcal{B}) \approx \mathcal{C}(\Xi, \mathbb{Z})_{\alpha_1}$. Then (6.2) gives

(6.4) $\quad 0 \to K_0(\mathcal{A})_{\alpha_1 \alpha_2} \xrightarrow{j_{2*} \circ j_{1*} \circ f_*^{-1}} K_0(\mathcal{A} \rtimes \mathbb{Z}^2) \xrightarrow{f_* \circ \mathrm{Ind}_1 \circ \mathrm{Exp}_2} K_0(\mathcal{A})^{\alpha_1 \alpha_2} \to 0.$

If the dynamical system $(\Xi, \mathbb{Z}^2)$ is minimal (which is the case for admissible model sets, see Theorem 3.8), the class of identity is the only nontrivial element in $K_0(\mathcal{A})^{\alpha_1 \alpha_2}$. Let ξ_1 be a lift of the identity $[1] \in K_0(\mathcal{A})^{\alpha_1 \alpha_2}$ in $K_0(\mathcal{A} \rtimes \mathbb{Z}^2)$, then:

LEMMA 6.11 ([**Elst**]). $\mathcal{T}_{\mathbb{P}}(\xi_1) = 1 \in \mathcal{T}_{\mathbb{P}}(K_0(\mathcal{A}))$.

REMARK 6.12. There is an alternative method to get (6.4) [**FH**], but Lemma 6.11 cannot be established just by considerations of isomorphisms because Ind_1 and Exp_2 must be computed explicitly.

Now, it is enough to notice that, because of trace invariance under α_1 and α_2, $\mathcal{T}_{\mathbb{P}}(K_0(\mathcal{A})_{\alpha_1\alpha_2}) = \mathcal{T}_{\mathbb{P}}(K_0(\mathcal{A}))$ to obtain

THEOREM 6.13 ([**Elst**]). $\mathcal{T}_{\mathbb{P}}(K_0(\mathcal{C}(\Xi) \rtimes_{\alpha_1\alpha_2} \mathbb{Z}^2)) = \mathcal{T}_{\mathbb{P}}(K_0(\mathcal{C}(\Xi))$.

Using Theorem 5.6, Theorem 6.1, Lemma 6.5 and Theorem 6.7, we get as a corollary the main result of this section

COROLLARY 6.2. *Let L be a two-dimensional admissible model set with acceptance domain M, M being a Borel set, let $(\mathbb{T}_M^{n+2}, \mathbb{R}^2)$ be its hull, $\mathbb{T}_M^n$ be the quotient of $\mathbb{R}_M^n$ by the $\mathbb{Z}^n$-action, and $\mathbb{P}$ be the Lebesgue measure on $\mathbb{T}_M^n$ normalized such that the measure of the acceptance domain M is one. Then*

$$\mathcal{T}_{\mathbb{P}*}(K_0(\mathcal{C}(\mathbb{T}_M^{n+2}) \rtimes \mathbb{R}^2)) = \mathcal{T}_{\mathbb{P}*}(K_0(\mathcal{C}(\mathbb{T}_M^n))) = \mathbb{P}(\mathcal{C}(\mathbb{T}_M^n, \mathbb{Z})),$$

and the IDOS *of Hamiltonians on two-dimensional quasicrystals takes values in the set $\mathbb{P}(\mathcal{C}(\mathbb{T}_M^n, \mathbb{Z})) \cap [0, 1]$.*

EXAMPLE 6.14 (The octagonal tiling). In the following, only model sets with $n = d = 2$ are considered, with $\mathcal{R} = \mathbb{Z}^4$. Moreover, the acceptance domain M is given by the π_2-projection of the unit hypercube of $\mathbb{R}^4$ with sides given by the canonical basis $\varepsilon_1, \ldots, \varepsilon_4$. Let $e_i = \pi_2(\varepsilon_i)$ for $i = 1, \ldots, 4$. Let $\mathbb{T}_M^2$ denote the quotient of $\mathbb{R}_M^2$ by the $\mathbb{Z}^2$-action. Given any $f \in \mathcal{C}(\mathbb{T}_M^2, \mathbb{Z})$, for $n \in \mathbb{N}$ the set $f^{-1}(\{n\})$ is both closed and open. Thus f is a finite $\mathbb{Z}$-linear combination of characteristic functions of clopen sets. Therefore the set of gap labels will be the $\mathbb{Z}$-module generated by the Lebesgue measure of clopen sets, provided the Lebesgue measure is normalized as to give measure one to M.

Let $\mathcal{F}$ be the family of affine hyperplanes in $\mathbb{R}^2 \times \mathbb{R}^2$ given by $\{F_i + a; a \in \mathcal{L} = \pi_2(\mathbb{Z}^4)\}$ where the F_i's are the hyperplanes parallel to the maximal faces of $\mathbb{R}^2 \times M$. For each i let u_i be a unit vector in $\mathbb{R}^2$ perpendicular to F_i. As outlined at the end of the Section 3.4, the M-topology can be equivalently defined as the coarsest topology for which the open half-planes $\{x \in \mathbb{R}^2; \langle u_i | x - a \rangle > 0\}$ (with $a \in \mathcal{L}$) are closed. $\mathbb{R}_M^2$ is nothing but the completion of $\mathbb{R}^2$ for this topology.

We present here a method to determine such elementary tiles which can directly be extended to higher dimensions [**Kel3**]. Let the set $\{F_{ij}(a, b) = (F_i + a) \cap (F_j + b); i, j \in [1, \ldots, 4], a, b \in \mathbb{Z}^4\}$. By projection on $\mathbb{R}^2$, this defines a family of points $\mathcal{F}^{(1)}$ endowed with an action of $\mathbb{Z}^4$. Thus, we consider $\mathcal{F}_{(0)} = \mathcal{F}^{(1)}/\mathbb{Z}^4$. This gives exactly the points we need to build elementary tiles in $\mathbb{T}_M^2$.

Now we study the *octagonal quasicrystal* [**Soc**]. It is obtained with $e_1 = (1, 0)$, $e_2 = (0, 1)$, $e_3 = (-\sqrt{2}/2, -\sqrt{2}/2)$, $e_4 = (\sqrt{2}/2, -\sqrt{2}/2)$ and with normalization $\|e_i\| = 1$. The acceptance domain is a regular octagon.

A combinatorial computation gives for this tiling: $\mathcal{F}_{(0)} = \{0, e_1/\sqrt{2}, e_3/\sqrt{2}\}$. This defines three types of intersection points. The last two types share the property of being intersection points of orthogonal lines only. Thus they generate *rectangles* and *triangles* as elementary tiles. Since $\mathcal{F}_{(0)}$ is not trivial, there are intersection points which are not projections of the lattice $\mathbb{Z}^4$. This is the reason why we cannot perform a triangulation of this tiling.

We only need to compute the measure of each of the rectangles and triangles, obtained by combining these three types of intersection points, to get

THEOREM 6.15 ([**BCL**]). *Let $\mathcal{A}_{oc}$ be the C^*-algebra associated to the octagonal quasicrystal. Let us denote by $\mathcal{T}_{\mathbb{P}}$ the trace associated to the Lebesgue measure on*

$\mathbb{R}^2$ and normalized such that the measure of the acceptance zone is 1. Then,

$$\mathcal{T}_{\mathbb{P}*}\Big(K_0(\mathcal{A}_{oc})\Big) = \frac{1/4\mathbb{Z} + \sqrt{2}/2\mathbb{Z}}{2(1+\sqrt{2})}$$
$$= \left\{ \frac{m+n\sqrt{2}}{8} \in \frac{\mathbb{Z}+\sqrt{2}\mathbb{Z}}{8} \text{ such that } m+n \text{ is even} \right\}.$$

References

[Agm] S. Agmon, *Lectures on exponential decay of solutions of second-order elliptic equations*, Math. Notes, vol. 29, Princeton Univ. Press, Princeton, 1982.

[AP] J. E. Anderson and I. F. Putnam, *Topological invariants for substitution tilings and their associated C^*-algebras*, Ergodic Theory Dynam. Systems **18** (1998), 509–537.

[AM] N. W. Ashcroft and N. D. Mermin, *Solid state physics*, Holt, Rinehart, and Winston, New York, 1976.

[BMS] M. Baake, R. V. Moody, and M. Schlottmann, *Limit-(quasi)periodic point sets as quasicrystals with p-adic internal spaces*, J. Phys. A **31** (1998), 5755–5765; math-ph/9901008.

[Bar] M. Barnsley, *Fractals everywhere*, Academic Press, Boston, MA, 1988.

[Bau] H. Bauer, *Maß- und Integrationstheorie*, 2nd ed., de Gruyter Lehrbuch, de Gruyter, Berlin, 1992.

[Bel1] J. Bellissard, *Schrödinger operators with almost periodic potential: an overview*, Mathematical Problems in Theoretical Physics (R. Schrader, ed.), Lecture Notes in Phys., vol. 153, Springer, Berlin, 1982, pp. 356–363.

[Bel2] _____, *K-theory of C^*-algebras in solid state physics*, Statistical Mechanics and Field Theory: Mathematical Aspects (T. C. Dorlas, ed.), Lecture Notes in Phys., vol. 257, Springer, Berlin, 1986, pp. 99–156.

[Bel3] _____, *C^*-algebras in solid state physics. 2D electrons in a uniform magnetic field*, Operator Algebras and Applications, vol. 2 (D. E. Evans and M. Takesaki, eds.), London Math. Soc. Lecture Note Series, vol. 136, Cambridge Univ. Press, Cambridge, 1988, pp. 49–76

[Bel4] _____, *Almost periodicity in solid state physics and C^*-algebras*, Mat.-Fys. Medd. Danske Vid. Selsk. **42** (1989), 35–75.

[Bel5] _____, *Spectral properties of Schrödinger operator with a Thue-Morse potential*, Number Theory and Physics (Les Houches, 1989) (J. M. Luck, P. Moussa, and M. Waldschmidt, eds.), Springer Proc. Phys., vol. 47, Springer, Berlin, 1990, pp. 140–150.

[Bel6] _____, *Gap labelling theorems for Schrödinger operators*, From Number Theory to Physics (M. Waldschmidt, P. Moussa, J. M. Luck, and C. Itzykson, eds.), Springer, Berlin, 1992, pp. 539–630.

[BBG] J. Bellissard, A. Bovier, and J.-M. Ghez. *Gap labelling theorems for one-dimensional discrete Schrödinger operators*, Rev. Math. Phys. **4** (1992), 1–37.

[BCL] J. Bellissard, E. Contensou, and A. Legrand, *K-théorie des quasi-cristaux, image par la trace: le cas du réseau octogonal*, C. R. Acad. Sci. Paris Sér. I Math. **327** (1998), 197–200.

[BLT] J. Bellissard, R. Lima, and D. Testard, *Almost periodic Schrödinger operators*, Mathematics + Physics, vol. 1 (L. Streit, ed.), World Scientific, Singapore, 1985, pp. 1–64.

[BS] J. Bellissard and E. Scoppola, *The density of states of almost periodic Schrödinger operators and the frequency module: a counterexample*, Commun. Math. Phys. **85** (1982), 301–308.

[BP] M. M. Benderskiĭ and L. A. Pastur, *The spectrum of the one-dimensional Schrödinger equation with random potential*, Math. Sborn. USSR **82** (1970), 273–284.

[Bla] B. Blackadar, *K-theory for operator algebras*, Math. Sci. Res. Inst. Publ., vol. 5, Springer, New York, 1986; 2nd ed., Cambridge Univ. Press, Cambridge, 1998.

[Blo] F. Bloch, *Über die Quantenmechanik der Elektronen in Kristallgittern*, Z. Phys. **52** (1928), 555–600.

[Boi] M. de Boissieu, *Quasicrystals: quasicrystalline order, atomic structure and phase transitions*, Lectures on Quasicrystals (F. Hippert and D. Gratias, eds.), EDP Sciences, Les Ulis, 1994, pp. 1–152.

[BHJ] M. Born, W. Heisenberg, and P. Jordan, *Zur Quantenmechanik*. II, Z. Phys. **35** (1925), 557–615.
[BSW] L. P. Bouckaert, R. Smoluchowski, and E. Wigner, *Theory of Brillouin zones and symmetry of wave functions in crystals*, Phys. Rev. **50** (1936), 58–67.
[BR] O. Bratteli and D. W. Robinson, *Operator algebras and quantum statistical mechanics*, vol. 1, 2nd ed., Springer, New York, 1987.
[Bri] L. Brillouin, *Les électrons libres dans les métaux et le rôle des reflexions de Bragg*, J. Phys. Radium **1** (1930), 377–400.
[BGR] L. G. Brown, P. Green, and M. A. Rieffel, *Stable isomorphism and strong Morita equivalence of C^*-algebras*, Pacific J. Math. **71** (1977), 349–363.
[Bru1] N. G. de Bruijn, *Sequences of zeros and ones generated by special production rules*, Nederl. Akad. Wetensch. Indag. Math. **43** (1981), 27–37.
[Bru2] _____*Algebraic theory of Penrose's nonperiodic tiling of the plane*. I, Nederl. Akad. Wetensch. Indag. Math. **43** (1981), 39–52; II, 53–66.
[CMS] L. A. Coburn, R. D. Moyer, and I. M. Singer, C^*-*algebra of almost periodic pseudodifferential operators*, Acta. Math. **139** (1973), 279–307.
[Co1] A. Connes, *Sur la théorie non commutative de l'intégration*, Algèbres d'opérateurs (Les Plans-sur-Bex, 1978) (A. Dold and B. Eckmann, eds.), Lecture Notes in Math., vol. 725, Springer, Berlin, 1979, pp. 19–143.
[Co2] _____, *An analogue of the Thom isomorphism for crossed products of a C^*-algebra by an action of $\mathbb{R}$*, Adv. in Math. **39** (1981), 31–55.
[Co3] _____, *A survey of foliation and operator algebras*, Operator Algebras and Applications, Part 2 (Kinston, 1980) (R. V. Kadison, ed.), Proc. Sympos. Pure Math., vol. 38, Amer. Math. Soc., Providence, RI, 1982, pp. 521–628.
[Co4] _____, *Noncommutative differential geometry*, Inst. Hautes Études Sci. Publ. Math. **62** (1986), 257–360.
[Co5] _____, *Géométrie non commutative*, InterÉditions, Paris, 1990.
[Co6] _____, *Noncommutative geometry*, Acad. Press., San Diego, CA, 1994.
[DT] B. E. J. Dahlberg and E. Trubowitz, *A remark on two-dimensional periodic potentials*, Comm. Math. Helv. **57** (1982), 130–134.
[DS] E. I. Dinaburg and Ya. G. Sinai, *The one-dimensional Schrödinger equation with a quasiperiodic potential*, Funkcional. Anal. i Priložen. **9** (1975), 8 21 (Russian); English transl., Funct. Anal. Appl. **9** (1975), 279–289.
[DK] M. Duneau and A. Katz, *Quasiperiodic patterns*, Phys. Rev. Lett. **54** (1985), 2688–2691.
[Els1] V. Elser, *Indexing problem in quasicrystal diffraction*, Phys. Rev. B **32** (1985), 4892–4898;
[Els2] _____, *Comment on "Quasicrystals: a new class of ordered structures"*, Phys. Rev. Lett. **54** (1985), 1730.
[Elst] A. van Elst, *Gap-labeling theorems for Schrödinger operators on the square and cubic lattice*, Rev. Math. Phys. **6** (1994), 319–342.
[Fes1] H. Feshbach, *Unified theory of nuclear reactions*, Ann. Phys. **5** (1958), 357–390.
[Fes2] _____, *A unified theory of nuclear reactions*. II, Ann. Phys. **19** (1962), 287–313.
[FH] A. H. Forrest and J. R. Hunton, *The cohomology and K-theory of commuting homeomorphisms of the Cantor set*, Ergodic Theory Dynam. Systems **19** (1999), 611–625.
[FHK1] A. H. Forrest, J. R. Hunton, and J. Kellendonk, *Projection quasicrystal*. I. *Toral rotations*, preprint 6.98, NTNU, Trondheim, 1998.
[FHK2] _____, *Cohomology of canonical projection tilings*, preprint no. 395, Sfb 288, TU Berlin, 1999.
[FHK3] _____, *Projection quasicrystal*. II. *Versus substitution tilings*, preprint no. 396, Sfb 288, TU Berlin, 1999.
[FK] A. H. Forrest and J. Kellendonk, 2000 (private communication).
[GK] F. Gähler and J. Kellendonk, *Cohomology groups for projection tilings in codimension 2*, Material Science and Engineering A (to appear).
[Gar] M. Gardner, *Extraordinary non-periodic tiling that enriches the theory of tiles*, Sci. Amer. **236** (1977), 110–121.
[Gin1] J. Ginibre, *Reduced density matrices of quantum gases*. I. *Limit of infinite volume*, J. Mathematical Phys. **6** (1965), 238–251.

[Gin2] J. Ginibre, *Some applications of functional integration in statistical mechanics*, Statistical Mechanics and Quantum Field Theory (C. DeWitt and R. Stora, eds.), Gordon and Breach, New York, 1971, pp. 327–427.

[GH] W. Gottschalk and G. A. Hedlund, *Topological dynamics*, Amer. Math. Soc. Colloq. Publ., vol. 36, Amer. Math. Soc., Providence, RI, 1955.

[Gre] F. P. Greenleaf, *Invariant means on topological groups and their applications*, Van Nostrand-Reinhold, New York, 1969.

[Gum] P. Gummelt, *Penrose tilings as coverings of congruent decagons*, Geom. Dedicata **62** (1996), 1–17.

[Hei1] W. Heisenberg, *Über quantentheoretische Umdeutung kinematischer und mechanischer Beziehungen*, Z. Phys. **33** (1925), 878–893.

[Hei2] W. Heisenberg, *Über den anschaulichen Inhalt der quantentheoretischen Kinematik und Mechanik*, Z. Phys. **43** (1927), 172–198.

[HM] B. Helffer and A. Mohamed, *Asymptotic of the density of states for the Schrödinger operator with periodic potential*, Duke Math. J. (to appear).

[HG] F. Hippert and D. Gratias (eds.), *Lectures on quasicrystals*, EDP Sciences, Les Ulis, 1994.

[Hof] A. Hof, *Diffraction by aperiodic structures*, The Mathematics of Long-Range Aperiodic Order (Waterloo, ON, 1995) (R. V. Moody, ed.), NATO ASI Ser. C: Math. Phys. Sci., vol. 489, Kluwer, Dordrecht, 1997, pp. 239–268.

[Hofs] D. Hofstadter, *The energy-levels of Bloch electrons in rational and irrational magnetic fields*, Phys. Rev. B **14** (1976), 2239–2249.

[JS] H. C. Jeong and P. J. Steinhardt, *Cluster approach for quasicrystals*, Phys. Rev. Lett. **73** (1994), 1943–1946.

[JM] R. Johnson and J. Moser, *The rotation number for almost periodic potentials*, Commun. Math. Phys. **84** (1982), 403–438.

[Jon] H. Jones, *The theory of Brillouin zones and electronic states in crystals*, Series in Physics, North-Holland, Amsterdam, 1960.

[KKL1] P. A. Kalugin, A. Y. Kitaev, and L. S. Levitov, $Al_{0.86}Mn_{0.14}$: *a six-dimensional crystal*, JETP Lett. **41** (1985), 145–149.

[KKL2] ———, *6-dimensional properties of* $Al_{0.86}Mn_{0.14}$ *alloy*, J. Physique Lett. **46** (1985), L601–L607.

[Kar] Yu. E. Karpeshina, *Perturbation theory for the Schrödinger operator with a periodic potential*, Lecture Notes in Math., vol. 1663, Springer, Berlin, 1997.

[KH] A. Katok and B. Hasselblatt, *Introduction to the modern theory of dynamical systems*, Cambridge Univ. Press, Cambridge, 1995.

[KD] A. Katz and M. Duneau, *Quasiperiodic patterns and icosahedral symmetry*, J. Physique **47** (1986), 181–196.

[KaGr] A. Katz and D. Gratias, *Tilings and quasicrystals*, Lectures on Quasicrystals (F. Hippert and D. Gratias, eds.), EDP Science, Les Ulis, 1994, pp. 187–264.

[Kel1] J. Kellendonk, *The local structure of tilings and their integer group of coinvariants*, Commun. Math. Phys. **187** (1997), 115–157.

[Kel2] ———, *Integer groups of coinvariants associated to octagonal tilings*, Operator Algebras and Their Applications (Waterloo, ON, 1994/1995) (P. Fillmore and J. Mingo, eds.) Fields Inst. Commun., vol. 13, Amer. Math. Soc., Providence, RI, 1997, pp. 155–169.

[Kel3] J. Kellendonk, 1999 (private communication).

[KP] J. Kellendonk and I. F. Putnam, *Tilings, C^*-algebras and K-theory*, Directions in Mathematical Quasicrystals (M. Baake and R. V. Moody, eds.), CRM Monogr. Ser., vol. 13, Amer. Math. Soc., Providence, RI, 2000, pp. 177–206, this volume.

[KM] W. Kirsch and F. Martinelli, *On the density of states of Schrödinger operators with random potentials*, J. Phys. A **15** (1982), 2139–2156.

[Kra1] P. Kramer, *Nonperiodic central space filling with icosahedral symmetry using copies of seven elementary cells*, Acta. Cryst. A **38** (1982), 257–264.

[Kra2] P. Kramer, *Quasicrystals: atomic coverings and windows are dual projects*, J. Phys. A **32** (1999), 5781–5793; math-ph/9904001.

[KN] P. Kramer and R. Neri, *On periodic and nonperiodic space fillings of E^m obtained by projections*, Acta. Cryst. A **40** (1984), 580–587; Erratum, **41** (1985), 619.

[LP] J. C. Lagarias and P. A. B. Pleasants, *Repetitive Delone sets and quasicrystals*, Ergodic Theory Dynam. Systems (to appear); math.DS/9909033.

[Lan]   O. E. Lanford III, *Entropy and equilibrium states in classical statistical mechanics*, Statistical Mechanics and Mathematical Problems (Seattle, 1971) (A. Lenard, ed.), Lecture Notes in Phys., vol. 20, Springer, Berlin, 1973, pp. 1–113.

[Le]   T. T. Q. Le, *Local rules for quasiperiodic tilings*, The Mathematics of Long-Range Aperiodic Order (Waterloo, ON, 1995) (R. V. Moody, ed.), NATO ASI Ser. C: Math. Phys. Sci., vol. 489, Kluwer, Dordrecht, 1997, pp. 331–366.

[LS]   D. Levine and P. J. Steinhardt, *Quasicrystals: a new class of ordered structures*, Phys. Rev. Lett. **53** (1984), 2477–2480.

[Mac]   A. Mackay, *Crystallography and the Penrose pattern*, Physica A **114** (1982), 609–613.

[Mey1] Y. Meyer, *Algebraic numbers and harmonic analysis*, North-Holland Math. Library, vol. 2, North-Holland, Amsterdam, 1972.

[Mey2] _____, *Quasicrystals, Diophantine approximations and algebraic numbers*, Beyond Quasicrystals (Les Houches, 1994) (F. Axel and D. Gratias, eds.), Springer, Berlin, 1995, pp. 3–16.

[Moo]   R. V. Moody, *Meyer sets and their duals*, The Mathematics of Long-Range Aperiodic Order (Waterloo, ON, 1995) (R. V. Moody, ed.), NATO ASI Ser. C: Math. Phys. Sci., vol. 489, Kluwer, Dordrecht, 1997, pp. 403–41.

[Mos]   J. Moser, *An example of a Schrödinger equation with an almost periodic potential and a nowhere dense spectrum*, Comm. Math. Helv. **56** (1981), 198–224.

[Moss] R. Mosseri, *Random tilings*, Lectures on Quasicrystals (F. Hippert and D. Gratias, eds.), EDP Science, Les Ulis, 1994, pp. 335–354.

[MS]   R. Mosseri and J. F. Sadoc, *Two and three dimensional nonperiodic networks obtained from self-similar tiling*, The Structure of Noncrystalline Materials 1982 (P. H. Gaskell, J. M. Parker, and E. A. Davis, eds.), Taylor and Francis, London, 1983, pp. 137–149.

[MRW]   P. S. Muhly, J. N. Renault, and D. P. Williams, *Equivalence and isomorphism for groupoid C^*-algebras*, J. Operator Theory **17** (1987), 3–22.

[Mur]   G. J. Murphy, *C^*-algebras and operator theory*, Academic Press, Boston, MA, 1990.

[NB]   S. Nakamura and J. Bellissard, *Low energy bands do not contribute to the quantum hall effect*, Commun. Math. Phys. **131** (1990), 283–305.

[Nak]   S. Nakao, *On spectral distribution for the Schrödinger operator with random potential*, Japan J. Math. (N.S.) **3** (1977), 111–139.

[NS]   D. Nelson and S. Sachev, *Statistical mechanics of pentagonal and icosahedral order in dense liquids*, Phys. Rev. B **32** (1985), 1480–1502.

[Par]   K. R. Parthasarathy, *Probability measures on metric spaces*, Probab. Math. Statist., vol. 3, Academic Press, New York, 1967.

[Pas]   L. A. Pastur, *Spectra of random selfadjoint operators*, Uspekhi. Mat. Nauk. **28** (1973), 3–66; English transl., Russian Math. Survey **28** (1973), 1–67.

[Ped]   G. K. Pedersen, *C^*-algebras and their automorphism groups*, London Math. Soc. Monogr., vol. 14, Academic Press, London, 1979.

[Pei]   R. Peierls, *Zur Theorie der elektrischen und thermischen Leitfähigkeit von Metallen*, Ann. Phys. **4** (1930), 121–148.

[Pen]   R. Penrose, *Pentaplexity: a class of nonperiodic tilings of the plane*, Math. Intelligencer **2** (1979/80), 32–37.

[PV]   M. Pimsner and D. Voiculescu, *Exact sequences for K-groups and Ext-group of certain cross-products of C^*-algebra*, J. Operator Theory **4** (1980), 93–118.

[Rad]   C. Radin, *Aperiodic tilings, ergodic theory and rotations*, The Mathematics of Long-Range Aperiodic Order (Waterloo, ON, 1995) (R. V. Moody, ed.), NATO ASI Ser. C: Math. Phys. Sci, vol. 489, Kluwer, Dordrecht, 1997, pp. 499–519.

[RW]   C. Radin and M. Wolff, *Space tilings and local isomorphism*, Geom. Dedicata **42** (1992), 355–360.

[RS]   M. Reed and B. Simon, *Methods of modern mathematical physics. I. Functional Analysis*, 2nd ed., Academic Press, San Diego, CA, 1980.

[Ren]   J. N. Renault, *A groupoid approach to C^*-algebras*, Lecture Notes in Math., vol. 793, Springer, Berlin, 1980.

[Rie1] M. A. Rieffel, *Morita equivalence for operator algebras*, Operator Algebras and Applications. Part 1 (Kingston, 1980) (R. V. Kadison, ed.), Proc. Symp. Pure Math., vol. 38, Amer. Math. Soc., Providence, RI, 1982, pp. 285–298.

[Rie2] _____, *Applications of strong Morita equivalence to transformation group C^*-algebras*, Operator Algebras and Applications. Part 1 (Kingston, 1980) (R. V. Kadison, ed.), Proc. Symp. Pure Math., vol. 38, Amer. Math. Soc., Providence, RI, 1982, pp. 299–310.

[Rue] D. Ruelle, *Statistical mechanics*: *Rigorous results*, Benjamin, New York, 1969; Addison-Wesley, Redwood City, CA, 1989 (reprint).

[Sak] S. Sakai, *C^*-algebras and W^*-algebras*, Ergeb. Math. Grenzgeb., vol. 60. Springer, Berlin, 1971.

[Sch] M. Schlottmann, *Cut-and-project sets in locally compact Abelian groups*, Quasicrystals and Discrete Geometry (Toronto, ON, 1995) (J. Patera, ed.), Fields Inst. Monogr., vol. 10, Amer. Math. Soc., Providence, RI, 1998, pp. 247–264.

[SBGC] D. Shechtman, I. Blech, D. Gratias, and J. W. Cahn, *Metallic phase with long-range orientational order and no translational symmetry*, Phys. Rev. Lett. **53** (1984), 1951–1953.

[SE] B. I. Shklovskii and A. L. Efros, *Electronic properties of doped semiconductors*, Springer, Berlin, 1984.

[Shu] M. Shubin, *The spectral theory and the index of elliptic operators with almost periodic coefficients*, Uspekhi Mat. Nauk **34** (1979), 95–135 (Russian); Russian Math. Survey **34** (1979), 109–157.

[Sim1] B. Simon, *Trace ideals and their applications*, London Math. Soc. Lecture Note Ser., vol. 35, Cambridge Univ. Press, Cambridge, 1979.

[Sim2] _____, *Almost periodic Schrödinger operators*, Adv. in Appl. Math. **3** (1982), 463–490.

[Sin] Ya. G. Sinaï, *Theory of phase transitions*: *rigorous results*, Internat. Ser. Natural Philos., vol. 108 Pergamon Press, Oxford, 1982.

[Sire] C. Sire, *Electronic spectrum of a $2D$ quasicrystal related to the octagonal quasi-periodic tiling*, Europhys. Lett. **10** (1990), 483–488.

[Skri] M. M. Skriganov, *The spectrum band structure of the three-dimensional Schrödinger operator with periodic potential*, Invent. Math. **80** (1985), 107–121.

[Sla] J. C. Slater, *Electrons in perturbed periodic lattices*, Phys. Rev. **76** (1949), 1592–1601.

[Soc] J. E. S. Socolar, *Simple octagonal and dodecagonal quasicrystals*, Phys. Rev. B **39** (1989), 10519–10551.

[Str1] M. J. O. Strutt, *Wirbelströme im elliptischen Zylinder*, Ann. Phys. **84** (1927), 485–506.

[Str2] _____ *Eigenschwingungen einer Saite mit sinusförmiger Massenverteilung*, Ann. Phys. **85** (1928), 129–136.

[Sut1] A. Sütő, *Singular continuous spectrum on a Cantor set of zero Lebesgue measure for the Fibonacci Hamiltonian*, J. Statist. Phys. **56** (1989), 525–531.

[Sut2] _____, *Spectra of some almost periodic operators*, Number Theory and Physics (Les Houches, 1989) (J. M. Luck, P. Moussa, and M. Waldschmidt, eds.), Springer Proc. Phys., vol. 47, Springer, Berlin, 1990, pp. 162–169.

[Tak] M. Takesaki, *Theory of operator algebras*. I, Springer, New York, 1979.

[Tsa] A. P. Tsai, *Developments in quasicrystals*: *new systems, single crystal and better structure perfection*, Quasicrystals (Tokyo, 1997) (S. Takeuchi and T. Fujiwara, eds.), World Scientific, Singapore, 1998, pp. 253–260.

[Weg] N. E. Wegge-Olsen, *K-theory and C^*-algebras*, Oxford Sci. Publ., Clarendon Press, Oxford, 1993.

[WS] E. Wigner and F. Seitz, *On the constitution of metallic sodium*, Phys. Rev. **43** (1933), 804–810.

[Zak] J. Zak, *Magnetic translation group*, Phys. Rev. A **134** (1964), 1602–1606.

I.R.S.A.M.C, Université Paul-Sabatier, 118, Route de Narbonne, F-31062 Toulouse Cedex, France
E-mail address: `jeanbel@irsamc2.ups-tlse.fr`

Institute for Theoretical Physics, University of Nijmegen, 6525 ED Nijmegen, The Netherlands
E-mail address: `danielh@sci.kun.nl`

I.R.S.A.M.C, Université Paul-Sabatier, 118, Route de Narbonne, F-31062 Toulouse Cedex, France
E-mail address: `marcz@irsamc2.ups-tlse.fr`

Quasicrystals, Parametric Density, and Wulff-Shape

Károly Böröczky, Jr., Uwe Schnell, and Jörg M. Wills

ABSTRACT. The growth of crystals and quasicrystals is based on various minimum energy principles, global ones such as the Gibbs-Curie-Wulff principle and local ones such as the Lennard-Jones principle. There is another approach based on the two well-known facts that minimum energy implies dense packings of atoms (of different size and with respect to the underlying periodic or quasiperiodic structure) and that energy flow acts in the boundary regions of crystals and quasicrystals. This approach, based on an appropriate density function, is a minimum volume principle, based on classical geometry and hence quite elegant. The essential point is that this minimum volume principle, i.e., maximal packing density, leads to a minimal Gibbs-Curie function. The approach is widely applicable and the results coincide with reality. This makes it a good additional model for crystals and quasicrystals. First we describe the crystal case and then we transfer the results to the case of quasicrystals.

1. Introduction

1.1. Structure and Shape. Crystals have been admired since earliest times. The flatness of their faces and their precise geometrical shapes distinguish them from other, in particular amorphous, forms in nature. Apparently nature here shows its atomic structure, and therefore many attempts were made to clarify this impression and to understand nature. Early scientists like Kepler (1611), Hooke (1665), Huyghens (1675) and others had the correct intuition that atomic dense packing is a fundamental property of condensed matter. So they investigated dense sphere packings in order to understand structure and shape of crystals.

A careful investigation of crystal structure began around 1830–1850 by Bravais, Hessel and Möbius (see [**Ha, Se1**]) and ended about 1890 with the famous list of Fedorov and Schoenflies of 32 crystal classes and the 230 space groups—a classification based on algebra and geometry only.

But it was not only until 1912, when Max von Laue showed the atomic structure of crystals via X-ray diffraction patterns, that the theoretical work was justified. Even quantum mechanics did not essentially influence this successful model.

2000 *Mathematics Subject Classification.* 52C23, 52C17, 05B40, 11H31, 82D25.
Research partially supported by OTKA 031984, Hungary.
Research partially supported by Erwin Schrödinger Institute, Vienna.

The short history of quasicrystals took the inverse course (see [**B, J, Se2**]). Their sensational discovery by Shechtman et al. in 1982 (published in 1984) via X-ray diffraction patterns was the very beginning of a tremendous number of publications for a better understanding of their structure and their 'forbidden' symmetries. The fact that R. Penrose had discovered his famous aperiodic tilings one decade before and that 3-dimensional tilings of similar type (Ammann-rhombs) were known, stimulated this research of structure. A very successful approach is the cut and project method, a purely geometric tool.

In both cases—crystals and quasicrystals—the structure can widely be described in mathematical terms without any physics and chemistry. This is due to the fact that the complicated physical and chemical background generates rigorous and clear long-range structures, periodic or quasiperiodic, which can be assumed to be infinite.

1.2. Shape of Crystals and Quasicrystals. As in the case of the structure, there was always the desire to explain also the shape of (large but finite) crystals and quasicrystals in mathematical terms only—or at least as far as possible. This was motivated by the previous arguments and by the fact that in spite of the great variety of shapes of crystals a careful observation shows that crystals of the same material and the same structure have something in common, e.g., the same set of angles and some face-relations.

A first highlight for crystal shapes was Haüy's 'Traité de minéralogie' (1784), based on geometric intuition. In the 17th, 18th and 19th century a general classification list (habit and combination of form) was developed, which can be found in all books of mineralogy. Of course pure descriptions do not give a deeper insight in crystal shape and growth. The basic problem for investigation of the shape is the boundary which cannot be neglected as in the case of structure, where infinitely large crystals and quasicrystals can be assumed. The influence of boundary effects complicates any investigation of crystals and quasicrystals or, with a well-known saying: 'The bulk was created by god, and the surface by the devil'.

In spite of these difficulties there is a highly developed theory for the local and global shape of crystals, and it is the aim of this contribution to introduce and summarize some of the concepts involved. This will be done in a unified fashion, to serve both crystals and quasicrystals.

1.3. Models for Shape and Growth. History of mathematically based models for crystals begins with the Gibbs-Curie principle of minimum energy (1878–1884) which leads to the Wulff-shape (1901), an excellent global model for crystal shape. This phenomenological approach was refined by Kossel and Stranski (1928/29) who developed a local atomic theory. A good description of these models is in [**Ha**]. Further refinements are due to Frank and Kasper (1958), and of course the Lennard-Jones potential and the Morse potential should be mentioned, which are very good models for clusters (see, e.g., [**HM**]).

All these models are based on energy arguments, and combined with the underlying geometric structure they give a good description of the great variety of crystal shapes.

In the short history of quasicrystals most efforts were made to understand their structure (e.g., [**B, BJS, BM, Br1, Br2, LP, Ni, P1, P2, Pl, VG1, VG2**]). Nevertheless there is a growing number of remarkable investigations concerning shape and growth of quasicrystals, and their problem of dense atomic packing (e.g., [**BN, H, JE, OA, OD, RiS, Sm**]). For details we refer to the survey in Section 3.

1.4. A New Model of Growth and Shape. All these approaches are based on minimum energy arguments, and all of them imply the geometric consequence of dense packings of atoms with respect to the underlying structure. Some rules of crystal shape and growth, e.g, by Miller and by Bravais reflect directly the principle of dense packings. The Lennard-Jones potential and the Morse potential can be interpreted as dense packing principles for soft balls, where the density is given by an energy measure. Here and in the following "soft" means that slight deformations of the balls are permitted.

So there is the question of whether there is a purely geometric approach based on a density measure, which simulates energy arguments and is a good model for a wide range of crystals and quasicrystals.

It is the aim of this survey to show that parametric density is an appropriate tool. Parametric density was introduced in [**W2**] for a joint mathematical theory of finite and infinite packings and in [**BHW1, BHW2**] such a theory was developed.

In fact, it turned out that dense sphere packings in the sense of parametric density are good models for crystals: *Small* ones for clusters (hard and soft packings), *long* ones (under the condition of online packing) for whiskers [**W5**], local packings for local growth on facets and, in particular, *large* packings for Wulff-shapes of crystals and quasicrystals [**BS2, BS1**]. The approach also works for packings of spheres of different size, hence for general crystals and quasicrystals, if the structure is prescribed [**S1, W6**]. Due to the mathematical methods (convexity), it does not work for nonconvex crystals or quasicrystals as, e.g., snowflakes, twins or drilled whiskers.

In this survey we only deal with Wulff-shapes of crystals and quasicrystals. Of course we spent some time to understand the crystallographic background, so we are grateful for the help of colleagues who gave useful hints and provided us with good references. We mention in particular W. Schwarz (Siegen) for clusters, P. Gritzmann (Munich) for whiskers, P. M. Gruber (Vienna) for Wulff-shape and Gibbs-Curie principle, and M. Baake, H. U. Nissen, N. Rivier and J. L. Verger-Gaugry for quasicrystals.

After this detailed introduction we describe in Section 2 the basic ideas and methods for crystals. In the main Section 3, these methods and its modification are applied to quasicrystals. The survey ends with a short outlook and open problems.

2. Crystals and the Wulff-Shape

2.1. Definitions. We define parametric density for the special case of hard ball packings, which is sufficient for most types of crystal and quasicrystal growth, but not for microclusters. For simplicity all balls have the same size, and because of scale-invariance have radius 1.

Let B^n denote the ball of radius 1 in Euclidean n-space E^n (in most applications $n = 2$ or $n = 3$). Let $C_k = \{c_1, \ldots, c_k\} \subset E^n$ such that $C_k + B^n$ is a packing of k nonoverlapping hard balls. Here $C_k + B^n$ means the vector sum. Further let conv C_k be the convex hull of C_k, i.e., the smallest convex set containing $c_1, \ldots, c_k$. Clearly, conv C_k is a convex polytope with at most k vertices. The *parametric density* of C_k is defined as

$$\delta(C_k, \varrho) = \frac{k}{V(\operatorname{conv} C_k + \varrho B^n)}, \tag{1}$$

where V denotes the volume an the parameter $\varrho > 0$ controls the influence of the boundary of the packing. Note that we consider here parametric point densities. Multiplying with the volume of a unit ball gives the equivalent notion of a volume density. Obviously the density is maximal for given k if and only if $V(\operatorname{conv} C_k + \varrho B^n)$ is minimal among all packing arrangements C_k (and fixed ϱ).

In this paper we consider point arrangements C_k which are contained in periodic sets as lattices (or a finite union of translates of a lattice, see Section 2.4) or quasiperiodic sets as model sets and vertex sets of tilings (see Section 3).

The underlying structure of any crystal is periodic with respect to some lattice L. The following considerations deal with the simplest case of one lattice. The more complicated and technical general case of multi-lattices can be obtained from the lattice case by superposition (for details see [**S1, W6**]). The density in (1) for expanding domains $\operatorname{conv} C_k$ tends to

$$\delta(L) = \frac{1}{\det L}$$

where $\det L$ is the determinant of L, which equals the volume of the fundamental domain (see, e.g., [**GL**]). For a lattice L in $\mathbb{R}^n$ we call a unit vector u *good* if the sublattice $L_u = (u^\perp \cap L)$ has dimension $n-1$, i.e., the orthogonal $(n-1)$-space has a positive density $\delta_L(u)$, which is defined analogously to $\delta(L)$. It is well known that these good vectors are exactly the vectors (normed to length 1) in the dual lattice $L^* = \{y \mid x \cdot y \in \mathbb{Z}, \forall x \in L\}$ and that $\delta_L(u) = 1/\det L_u$ (see [**GL**]).

A convex polytope is the bounded intersection of finitely many halfspaces, or equivalently, the convex hull of finitely many points. We also allow nonconvex polytopes, i.e., finite unions of convex polytopes. The boundary of a polytope consists of finitely many bounded convex $(n-1)$-polytopes, called *facets*. If each vertex of the polytope P is a point of L then P is called an L-polytope, and each facet unit normal is a good direction in this case.

2.2. Surface Energy. Consider a polytope P with facets F_i such that the exterior unit normal u_i to F_i is good for each facet. Then the Gibbs-Curie energy of P with respect to a parameter $\varrho > 0$ is defined as

$$E_L(P, \varrho) = \sum_i \left(\varrho - \frac{\delta_L(u_i)}{2\delta(L)} \right) \cdot |F_i| \tag{2}$$

where $|\cdot|$ is the $(n-1)$-content. The sum can be interpreted as a weighted surface area, where the weights of the $|F_i|$ are free surface energies. They are all $< \varrho$, but come arbitrarily close to ϱ, hence ϱ can be interpreted as the **free upper surface energy** (fuse). Actually, all the results and definitions in this section can be extended to any domain whose boundary is a Lipschitz $(n-1)$-manifold but we avoid technicalities in this survey.

According to Wulff, a crystal prefers a shape which minimizes the Gibbs-Curie surface energy if the volume is given (see [**Ha, RW**] for an explanation). In order to have only positive summands in the Gibbs-Curie energy (2), we assume that

$$\varrho > \varrho_0 = \max_{u \text{ good}} \frac{\delta_L(u)}{2\delta(L)}. \tag{3}$$

As in the classical Wulff construction we define the Wulff-shape

$$(4) \qquad W(L, \varrho) = \left\{ x \in \mathbb{R}^n \; \middle| \; \forall u \text{ good}, u \cdot x \leq \varrho - \frac{\delta_L(u)}{2\delta(L)} \right\}.$$

Although $W(L, \varrho)$ is formally the intersection of infinitely many halfplanes it is a convex polytope. This follows from Diophantine approximation (see [**W4**], for a more general and elementary proof see [**S1**]). Note that if $\varrho \to \infty$, then the Wulff-shape tends to a ball. In any case, $W(L, \varrho)$ is always the optimal shape:

THEOREM 1 (Wulff, 1901). *Assume P to be a polytope (convex or nonconvex) such that each facet unit normal is a good direction and $V(P) = V(W)$ where $W = W(L, \varrho)$) for some $\varrho > \varrho_0$. Then $E_L(P, \varrho) \geq E_L(W, \varrho)$, and equality holds if and only if $P = W$.*

Wulff gave no complete proof of his result, which was later done by many authors. The most famous proof is by Max von Laue [**L**]. The first geometric proof for convex P appeared in 1943 (see [**D**]) and the shortest geometric proof is by P. M. Gruber (unpublished). The result can be extended to nonconvex polytopes using Busemann's notion of surface area (see [**Th**]).

Theorem 1 explains why crystals prefer convex polytopal shapes, and why the "dense" directions are preferred for the facets.

2.3. Density Deviation. Next we describe how the optimality of the Wulff shape can be modelled using the theory of ball packings and parametric density. For simplicity we first describe the simplest case of lattice packing, which is a model for metals and noble gases only. The underlying lattice L is given by crystallographic facts. Then we have the following characterization of the Wulff shape (see [**BB2**]):

THEOREM 2. *Let ϱ be suitably large (say $\varrho > \varrho_0$ if $n = 3$) and assume that $C_k \subset L$ maximizes the parametric density for given k. Then*

$$\lim_{k \to \infty} \frac{\operatorname{conv} C_k}{\lambda_k} = W(L, \varrho),$$

where λ_k is of order $k^{1/n}$.

The proof relies on a number theoretic formula counting the lattice points in large bodies [**BB1**]. The condition $\varrho > \varrho_0$ is essential in Theorem 2, since if $\varrho < \varrho_0$ then the optimal arrangement is lower dimensional. For example, if ϱ is small, then the optimal C_k is formed by collinear points (see [**BHW1, BHW2**]). Moreover, $\varrho > \varrho_0$ coincides with the interpretation of $\varrho - \varrho_0$ as a free surface energy, cf. Section 2.2, Equations (2) and (3).

Densest individual packings, i.e., atom by atom, can have substantial global changes, if only one atom is added, which contradicts physical reality and intuition. It is much easier and closer to reality to investigate the average crystal growth rather than the growth for each individual atom with all its irregularities. This approach leads to the density deviation and is explained in the following. In fact, this averaging process was the first approach to obtain Wulff-shapes and was developed in [**W4**] and [**W3**] for the simplest case of lattice packings of spheres of the same size. Of course it leads to the same Wulff-shapes as in the individual approach [**BB2**]. We sketch the idea:

Let P be a polytope with facets F_i and exterior unit normals u_i in good directions. Then we consider, as an averaging process, the growing shells $\lambda P, \lambda \in \mathbb{N}$.

With $k = \text{card}(\lambda P \cap L)$ and (1) we have
$$\delta(\lambda P, \varrho) = \frac{\text{card}(\lambda P \cap L)}{V(\lambda P + \varrho B^n)}.$$

This is the quotient of two well-known polynomials. The lattice point enumerator $\text{card}(\lambda P \cap L)$ is the Ehrhart polynomial, (see [**GL**]):

(5) $\quad \text{card}(\lambda P \cap L) = \delta(L) V(P) \cdot \lambda^n + \frac{1}{2} \left(\sum_i \delta_L(u_i) \cdot |F_i| \right) \cdot \lambda^{n-1} + O(\lambda^{n-2}).$

In the denominator, it follows from the Steiner-formula for the volume of the parallel body (see [**SY**]) that

(6) $\quad V(\lambda P + \varrho B^n) = V(P) \cdot \lambda^n + \varrho \left(\sum_i |F_i| \right) \cdot \lambda^{n-1} + O(\lambda^{n-2}).$

Polynomial division leads to the Laurent expansion:
$$\delta(\lambda P, \varrho) = \delta(L) + \lambda^{-1} (V(P))^{-1/n} \Delta_L(P, \varrho) + O(\lambda^{-2}),$$
where

(7) $\quad \Delta_L(P, \varrho) = \lim_{\lambda \to \infty} V(\lambda P)^{1/n} \cdot \left(\delta(L) - \delta(\lambda P, \varrho) \right),$

is called the *density deviation*. It can be considered as a second order density or as a derivative at infinity. Using (5) and (6) we obtain

(8) $\quad \Delta_L(P, \varrho) = \delta(L) \cdot V(P)^{\frac{1-n}{n}} \sum_i \left(\varrho - \frac{\delta_L(u_i)}{2\,\delta(L)} \right) \cdot |F_i|.$

It follows that the geometric notion of the density deviation is equivalent to the Gibbs-Curie surface energy. Observe that $\Delta(P, \varrho) < \Delta(Q, \varrho)$ implies that the parametric density of λP is larger than that of λQ for large λ. The concept of density deviation (or equivalently the surface energy) has been used for extensions to general crystals (see Section 2.4) and to nonperiodic sets (see Section 3).

2.4. General Crystals and Wulff Shapes. In the previous section, the simplest case of lattice packings was described. But most crystals are built by several types of atoms or, as in the case of the diamond, the underlying structure is no mathematical lattice but a periodic structure, i.e., the union of finitely many translates of a lattice. So the atoms, even of one sort, may form a multi-orbital structure, where an orbit is meant with respect to the underlying crystallographic group. These more general structures, i.e., periodic packings of balls in different orbits and possibly of different size, were first considered in [**S1**] and later in [**W6**]. The idea is the same as described, again with an appropriate density deviation, but the technical details are much more complicated than in the 1-atomic case. The essential point is that in all examples chosen in [**S1**], namely diamond, NaCl and ZnS, the obtained results coincide with reality for suitably chosen parameter ϱ, i.e., the range of possible Wulff-shapes can be realized by the choice of ϱ. The basic shapes are obtained for ϱ slightly larger than ϱ_0 (see (3)). With the help of a computer program (unpublished) U. Schnell determined also the Wulff-shape for pyrite FeS_2, which is for certain ϱ a nonregular dodecahedron (see Figure 1).

Further we mention that in [**S2**] with this model an explanation was given for the fact that most of the noble gases crystallize in the face centred cubic lattice fcc and not in the hexagonal closest packing hcp or related periodic structures.

 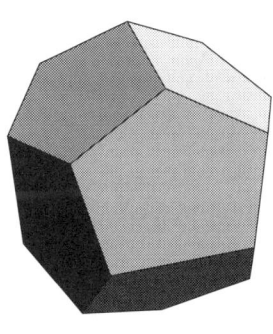

FIGURE 1. Pyrite, the natural shape and the Wulff-shape obtained by parametric density and a computer program

The Lennard-Jones potential does not reflect this fact since the hcp structure has a slightly lower energy per particle than fcc ([**Wo, BR**]).

3. Quasicrystals

3.1. Model Sets. A model set is as a projection of points of a higher-dimensional lattice L into an irrational subspace E. Model sets show a similar kind of quasiperiodicity and quasisymmetry as the quasicrystals. So they are an adequate mathematical model for the structure of quasicrystals. The construction is based on the cut and project method, which can be described as follows:

(a) Assume that L is a lattice in $\mathbb{R}^{n+m}$ and E (the so-called physical space) is a linear n-flat such that the origin is the only lattice point in E.
(b) Fix an open Jordan measurable subset (also called Riemann integrable, see for instance [**Sch2**] in this volume) Ω of $E_\perp$. This set is called a *window*, and the model set $\Lambda \subset E$ is the projection of the points of $(\Omega + E) \cap L$ into E; namely, the points whose projection into $E_\perp$ is contained in Ω.

A model set is called primitive, if E is not contained in any lattice $(m + n − 1)$-plane; or equivalently, the projection of L into $E_\perp$ is dense. Note that in the early literature (say [**M1**]), the term "model set" is used for primitive model sets.

To illustrate this method, we consider the one-dimensional example pictured in Figure 2. Here, $L = \mathbb{Z}^2$ and E is a line of irrational slope. The lattice points contained in the strip project to a one-dimensional nonperiodic set.

The model set Λ has some long-range order. For example, balls of fixed radius T contain only finitely many different configurations of points of Λ up to translation. In addition, there exists an R depending on T such that any ball of radius R in E contains a translate of each configuration, and hence Λ is *repetitive* or *minimal*. On the other hand, Λ is not periodic with respect to any vector, because E is irrational.

An important property of model sets is that they can have symmetries which are impossible for lattices. If both L and Ω are invariant under the action of a group, then Λ is also invariant. This way model sets can be constructed with five-fold symmetry (e.g., the vertices of the Penrose tiling, $L = \mathbb{Z}^5$ or $L = A_4$), or with the

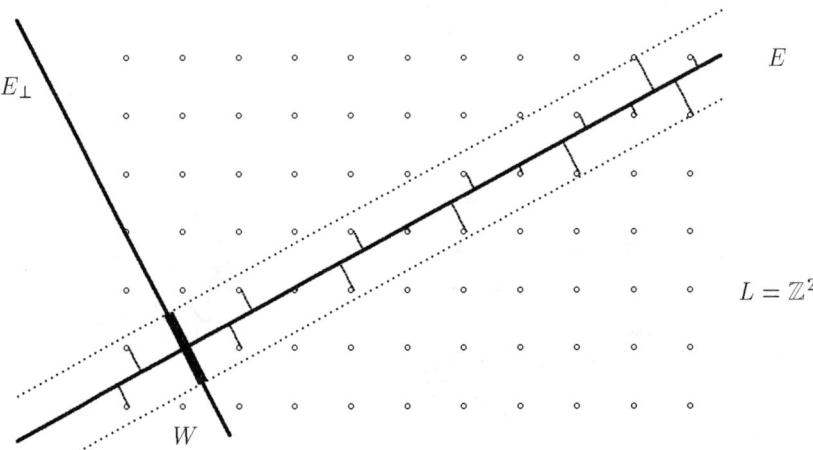

FIGURE 2. A one-dimensional example of a model set obtained by the cut and project method.

symmetries of the regular icosahedron (e.g., $L = \mathbb{Z}^6$ or $L = D_6$). These forbidden symmetries were the crucial point of Shechtman's discovery of quasicrystals. For 3D quasicrystals with icosahedral symmetry see [**BN, NWBC**]. Some more examples are discussed below in detail.

The discussion below shows that the analogues of the results about crystals seem to hold for a large class of model sets. Clearly more examples are desirable, in order to check how far reaching the method is and under which conditions it possibly needs some corrections.

Let K be a Jordan measurable subset with nonempty interior in E. If Λ is primitive and λ is large then

$$\text{card}(\lambda K \cap \Lambda) \sim \frac{|\Omega|}{\det L} \cdot V(\lambda K), \tag{9}$$

where $|\Omega|$ denotes the m-dimensional content of the window Ω (see [**Sch1, Ho, BS1**]). So it is natural to define

$$\delta(\Lambda) = \frac{|\Omega|}{\det L}.$$

Now if Λ is the disjoint union of the primitive model sets $\Lambda_1, \ldots, \Lambda_k$ then define $\delta(\Lambda) = \sum_{i=1}^{k} \delta(\Lambda_i)$. For example, the vertices of the Penrose tiling is the disjoint union of four primitive model sets (see Section 3.2). In any case,

$$\delta(\Lambda) = \lim_{\lambda \to \infty} \frac{\text{card}(\lambda K \cap \Lambda)}{V(\lambda K)}.$$

In order to define the Wulff shape, we need the lower dimensional densities. Therefore assume that if an affine subspace in $E_\perp$ intersects the interior of the window Ω, then the intersection is Jordan measurable inside the affine subspace. Call a unit vector $u \in E$ *good* if $u^\perp \cap L$ contains an at least n-dimensional lattice $L(u)$ such that the linear hull of $L(u)$ intersects E in an $(n-1)$-dimensional subspace.

Then $(x + u^\perp) \cap \Lambda$ is an $(n-1)$-dimensional model set for any $x \in \Lambda$, and define
$$\delta_\Lambda(u) = \sup_{x \in E} \delta\big((x + u^\perp) \cap \Lambda\big).$$
Here we can take only supremum because the density of the model subset with dimension $n-1$ depends on the content of the intersection of the window Ω and $x + \lin L(u)$ (see the examples in Sections 3.2 and 3.3).

Now the Gibbs-Curie energy and the Wulff shape can be defined analogously to the periodic case. Let Λ be a model set and
$$\varrho > \varrho_0 = \max_{u \text{ good}} \frac{\delta_\Lambda(u)}{2\delta(\Lambda)}.$$
If for a polytope P, the exterior unit normal u_i is good for each facet F_i then the Gibbs-Curie energy of P is
$$E_\Lambda(P, \varrho) = \sum_i \left(\varrho - \frac{\delta_\Lambda(u_i)}{2\delta(\Lambda)}\right) \cdot |F_i|.$$
In addition, we define the "quasi-crystallographic" Wulff shape as
$$W(\Lambda, \varrho) = \left\{x \in \mathbb{R}^n \mid \forall u \text{ good}, u \cdot x \leq \varrho - \frac{\delta_\Lambda(u)}{2\delta(\Lambda)}\right\}.$$
These definitions are analogous to those for the periodic case ((3), (2), and (4)), but it is still not really understood when $W(\Lambda, \varrho)$ is a convex polytope.

CONJECTURE 1. *If E can be defined in a basis of L using an algebraic extension of $\mathbb{Q}$, then $W(\Lambda, \varrho)$ is a convex polytope for $\varrho > \varrho_0$.*

The existing methods verify the conjecture for quadratic extensions (see [**BS2, BS1**] and Sections 3.2 and 3.3).

The proof of the Wulff theorem applies also to this case. In particular,

THEOREM 3. *If $W(\Lambda, \varrho)$ is a polytope and P is another polytope, with $V(P) = V\big(W(\Lambda, \varrho)\big)$, such that each exterior unit normal is good, then*
$$E(P, \varrho) \geq E(W(\Lambda, \varrho), \varrho)$$
with equality if and only if P is congruent with $W(\Lambda, \varrho)$.

Note that the definition of Gibbs-Curie energy can be extended to any body in E with Lipschitz boundary, and Theorem 3 still holds in this extended context. There are various examples of quasicrystals, where the shape predicted by our construction coincides with reality (see Figures 3 and 4, and Section 3.3).

Next we discuss how parametric density of subsets of model sets is related to quasicrystal growth. The crucial point is to find the analogue of the formula (5) of Ehrhart. In general, we only have the following result (see [**BS1**]):

THEOREM 4. *Let Λ be a primitive model set which can be defined in a basis of L using an algebraic extension of degree q of $\mathbb{Q}$. If K and the window Ω is a polytope or a convex body, and λ is large then*
$$\card(\lambda K \cap \Lambda) = \delta(\Lambda) V(K) \cdot \lambda^n + O\big(\lambda^{n - 1/q(n+m)} (\log \lambda)^{1/q}\big).$$

Unfortunately, the exponent of λ most probably can not be brought down to $n-1$ in the error term (not even for quadratic extensions). So we consider model sets with "deflation and inflation". Using say local matching rules, one can obtain down scaled copies of Λ by a factor ϑ^{-1} for many model sets. This process is called deflation. Now inflation is the inverse process, and ϑ is called the *inflation multiplier*

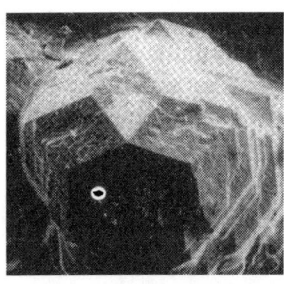 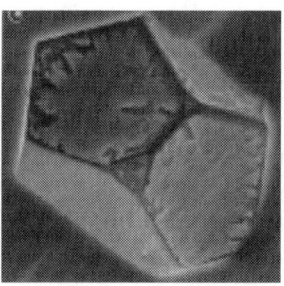 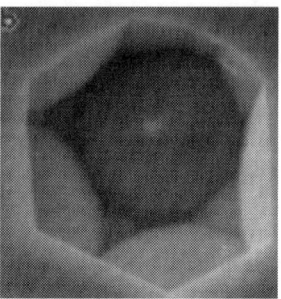

FIGURE 3. Scanning electron microscopy images of single grains of real quasicrystals.

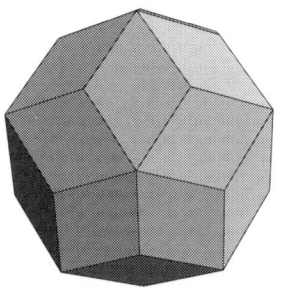 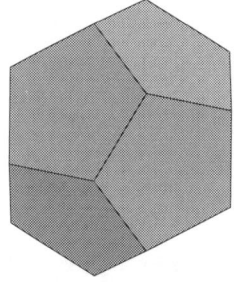 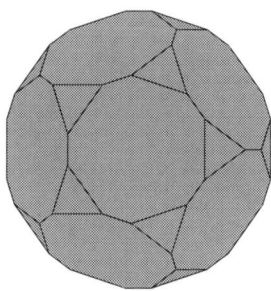

FIGURE 4. Wulff-shapes obtained by parametric density and a computer program.

of Λ if $\vartheta\Lambda \subset \Lambda$. Then ϑ is an algebraic number whose degree divides $n + m$. In addition, ϑ is Pisot number; namely, the absolute value of each of its conjugates is less than 1. Actually, there exists a related nonsingular linear transformation M of $\mathbb{R}^{n+m}$ such that M acts on E by multiplication by ϑ. If Λ is primitive then even $M(L) \subset L$ and $M(W) \subset W$. All these properties can be found in [**M2**, Theorem 6] and [**Lag**, Theorem 4.1] and see [**BDGPS**] for the properties of Pisot numbers. A beautiful example of these phenomena is the Penrose tiling discussed in the next section.

In spite of the restrictive algebraic conditions, many model sets occurring in the applications do possess inflation.

It is natural to call a polytope a Λ-*polytope* if each vertex is in Λ. Note that most probably, each unit facet normal is a good direction in this case.

CONJECTURE 2. *If Λ has an inflation multiplier ϑ and P is a Λ-polytope, then*

$$\mathrm{card}(\vartheta^k P \cap \Lambda) = \delta(\Lambda)V(P)\cdot \vartheta^{n\cdot k} + \frac{1}{2}\left(\sum_i \delta_\Lambda(u_i)\cdot |F_i|\right)\cdot \vartheta^{(n-1)\cdot k} + O\bigl(\vartheta^{(n-2)\cdot k}\bigr).$$

This formula is the direct analogue to Ehrhart's formula (5), except that the factor of dilation grows exponentially. In [**BS2**] the conjecture is verified for the Penrose tiling (see Section 3.2).

QUASICRYSTALS, PARAMETRIC DENSITY, AND WULFF-SHAPE 269

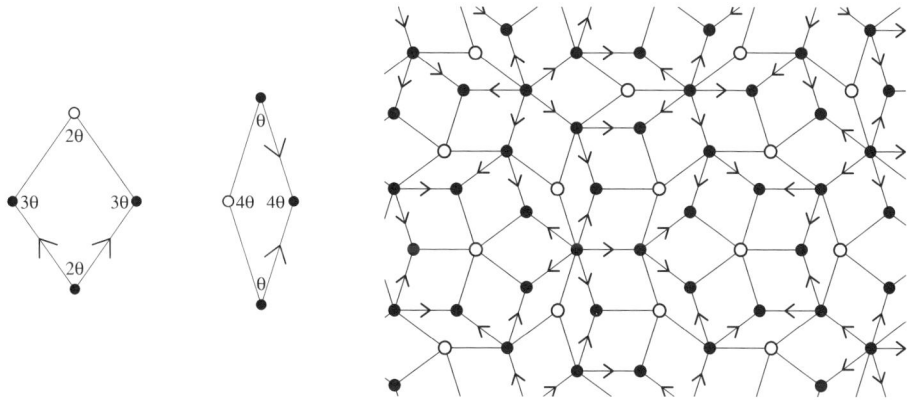

FIGURE 5. The Penrose tiling is obtained as a tiling of two rhombs together with matching rules.

If Λ is a model set with inflation multiplier ϑ, and P is a Λ-polytope, then we define the density deviation analogous to (7):

$$(10) \qquad \Delta_\Lambda(P, \varrho) = \lim_{k \to \infty} V(\vartheta^k P)^{1/n} \cdot \left(\delta(\Lambda) - \frac{\mathrm{card}(\vartheta^k P \cap \Lambda)}{V(\vartheta^k P + \varrho B^n)} \right).$$

Now let us assume that Conjecture 2 holds. Then using (6), it follows analogously to (8) that

$$\Delta_\Lambda(P, \varrho) = \delta(\Lambda) V(P)^{(1-n)/n} \cdot E_\Lambda(P, \varrho),$$

and Conjecture 1 yields that the density deviation is minimized by the Wulff shape. Therefore, similarly as for crystals, parametric density of subsets of model sets and shape of quasicrystals are intimately connected.

3.2. Penrose Tiling. Penrose tilings and their analogues show a similar kind of quasi-periodicity and quasi-symmetry as the quasicrystals which were, in fact, discovered ten years later.

First we consider the classical 2-dimensional Penrose tiling as a tiling, and we come back to the interpretation of the vertices as a model set at the end. The material of this section is explained in [**BS2**] in more detail. The results indicate that the shape of quasicrystals might be explained in terms of parametric density.

The 2-dimensional Penrose tiling is obtained as a tiling of two rhombs, a thick one with angles $2\pi/5$ and $3\pi/5$, and a thin one with angles $\pi/5$ and $4\pi/5$, with certain matching rules, indicated by the black and white points and the arrows (see Figure 5).

Further it can be produced by the process of inflation. In one step of inflation the rhombs are decomposed into half-rhombs (see Figure 6).

These are rearranged to obtain new rhombs and then multiplied by the factor of the golden ratio $\tau = (\sqrt{5}+1)/2 = 1.618\ldots$. Then the new rhombs have the original size again. Iteration of this procedure leads to a sequence of space filling patches of the same shape (see Figure 7).

The family of vertices of the Penrose rhombs is denoted by P and a union R of rhombs is called a P-set. As one can see in Figure 8, it may be nonconvex.

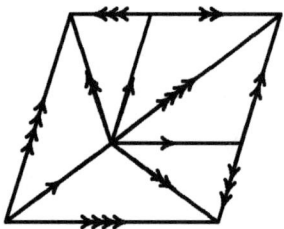 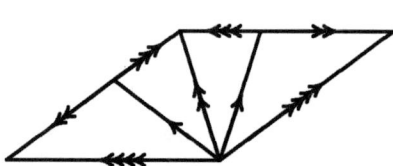

FIGURE 6. Inflation rules for the Penrose tiling.

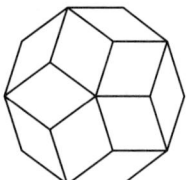 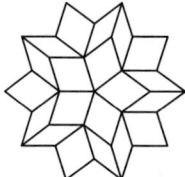 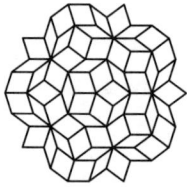 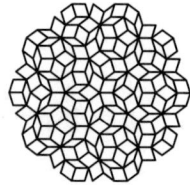

FIGURE 7. Spacefilling sequence obtained by inflation.

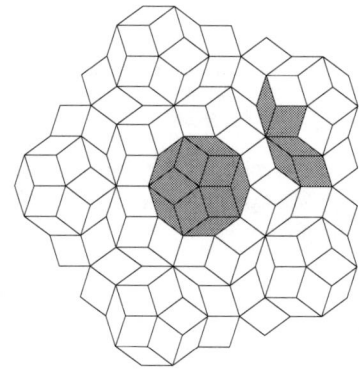

FIGURE 8. The regular decagon is a P-set. P-sets may be nonconvex.

Further, let R_k be the resulting set after k steps of inflation. By a careful analysis of the inflation rules, we can give a formula for the number $\mathrm{card}(P \cap R_k)$ of P-points in R_k for a P-set R.

THEOREM 5. *Let R be a P-set. Then*

$$\mathrm{card}(P \cap R_k) = \mathrm{card}(P \cap \tau^k R) = \delta V(\tau^k R) + \delta'/2 F(\tau^k R) + O(1),$$

where $\delta = 2(\tau + 1) \cdot \sin \pi/5 \sim 3.077$, $\delta' = (2\tau + 1)/\sqrt{10} \sim 1.339$ and P denotes the surface area (perimeter).

The constant δ is the 2-dimensional density of P-points in E and δ' is the density along the five dense directions of the Penrose tiling (parallel to the diagonals of the regular decagon in Figure 8, see also Figure 9).

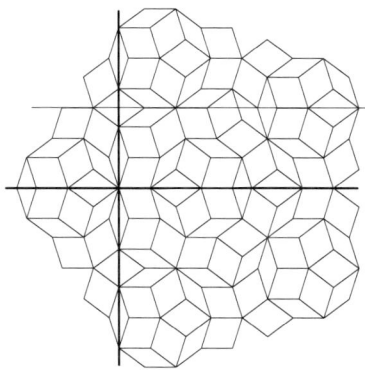

FIGURE 9. 1-dimensional densities. The horizontal line is one of the 10 densest directions, the vertical line one of the 10 second best.

This formula is analogous to Ehrhart's formula for the lattice point enumerator (5), and verifies Conjecture 2 for P-sets as shown in the following.

The Penrose tiling contains a regular decagon T which is the union of ten rhombs, five of each type (see Figure 8).

For P-sets we can show the following isoperimetric inequality:

THEOREM 6. *Let R be a P-set. Then*
$$V(R) \leq \sin\left(\frac{\pi}{5}\right) \cdot \frac{\tau+1}{20} F(R)^2,$$
with equality if and only if R is the regular decagon.

The density deviation for P-sets is defined by
$$\Delta(R, \varrho) = \lim_{n \to \infty} \left(V(R_n)\right)^{1/2} \cdot \left(\delta - \frac{\operatorname{card}(P \cap R_n)}{V(R_n + \varrho B^2)}\right).$$

Now $\Delta(R, \varrho)$ can be calculated using (6) and Theorem 5 to obtain

(11) $$\Delta(R, \varrho) = \delta\left(\varrho - \frac{\delta'}{2\delta}\right) F(R) V(R)^{-1/2}.$$

An immediate consequence of Theorem 6 is

COROLLARY 1. *Let R be a P-set. Then*
$$\Delta(R, \varrho) \geq \Delta(T, \varrho),$$
with equality if and only if R is the regular decagon T.

Now we turn to the representation of the vertices of the Penrose tiling as a model set. In this case $L = \mathbb{Z}^5$. The cyclic group of order five has a natural action on the orthonormal basis of $\mathbb{Z}^5$. The invariant subspaces of this action is the line spanned by the vector $v_0 = (1,1,1,1,1)$ and two 2-planes. Let E be one of the 2-planes.

Finally, the window Ω is the projection of the unit cube into $E_\perp$, which is Kepler's rhombic triacontahedron. In particular v_0 is orthogonal to E.

There exist four other 2-planes parallel to E which contain some lattice point and intersect Ω. This yields that P is the union of four primitive model sets, and each of them is defined using a lattice isomorphic to the four-dimensional lattice A_4. For a representation of the Penrose tiling constructed via 4-space see [**BKSZ**].

The set of good directions can be characterized similarly as in the 3D case (see Sectio 3.3), so we do not discuss it here. The Gibbs-Curie energy and the Wulff shape can be defined as in Section 3.1, and we have

THEOREM 7. (a) If $\varrho > \varrho_0$, then $W(P, \varrho)$ is a convex polygon, and hence the Wulff shape $W(P, \varrho)$ optimizes the Gibbs-Curie energy.
(b) If $\varrho_0 < \varrho < 6\varrho_0$, then $W(P, \varrho)$ is a regular decagon.

For increasing ϱ the Wulff-shape $W(P, \varrho)$ gets more and more edges (e. g. in the next step twenty edges parallel to the densest and the second densest lines) and for $\varrho \to \infty$ it tends to a circle.

For the vertices of the Penrose tiling, the inflation multiplier introduced in section 3.1 is $\vartheta = \tau$, corresponding to the matrix

$$M = \frac{1}{2} \begin{pmatrix} 1 & 1 & -1 & -1 & 1 \\ 1 & 1 & 1 & -1 & -1 \\ -1 & 1 & 1 & 1 & -1 \\ -1 & -1 & 1 & 1 & 1 \\ 1 & -1 & -1 & 1 & 1 \end{pmatrix}.$$

So Conjecture 2 is verified for P-sets by Theorem 5.

3.3. 3D Quasicrystals. In this section, we consider 3D quasicrystals following [**BS1**]. For most real quasicrystals the lattice L is the 6-dimensional integer lattice $\mathbb{Z}^6$ or the face-centred hypercubic lattice $D_6 = \{x \in \mathbb{Z}^6 \mid x_1 + \cdots + x_6 \text{ even}\}$. Both lattices are invariant under the natural action of the icosahedral group Y_h. In both cases, one common choice of the window Ω is the projection of the interior of the projection of the 6-dimensional unit cube into $E_\perp$. Then Ω is Kepler's rhombic triacontahedron with edgelength $1/\sqrt{2}$ and volume $V(\Omega) = 2\sqrt{2}\sin(\pi/5)(\tau + 1) = 4.35\ldots$. We denote by Λ both model sets, which are primitive in this case. These two 3-dimensional irrational subspaces E and $E_\perp$ can be represented as image and kernel of the matrix

$$\begin{pmatrix} \sqrt{5} & -1 & -1 & -1 & -1 & -1 \\ -1 & \sqrt{5} & -1 & 1 & 1 & -1 \\ -1 & -1 & \sqrt{5} & -1 & 1 & 1 \\ -1 & 1 & -1 & \sqrt{5} & -1 & 1 \\ -1 & 1 & 1 & -1 & \sqrt{5} & -1 \\ -1 & -1 & 1 & 1 & -1 & \sqrt{5} \end{pmatrix}.$$

This construction goes back to [**K, KN**]. For the surface energy of periodic crystals, the density of 2-planes plays an important role. So we consider here densities of 2-planes (see also [**MP**]). For planar model sets, of course, we consider densities of lines. The horizontal line in Figure 9 indicates one of the 10 densest directions and the vertical line one of the 10 second best directions in the Penrose tiling. As the thin horizontal line shows, parallel lines may have different densities.

Here the set of good directions as defined in Section 3.1 can be characterized by

(12) $$G = \{(R + \tau RH) \cdot x \mid x \in \mathbb{Z}^6\} \subset E$$

where R and H are certain (6×6)-integer matrices.

For $u = a + \tau b \in G$ with $a, b \in \mathbb{Z}^6$ there is a corresponding (or conjugate) vector $u' = b - \tau a \in E_\perp$ such that u and u' together span a rational subspace. Let $L_2(u) = \text{lin}(u, u') \cap L$. With an argument from uniform distribution we can determine the maximal density $\delta_2(u)$ of a plane orthogonal to u.

$$\delta_2(u) = \frac{q(u)}{\det L \cdot \det L_2(u)}. \tag{13}$$

where $q(u) = V_2(\Omega \cap E_2(u'))$ and $E_2(u') \subset E_\perp$ is orthogonal to u'.

By Diophantine approximation it can be shown that there is a sequence of parallel lines orthogonal to $u \in G$, whose densities tend to this supremum $\delta_2(u)$.

A Λ-polytope P in E is characterized by the fact that the exterior unit normal facet vectors are contained in G. Then one can introduce the Gibbs-Curie surface energy for G-polytopes and the Wulff shape as in Section 3.1.

THEOREM 8. *If $\varrho > \varrho_0$ then $W(\Lambda, \varrho)$ is a polytope, and hence the Wulff shape $W(\Lambda, \varrho)$ optimizes the Gibbs-Curie energy.*

  (a) *If $L = \mathbb{Z}^6$ and $\varrho_0 < \varrho < 2\varrho_0$, then $W(\Lambda, \varrho)$ is a regular rhombic triacontahedron.*
  (b) *If $L = D_6$ and $\varrho_0 < \varrho < 1.21\varrho_0$, then $W(\Lambda, \varrho)$ is a regular dodecahedron.*

If ϱ tends to ∞, then the Wulff-shape tends to a ball and in general it is hard to compute. The shapes in parts (a) and (b) in fact appear in reality as the scanning electron microscopy images in Figure 3 show. The process of inflation can be applied also in these cases (see [**VG1**] or [**OA**]). The inflation multiplier introduced in Section 3.1 is $\vartheta = \tau$, and say for D_6, the corresponding matrix is

$$M = \frac{1}{2} \begin{pmatrix} 1 & 1 & 1 & 1 & 1 & 1 \\ 1 & 1 & 1 & -1 & -1 & 1 \\ 1 & 1 & 1 & 1 & -1 & -1 \\ 1 & -1 & 1 & 1 & 1 & -1 \\ 1 & -1 & -1 & 1 & 1 & 1 \\ 1 & 1 & -1 & -1 & 1 & 1 \end{pmatrix}.$$

It is unknown whether Conjecture 2 holds in this case.

4. Concluding Remarks

In the previous chapter the successful approach to quasicrystals was shown. But there is still much less known than in the case of crystals. Besides providing a survey of the recent developments, the main purpose of this paper is to initiate further research:

- The polytopality of the Wulff-shape corresponding to a model set Λ (see Conjecture 1) is essential for the theory, but it has been only established if Λ can be defined in a quadratic extension of $\mathbb{Q}$. On the other hand, say planar model sets with seven-fold symmetry can be only defined using some cubic extensions of $\mathbb{Q}$ (see [**BJS**]).
- Although density deviation is an averaging process, the Wulff shape for quasicrystals does not take into consideration the possible randomness of the underlying structure yet as described in [**BM**] for model sets with entropy.
- It would be desirable to transfer the results for online packings and local growth (see [**W5**]) from crystals to quasicrystals.

- In our investigations we considered the simplest case of equal atoms centred in the points of the model set. But clearly one can also choose subsets and introduce weights for different atoms as was done in the periodic case (see [**S1**]).
- It is a basic problem to find a good asymptotic formula for the number of the points of a given model set in an expanding domain (see Conjecture 2).

Independent of these remarks there is a general conclusion that the geometric approach via parametric density is a quite general and realistic model for the shape of crystals and quasicrystals.

References

[AM] N. W. Ashcroft and N. D. Mermin, *Solid state physics*, Holt, Rinehart and Winston, New York, 1976.

[B] M. Baake, *A guide to mathematical quasicrystals*, Quasicrystals (J.-B. Suck, M. Schreiber, and P. Häussler, eds.), Springer, Berlin, 2001 (to appear); math-ph/9904014.

[BJS] M. Baake, D. Joseph, and M. Schlottmann, *The root lattice D_4 and planar quasilattices with octagonal and dodecagonal symmetry*, Internat. J. Modern Phys. B **5** (1991), 1927–1953; mp_arc/00-255.

[BKSZ] M. Baake, P. Kramer, M. Schlottmann, and D. Zeidler, *Planar patterns with fivefold symmetry as sections of periodic structures in 4-space*, Internat. J. Modern Phys. B **4** (1990), 2217–2268.

[BM] M. Baake and R. V. Moody, *Diffractive point sets with entropy*, J. Phys. A **31** (1998), 9023–9039.

[BN] C. Beeli and H. U. Nissen, *Growth morphology of icosahedral Al-Mn-Pd single quasicrystals*, Philos. Mag. B **68** (1993), 487–512.

[BDGPS] M.-J. Bertin, A. Decomps-Guilloux, M. Grandet-Hugot, M. Pathiaux-Delefosse, and J.-P. Schreiber, *Pisot and Salem numbers*, Birkhäuser, Basel, 1992.

[BB1] U. Betke and K. Böröczky, Jr., *Asymptotic formulae for the lattice point enumerator*, Canad. J. Math. **51** (1999), 225–249.

[BB2] _____, *Finite lattice packings and the Wulff-shape*, Mathematika (to appear).

[BHW1] U. Betke, M. Henk, and J. M. Wills, *Finite and infinite packings*, J. Reine Angew. Math. **453** (1994), 165–191.

[BHW2] _____, *Sausages are good packings*, Discrete Comput. Geom. **13** (1995), 297–311.

[BR] B. Borden and C. Radin, *The crystal structure of noble gases*, J. Chem. Phys. **75** (1981), 2012–2013.

[BS1] K. Böröczky, Jr., and U. Schnell, *Wulff shape for nonperiodic arrangements*, Lett. Math. Phys. **45** (1998), 81–94.

[BS2] _____, *Quasicrystals and the Wulff-shape*, Discrete Comput. Geom. **21** (1999), 421–436.

[Br1] N. G. de Bruijn, *Algebraic theory of Penrose's nonperiodic tilings in the plane*. I, Nederl. Akad. Wetensch. Indag. Math. **43** (1981), 39–52; II, 53–66.

[Br2] _____, *Updown generation of Penrose patterns*, Indag. Math. (N.S.) **1** (1990), 201–219.

[D] A. Dinghas, *Über einen geometrischen Satz von Wulff für die Gleichgewichtsform von Kristallen*, Z. Kristallogr., Mineral. Petrogr. **105** (1944), 304–314.

[DKS] R. Dobrushin, R. Kotecký, and S. Shlosman, *Wulff construction: a global shape from local interaction*, Transl. Math. Monogr., vol. 104, Amer. Math. Soc., Providence, RI, 1992.

[GL] P. M. Gruber and C. G. Lekkerkerker, *Geometry of numbers*, 2nd ed., North-Holland Math. Library, vol. 37, North-Holland, Amsterdam, 1987.

[GS] B. Grünbaum and G. C. Shephard, *Tilings and patterns*, Freeman, San Francisco, 1986.

[Ha] P. Hartman, (ed.), *Crystal growth: an introduction*, North-Holland, Amsterdam, 1973.

[H] C. Henley, *Sphere packings and local environments in Penrose tilings*, Phys. Rev. B **34** (1986), 797–816.

[HM] M. R. Hoare and J. A. McInness, *Morphology and statistical statics of simple microclusters*, Adv. in Phys. **32** (1983), 791–821.

[Ho]   A. Hof, *Diffraction by aperiodic structures*, The Mathematics of Long-Range Aperiodic Order (Waterloo, ON, 1995) (R. V. Moody, ed.), NATO ASI Ser. C: Math. Phys. Sci., vol. 489, Kluwer, Dordrecht, 1997, pp. 239–268.

[J]   C. Janot, *Quasicrystals. A primer*, 2nd ed., Monographs on the Physics and Chemistry of Materials, Clarendon Press, Oxford, 1994.

[JE]   D. Joseph and V. Elser, *A model of quasicrystal growth*, Phys. Rev. Lett. **79** (1997), 1066–1069.

[K]   P. Kramer, *Nonperiodic central space filling with icosahedral symmetry using copies of seven elementary cells*, Acta Cryst. A **38** (1982), 257–264.

[KN]   P. Kramer and R. Neri, *On periodic and nonperiodic space fillings of $\mathbb{E}^m$ obtained by projection*, Acta Cryst. A **40** (1984), 580–587; Erratum, **41** (1985), 619.

[L]   M. von Laue, *Der Wulffsche Satz für die Gleichgewichtsform von Kristallen*, Z. Kristallogr., Mineral. Petrogr. **105** (1943), 124–133.

[Lag] J. C. Lagarias, *Geometric models for quasicrystals I. Delone sets of finite type*, Discrete Comput. Geom. **21** (1999), 161–191.

[LP]   W. F. Lunnon and P. A. B. Pleasants, *Quasicrystallographic tilings*, J. Math. Pures Appl. (9) **66** (1987), 217–263.

[M1]   Y. Meyer, *Algebraic Numbers and Harmonic Analysis*, North-Holland Math. Library, vol. 2, North-Holland, Amsterdam, 1972.

[M2]   _____, *Quasicrystals, Diophantine approximation and algebraic numbers*, Beyond Quasicrystals (Les Houches, 1994) (F. Axel and D. Gratias, eds.), Springer, Berlin, 1995, pp. 3–16.

[MP]   R. V. Moody and J. Patera, *Densities, minimal distances, and coverings of quasicrystals*, Commun. Math. Phys. **195** (1998), 613–626.

[Ni]   K. Niizeki, *A classification of special points of icosahedral quasilattices*, J. Phys. A **22** (1989), 4295–4302.

[NWBC] H. U. Nissen, R. Wessiken, C. Beeli, and A. Csanády, *Al-Mn quasicrystal aggregates with icosahedral morphological symmetry*, Philos. Mag. B **57** (1988), 587–597.

[OD]   C. Oguey and M. Duneau, *The question of densities in modelling AlMnSi quasicrystals*, Europhys. Lett. **7** (1988), 49–54.

[OA]   Z. Olami and S. Alexander, *Quasiperiodic packing densities*, Phys. Rev. B **37** (1988), 3973–3978.

[OK]   Z. Olami and M. Kleman, *A two-dimensional aperiodic dense tiling*, J. Physique **50** (1989), 19–33.

[P1]   R. Penrose, *The role of aesthetics in pure and applied mathematical research*, Bull. Inst. Math. Appl. **10** (1974), 266–271.

[P2]   _____, *Pentaplexity*, Eureka **39** (1978), 16–22.

[Pl]   P. A. B. Pleasants, *The construction of quasicrystals with arbitrary symmetry group*, Quasicrystals (C. Janot and R. Mosseri, eds.), World Scientific, Singapore, 1995, pp. 22–30.

[RaS] C. Radin and L. Sadun, *The isoperimetric problem for pinwheel tilings*, Commun. Math. Phys. **177** (1996), 255–263.

[RiS] N. Rivier and J. F. Sadoc, *Polymorphism and disorder in a close-packed structure*, Europhys. Lett. **7** (1988), 523–528.

[RW]   C. Rottman and M. Wortis, *Statistical mechanics of equilibrium crystal shapes: Interfacial phase diagrams and phase transitions*, Phys. Reports **103** (1984), 59–79.

[SY]   J. R. Sangwine-Yager, *Mixed volumes*, Handbook of Convex Geometry (P. M. Gruber and J. M. Wills, eds.), North-Holland, Amsterdam, 1993, pp. 43–71.

[Sch1] M. Schlottmann, *Cut-and-project sets in locally compact Abelian groups*, Quasicrystals and Discrete Geometry (Toronto, ON, 1995) (J. Patera, ed.), Fields Inst. Monogr., vol. 10, Amer. Math. Soc., Providence, RI, 1998, pp. 247–264.

[Sch2] _____, *Generalized model sets and dynamical systems*, Directions in Mathematical Quasicrystals (M. Baake and R. V. Moody, eds.), CRM Monogr. Ser., vol. 13, Amer. Math. Soc., Providence, RI, 2000, pp. 143–159, this volume.

[S1]   U. Schnell, *Periodic sphere packings and the Wulff-shape*, Beiträge Algebra Geom. **40** (1999), 125–140.

[S2]   _____, *FCC versus HCP via parametric density*, Discrete Math. **211** (2000), 269–274.

[Se1]   M. Senechal, *Brief history of geometrical crystallography*, in: Historical Atlas of Crystallography (J. Lima-de-Faria, ed.), Kluwer, Dordrecht, 1990, pp. 43–59.
[Se2]   _____, *Quasicrystals and geometry*, Cambridge Univ. Press, Cambridge, 1995; 1996 (corrected reprint).
[Sm]   A. P. Smith, *The sphere packing problem in quasicrystals*, J. Noncryst. Solids **153 & 154** (1993), 258–263.
[Th]   A. C. Thompson, *Minkowski geometry*, Encyclopedia Math. Appl., vol. 63, Cambridge Univ. Press, Cambridge, 1996.
[VG1]   J.-L. Verger-Gaugry, *Approximate icosahedral periodic tilings with pseudo-icosahedral symmetry in reciprocal space*, J. Physique **49** (1988), 1867–1874.
[VG2]   _____, *On a generalization of the Hermite constant*, Period. Math. Hungar. **34** (1997), 153–164.
[W1]   J. M. Wills, *A quasicrystalline sphere-packing with unexpected high density*, J. Physique **51** (1990), 1061–1064.
[W2]   _____, *Finite sphere packings and sphere coverings*, Rend. Sem. Mat. Messina Ser. II **2** (1993), 91–97.
[W3]   _____, *Lattice packings of spheres and Wulff-shape*, Mathematika **43** (1996), 229–236.
[W4]   _____, *On large lattice packings of spheres*, Geom. Dedicata **65** (1997), 117–126.
[W5]   _____, *Parametric density, online-packings and crystal growth*, II International Conference in "Stochastic Geometry, Convex Bodies and Empirical Measures" (Agrigento, 1996) (M. I. Stoka, ed.) Rend. Circ. Mat. Palermo (2) Suppl., vol. 50, Circolo Matematico di Palermo, Palermo, 1997, pp. 413–424.
[W6]   _____, *The Wulff-shape of large periodic packings*, Discrete Mathematical Chemistry (P. Hansen, P. Fowler, and M. Zheng, eds.), DIMACS Ser. Discrete Math. Theoret. Comput. Sci., vol. 51, Amer. Math. Soc., Providence, RI, 2000, pp. 367–375.
[Wo]   L. V. Woodcock, *Entropy difference between the face-centered cubic and hexagonal close-packed crystal structures*, Nature **385** (1997), 141–143.

Rényi Institute of Mathematics, 1364 Budapest, Pf. 127, Hungary
E-mail address: carlos@renyi.hu

Mathematisches Institut, Universität Siegen, 57068 Siegen, Germany
E-mail address: schnell@mathematik.uni-siegen.de
E-mail address: wills@mathematik.uni-siegen.de

Gordon-Type Arguments in the Spectral Theory of One-Dimensional Quasicrystals

David Damanik

ABSTRACT. We review the recent developments in the spectral theory of discrete one-dimensional Schrödinger operators with potentials generated by substitutions and circle maps. We discuss how occurrences of local repetitive structures allow for estimates of generalized eigenfunctions. Among the recent applications of this general approach are almost sure and uniform results on the absence of eigenvalues as well as continuity of the spectral measures with respect to Hausdorff measures.

Introduction

In 1983, two groups independently proposed and investigated a simple model, a discrete one-dimensional Schrödinger operator with a potential taking only two values, which was shown to exhibit unexpected and spectacular behavior. Kohmoto et al. [**KKT**] and Ostlund et al. [**OPRSS**] employed dynamical systems methods to study the scaling properties that were embodied in their model problem. It was argued that the eigenfunctions display a critical behavior in that they are neither extended nor localized. Moreover, this critical behavior seemed to be a universal feature of the model being quite independent of a modulation of the strength of the potential values. This was in sharp contrast to another popular model, the Harper operator, which exhibits such critical eigenfunctions only for some fixed modulation which in fact represents a sharp transition from extended to localized states.

One year later, in 1984, Shechtman et al. published their discovery of a structure which has long range order without being globally translation invariant [**SBGC**]. This discovery was essentially the birth of a whole new field, the investigation of quasicrystals, structures whose existence has been absolutely unexpected up to then, and it triggered intensive research activities including a reconsideration of the nature and definition of order, compare, for example, [**Ba99a**].

Further structures with this property were soon discovered and on the theoretical side, models were proposed which reflect the observed phenomena. Chief among these were quasiperiodic structures that are constructed by a cut-and-project mechanism, which basically projects an ordered structure from a higher-dimensional

---

2000 *Mathematics Subject Classification.* Primary 81Q10; Secondary 47B80, 52C23.

The author was supported by the German Academic Exchange Service (DAAD) through Hochschulsonderprogramm III (Postdoktoranden).

©2000 David Damanik

space to the physical space. In 1986, Luck and Petritis showed in [**LP86**] that the model proposed by Kohmoto et al. and Ostlund et al. is naturally associated to such a cut-and-project structure. The features of this model were then generalized to related models by Kalugin et al. [**KKL86**] and Gumbs and Ali [**GA**]. All these works concluded the critical behavior of eigenfunctions as well as the nowhere dense structure of the set of allowed energies which was in fact claimed to be a set of zero Lebesgue measure (compare in particular [**KO**]). Also in 1986, using symbolic dynamics methods, Casdagli presented a fine analysis of the Fibonacci dynamical system which further supported this claim [**Ca**].

In a 1987 paper, Sütő pursued a rigorous study of these phenomena for the basic original model, the *Fibonacci Hamiltonian* [**Sü87**]. He was able to confirm parts of the observations, namely delocalized states and thus absence of point spectrum, and, for sufficiently large modulation of the potential, nowhere dense structure of the spectrum. He complemented these results in 1989 by showing that the spectrum is indeed always a set of measure zero [**Sü89**]. Absence of absolutely continuous spectrum then follows immediately. In the same year, Bellissard et al. obtained these results for a more general class by essentially the same strategy [**BIST**]. It was thus rigorously established that those operators exhibit purely singular continuous zero-measure spectrum, reflecting the fact that they model one-dimensional structures being intermediate between periodic (leading to absolutely continuous spectrum) and disordered (leading to pure point spectrum).

From a mathematical point of view, the occurrence of purely singular continuous spectrum, however, was still considered to be a curiosity joining the "constructed" toy model examples by Pearson [**Pe78**]. Only in the mid-90's did Simon and co-workers aim at a conceptual understanding of singular continuous spectral measures. In a series of papers [**Si95a, DMS, JS, DJLS, SS, Si96a, Si96b**], they not only exhibited many new examples of operators with this spectral type, they even showed that the occurrence of purely singular continuous spectrum is generic in an appropriate sense.

Moreover, in the course of this decade, building upon the landmark papers mentioned above, the occurrence of purely singular continuous spectrum has been shown to be a universal feature of operators associated to structures displaying aperiodicity at an intermediate level. To review these developments is the purpose of the present article. In 1994, Sütő contributed a review in similar spirit to a Les Houches winter school [**Sü95**]. We therefore put particular emphasis on ideas and approaches that were introduced since then. Moreover, we shall center these approaches around one core idea which is due to Gordon and dates back to the 70's [**Go76**]. It was recently realized that the philosophy embodied in Gordon's paper may serve as a universal tool in the theory as the necessary input, local repetitive structures, is always present in the proposed models. For further background and an introduction to the relevant basics in operator theory, we refer the reader to Sütő's article and, as further introductory reading, we want to mention [**BG95, D97**].

Our main goal here is to explain what quantities to look at when studying spectral properties of one-dimensional quasicrystal models, to present strategies and methods of their investigation which either have been successfully applied in the past or which seem promising when tackling some of the open problems, and to list some of these problems which appear to be important and interesting.

The organization is as follows. We start by presenting, in Section 1, essential parts of the theory of ergodic families of discrete one-dimensional Schrödinger operators which provides a very useful framework to work within due to strong general results and proof strategies. Section 2 presents two classes of such families which we will focus on, both being natural generalizations of the family of Fibonacci operators. It is shown how these classes fit into the framework, and known results for these families are recalled. The fundamental Gordon idea, upon which the core message of this paper is based, is discussed in Section 3. A certain variant of the Gordon method suggests investigating traces of some unimodular 2×2-matrices associated to an operator. Section 4 explains how useful bounds on these traces can be obtained by studying a dynamical system which is induced by hierarchical structures in the potential value arrangement. A recently introduced general strategy for obtaining results that hold for all members of a family of operators is presented in Section 5. It basically emphasizes a combinatorial point of view as opposed to measure theoretical type of arguments suggested by ergodic theory. Using this method, one may investigate quantities that do not behave nicely under very weak perturbations, such as the point spectrum. Consequently, Section 6 deals with results on the absence of point spectrum. We present three types of results which are of increasing completeness, the most complete, of course, being uniform results. The latter is shown to be accessible by combining the Gordon method, the bounds on the traces, and the combinatorial point of view. Parts of the results obtained in the proofs of absence of eigenvalues can in fact be used to show that the spectrum has Lebesgue measure zero as explained in Section 7. Section 8 is concerned with transport properties of one-dimensional quasicrystals and hence with the unitary groups generated by the operators. We discuss recent results and possible ways to obtain bounds on the time evolution. Finally, we present some open problems in Section 9.

1. Ergodic Families of Schrödinger Operators

In this section we recall the concept of ergodic families of Schrödinger operators which has proved to provide a convenient framework for the operators that are of interest in this article. For more detailed presentations we refer the reader to the books by Carmona-Lacroix [**CL**] and Cycon et al. [**CFKS**]. In particular, every application of the fundamental Kotani theory requires this framework, and this theory is at the heart of all results concerning absence of absolutely continuous spectrum and the zero-measure property for the operators under consideration. Our presentation does not strive for greatest generality but rather for an appropriate notion of ergodic family which comprises all the examples we want to discuss here.

DEFINITION 1.1. Let Ω be a compact metric space and let $T \colon \Omega \to \Omega$ be a homeomorphism. The pair (Ω, T) is called a *topological dynamical system*. Given some $\omega \in \Omega$, the set $\{T^n \omega : n \in \mathbb{Z}\}$ is called the *orbit* of ω. Denote by $\mathcal{B}$ the Borel σ-algebra of Ω. A Borel probability measure μ is called *stationary* if $\mu(T(B)) = \mu(B)$ for every $B \in \mathcal{B}$. A Borel set B is called *shift invariant* if $T(B) = B$. A stationary measure is called *ergodic* if any shift invariant set has measure zero or one. The topological dynamical system (Ω, T) is called *minimal* if the orbit of every $\omega \in \Omega$ is dense. It is called *uniquely ergodic* if there exists a unique ergodic measure, and it is called *strictly ergodic* if it is both minimal and uniquely ergodic.

It is in fact well known that if there is a unique stationary measure μ, then μ is necessarily ergodic [**W**]. The two examples below will be of major importance in what follows.

EXAMPLE 1.2. Let $\Omega = \mathbb{T} \simeq [0,1)$ and $T\omega = \omega + \alpha \mod 1$ for some irrational $\alpha \in (0,1)$. It is well known that the Lebesgue measure on $\mathbb{T}$ is the unique stationary measure (i.e., the system is uniquely ergodic) and that every orbit is dense (i.e., the system is minimal). Hence, (Ω, T) is strictly ergodic.

Before turning to the next example let us introduce some notation.

DEFINITION 1.3. Let $\mathcal{A} = \{a_1, \ldots, a_s\}$ be a finite set, called *alphabet*. Endow $\mathcal{A}$ with the discrete topology. The a_i are called *symbols* or *letters*, the elements of $\mathcal{A}^* = \bigcup_{k \geq 1} \mathcal{A}^k$ are called *words*. We denote by $|v|$ the *length* of a word $v \in \mathcal{A}^*$ (i.e., $|v| = l$ if $v \in \mathcal{A}^l$). For $v, w \in \mathcal{A}^*$, $\#_v(w)$ denotes the number of occurrences of v in w (e.g., $\#_{aa}(aabaaa) = 3$). Let $\mathcal{A}^{\mathbb{N}}$, $\mathcal{A}^{\mathbb{Z}}$ denote the sets of one-sided and two-sided infinite sequences, called *infinite words*, over $\mathcal{A}$, both being equipped with product topology which is easily seen to be a metric topology. Given a finite or infinite word w, a finite word v is called a *subword* or *factor* of w if there are (finite or infinite) words r, s such that $w = rvs$, with the obvious definition of concatenation of words. We define $F_w = \{y : y \text{ is a factor of } w\}$ and $F_w(n) = F_w \cap \mathcal{A}^n$, $n \in \mathbb{N}$. The *complexity function* $p_w \colon \mathbb{N} \to \mathbb{N}_0$ corresponding to w is given by $p_w(n) = |F_w(n)|$, $|\cdot|$ denoting cardinality.

EXAMPLE 1.4. Let $\mathcal{A}$ be an alphabet. The shift T on $\mathcal{A}^{\mathbb{Z}}$ is defined by $(T\psi)_n = \psi_{n+1}$ for $\psi \in \mathcal{A}^{\mathbb{Z}}$. Let $\Omega \subseteq \mathcal{A}^{\mathbb{Z}}$ be closed and invariant under T, that is, $T\Omega = \Omega$. The topological dynamical system (Ω, T) is called a *subshift*. Let us consider subshifts that are generated as follows. Given some $\psi \in \mathcal{A}^{\mathbb{N}}$, we define Ω_ψ to be the set of two-sided sequences having all their subwords occur in ψ, that is,

$$\Omega_\psi = \{\omega \in \mathcal{A}^{\mathbb{Z}} : F_\omega \subseteq F_\psi\}.$$

It is clear that Ω_ψ is closed and invariant. Moreover, unique ergodicity and strict ergodicity of (Ω_ψ, T) can be characterized in terms of frequencies of subwords as follows (cf. [**Q**]). The subshift is uniquely ergodic if and only if for every $v \in F_\psi$ there is a number $d_\psi(v) \geq 0$, the *frequency of v in ψ*, such that for every $k \in \mathbb{N}$, we have

$$\frac{\#_v(\psi_k \ldots \psi_{k+n-1})}{n} \longrightarrow d_\psi(v)$$

as $n \to \infty$, uniformly in k. Minimality of (Ω, T) is equivalent to the fact that every word in F_ψ occurs infinitely often in ψ and the gaps between consecutive occurrences are bounded by a constant depending on the word, that is, the occurrences of every word are relatively dense. Thus, the subshift is strictly ergodic if and only if it is uniquely ergodic with strictly positive frequencies, that is, $d_\psi(v) > 0$ for all $v \in F_\psi$. In the uniquely ergodic case, the unique stationary measure μ obeys

(1.1) $$\mu(\{\omega \in \Omega : \omega_m \ldots \omega_{m+|v|-1} = v\}) = d_\psi(v)$$

for every $v \in F_\psi$ and every $m \in \mathbb{Z}$. That is, the measure of some cylinder set can be determined by studying the frequency of the defining word.

DEFINITION 1.5. Given a topological dynamical system (Ω, T), an ergodic measure μ, and a measurable function $g \colon \Omega \to \mathbb{R}$, one defines for each $\omega \in \Omega$ a two-sided infinite sequence $V_\omega : \mathbb{Z} \to \mathbb{R}$ by $V_\omega(n) = g(T^n \omega)$. This gives rise to a discrete one-dimensional Schrödinger operator H_ω on $\ell^2(\mathbb{Z})$ which acts on some $\phi \in \ell^2(\mathbb{Z})$ by

$$(H_\omega \phi)(n) = \phi(n+1) + \phi(n-1) + V_\omega(n)\phi(n).$$

The family $(H_\omega)_{\omega \in \Omega}$ is called an *ergodic family of Schrödinger operators*.

The striking fundamental result, which motivates the choice of this framework even for deterministic models such as the ones we consider in this paper, is the following result which essentially says that the spectrum and the spectral type are deterministic up to sets of measure zero.

THEOREM 1.6 (Pastur, Kunz-Souillard). *Let $(H_\omega)_{\omega \in \Omega}$ be an ergodic family of Schrödinger operators. Then there exist sets $\Omega_0 \subseteq \Omega$, $\Sigma, \Sigma_{\mathrm{pp}}, \Sigma_{\mathrm{sc}}, \Sigma_{\mathrm{ac}} \subseteq \mathbb{R}$ such that $\mu(\Omega_0) = 1$ and $\sigma(H_\omega) = \Sigma$, $\sigma_{\mathrm{pp}}(H_\omega) = \Sigma_{\mathrm{pp}}$, $\sigma_{\mathrm{sc}}(H_\omega) = \Sigma_{\mathrm{sc}}$, $\sigma_{\mathrm{ac}}(H_\omega) = \Sigma_{\mathrm{ac}}$ for every $\omega \in \Omega_0$.*

A proof of this result can be found in [**Pa, KS**]. However, for a discussion of this result and most of what follows in this section, the reader could also consult the books [**CL, CFKS**]. In fact, an additional assumption even allows for a strengthening of Theorem 1.6.

DEFINITION 1.7. An ergodic family of Schrödinger operators $(H_\omega)_{\omega \in \Omega}$ is called *minimal* if for each pair $\omega_1, \omega_2 \in \Omega$, the sequence V_{ω_1} is a pointwise limit of translates of V_{ω_2}.

REMARK 1.8. Note that if the family is generated by a dynamical system as defined in Example 1.4 along with $g(\omega) = f(\omega_0)$ where $f : \mathcal{A} \to \mathbb{R}$ is arbitrary, then minimality of the family follows from minimality of the dynamical system (Ω, T).

THEOREM 1.9. *Let $(H_\omega)_{\omega \in \Omega}$ be a minimal ergodic family of Schrödinger operators. Then, we have $\sigma(H_\omega) = \Sigma$, $\sigma_{\mathrm{ac}}(H_\omega) = \Sigma_{\mathrm{ac}}$ for every $\omega \in \Omega$.*

The statement about the spectrum is already part of the folklore and is essentially contained in [**RS80**]. The result on the absolutely continuous spectrum was recently obtained by Last and Simon [**LS**]. Thus, given a minimal ergodic family, one can pick any member of the family when studying the spectrum or the absolutely continuous spectrum. This is a clear motivation for embedding even a deterministic model into this framework since it may well be that another member of the family, not the one we started with, is easier to study. Let us remark that minimality does not imply constancy of $\sigma_{\mathrm{pp}}(H_\omega)$ or $\sigma_{\mathrm{sc}}(H_\omega)$ as there are explicit counterexamples [**JS**].

Let us now turn to a beautiful theory which has been termed *Kotani theory*. The results we shall describe below indeed form the core of much of the theory of ergodic families of Schrödinger operators in one dimension and certainly provide a basis for most results on Fibonacci-type operators which have been obtained so far. Given a family $(H_\omega)_{\omega \in \Omega}$, it is often very useful to consider the associated eigenvalue equation in difference sense, that is,

(1.2) $$\phi(n+1) + \phi(n-1) + V_\omega(n)\phi(n) = E\phi(n),$$

where $E \in \mathbb{C}$ and ϕ is just required to be a two-sided sequence, $\phi : \mathbb{Z} \to \mathbb{C}$. Some of the most useful tools in one-dimensional Schrödinger operator theory are results that establish a link between the behavior of solutions to (1.2) and spectral properties of the operator since the former are to some extent relatively easy to investigate. In fact, Kotani theory provides a link of this kind in the case of the ergodic framework. Connections in the deterministic case have been found by

Gilbert, Pearson, and Khan [**GP, Gi, KP**] (see Jitomirskaya-Last [**JL96, JL99a, JL99b**] for an extension of the results and a simplification of the proof) and by Last and Simon [**LS**]. Let us recall a standard reformulation of (1.2). Given a two-sided sequence ϕ we define $\Phi \colon \mathbb{Z} \to \mathbb{C}^2$ by

$$\Phi(n) = \begin{pmatrix} \phi(n+1) \\ \phi(n) \end{pmatrix}.$$

Defining

$$T_{E,\omega}(n) = \begin{pmatrix} E - V_\omega(n) & -1 \\ 1 & 0 \end{pmatrix},$$

$$M_{E,\omega}(n) = \begin{cases} T_{E,\omega}(n) \times \cdots \times T_{E,\omega}(1), & n \geq 1, \\ I, & n = 0, \\ T_{E,\omega}(n+1)^{-1} \times \cdots \times T_{E,\omega}(0)^{-1}, & n \leq -1, \end{cases}$$

one may easily check that

ϕ solves (1.2) $\iff$ Φ solves $\Phi(n) = M_{E,\omega}(n)\Phi(0)$ for every $n \in \mathbb{Z}$.

The matrices $M_{E,\omega}(\cdot)$ are called *transfer matrices*. They have determinant 1 since they are products of the *elementary transfer matrices* $T_{E,\omega}(\cdot)$ which obviously have determinant 1. The linear space of solutions to (1.2) for fixed E is two-dimensional, as can be seen from the above relation. Consider the two solutions $\phi_{1,2}$ induced by the initial conditions $\phi_1(0) = \phi_2(1) = 0$, $\phi_1(1) = \phi_2(0) = 1$. Then we also have

$$M_{E,\omega}(n) = \begin{pmatrix} \phi_1(n+1) & \phi_2(n+1) \\ \phi_1(n) & \phi_2(n) \end{pmatrix}.$$

Thus, the matrix contains all information about $\phi_{1,2}$ and hence all solutions in a pointwise way. In particular, bounds on $\|M_{E,\omega}(n)\|$, for example, yield bounds on $\|\Phi(n)\|$ for all solutions. It turns out to be useful to distinguish between exponential and sub-exponential growth of $\|M_{E,\omega}(n)\|$. Thus, we let

$$\gamma_{\omega,\pm}(E) = \lim_{n \to \pm\infty} \frac{1}{n} \ln \|M_{E,\omega}(n)\|,$$

provided the limit exists. Regarding existence of this limit, the following has been obtained in [**FK**].

THEOREM 1.10 (Furstenberg-Kesten). *For every $E \in \mathbb{C}$, there exist $\Omega_E \subseteq \Omega$ and $\gamma(E) \in \mathbb{R}$ such that $\mu(\Omega_E) = 1$ and for every $\omega \in \Omega_E$, $\gamma_{\omega,\pm}(E)$ exist and are equal to $\gamma(E)$, that is, $\gamma_{\omega,+}(E) = \gamma_{\omega,-}(E) = \gamma(E)$.*

The number $\gamma(E)$ is called the *Lyapunov exponent*.

THEOREM 1.11 (Osceledec). *Suppose that for some $E \in \mathbb{C}$, $\gamma(E) > 0$. Then, for every $\omega \in \Omega_E$, there exist solutions ϕ_d^+, ϕ_d^- of $H_\omega \phi = E\phi$ such that $\phi_d^\pm$ decays exponentially at $\pm\infty$, respectively, at the rate $-\gamma(E)$. Moreover, every solution which is linearly independent of ϕ_d^+ (resp., ϕ_d^-) grows exponentially at $+\infty$ (resp., $-\infty$) at the rate $\gamma(E)$.*

See [**LS, O, Ru**]. Thus, in the case of a positive Lyapunov exponent, one has a complete understanding of the asymptotics of the solutions at infinity.

Kotani theory now establishes a link between the Lyapunov exponent and the absolutely continuous spectrum. Define

$$A = \{E \in \mathbb{R} : \gamma(E) = 0\}.$$

The *essential closure* $\overline{S}^{\text{ess}}$ of a set $S \subseteq \mathbb{R}$ is defined by
$$\overline{S}^{\text{ess}} = \{E \in \mathbb{R} : |(E - \varepsilon, E + \varepsilon) \cap S| > 0 \ \forall \varepsilon > 0\},$$
where $|\cdot|$ denotes Lebesgue measure. In particular, $\overline{S}^{\text{ess}} = \emptyset$ for every set S of zero Lebesgue measure.

THEOREM 1.12 (Ishii-Pastur-Kotani). $\Sigma_{\text{ac}} = \overline{A}^{\text{ess}}$.

For a proof of the inclusion "$\subseteq$" the reader may consult [**I, Pa**] (see also [**Bu**] for an alternative proof using Gilbert-Pearson theory), the opposite inclusion has been treated in [**Ko84**] (see [**Si83**] for an adaptation to the discrete case). The following corollary, obtained in [**Ko89**], to the proof given in [**Ko84, Si83**] is of great interest to us since all the potentials we shall be dealing with take only finitely many values. Moreover, the additional assumption of aperiodicity is rather non-restrictive since there is a well-established theory treating the case of periodic potentials; see, for example, [**RS78**].

THEOREM 1.13 (Kotani). *If the potentials* V_ω *are aperiodic and take only finitely many values, then* $|A| = 0$. *In particular,* $\Sigma_{\text{ac}} = \emptyset$.

2. Models Generated by Circle Maps and Primitive Substitutions

In this section we present the two classes of ergodic families we shall discuss in the sequel. Both classes are natural extensions of different aspects of the Fibonacci model, one generalizing its quasiperiodicity (models generated by circle maps) or, when restricted to a subclass, its word complexity properties (models generated by Sturmian sequences), the other one generalizing its self-similar structure (models generated by primitive substitutions). We show how they fit into the general framework presented in the preceding section and recall known results for these classes. For the sake of brevity we introduce the following notions.

DEFINITION 2.1. A family $(H_\omega)_{\omega \in \Omega}$ is called *EFA* if it is an ergodic family of Schrödinger operators such that the potentials V_ω are aperiodic and take only finitely many values. It is called *MEFA* if it is EFA and minimal.

With this convention at hand we infer from the above results that EFA (MEFA) families exhibit almost surely (uniformly) purely singular spectrum, that is, $\Sigma = \Sigma_{\text{pp}} \cup \Sigma_{\text{sc}}$. More precisely, the set A associated to such a family has zero Lebesgue measure. Furthermore, MEFA families have constant spectrum. In fact, all the classes of families $(H_\omega)_{\omega \in \Omega}$ we shall present and discuss now will be MEFA families, so let us bear in mind that when studying the spectral type, we only need to distinguish between point spectrum and singular continuous spectrum since all these families have empty absolutely continuous spectrum for all H_ω, $\omega \in \Omega$; thus the latter does not present any issue at all.

Circle map models. A circle map model is parameterized by three parameters, namely, an irrational *rotation number* $\alpha \in (0,1)$, an *interval length* $\beta \in (0,1)$, and a *coupling constant* $\lambda \in \mathbb{R}\backslash\{0\}$. There are two ways to choose Ω, T, g, both of which have been used in the past. Their mutual relation is, for example, given in [**DL99c**]. The first way follows Example 1.2. Thus, let $\Omega = \mathbb{T} \simeq [0,1)$ and $T: \Omega \to \Omega$, $\omega \mapsto \omega + \alpha \mod 1$. As noted above, (Ω, T) is strictly ergodic with the Lebesgue measure on $\mathbb{T}$ as unique ergodic measure μ. Let g be given by $g(\omega) = \lambda \cdot \chi_{[1-\beta,1)}(\omega)$. This yields potentials $V_\omega(n) = \lambda \cdot \chi_{[1-\beta,1)}(\alpha n + \omega \mod 1)$. The other possibility of associating a family of potentials with the one-sided sequence $v_{\alpha,\beta,\theta}(n) = \chi_{[1-\beta,1)}(n\alpha + \theta \mod 1)$, $n \in \mathbb{N}$, and follows the lines of Example 1.4. Define $\Omega_{\alpha,\beta} = \Omega_{v_{\alpha,\beta,\theta}}$. The notation is justified since the subshift does not depend on θ. It was shown by Hof in [**H**] that the dynamical system $(\Omega_{\alpha,\beta}, T)$ is strictly ergodic. The function g generating the potentials is in this case given by $g(\omega) = f(\omega_0)$, where $f(0) = 0$, $f(1) = \lambda$. In the case $\alpha = \beta$ we write $v_{\alpha,\theta}$ instead of $v_{\alpha,\alpha,\theta}$ and Ω_α instead of $\Omega_{\alpha,\alpha}$, and we call the dynamical system (Ω_α, T) as well as the resulting potentials *Sturmian*. Sturmian potentials have an interesting combinatorial property. Consider a one-sided sequence s. Recall that the complexity function $p_s(n)$ counts the number of factors of length n in s. One can show (see [**Lo83, Lo99**] for this and much more on combinatorics on words in general and Sturmian sequences in particular, compare also [**LP92**]) that $p_s(n) \leq n$ for some n implies that s is ultimately periodic. Thus, any non-ultimately periodic sequence s obeys $p_s(n) \geq n+1$ for every n. Sequences s having $p_s(n) = n+1$ for all n are called *Sturmian*. The terminology is now motivated by the fact that every $v_{\alpha,\theta}$ is a Sturmian sequence and every $\{0,1\}$-valued Sturmian sequence coincides with the restriction of some element of an appropriate Ω_α to $\mathbb{N}$.

Our discussion of this relation involving one-sided sequences may appear somewhat awkward, but the corresponding relation for two-sided sequences is not true, that is, there exist $\{0,1\}$-valued two-sided sequences s with complexity function obeying $p_s(n) = n+1$ which, however, do not belong to some Ω_α (consider, e.g., the sequence s defined by $s_n = 0$ for $n < 0$ and $s_n = 1$ for $n \geq 0$), compare [**CH**] for a complete characterization.

Let us note the following concerning families generated by circle maps.

PROPOSITION 2.2. *Both ways of generating an operator family $(H_\omega)_{\omega \in \Omega}$ corresponding to the parameters α, β, λ induce MEFA families.*

For fixed parameter values, the operator family induced by the dynamical system on the torus is contained in the operator family obtained from the subshift. Interestingly, the latter family is strictly larger even though both families are minimal [**DL99c**]! The Sturmian model generated by the golden mean $\alpha = \frac{1}{2}(\sqrt{5}-1)$ is called the *Fibonacci model*.

Models generated by primitive substitutions. Let $\mathcal{A}$ be an alphabet. A *substitution* S is a map $S \colon \mathcal{A} \to \mathcal{A}^*$. S can be extended morphically to $\mathcal{A}^*$ (resp., $\mathcal{A}^\mathbb{N}$) by $S(b_1 \ldots b_n) = S(b_1) \ldots S(b_n)$ (resp., $S(b_1 b_2 b_3 \ldots) = S(b_1)S(b_2)S(b_3)\ldots$). S is called *primitive* if there exists $k \in \mathbb{N}$ such that for every $a \in \mathcal{A}$, $S^k(a)$ contains every symbol from $\mathcal{A}$. Prominent examples of primitive substitutions are given by

$a \mapsto ab, b \mapsto a$	Fibonacci,
$a \mapsto ab, b \mapsto aa$	period doubling,
$a \mapsto ab, b \mapsto aaa$	binary non-Pisot,
$a \mapsto ab, b \mapsto ba$	Thue-Morse,
$a \mapsto ab, b \mapsto ac, c \mapsto db, d \mapsto dc$	Rudin-Shapiro.

A fixed point $u \in \mathcal{A}^\mathbb{N}$ of S is called *substitution sequence*. The existence of such a fixed point is ensured by the following conditions,

- there exists a letter $a \in \mathcal{A}$ such that the first letter of $S(a)$ is a,
- $\lim_{n \to \infty} |S^n(a)| = \infty$,

which are easily seen to hold for a suitable power of S if S is primitive. Without loss of generality (since any power of S is primitive if S is primitive), we assume this power to be equal to one. In this case, $u = \lim_{n\to\infty} S^n(a)$ exists and is a substitution sequence. Define Ω by $\Omega = \Omega_u$. If S is primitive, Ω does not depend on the choice of the substitution sequence. The subshift (Ω, T) is called the *substitution dynamical system associated to S* and it is strictly ergodic [**Q**]. Choose some function $f : \mathcal{A} \to \mathbb{R}$ and define $g(\omega) = f(\omega_0)$.

PROPOSITION 2.3. *Suppose that the substitution S is primitive, the substitution sequence u is not ultimately periodic, and the function f takes at least two values. Then the induced operator family $(H_\omega)_{\omega \in \Omega}$ is MEFA.*

In case of the Fibonacci substitution the subshift is equivalent to the Sturmian subshift corresponding to $\alpha = \frac{1}{2}(\sqrt{5}-1)$ via $a \mapsto 1$, $b \mapsto 0$. Thus the corresponding families of operators coincide (up to a spectral shift).

These two classes of MEFA families have been studied extensively in the past; see [**BIST, BIT, D98a, DKL, DL99a, DL99b, DP86, IT, IRT, HKS, J91, Ka, Ra, Sü87, Sü89**] and [**Be, BBG91, BBG92, BG93, D98b, D98c, D99b, DP91, H, HKS**] for some important contributions in the case of circle map models and substitution models, respectively. The results comprise in particular singular continuity of spectral measures, zero Lebesgue measure of the spectrum, Gap labeling via K-theory, opening of the gaps at low coupling, continuity of gap boundaries with respect to the rotation number, and uniform existence of the Lyapunov exponent for large subclasses.

3. Pointwise Methods and Variants of the Gordon Criterion

This section is concerned with methods in the spectral theory of some fixed Schrödinger operator with particular emphasis on several variants of an idea originally due to Gordon [**Go76**]. The methods we present can be applied, for example, to a fixed member of some ergodic family $(H_\omega)_{\omega \in \Omega}$.

Consider a bounded function $V : \mathbb{Z} \to \mathbb{R}$ and the associated Schrödinger operator

(3.1) $$(H\phi)(n) = \phi(n+1) + \phi(n-1) + V(n)\phi(n)$$

along with the difference equation

(3.2) $$\phi(n+1) + \phi(n-1) + V(n)\phi(n) = E\phi(n),$$

where $E \in \mathbb{C}$. Similarly to the above, we introduce a reformulation of (3.2) in terms of *transfer matrices* $M_E(n)$,

$$\Phi(n) = \begin{pmatrix} \phi(n+1) \\ \phi(n) \end{pmatrix},$$

$$T_E(n) = \begin{pmatrix} E - V(n) & -1 \\ 1 & 0 \end{pmatrix},$$

$$M_E(n) = \begin{cases} T_E(n) \times \cdots \times T_E(1), & n \geq 1, \\ I, & n = 0, \\ T_E(n+1)^{-1} \times \cdots \times T_E(0)^{-1}, & n \leq -1. \end{cases}$$

Let us discuss "Gordon-type arguments," that is, the exploitation of local repetitions. Recall that transfer matrices have determinant 1, independently of the potential V, the energy E, and the site n. Thus, by the Cayley-Hamilton theorem, the following universal equation holds,

(3.3) $$M_E(n)^2 - \operatorname{tr}(M_E(n))M_E(n) + I = 0.$$

Suppose now that V repeats its values on the interval $\{1, \ldots, n\}$ once, that is,

(3.4) $$V(j) = V(j+n), \quad 1 \leq j \leq n.$$

Due to the fact that the definition of transfer matrices is local, we infer that for any energy E, we have

(3.5) $$M_E(n)^2 = M_E(2n).$$

Plugging this into (3.3), we obtain

(3.6) $$M_E(2n) - \operatorname{tr}(M_E(n))M_E(n) + I = 0.$$

Now consider any initial vector $\Phi(0)$, which we may assume to be normalized, that is, $\|\Phi(0)\| = 1$. We apply (3.6) to $\Phi(0)$ and get

(3.7) $$\Phi(2n) - \operatorname{tr}(M_E(n))\Phi(n) + \Phi(0) = 0.$$

Since $\Phi(0)$ has norm 1, either $\Phi(2n)$ or $\operatorname{tr}(M_E(n))\Phi(n)$ has to have norm at least $\frac{1}{2}$. Thus,

(3.8) $$\max(\|\Phi(n)\|, \|\Phi(2n)\|) \geq \frac{1}{2} \min\left(1, \frac{1}{|\operatorname{tr}(M_E(n))|}\right).$$

If we can find, for some fixed energy E, a sequence $n_k \to \infty$ such that the potential repeats the values on $\{1, \ldots, n_k\}$ once and the sequence $\operatorname{tr}(M_E(n_k))$ remains bounded, then the right-hand side of (3.8) is strictly bounded away from zero on this sequence of sites for every initial vector! Thus, in this case no solution tends to zero and, in particular, E is not an eigenvalue of H. Let us summarize this in the following lemma.

LEMMA 3.1 (Two-block method). *Fix a potential V and an energy E. Suppose there is a sequence $n_k \to \infty$ and some $1 \leq C < \infty$ such that we have for every k,*

(1) $V(j) = V(j + n_k)$, $1 \leq j \leq n_k$,
(2) $|\operatorname{tr}(M_E(n_k))| \leq C$.

Then, E is not an eigenvalue of H and no solution of (3.2) tends to zero at $+\infty$. More precisely, for every k, every solution obeys

$$\max(\|\Phi(n_k)\|, \|\Phi(2n_k)\|) \geq \frac{1}{2C}.$$

REMARK 3.2. Of course, there are obvious variations on this idea. First of all, it is not important that the squares are aligned at the origin; any other site will do. Similarly, the squares can also be aligned to their right side and one can work on the left half-line. In this case one gets that, once the modified conditions are satisfied, no solution tends to zero at $-\infty$ with similar uniform lower bound for the norms.

But what if a study of transfer matrix traces is not feasible? The answer is simple, just find another block of repetition! The key ingredient in the above argument is the three-term expression in (3.6) which, given largeness of one term, yields largeness of at least one of the others. Now suppose that we have a repetition of $V(1), \ldots, V(n)$ and that the trace of $M_E(n)$ is large. Even if we infer "largeness" of the middle term, this may only be due to the trace but not to the matrix (resp., the vector after application of the equation to an initial vector). This complication does not occur for the other terms. So, in case of a large trace, try to find a repetition of the potential values from 1 to n to the left and, if successful, apply (3.6) to $\Phi(-n)$ but retain normalization at the origin! This yields the equation

$$\Phi(-n) - \operatorname{tr}(M_E(n))\Phi(0) + \Phi(n) = 0. \tag{3.9}$$

Now, the middle term is large (a large factor times a normalized vector) and, again, this says that at least one of the other vectors has to be large. Quantitatively, we have the following lemma.

LEMMA 3.3 (Three-block method). *Fix a potential V. Suppose there is a sequence $n_k \to \infty$ such that we have for every k,*

$$V(j - n_k) = V(j) = V(j + n_k), \quad 1 \leq j \leq n_k.$$

Then for every energy E, we have that E is not an eigenvalue of H and no solution of (3.2) tends to zero at both $\pm\infty$. More precisely, for every k, every solution obeys

$$\max(\|\Phi(-n_k)\|, \|\Phi(n_k)\|, \|\Phi(2n_k)\|) \geq \frac{1}{2}.$$

REMARK 3.4. In general it may well happen that for some energy E, every solution decays at either $+\infty$ or $-\infty$. This happens, for example, for energies outside the spectrum of H. Thus, even if one has cubes rather than only squares, the additional investigation of transfer matrix traces pays off in the form of a stronger conclusion.

Lemma 3.3 is very close to the original Gordon result [**Go76**] (see also [**CFKS**]) which, however, requires another block of repetition. It was stated and proved in this form by Delyon and Petritis [**DP86**]. Lemma 3.1 was proved by Sütő in [**Sü87**].

4. Trace Map Characterization of the Spectrum

The two-block version of the Gordon criterion presented in Lemma 3.1 suggests investigating both repetitive structures in the potential and traces of transfer matrices when trying to obtain bounds on solutions of the eigenvalue equation. In the case of Sturmian models or models generated by primitive substitutions, transfer matrix traces can be investigated by studying a (generalized) dynamical system, the *trace map*, which is induced by hierarchical structures in the potentials. These are by definition present in potentials generated by substitutions and their presence in the Sturmian case can be exhibited using continued fraction expansion theory. The present section is concerned with a discussion of the correspondence between the dynamics of trace maps and the spectra of operators from the two classes. For further information on trace maps, we refer the reader to [**AP, BGJ, BR, KN, PWW, Ro, RB**].

The introduction of a trace map follows a universal program which can be summarized as follows.

(1) Exhibit a sequence of generating words obeying recursive relations.
(2) Consider transfer matrices as being associated to finite words rather than to infinite sequences from the subshift.
(3) Translate the recursive relations to the level of transfer matrices.
(4) Pass to the traces of these matrices using suitable identities for unimodular 2×2-matrices.

Among the models we are interested in, trace maps have been found for all Sturmian models and all substitution models. Let us indicate how to establish the above steps in these two cases.

Given some one-sided sequence ψ such that the associated subshift (Ω_ψ, T) is minimal, it is easy to check that we have (this is essentially Gottschalk's theorem, see [**Pe83**])

(4.1) $$\Omega_\psi = \{\omega \in \mathcal{A}^{\mathbb{Z}} : F_\omega = F_\psi\}.$$

In the case of ψ being a substitution sequence associated to some primitive substitution S, ψ has the form $\psi = \lim_{n\to\infty} S^n(a)$ for a suitable $a \in \mathcal{A}$. In particular, we have

(4.2) $$\Omega_\psi = \{\omega \in \mathcal{A}^{\mathbb{Z}} : \forall w \in F_\omega \, \exists n \in \mathbb{N} \text{ such that } w \in F_{S^n(a)}\}.$$

Thus, the words $S^n(a)$ entirely determine the hull. Due to primitivity, any other sequence of the form $S^n(b)$, $b \in \mathcal{A}$, can be used. The set of words $S^n(b)$ where n ranges over $\mathbb{N}_0$ and b ranges over $\mathcal{A}$ therefore serves as a good basis for a study of the local properties of the potential value arrangements. Moreover, among these words the presence of recursive relations is immediate from the substitution rule,

(4.3) $$S^n(b) = S^{n-1}(S(b)) = S^{n-1}(c_1 \ldots c_k) = S^{n-1}(c_1) \ldots S^{n-1}(c_k).$$

Note that the concrete expression, that is, the way to pass from the words $S^{n-1}(c_1) \ldots S^{n-1}(c_k)$ to the word $S^n(b)$, is n-independent.

EXAMPLE 4.1. Let us consider the Fibonacci substitution S_F which acts as $S_F(a) = ab$, $S_F(b) = a$. The recursive relations are given by

(4.4) $$S_F^n(a) = S_F^{n-1}(a) S_F^{n-1}(b), \quad S_F^n(b) = S_F^{n-1}(a).$$

Let us now consider a Sturmian subshift Ω_α. Consider the continued fraction expansion of α (for general information on continued fractions, see, e.g., [**Kh, Pe54**]),

(4.5) $$\alpha = \cfrac{1}{a_1 + \cfrac{1}{a_2 + \cfrac{1}{a_3 + \ldots}}}$$

with uniquely determined $a_n \in \mathbb{N}$. The associated rational approximants p_n/q_n obey

(4.6) $$p_0 = 0, \quad p_1 = 1, \quad p_n = a_n p_{n-1} + p_{n-2},$$

(4.7) $$q_0 = 1, \quad q_1 = a_1, \quad q_n = a_n q_{n-1} + q_{n-2}.$$

To make things formally similar to the substitution case, define words s_n over the alphabet $\mathcal{A} = \{0, 1\}$ by

(4.8) $$s_n = v_{\alpha,0}(1) \ldots v_{\alpha,0}(q_n)$$

and the one-sided sequence c_α by

(4.9) $$c_\alpha = \lim_{n\to\infty} s_n.$$

Of course, c_α is nothing else than $v_{\alpha,0}$ restricted to $\mathbb{N}$. Note that the words s_n have length q_n. The equation (4.7) now has the following analog on word level,

(4.10) $$s_n = s_{n-1}^{a_n} s_{n-2}.$$

The equation holds in this form for $n \geq 3$. The correct initial conditions are recovered by

$$(4.11) \qquad s_{-1} = 1, \quad s_0 = 0, \quad s_1 = s_0^{a_1 - 1} s_{-1},$$

and with these definitions (4.10) also holds for all $n \geq 2$. This can be proved by using the fact that continued fraction approximants $\frac{p_n}{q_n}$ provide the best possible approximation to α; see [**BIST**] for details. The following useful formula can be deduced from (4.10).

PROPOSITION 4.2. *For each $n \geq 2$,*

$$(4.12) \qquad s_n s_{n+1} = s_{n+1} s_{n-1}^{a_n - 1} s_{n-2} s_{n-1}.$$

PROOF. $s_n s_{n+1} = s_n s_n^{a_{n+1}} s_{n-1} = s_n^{a_{n+1}} s_n s_{n-1} = s_n^{a_{n+1}} s_{n-1}^{a_n} s_{n-2} s_{n-1} = s_{n+1} s_{n-1}^{a_n - 1} s_{n-2} s_{n-1}.$ $\square$

Again we have found a sequence of words obeying recursive relations tending to a limit word c_α such that the hull can be written in the following form,

$$(4.13) \qquad \Omega_\alpha = \{\omega \in \mathcal{A}^{\mathbb{Z}} : F_\omega = F_{c_\alpha}\}.$$

The next step is to pass from the recursively generated words to the transfer matrices associated to these words. Recall that given a subshift over some alphabet $\mathcal{A}$, the potentials V_ω were generated by replacing the symbols in the sequences by real numbers according to some function $f \colon \mathcal{A} \to \mathbb{R}$. Thus, one can study local properties of the potentials by introducing the following energy-indexed representation of words as $\mathrm{SL}(2, \mathbb{R})$ matrices. Fix some real energy E and define the map $M_E \colon \mathcal{A} \to \mathrm{SL}(2, \mathbb{R})$ by

$$(4.14) \qquad M_E(a) = \begin{pmatrix} E - f(a) & -1 \\ 1 & 0 \end{pmatrix}$$

Extend this mapping to $\mathcal{A}^*$ by

$$(4.15) \qquad M_E(a_1 \ldots a_n) = M_E(a_n) \times \cdots \times M_E(a_1).$$

For the above two classes, the recursions (4.3) and (4.10) then extend to the associated matrices in a straightforward way. We obtain, for example, in the Sturmian case

$$(4.16) \qquad M_n = M_{n-2} M_{n-1}^{a_n},$$

where we set $M_n = M_E(s_n)$. Note that due to (4.15), the order of the factors has been reversed. Depending on the explicit form of a substitution S, we have a similar analog in this case. In principle, we would like to apply the matrix trace to these equations and to study the dynamics, that is, the limit $n \to \infty$, for the traces. This is motivated by a simple argument which we will discuss in a moment. However, since the trace is not multiplicative, this transition is not as straightforward as the transition from words to matrices. This can be remedied by using appropriate identities to break down powers and by extending the set of underlying variables (from a set of size $|\mathcal{A}|$ to a larger, but still finite, set). First of all, all the powers can be broken down to one just by using the characteristic equation of unimodular matrices M, that is,

$$(4.17) \qquad M^2 = \mathrm{tr}(M) M - I.$$

Moreover, one can pass to products not having multiple occurrences of factors by using the equation (see [**KN**])

(4.18) $$\mathrm{tr}(MNMO) = \mathrm{tr}(MN)\mathrm{tr}(MO) + \mathrm{tr}(NO) - \mathrm{tr}(N)\mathrm{tr}(O).$$

This set of remaining possible products, the enlarged alphabet $\mathcal{E}$, has cardinality bounded by [**KN**] (see [**AB, ABG**] for improvements)

(4.19) $$\sum_{l=1}^{|\mathcal{A}|} \frac{|\mathcal{A}|!}{l(|\mathcal{A}|-l)!}.$$

Although one may obtain messy expressions, this method generates polynomial expressions for the traces of these products of level n in terms of the traces on level $n-1$. In case of a two-letter alphabet $\mathcal{A} = \{a,b\}$ one may choose $\mathcal{E} = \{a,b,ab\}$.

EXAMPLE 4.3. In the Fibonacci case we infer from (4.4) that $x_n(E) = \mathrm{tr}(M_E(S_F^n(a)))$, $y_n(E) = \mathrm{tr}(M_E(S_F^n(b)))$, $z_n(E) = \mathrm{tr}(M_E(S_F^n(a))M_E(S_F^n(b)))$ obey

$$x_n(E) = z_{n-1}(E),$$
$$y_n(E) = x_{n-1}(E),$$
$$z_n(E) = x_{n-1}(E)z_{n-1}(E) - y_{n-1}(E).$$

In this example we do not even need the enlarged alphabet since $z_n(E) = x_{n+1}(E)$. Thus, one has the equivalent recursion

$$x_n(E) = x_{n-1}(E)x_{n-2}(E) - x_{n-3}(E)$$

involving only the x-variables.

Now why should one expect a connection between the trace map and the spectrum Σ? By the combinatorial subshift definitions (4.1) and (4.13) and by the recursions (4.3) and (4.10), any operator H_ω is a strong limit of operators H_n with periodic potentials where the periods have length $|S^n(a)|$ (resp., q_n) and the potential values are obtained by applying f to the symbols in $S^n(a)$ (resp., s_n). The ω-dependence of these operators is solely reflected in the locations of the periods, the actual periodic blocks are the same. Write $x_n(E)$ for $\mathrm{tr}(M_E(S^n(a)))$ (resp., $\mathrm{tr}(M_n)$). By the general periodic theory [**RS78**] we have

(4.20) $$\sigma(H_n) = \{E : |x_n(E)| \le 2\}.$$

Thus, the strong approximation gives (see [**RS80**])

(4.21) $$\Sigma = \sigma(H_\omega) \subseteq \bigcap_{k \in \mathbb{N}} \overline{\bigcup_{n \ge k} \{E : |x_n(E)| \le 2\}}.$$

If one can prove that $\bigcup_{n \ge k}\{E : |x_n(E)| \le 2\}$ is closed, then one ends up with a nice (and useful!) property of energies in the spectrum, namely, boundedness of traces on a subsequence (even with a uniform bound),

(4.22) $$\Sigma \subseteq \bigcap_{k \subset \mathbb{N}} \bigcup_{n \ge k} \{E : |x_n(E)| \le 2\} = \{E : |x_n(E)| \le 2 \text{ for infinitely many } n\}.$$

It turns out that the trace map is a good tool to establish such a property. In many cases it can even be shown that equality holds in (4.22). Namely, building upon Sütő [**Sü87**], the following has been shown for Sturmian models by Bellissard et al. [**BIST**].

THEOREM 4.4 (Trace map characterization of spectra of Sturmian models). *Let* $(H_\omega)_{\omega \in \Omega}$ *be a Sturmian family corresponding to* α, λ. *Define* $C_\lambda = 2 + \sqrt{8 + \lambda^2}$. *Then, we have*

$$\Sigma = \{E : |x_n(E)| \leq 2 \text{ for infinitely many } n\} = \{E : |x_n(E)| \leq C_\lambda \forall n\}.$$

Under certain assumptions, so-called *semi-primitvity* of the trace map and the occurrence of a square in the substitution sequence, which were shown to be satisfied by many prominent examples including Fibonacci, period doubling, binary non-Pisot, and Thue-Morse, Bovier and Ghez obtained a trace map characterization of the spectrum in the primitive substitution case. We refer the reader to their paper [**BG93**] and also its precursors [**Be**, **BBG91**] for a precise statement of the results. We note, however, that they do not obtain boundedness of orbits for energies from the spectrum in their general setting but rather boundedness on a subsequence. In this sense the trace maps associated to primitive substitution models exhibit only a partial analogy to Theorem 4.4.

5. Partitions and Uniform Results

In this section we focus on the local structures of sequences in a subshift which is induced by a recursively generated infinite word. As we saw above, this class comprises subshifts generated by substitution sequences and Sturmian words. We exhibit a uniform combinatorial property for the elements in the subshift. We then discuss how to employ this uniform combinatorial property to obtain uniform spectral properties.

To keep the notation simple we shall focus on Sturmian subshifts. However, the reader may easily verify that the ideas we present here apply to substitution models equally well. Fix some irrational $\alpha \in (0,1)$. Recall the description of a Sturmian hull Ω_α in terms of a subword definition (4.13). From this, one sees that the elements of Ω_α have a uniform combinatorial property, namely, they have the same set of factors which is equal to the set of factors of c_α. Now, apart from its subword structure, c_α has an additional structure. Namely, it is built from blocks s_n which obey a recursive relation. From (4.9) and (4.10) one can infer that for each $n \in \mathbb{N}$, c_α can be decomposed into blocks of type s_n and s_{n-1}. This decomposition is even unique. Moreover, in this decomposition, blocks of type s_{n-1} are always isolated and blocks of type s_n have multiplicity a_{n+1} or $a_{n+1} + 1$. It is now natural to ask whether this structure is inherited by the elements of Ω_α. After all, as was discussed in the previous section, we already know that these blocks have useful properties in that the transfer matrices associated to them can be analyzed. It turns out that all these properties persist when passing from c_α to $\omega \in \Omega_\alpha$ [**DL99a**].

PROPOSITION 5.1. *Let* $\omega \in \Omega_\alpha$. *Then for every* $n \in \mathbb{N}$, *there exists a unique decomposition of* ω *into blocks of type* s_n *and* s_{n-1}. *In this decomposition, the multiplicity of each occurrence of* s_n *(resp., s_{n-1}) is a_{n+1} or $a_{n+1} + 1$ (resp., 1).*

This decomposition is called *n-partition of ω*. The above result indicates that to a certain extent, the members of Ω_α are equally well accessible to the pointwise methods introduced above. Indeed, in order to establish spectral properties that cannot be studied by pointwise convergence uniformly, one may apply a pointwise criterion to each ω separately. From Proposition 5.1 we learn that local repetitive structures abound in Sturmian potentials. For instance, we know that for every n, each s_{n-1}-block in the n-partition is followed and preceded by at least a_{n+1} copies of s_n, respectively.

Qualitatively, we have the same phenomena in hulls generated by primitive substitution hulls, the specific properties, however, depending on the concrete substitution given.

6. Absence of Eigenvalues: Locating Squares and Cubes

We now turn to results on absence of point spectrum for models generated by circle maps and primitive substitutions. With one exception—the paper by Hof et al. [**HKS**] which employs a criterion in similar spirit using palindromes rather than powers (see [**Ba99b**] for extensions)—virtually all the known results were obtained by using variants of the Gordon method. In this section we show how the methods and tools presented in the three preceding sections can be combined to yield these results.

Let us begin by discriminating between the types of results that have been obtained. Given a family $(H_\omega)_{\omega \in \Omega}$ such that the underlying dynamical system (Ω, T) is a strictly ergodic subshift over some alphabet $\mathcal{A}$ with unique invariant measure μ, define
$$\Omega_c = \{\omega \in \Omega : \sigma_{\mathrm{pp}}(H_\omega) = \emptyset\}.$$
We say that absence of eigenvalues for $(H_\omega)_{\omega \in \Omega}$ is *generic* if Ω_c is a dense G_δ (i.e., a countable intersection of open sets which is dense in Ω), *almost sure* if Ω_c has full μ-measure, and *uniform* if $\Omega_c = \Omega$. Let us recall some general arguments that are useful in this context. First of all, to establish a generic result it is sufficient to exclude eigenvalues for just one $\omega \in \Omega$.

PROPOSITION 6.1. *If Ω_c is non-empty, then it is a dense G_δ.*

PROOF. Simon has shown that Ω_c is a G_δ [**Si95a**]. If Ω_c is not empty, then it contains an entire orbit which is dense by minimality. □

Next, by invariance Ω_c has μ-measure 0 or 1. In order to establish an almost sure result, it therefore suffices to bound the measure of Ω_c from below by a positive number. Here is a more elaborate version of this which is useful in connection with local investigations.

PROPOSITION 6.2. *Suppose $G(n)$, $n \in \mathbb{N}$, are Borel sets such that*
(1) $\limsup_{n \to \infty} G(n) \subseteq \Omega_c$,
(2) $\limsup \mu(G(n)) > 0$.
Then, $\mu(\Omega_c) = 1$.

PROOF. The assertion follows from
$$\mu\Big(\limsup_{n \to \infty} G(n)\Big) \geq \limsup_{n \to \infty} \mu\big(G(n)\big),$$
which is readily verified. □

As pointed out earlier, one cannot expect a general way to establish uniform absence of eigenvalues for the models we consider here by inspecting a set of ω's which is strictly smaller than Ω. To a certain extent, this can be understood in view of the discreteness of the potential values and the well-known and heavily studied sensitivity of point spectrum with respect to rank one perturbations [**SW**,

DMS, Go94, Si95b]. Thus one is led to consider each ω individually and to apply pointwise methods.

It seems interesting to note that a study of the eigenvalue problem motivates three different viewpoints, namely, topological arguments for generic results, measure-theoretical arguments for almost sure results, and combinatorial arguments for uniform results.

Let us now combine these general strategies with the Gordon-type criteria from Section 4. We will treat generic, almost sure, and uniform results separately.

6.1. Generic Results. We immediately deduce as a first application a criterion for generic absence of eigenvalues as follows.

PROPOSITION 6.3. *Suppose there exists* $\omega \in \Omega$ *such that* $V = V_\omega$ *obeys the assumption of either Lemma 3.1 or Lemma 3.3. Then, Ω_c is a dense G_δ.*

A single element with Gordon-type symmetries was found in the Fibonacci case [**Sü87**], in the general Sturmian case [**BIST, D98a**], and for a class of substitution models including period doubling and binary non-Pisot [**D98d**]. We also want to remark that the work by Hof et al. provides a method to prove generic absence of eigenvalues by studying palindromic structures in the potentials [**HKS**]. However, their method seems to be restricted to sets of measure zero [**DZ, D99b**] and is thus not able to establish almost sure or uniform results.

6.2. Almost Sure Results. Similarly, one gets a criterion for almost sure absence of eigenvalues as follows. We start with the three-block method. Define

$$G(n) = \{\omega \in \Omega : V_\omega(k-n) = V_\omega(k) = V_\omega(k+n), 1 \le k \le n\}.$$

Obviously, the $G(n)$ are Borel sets since they are finite unions of cylinder sets. Combining Lemma 3.3 and Proposition 6.2, we obtain the following proposition.

PROPOSITION 6.4. *Suppose*

$$(6.1) \qquad \limsup_{n \to \infty} \mu\big(G(n)\big) > 0.$$

Then, $\mu(\Omega_c) = 1$.

Moreover, $\mu\big(G(n)\big)$ can be estimated by inspecting frequencies of cubes of length $3n$ due to equation (1.1). An argument that is often useful is the following.

LEMMA 6.5. *Suppose $\Omega = \Omega_\psi$ and there is a fourth power v^4 occurring in ψ such that $|v| = n$. Then,*

$$(6.2) \qquad \mu\big(G(n)\big) \ge n d_\psi(v^4).$$

In particular, the assumption of Proposition 6.4 is satisfied if one can find a constant $B > 0$ and a sequence of words v_k with $|v_k| = n_k \to \infty$ as $k \to \infty$ such that for all k, $v_k^4 \in F_\psi$ and

$$(6.3) \qquad d_\psi(v_k^4) \ge \frac{B}{n_k}.$$

This criterion has been applied to circle map models as well as substitution models. In [**DP86**], the following theorem for circle map models has been proved.

THEOREM 6.6 (Delyon-Petritis). *Let $\Omega_{\alpha,\beta}$ be a circle map hull. Suppose that the coefficients a_n in the continued fraction expansion of α obey*

$$(6.4) \qquad \limsup_{n \to \infty} a_n \ge 5.$$

Then for every coupling constant λ, eigenvalues are almost surely absent.

The condition (6.4) has been slightly relaxed by Kaminaga in [**Ka**].

THEOREM 6.7 (Kaminaga). *Let $\Omega_{\alpha,\beta}$ be a circle map hull. Suppose that the coefficients a_n in the continued fraction expansion of α obey*

(6.5) $$\limsup_{n\to\infty} a_n \geq 4.$$

Then for every coupling constant λ, eigenvalues are almost surely absent.

For substitution models the criterion in [**D98c**] reads as follows.

THEOREM 6.8. *Let u be a fixed point of a primitive substitution S. Suppose u has a fourth power as a factor. Then for the associated operator family $(H_\omega)_{\omega\in\Omega}$, we have $\mu(\Omega_c) = 1$.*

The criterion applies in particular to the binary non-Pisot case. Moreover, the argument in the proof can be modified to include the period doubling case, too (see [**D98b**] for a slightly more direct proof in this case).

Turning now to the two-block method, we can modify the above steps as follows. Define for $n \in \mathbb{N}$ and $C < \infty$,

$$G'(n,C) = \{\omega \in \Omega : V_\omega(k) = V_\omega(k+n), 1 \leq k \leq n, |\mathrm{tr}(M_{E,\omega}(n))| \leq C\ \forall E \in \Sigma\}.$$

Again the $G'(n,C)$ are finite unions of cylinder sets and hence Borel sets. Lemma 3.1 and Proposition 6.2 now imply the following.

PROPOSITION 6.9. *Suppose there exists $C < \infty$ such that*

(6.6) $$\limsup_{n\to\infty} \mu(G'(n,C)) > 0.$$

Then, $\mu(\Omega_c) = 1$.

We have the following criterion which is similar to Lemma 6.5.

LEMMA 6.10. *Suppose $\Omega = \Omega_\psi$ and there is a cube v^3 occurring in ψ such that $|v| = n$ and $|\mathrm{tr}(M_E(v))| \leq C$ for every $E \in \Sigma$. Then,*

(6.7) $$\mu(G'(n,C)) \geq n d_\psi(v^3).$$

In particular, the assumption of Proposition 6.9 is satisfied if one can find constants $B > 0$ and $C < \infty$ and a sequence of words v_k with $|v_k| = n_k \to \infty$ as $k \to \infty$ such that $v_k^3 \in F_\psi$,

(6.8) $$|\mathrm{tr}(M_E(v_k))| \leq C \quad \forall E \in \Sigma,$$

and

(6.9) $$d_\psi(v_k^3) \geq \frac{B}{n_k}.$$

These criteria are particularly useful in the Sturmian case since we have a uniform trace bound for fixed λ; compare Theorem 4.4. Moreover, Proposition 5.1 allows one to estimate frequencies of s_n^3 using (4.12). Putting this together, one obtains the following result.

THEOREM 6.11 (Kaminaga). *Let Ω_α be a Sturmian hull. Then for every λ, $\mu(\Omega_c) = 1$.*

The proof given in [**Ka**] does not use this two-block argument but rather a more elaborate three-block argument. However, using Proposition 5.1 one may also find suitably positioned cubes. We want to point out that the use of the two-block argument yields additional information; compare Remark 3.4.

6.3. Uniform Results.
This last theorem can even be strengthened. In fact, uniform absence of eigenvalues was recently established for all Sturmian hulls.

THEOREM 6.12. *Let Ω_α be a Sturmian hull. Then for every λ, $\Omega_c = \Omega_\alpha$.*

The proof shows that for all λ, α, Lemma 3.1 is applicable to every V_ω [**DL99a, DKL**]. This is the only uniform singular continuity result in this context that is known so far.

Table 1 summarizes the known results on absence of eigenvalues with reference to the respective first proof.

TABLE 1.

Model	generic	almost sure	uniform
Circle maps (every λ, α, $\beta = \alpha$)	[**BIST**]	[**Ka**]	[**DL99a, DKL**]
Circle maps (every λ, β, a.e. α)	[**DP86**]	[**DP86**]	open
Circle maps (every λ, α, β)	[**HKS**]	open	open
Fibonacci substitution	[**Sü87**]	[**Ka**]	[**DL99a**]
Period doubling substitution	[**BBG91**]	[**D98b**]	open
Binary non-Pisot substitution	[**HKS**]	[**D98c**]	open
Thue-Morse substitution	[**DP91**]	open	open
Rudin-Shapiro substitution	open	open	open

7. Zero-Measure Spectrum

In this section we show how the fact that the spectrum has zero Lebesgue measure can be proved by using the lower bounds that were established when proving absence of eigenvalues. In this sense, the zero-measure property is merely a corollary to a proof of absence of eigenvalues if the latter is based on the two-block method. In fact, this type of argument virtually recovers all the known results on zero-measure spectrum. A more comprehensive discussion of this simple but somewhat surprising fact is given in [**DL99c**].

Recall that A denotes the set of energies where the averaged Lyapunov exponent vanishes and that it has zero Lebesgue measure for the operators under study. The standard way of proving the zero-measure property is to show that the spectrum Σ is contained in this set,

(7.1) $$\Sigma \subseteq A.$$

This can be done with the two-block method as follows. Recall that for every E, there exists a full measure set Ω_E such that for $\omega \in \Omega_E$, the pointwise Lyapunov exponent $\gamma_{\omega,+}(E)$ exists and its value is given by $\gamma(E)$. In fact, it has been shown that for $E \in \Sigma$, $\Omega_E = \Omega$ for all substitution models [**H**] and all Sturmian models [**DL99b**]. So to prove (7.1) it is sufficient to show that the two-block method is applicable for some $\omega \in \Omega$ since it yields that no solution is decaying at $+\infty$, whereas, by the Osceledec result Theorem 1.11, a positive Lyapunov exponent would give rise to a solution which is exponentially decaying at $+\infty$! This simple argument can be applied to all Sturmian models [**DL99c**] (thus recovering the main result from [**BIST**]) and it can also be used in a slightly modified form (see, e.g., [**D99a**]) to recover and elucidate results of Bellissard et al. [**BBG91**] and Bovier and Ghez [**BG93**] for substitution models including period doubling, Thue-Morse, and binary non-Pisot.

The main point here is to emphasize that the occurrence of zero-measure and thus Cantor spectrum is natural, given the Kotani result and hierarchical structures in the potential, the latter leading to a trace map characterization of the spectrum. The additional input of the occurrence of squares is unavoidable in the case of Sturmian models and substitution models over a two-letter alphabet, and it appears as a natural further assumption in the result of Bovier and Ghez for substitution models on larger alphabets.

Table 2 summarizes the known results on zero-measure spectrum together with references to their proofs.

TABLE 2.

Model	zero-measure spectrum
Circle maps (every λ, α, $\beta = \alpha$)	[BIST]
Circle maps (every λ, α, $\beta \neq \alpha$)	open
Fibonacci substitution	[Sü89]
Period doubling substitution	[BBG91]
Binary non-Pisot substitution	[BG93]
Thue-Morse substitution	[BBG91]
Rudin-Shapiro substitution	open

8. Quantum Dynamics

How would such spectral properties show up in physical systems (resp., observables)? To give a first hint, we are now concerned with the transport properties of one-dimensional quasicrystal models, that is, with the long time behavior of the unitary groups generated by the operators under study. We have seen that the presence of purely singular continuous spectrum seems to be the rule for circle map and substitution Hamiltonians. Consequently, we discuss recent ideas and results concerning the dynamics of operators with purely singular continuous spectra. An analysis of the quantum dynamics naturally consists of two parts. On the one hand, one tries to identify crucial characteristics of singular continuous spectral measures which enable one to obtain bounds on the dynamics. Certain dimensions, such as Hausdorff dimension and packing dimension, have proved to be useful in this context. On the other hand, one seeks methods to study these dimensions which apply to the models of interest. Investigations in these directions are still in their early stages. In particular the results for concrete operators are very limited as the reader will notice. This, however, should be seen as a challenge.

Let us first recall parts of the general theory. Suppose H is a self-adjoint operator in $\ell^2(\mathbb{Z}^d)$ and $\psi \in \ell^2(\mathbb{Z}^d)$ with $\|\psi\| = 1$. The spectral measure μ_ψ of ψ is uniquely defined by

$$\langle \psi, f(H)\psi \rangle = \int_\mathbb{R} f(x)\, d\mu_\psi(x)$$

for any measurable function f. We are interested in the long time behavior of

$$\psi(t) = e^{-itH}\psi.$$

In the singular continuous regime it is convenient to consider time averaged quantities as suggested by the RAGE theorem [**RS79**], a common quantity being the Cesaro mean of the moments of order p of the position operator for ψ, that is,

$$\langle\!\langle |X|^p \rangle\!\rangle(T) = \frac{1}{T} \int_0^T \langle \psi(t), |X|^p \psi(t) \rangle \, dt,$$

where $|X|^p$ is given by

$$|X|^p = \sum_{n \in \mathbb{Z}^d} |n|^p \langle \delta_n, \cdot \rangle \delta_n$$

with the standard orthonormal basis $(\delta_n)_{n \in \mathbb{Z}^d}$ of $\ell^2(\mathbb{Z}^d)$. Several authors have established lower bounds on $\langle\!\langle |X|^p \rangle\!\rangle(T)$ in terms of certain continuity properties of μ_ψ. Typical bounds provide a power law behavior where the power depends on the moment, the continuity (measured by some $\alpha \in (0,1)$), and the space dimension in the following way,

(8.1) $$\langle\!\langle |X|^p \rangle\!\rangle(T) > C_{\psi,p} T^{p\alpha/d}.$$

The first type of result in this direction is due to Guarneri [**Gu**] and Combes [**Co**]. It requires uniform α-Hölder continuity of μ_ψ, that is, $\mu_\psi(I) < C|I|^\alpha$ for every interval I with $|I| < 1$, $|\cdot|$ denoting Lebesgue measure. It was extended by Last in [**La96**] to measures with non-trivial α-continuous component, that is, μ_ψ which are not supported on a set of zero h^α measure, where h^α denotes the α-dimensional Hausdorff measure; see also [**BCM**] and [**BT**]. A recent result by Guarneri and Schulz-Baldes relaxes the requirement that the bound holds for all times. They are able to prove a similar bound for a sequence of time scales $T_n \to \infty$ in terms of the packing dimension of μ_ψ which sometimes gives a better exponent; see [**GS**] for details. Similar upper bounds purely in terms of Hausdorff dimensional properties of μ_ψ cannot hold true due to an example in [**DJLS**] which shows that even a pure point measure μ_ψ can give rise to a growth rate of $\langle\!\langle |X|^2 \rangle\!\rangle(T)$ which is arbitrarily close to ballistic (compare, however, [**Si90**]).

In the one-dimensional Schrödinger operator case, Jitomirskaya and Last developed a beautiful way to study such dimensional properies of spectral measures [**JL96, JL99a, JL99b**]. Their method is in fact an extension of the Gilbert-Pearson theory [**GP, Gi, KP**]. It consists of studying the lim inf of

$$\frac{\|\phi_1\|_L^{2-\alpha}}{\|\phi_2\|_L^\alpha}$$

as L tends to infinity, where $\phi_{1,2}$ are solutions of (3.2) with "orthogonal" boundary conditions at the origin and

(8.2) $$\|\phi\|_L = \left(\sum_{n=1}^{\lfloor L \rfloor} |\phi(n)|^2 + (L - \lfloor L \rfloor)|\phi(\lfloor L \rfloor + 1)|^2 \right)^{1/2}.$$

This approach has been applied to Sturmian models in [**JL96, JL99b, D98a, DKL**]. Those works obtain the bound (8.1) for all elements in the hull in the case where the rotation number has bounded density. It holds for all initial states ψ with a positive α which depends on the rotation number and the coupling constant. The proof essentially consists of three steps. One first linearizes the trace map in order to prove a power-law *upper* bound on (8.2), uniformly for all solutions (see [**IRT, IT**]). Then one proves a similar uniform power-law *lower* bound. Interestingly, in this step a Gordon-type argument is the key ingredient. Finally, one essentially employs the maximum principle to infer the desired property of the whole-line problem from the analysis of the half-line solution behavior.

9. Open Problems

In this concluding section we list some open problems. We state explicit questions as well as vague directions that seem interesting and important.

9.1. A Constructive Proof of Kotani's Result.

Theorem 1.13, essentially a corollary to Kotani theory, is right at the heart of most of the results presented and discussed in this paper. Specifically, proofs of absence of absolutely continuous spectrum and zero-measure spectrum were possible only after Kotani published this theorem in 1989. Its value to the known results therefore cannot be overrated. However, we feel that an alternative way of understanding these results would be extremely interesting. Kotani's proof of the fact that the set of energies where the Lyapunov exponent vanishes has Lebesgue measure zero is indirect and inconstructive. It does not give any further information as to why the statement of the theorem is true. On the one hand, it would be nice to have an intuitive understanding of the very uniform absence of absolutely continuous spectrum. This phenomenon, of course, relies heavily on the fact that the potentials take only finitely many values. For example, circle map potentials with the discontinuous characteristic function replaced by a smoother function f seem to exhibit a much smoother transition from absolutely continuous spectrum through singular continuous spectrum to pure point spectrum as the coupling constant λ ranges from 0 through finite values to ∞; compare [**J99**] for the completion of the proof of this phenomenon in the case of the almost Mathieu operator (i.e., $f = \cos$) and [**J95, La95**] for the state of the almost Mathieu art as of 1994. In particular, absolutely continuous spectrum is present in this case for non-zero λ. On the other hand, it would be interesting to investigate the Hausdorff dimension of the spectrum. Kotani's proof does not provide any clue how to tackle this problem in the general case. However, in the 1987 Sütő paper [**Sü87**], a constructive proof of Cantor spectrum was given in the Fibonacci case at large coupling ($\lambda > 4$). Extending this approach, Raymond found in [**Ra**] a way to obtain upper bounds on the Hausdorff dimension in this case. His method should extend to some rotation numbers α other than the golden mean, but already the proof in the Fibonacci case requires considerable effort. The study of the dimension is not only interesting from a purely mathematical perspective. Killip et al. have shown that Raymond's study can be used to establish upper bounds on the dynamics in the Fibonacci case [**KKL99**]. We want to stress, however, that these results are limited to large λ, one particular α, and $\beta = \alpha$. This brings us to the next direction of possible future research activity.

9.2. Spectral Dimensions and Quantum Dynamics.

As discussed in an earlier section, there has been considerable progress in the understanding of the dynamics of a Schrödinger time evolution in the presence of peculiar spectral measures, such as purely singular continuous measures with multifractal structures. Such multifractal behavior is expected to be present for Fibonacci-type operators. Results in this direction for concrete operators, however, are extremely limited, and it will be worthwhile to pursue such investigations. Let us list some questions and problems.

(1) Develop methods that allow for the investigation of dimensional properties of spectral measures.
(2) Find sufficient criteria for non-trivial dynamical upper bounds.
(3) Study dimensions and dynamics for circle map and substitution Hamiltonians.
(4) Is it possible to extend the α-continuity result for Sturmian models with bounded density rotation numbers to a larger class or is there in fact a delicate dependence of the transport properties of the operator on the Diophantine properties of the rotation number?
(5) A partial answer to the above question could be obtained by an extension of Raymond's bound on the Hausdorff dimension of the spectrum to rotation numbers other than the golden mean since this dimension provides a natural upper bound on the dimension of continuity of the spectral measure. Is it possible, for example, to prove that the Hausdorff dimension of the spectrum is zero for, say, Liouville rotation number and large coupling constant?

9.3. Complexity and Spectral Theory. Consider two-sided sequences over a two-letter alphabet, that is, elements of $\mathcal{A}^{\mathbb{Z}}$, where $\mathcal{A} = \{x_1, x_2\} \subseteq \mathbb{R}$. Given such a sequence s, we ask what properties of s are crucial in determining the spectral type of the associated Schrödinger operator, $\Delta + s$ in $\ell^2(\mathbb{Z})$, where Δ denotes the discrete Laplacian. It is clear that any spectral type can occur. A possible point of view, namely that combinatorial properties of s might discriminate between the several spectral types, is discussed in this subsection.

Recall the complexity function p_s which measures the subword complexity of some sequence s, that is, $p_s(n)$ equals the number of subwords in s having length n. The condition $p_s(n) \le n$ for some n is equivalent to p_s being bounded and s being periodic. Moreover, the aperiodic sequences s of minimal complexity (i.e., $p_s(n) = n+1$ for every n) are essentially just the circle map sequences with $\alpha = \beta$ irrational. Restricting our attention to sequences which are *recurrent* in the sense that every subword occurs infinitely often, we can formulate two surprising implications.

(1) p_s bounded $\implies \Delta + s$ has purely absolutely continuous spectrum,
(2) $p_s(n) = n + 1 \implies \Delta + s$ has purely singular continuous spectrum.

As we saw above, sequences generated by primitive substitutions tend to also give rise to operators with purely singular continuous spectrum. This fits very nicely into this picture since their combinatorial complexity is also at the bottom of the hierarchy: It is always bounded by a linear function [**Q**].

On the other complexity extreme, it seems that we encounter a tendency to pure point spectrum. This can be argued as follows. Put some non-trivial probability measure ν on $\mathcal{A}$ (i.e., assign the probability $p \in (0, 1)$ to one letter and $1 - p$ to the other) and consider the product measure $\bigotimes_{n \in \mathbb{Z}} \nu$ on $\mathcal{A}^{\mathbb{Z}}$. Now it is known that almost every sequence with respect to this measure leads to an operator with pure point spectrum (by results of Carmona et al. on localization for Bernoulli Anderson models [**CKM**]). On the other hand, almost every sequence s has every word from $\mathcal{A}^*$ as a subword and thus has complexity function $p_s(n) = 2^n$. We observe that as complexity is increased, the spectral measures become more singular. This raises several questions.

(1) Is there a direct proof of the above observations?

(2) Is there a sharp transition from purely singular continuous spectrum to pure point spectrum?
(3) Are there examples of potentials in $\mathcal{A}^{\mathbb{Z}}$ where several spectral types coexist?

9.4. The Eigenvalue Problem. Although there has been considerable effort to study the eigenvalue problem for one-dimensional circle map and substitution models, we feel that there is still room for improving our understanding of this problem. It is extremely puzzling that no counterexample is known to the apparent tendency that the spectrum is purely singular continuous, and yet one is not able to prove absence of eigenvalues for the entire class of potentials. Instead one is currently able to deduce the desired result from certain local symmetries such as powers or palindromes. However, these symmetries are not always present, as the example of the Rudin-Shapiro substitution shows. In fact, this example epitomizes our lack of understanding in that it defies almost all known and well-established approaches. Apart from the absence of absolutely continuous spectrum, essentially nothing is known about this model. Ironically, this again hints at the power of the inconstructive Kotani result. This leads us to the following concrete problems.

(1) Find an example with non-empty point spectrum or prove absence of eigenvalues for all models.
(2) More modestly, find new ways to prove absence of eigenvalues or to prove presence thereof.
(3) Concretely, study the Rudin-Shapiro case.
(4) Less concretely, try to prove hierarchical structures in the eigenfunctions using hierarchical structures in the potentials (e.g., for Sturmian models or substitution models) which prevent them from being square-summable. In doing so one would also gain important insight into dimensional issues using, for example, the Jitomirskaya-Last theory. The reader may take a look at the very interesting paper [**J91**] by Jitomirskaya which studies circle map potentials for certain parameter values from this point of view.
(5) Finally, since the Thue-Morse substitution is among the most prominent primitive substitutions, it would be nice to prove almost sure absence of eigenvalues also in this case.

9.5. Multi-Dimensional Models. Although the focus of this paper is on one-dimensional models, we also want to address the problem of extending some of the known results to analogous models in higher dimensions. We hope to have demonstrated that a considerable amount of knowledge and results has been accumulated for one-dimensional Schrödinger operators with circle map or substitution potentials. It is somewhat striking that only very little is known for their multi-dimensional analogs. This is, of course, due to the fact that most results were proved by using the transfer matrix formalism which is a purely one-dimensional concept. However, we believe that some results, such as absence of eigenvalues for certain models, should extend to higher dimensions, and that these extensions should be among the major future objectives in this field.

References

[AP]   J.-P. Allouche and J. Peyrière, *Sur une formule de récurrence sur les traces de produits de matrices associés a certaines substitutions*, C. R. Acad. Sci. Paris Sér. II Méc. Phys. Chim. Sci. Univers Sci. Terre **302** (1986), 1135–1136.

[AB]   Y. Avishai and D. Berend, *Trace maps for arbitrary substitution sequences*, J. Phys. A **26** (1993), 2437–2443.

[ABG] Y. Avishai, D. Berend, and D. Glaubman, *Minimum-dimension trace maps for substitution sequences*, Phys. Rev. Lett. **72** (1994), 1842–1845.

[Ba99a] M. Baake, *A guide to mathematical quasicrystals*, Quasicrystals (J.-B. Suck, M. Schreiber, and P. Häussler, eds.), Springer, Berlin, 2001 (in press); math-ph/9901014.

[Ba99b] _____, *A note on palindromicity*, Lett. Math. Phys. **49** (1999), 217–227.

[BGJ] M. Baake, U. Grimm, and D. Joseph, *Trace maps, invariants, and some of their applications*, Internat. J. Modern Phys. B **7** (1993), 1527–1550.

[BR]   M. Baake and J. A. G. Roberts, *Reversing symmetry group of $GL(2,\mathbb{Z})$ and $PGL(2,\mathbb{Z})$ matrices with connections to cat maps and trace maps*, J. Phys. A **30** (1997), 1549–1573.

[BCM] J. M. Barbaroux, J. M. Combes, and R. Montcho, *Remarks on the relation between quantum dynamics and fractal spectra*, J. Math. Anal. Appl. **213** (1997), 698–722.

[BT]   J. M. Barbaroux and S. Tcheremchantsev, *Universal lower bounds for quantum diffusion*, J. Funct. Anal. **168** (1999), 327–354.

[Be]   J. Bellissard, *Spectral properties of Schrödinger's operator with a Thue-Morse potential*, Number Theory and Physics (Les Houches, 1989) (J. M. Luck, P. Moussa, and M. Waldschmidt, eds.), Springer Proc. Phys., vol. 47, Springer, Berlin, 1990, pp. 140–150.

[BBG91] J. Bellissard, A. Bovier, and J.-M. Ghez, *Spectral properties of a tight binding Hamiltonian with period doubling potential*, Commun. Math. Phys. **135** (1991), 379–399.

[BBG92] _____, *Gap labelling theorems for one-dimensional discrete Schrödinger operators*, Rev. Math. Phys. **4** (1992), 1–37.

[BIST] J. Bellissard, B. Iochum, E. Scoppola, and D. Testard, *Spectral properties of one-dimensional quasicrystals*, Commun. Math. Phys. **125** (1989), 527–543.

[BIT] J. Bellissard, B. Iochum, and D. Testard, *Continuity properties of the electronic spectrum of 1D quasicrystals*, Commun. Math. Phys. **141** (1991), 353–380.

[BG93] A. Bovier and J.-M. Ghez, *Spectral properties of one-dimensional Schrödinger operators with potentials generated by substitutions*, Commun. Math. Phys. **158** (1993), 45–66; Erratum, Commun. Math. Phys. **166** (1994), 431–432.

[BG95] _____, *Remarks on the spectral properties of tight-binding and Kronig-Penney models with substitution sequences*, J. Phys. A **28** (1995), 2313–2324.

[Bu]   D. Buschmann, *A proof of the Ishii-Pastur theorem by the method of subordinacy*, Univ. Iagel. Acta Math. **34** (1997), 29–34.

[CKM] R. Carmona, A. Klein, and F. Martinelli, *Anderson localization for Bernoulli and other singular potentials*, Commun. Math. Phys. **108** (1987), 41–66.

[CL]   R. Carmona and J. Lacroix, *Spectral theory of random Schrödinger operators*, Probab. Appl., Birkhäuser, Boston, MA, 1990.

[Ca]   M. Casdagli, *Symbolic dynamics for the renormalization group of a quasiperiodic Schrödinger equation*, Commun. Math. Phys. **107** (1986), 295–318.

[Co]   J.-M. Combes, *Connections between quantum dynamics and spectral properties of time-evolution operators*, Differential Equations with Applications to Mathematical Physics (W. F. Ames, E. M. Harrel, II, and J. V. Herod, eds.), Math. Sci. Engrg., vol. 192, Academic Press, Boston, MA, 1993, pp. 59–68.

[CH]   E. M. Coven and G. A. Hedlund, *Sequences with minimal block growth*, Math. Systems Theory **7** (1973), 138–153.

[CFKS] H. L. Cycon, R. Froese, W. Kirsch, and B. Simon, *Schrödinger operators, with application to quantum mechanics and global geometry*, Texts Monogr. Phys., Springer, Berlin, 1987.

[D97] D. Damanik, *Schrödinger operators with potentials generated by primitive substitutions: An invitation*, Univ. Iagel. Acta Math. **34** (1997), 45–56.

[D98a] _____, *α-continuity properties of one-dimensional quasicrystals*, Commun. Math. Phys. **192** (1998), 169–182.

[D98b] _____, *Singular continuous spectrum for the period doubling Hamiltonian on a set of full measure*, Commun. Math. Phys. **196** (1998), 477–483.

[D98c] _____, *Singular continuous spectrum for a class of substitution Hamiltonians*, Lett. Math. Phys. **46** (1998), 303–311.

[D98d] _____, *Singulärstetiges Spektrum für Substitutionsoperatoren*, Ph.D. thesis, Univ. Frankfurt, Germany, 1998.

[D99a] _____, *Substitution Hamiltonians with bounded trace map orbits*, J. Math. Anal. Appl. (to appear).

[D99b] _____ (in preparation).

[DKL] D. Damanik, R. Killip, and D. Lenz, *Uniform spectral properties of one-dimensional quasicrystals. III. α-continuity*, Commun. Math. Phys. **212** (2000), 191–204; math-ph/9910017.

[DL99a] D. Damanik and D. Lenz, *Uniform spectral properties of one-dimensional quasicrystals. I. Absence of eigenvalues*, Commun. Math. Phys. **207** (1999), 687–696.

[DL99b] _____, *Uniform spectral properties of one-dimensional quasicrystals. II. The Lyapunov exponent*, Lett. Math. Phys. **50** (1999), 245–257; math-ph/9905008, mp_arc/99-184.

[DL99c] _____, *Half-line eigenfunction estimates and singular continuous spectrum of zero Lebesgue measure*, math.SP/9905099.

[DZ] D. Damanik and D. Zare, *Palindrome complexity bounds for primitive substitution sequences*, Discrete Math. (to appear); preprint (1999).

[DJLS] R. del Rio, S. Jitomirskaya, Y. Last, and B. Simon, *Operators with singular continuous spectrum. IV. Hausdorff dimensions, rank one perturbations, and localization*, J. Analyse Math. **69** (1996), 153–200.

[DMS] R. del Rio, N. Makarov and B. Simon, *Operators with singular continuous spectrum, II. Rank one operators*, Commun. Math. Phys. **165** (1994), 59–67.

[DP86] F. Delyon and D. Petritis, *Absence of localization in a class of Schrödinger operators with quasiperiodic potential*, Commun. Math. Phys. **103** (1986), 441–444.

[DP91] F. Delyon and J. Peyrière, *Recurrence of the eigenstates of a Schrödinger operator with automatic potential*, J. Statist. Phys. **64** (1991), 363–368.

[FK] H. Furstenberg and H. Kesten, *Products of random matrices*, Ann. Math. Statist. **31** (1960), 457–469.

[Gi] D. J. Gilbert, *On subordinacy and analysis of the spectrum of Schrödinger operators with two singular endpoints*, Proc. Roy. Soc. Edinburgh Sect. A**112** (1989), 213–229.

[GP] D. J. Gilbert and D. B. Pearson, *On subordinacy and analysis of the spectrum of one-dimensional Schrödinger operators*, J. Math. Anal. Appl. **128** (1987), 30–56.

[Go76] A. Gordon, *On the point spectrum of the one-dimensional Schrödinger operator*, Uspehi Mat. Nauk. **31** (1976), 257–258.

[Go94] _____, *Pure point spectrum under 1-parameter perturbations and instability of Anderson localization*, Commun. Math. Phys. **164** (1994), 489–505.

[Gu] I. Guarneri, *Spectral properties of quantum diffusion on discrete lattices*, Europhys. Lett. **10** (1989), 95–100.

[GS] I. Guarneri and H. Schulz-Baldes, *Lower bounds on wave packet propagation by packing dimensions of spectral measures*, Math. Phys. Electron. J. **5** (1999), no. 1.

[GA] G. Gumbs and M. K. Ali, *Dynamical maps, Cantor spectra, and localization for Fibonacci and related quasiperiodic lattices*, Phys. Rev. Lett. **60** (1988), 1081–1084.

[H] A. Hof, *Some remarks on discrete aperiodic Schrödinger operators*, J. Statist. Phys. **72** (1993), 1353–1374.

[HKS] A. Hof, O. Knill, and B. Simon, *Singular continuous spectrum for palindromic Schrödinger operators*, Commun. Math. Phys. **174** (1995), 149–159.

[IRT] B. Iochum, L. Raymond, and D. Testard, *Resistance of one-dimensional quasicrystals*, Phys. A **187** (1992), 353–368.

[IT] B. Iochum and D. Testard, *Power law growth for the resistance in the Fibonacci model*, J. Statist. Phys. **65** (1991), 715–723.

[I] K. Ishii, *Localization of eigenstates and transport phenomena in the one dimensional disordered system*, Progr. Theoret. Phys. Suppl. **53** (1973), 77–138.

[J91] S. Jitomirskaya, *Singular spectral properties of a one-dimensional discrete Schrödinger operator with quasiperiodic potential*, Dynamical Systems and Statistical Mechanics

(Moscow, 1991) Adv. Soviet Math., vol. 3, Amer. Math. Soc., Providence, RI, 1991, 215–254.

[J95] _____, *Almost everything about the almost Mathieu operator.* II, XI International Congress of Mathematical Physics (Paris, 1994) (D. Iagolnitzer, ed.), International Press, Cambridge, MA, 1995, pp. 373–382.

[J99] _____, *Metal-insulator transition for the almost Mathieu operator*, Ann. of Math. (2) **150** (1999), 1159–1175.

[JL96] S. Jitomirskaya and Y. Last, *Dimensional Hausdorff properties of singular continuous spectra*, Phys. Rev. Lett. **76** (1996), 1765–1769.

[JL99a] _____, *Power law subordinacy and singular spectra.* I. *Half-line operators*, Acta Math. **183** (1999), 171–189.

[JL99b] _____, *Power law subordinacy and singular spectra,* II. *Line operators*, Commun. Math. Phys. **211** (2000), 643–658.

[JS] S. Jitomirskaya and B. Simon, *Operators with singular continuous spectrum.* III. *Almost periodic Schrödinger operators*, Commun. Math. Phys. **165** (1994), 201–205.

[KKL86] P. A. Kalugin, A. Yu. Kitaev, and L. S. Levitov, *Electron spectrum of a one-dimensional quasicrystal*, Soviet Phys. JETP **64** (1986), 410–415.

[Ka] M. Kaminaga, *Absence of point spectrum for a class of discrete Schrödinger operators with quasiperiodic potential*, Forum Math. **8** (1996), 63–69.

[KP] S. Khan and D. B. Pearson, *Subordinacy and spectral theory for infinite matrices*, Helv. Phys. Acta **65** (1992), 505–527.

[Kh] A. Khintchine, *Continued fractions*, Dover, Mineola, NY, 1997.

[KKL99] R. Killip, A. Kiselev, and Y. Last, in preparation.

[KKT] M. Kohmoto, L. P. Kadanoff, and C. Tang, *Localization problem in one dimension: Mapping and escape*, Phys. Rev. Lett. **50** (1983), 1870–1872.

[KO] M. Kohmoto and Y. Oono, *Cantor spectrum for an almost periodic Schrödinger equation and a dynamical map*, Phys. Lett. A **102** (1984), 145–148.

[KN] M. Kolár and F. Nori, *Trace maps of general substitutional sequences*, Phys. Rev. B **42** (1990), 1062–1065.

[Ko84] S. Kotani, *Ljapunov indices determine absolutely continuous spectra of stationary random one-dimensional Schrödinger operators*, Stochastic Analysis (Katata/Kyoto, 1982) (K. Itô, ed.), North-Holland Math. Library, vol. 32, North-Holland, Amsterdam, 1984, pp. 225–247.

[Ko89] _____, *Jacobi matrices with random potentials taking finitely many values*, Rev. Math. Phys. **1** (1989), 129–133.

[KS] H. Kunz and B. Souillard, *Sur le spectre des opérateurs aux différences finies aléatoires*, Commun. Math. Phys. **78** (1980), 201–246.

[La95] Y. Last, *Almost everything about the almost Mathieu operator.* I, XI International Congress of Mathematical Physics (Paris, 1994) (D. Iagolnitzer, ed.), International Press, Cambridge, MA, 1995, pp. 366–372.

[La96] _____, *Quantum dynamics and decompositions of singular continuous spectra*, J. Funct. Anal. **142** (1996), 406–445.

[LS] Y. Last and B. Simon, *Eigenfunctions, transfer matrices, and absolutely continuous spectrum of one-dimensional Schrödinger operators*, Invent. Math. **135** (1999), 329–367.

[Lo83] M. Lothaire (ed.), *Combinatorics on words*, 2nd ed., Cambridge Math. Lib., Cambridge Univ. Press, Cambridge, 1997.

[Lo99] _____, *Algebraic combinatorics on words*, in preparation.

[LP86] J. M. Luck and D. Petritis, *Phonon spectra in one-dimensional quasicrystals*, J. Statist. Phys. **42** (1986), 289–310.

[LP92] W. F. Lunnon and P. A. B. Pleasants, *Characterization of two-distance sequences*, J. Austral. Math. Soc. A **53** (1992), 198–218.

[O] V. Osceledec, *A multiplicative ergodic theorem. Lyapunov characteristic numbers for dynamical systems*, Trans. Moscow Math. Soc. **19** (1968), 197–231.

[OPRSS] S. Ostlund, R. Pandit, D. Rand, H. J. Schellnhuber, and E. D. Siggia, *One-dimensional Schrödinger equation with an almost periodic potential*, Phys. Rev. Lett. **50** (1983), 1873–1877.

[Pa]   L. Pastur, *Spectral properties of disordered systems in the one-body approximation*, Commun. Math. Phys. **75** (1980), 179–196.
[Pe78]   D. B. Pearson, *Singular continuous measures in scattering theory*, Commun. Math. Phys. **60** (1978), 13–36.
[Pe54]   O. Perron, *Die Lehre von den Kettenbrüchen. I. Elementare Kettenbrüche*, Teubner, Stuttgart, 1954.
[Pe83]   K. Petersen, *Ergodic theory*, Cambridge Stud. Adv. Math., vol. 2, Cambridge Univ. Press, Cambridge, 1983.
[PWW]   J. Peyrière, Z. Y. Wen, and Z. X. Wen, *Polynômes associés aux endomorphismes de groupes libres*, Enseign. Math. (2) **39** (1993), 153–175.
[Q]   M. Queffélec, *Substitution dynamical systems—Spectral analysis*, Lecture Notes in Math., vol. 1284, Springer, Berlin, 1987.
[Ra]   L. Raymond, *A constructive gap labelling for the discrete Schrödinger operator on a quasiperiodic chain*, preprint (1997).
[RS80]   M. Reed and B. Simon, *Methods of modern mathematical physics. I. Functional analysis*, 2nd ed., Academic Press, San Diego, CA, 1980.
[RS79]   _____, *Methods of modern mathematical physics. III. Scattering theory*, Academic Press, New York, 1979.
[RS78]   _____, *Methods of modern mathematical physics. IV. Analysis of operators*, Academic Press, New York, 1978.
[Ro]   J. A. G. Roberts, *Escaping orbits in trace maps*, Physica A **228** (1996), 295–325.
[RB]   J. A. G. Roberts and M. Baake, *Trace maps as 3D reversible dynamical systems with an invariant*, J. Statist. Phys. **74** (1994), 829–888.
[Ru]   D. Ruelle, *Ergodic theory of differentiable dynamical systems*, Inst. Hautes Études Sci. Publ. Math. **50** (1979), 27–58.
[SBGC]   D. Shechtman, I. Blech, D. Gratias, and J. V. Cahn, *Metallic phase with long range orientational order and no translational symmetry*, Phys. Rev. Lett. **53** (1984), 1951–1953.
[Si83]   B. Simon, *Kotani theory for one dimensional stochastic Jacobi matrices*, Commun. Math. Phys. **89** (1983), 227–234.
[Si90]   _____, *Absence of ballistic motion*, Commun. Math. Phys. **134** (1990), 209–212.
[Si95a] _____, *Operators with singular continuous spectrum. I. General operators*, Ann. of Math. (2) **141** (1995), 131–145.
[Si95b] _____, *Spectral analysis of rank one perturbations and applications*, Mathematical Quantum Theory. II. Schrödinger Operators (Vancouver, BC, 1993) (J. Feldman, R. Froese, and L. M. Rosen, eds.) CRM Proc. Lecture Notes, vol. 8, Amer. Math. Soc., Providence, RI, 1995, pp. 109–149.
[Si96a] _____, *Operators with singular continuous spectrum. VI. Graph Laplacians and Laplace-Beltrami operators*, Proc. Amer. Math. Soc. **124** (1996), 1177–1182.
[Si96b] _____, *Operators with singular continuous spectrum. VII. Examples with borderline time decay*, Commun. Math. Phys. **176** (1996), 713–722.
[SS]   B. Simon and G. Stolz, *Operators with singular continuous spectrum. V. Sparse potentials*, Proc. Amer. Math. Soc. **124** (1996), 2073–2080.
[SW]   B. Simon and T. Wolff, *Singular continuous spectrum under rank one perturbations and localization for random Hamiltonians*, Commun. Pure Appl. Math. **39** (1986), 75–90.
[Sü87]   A. Sütő, *The spectrum of a quasiperiodic Schrödinger operator*, Commun. Math. Phys. **111** (1987), 409–415.
[Sü89]   _____, *Singular continuous spectrum on a Cantor set of zero Lebesgue measure for the Fibonacci Hamiltonian*, J. Statist. Phys. **56** (1989), 525–531.
[Sü95]   _____, *Schrödinger difference equation with deterministic ergodic potentials*, Beyond Quasicrystals (Les Houches, 1994) (F. Axel and D. Gratias, eds.) Springer, Berlin, 1995, pp. 481–549.
[W]   P. Walters, *An introduction to ergodic theory*, Grad. Texts in Math., vol. 79, Springer, New York, 1982.

Fachbereich Mathematik, Johann Wolfgang Goethe-Universität, D-60054 Frankfurt, Germany

Current address: Department of Mathematics 253-37, California Institute of Technology, Pasadena, CA 91125, USA

E-mail address: `damanik@its.caltech.edu`

Centre de Recherches Mathématiques
CRM Monograph Series
Volume **13**, 2000

The Planar Dimer Model with Boundary: A Survey

Richard Kenyon

1. Introduction

A *dimer covering* of a finite graph is a perfect matching of the graph, that is, a set of edges with the property that every vertex is contained in a unique edge. The *dimer model* is the statistical model dealing with the set of all dimer coverings of a graph.

Kasteleyn [**K1**] and Temperley and Fisher [**TF**] initiated the study of the dimer model by showing how to *count exactly* the number of perfect matchings of an $m \times n$ grid (when at least one of m and n is even). The number is the square root of the determinant of an $mn \times mn$ matrix, and is explicitly:

$$\left[\prod_{j=1}^{m}\prod_{k=1}^{n}\left(2\cos\left(\frac{\pi j}{m+1}\right) + 2i\cos\left(\frac{\pi k}{n+1}\right)\right)\right]^{1/2}.$$

Later, Kasteleyn [**K2**] showed how to efficiently count the number of perfect matchings of any finite planar graph: the number is the square root of the determinant of a matrix closely related to the adjacency matrix of the graph.

Since that time a great deal of work has been done on the physical aspects of the dimer model (see, e.g., [**NYB, FW, Hen, T**]). In the early 1990's some intriguing combinatorial results brought a renewed mathematical interest in the model. These results touched on the influence of the *boundary* on a random dimer covering of a bounded graph [**EKLP, CEP, CKP, Pr2, DMB**]. In this paper we would like to give a short survey of some of these new results. We should note that some preliminary results along these lines appeared somewhat earlier: in [**Els**] and [**GCZ**].

In this paper we concentrate on the mathematical point of view. The reader interested in the more physical aspects of dimers should consult the review of Nagle [**NYB**], works of Kasteleyn [**K1, K3, K2**], Fisher [**Fi**], Fan and Wu [**FW**], McCoy and Wu [**MCW**], Henley [**Hen**], Richard et al. [**RHHB**] and others.

Figure 1 shows four planar tilings. The first is a tiling with *dominos*, that is, 1×2 rectangles in two orientations. The second is a tiling with lozenges, where a *lozenge* is a rhombus with a 60° angle and sides of length 1 (*losange* is the French word for rhombus). Lozenges come in three orientations. The third example is a special kind of tiling with squares and isosceles right triangles called "diabolo

---

2000 *Mathematics Subject Classification.* Primary 82B20; Secondary 60K35.

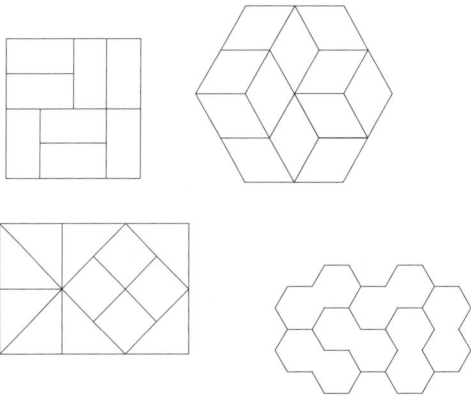

FIGURE 1. Four tiling problems.

tilings" (see exact definition below), and the last is a tiling with "bibones" which are pairs of hexagons joined along an edge. Bibones also come in three orientations.

In the above tilings the tiles are required to meet edge-to-edge (a domino is assumed to have a vertex on the center of its long edge) except in the case of the diabolo tilings, where we also allow a triangle to intersect two squares along its hypotenuse (we don't allow two triangles to meet along their hypotenuses in a "staggered" fashion).

Each of these tiling models is in fact a dimer model in disguise. See Figure 2. The reason for introducing them as tilings is simply a matter of taste. Apparently mathematicians prefer to talk about tilings, physicists about dimers, and computer scientists about perfect matchings! The diabolo tilings were coined by Jim

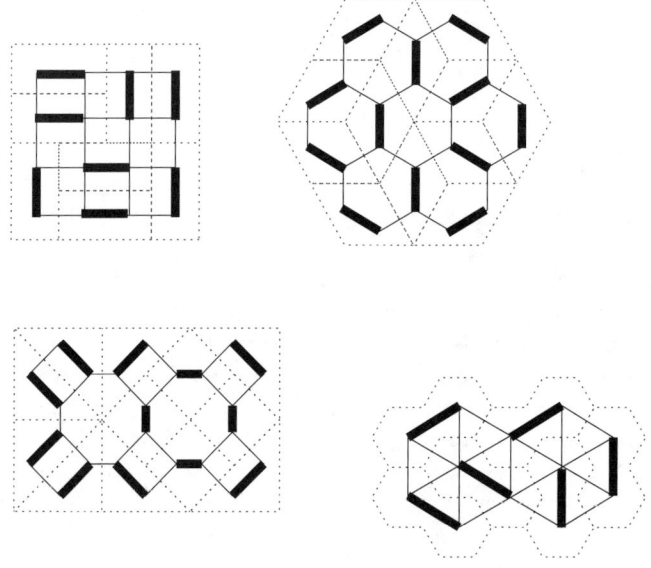

FIGURE 2. Matchings corresponding to the tilings of Figure 1.

Propp (although he has subsequently expressed reservations about this term, we will stick to it here). Each tile is a union of two isosceles-right triangles in a certain isosceles-right triangulation of the plane, whose dual is the "square-octagon" lattice (a periodic graph whose faces are squares and octagons). The word "bibone" comes from [**KR**] where it was used in analogy with terminology of Thurston [**Th**] who used the term "tribone" to denote a line of three adjacent vertices in the triangular lattice (dually, a line of three hexagons).

Another well-known statistical model which is a planar dimer model in disguise is the Ising model (with no external field) on the square grid $\mathbb{Z}^2$. This fact was first made explicit in Fisher [**Fi**] (although some credit should go to Kasteleyn [**K1**]). Recall that configurations in the Ising model are assignments of spins $\{+,-\}$ to points in $\mathbb{Z}^2$; each configuration has an *energy* which we may take to be proportional to the number of pairs of adjacent vertices with differing spins.

In Figure 3, we put the spins rather on the faces of the grid so that for each pair of neighboring faces with differing spins there is a boldface edge; the set of boldface edges then forms a subgraph which has the property that at each vertex it has even degree. Conversely, any subgraph of the grid which has even degree at each vertex corresponds to exactly two different spin configurations (one may choose arbitrarily the spin of the face at the origin; then the remaining spins are determined). Figure 4 then shows the bijection between the Ising model and the dimer model on a related graph: one replaces each vertex in the Ising model with a "butterfly" graph. There is a way to assign energies to dimer configurations so that this bijection preserves energies; see Section 1.1. The resulting planar graph is called the *Fisher lattice*.

Other well-known statistical models such as the 6-vertex and 8-vertex model have similar planar dimer versions under certain restrictions on the vertex energies: see [**FW**] or [**Bax**].

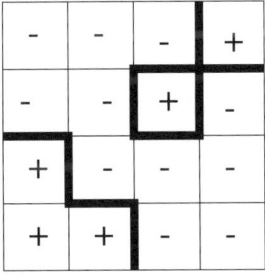

FIGURE 3. The Ising model.

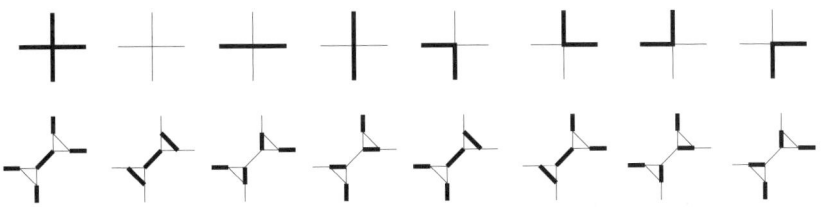

FIGURE 4. Matchings corresponding to the Ising model.

1.1. Defining the Measure.
For a finite planar graph G let $M(G)$ be the set of all its perfect matchings. We assume that the graph G comes equipped with an *energy* associated to each edge, which is the energy we must furnish to put a dimer on that edge. This energy may be positive or negative. A perfect matching then has associated to it an energy which is the sum of the energies for each of its *matched* edges.

Mathematically these energies can be thought of as simply a convenient device used to define the *measure* on $M(G)$ (see [**Ke**]). The measure is defined using *Gibbs' axiom* (hence it is called a Gibbs measure), which says that the probability of a configuration is proportional to the exponential of its energy, that is, if a configuration C has energy $E(C)$ then its probability is $e^{-\beta E(C)}/Z$, where Z is the constant of proportionality necessary to make the measure a probability, and β is a quantity independent of C, which in physics is equal to $1/(kT)$, k being Boltzmann's constant and T being the temperature. The function Z is called the *partition function* of the system and is by definition the sum over all configurations C of $e^{-\beta E(C)}$.

As we have defined it the Gibbs measure depends on T. As T decreases the configurations with lower energy have higher probability. In fact, in the limit $T = 0$ only the configurations of lowest energy have nonzero probability. These are called the *ground states*. On the other hand, in the limit as $T \to \infty$ all configurations have the same probability, so the problem just becomes one of counting the number of configurations. Recall that this was for finite graphs. Usually the situation is not quite so simple: one is most often interested in letting the size of the system tend to ∞ at the same time as the temperature tends to zero or infinity. Fortunately we won't have to deal with these issues here.

Notationally it is often convenient to shortcut the exponentiation step and assign to an edge of energy E an *activity* equal to $e^{-\beta E}$. Then the probability of a configuration is proportional to the *product* of the activities of its matched edges. Often the activities are referred to simply as *weights*.

In the Ising model each spin configuration has an energy equal to the number of neighbors of differing spins. This is the number of bold edges in Figure 3. In the corresponding dimer model we assign the energies of the edges in the butterfly to be zero, and the other edges energy 1. The corresponding Gibbs measures are now mapped to one another under the bijection we described earlier.

1.2. The Boundary.
There are some natural questions to ask right from the start about these Gibbs measures on $M(G)$. For each finite graph the measure is a sum of a finite number of point masses (since the number of configurations is finite). So perhaps the first question should be regarding the convergence of the measures when the graphs become large. Put differently, if we take a large graph G how does the measure depend on G?

In all the cases we are considering the graphs are "periodic": they are subgraphs of an infinite graph on which $\mathbb{Z}^2$ acts via translations (with finite quotient). So by "large" graph G we should mean one which contains a large neighborhood of the origin in $\mathbb{Z}^2$. Then the question really becomes one about the choice of boundary of G. Does the choice of boundary outside a large ball in G significantly affect what the measure looks like near the origin in $\mathbb{Z}^2$? Remember that we are only dealing with complete dimer coverings of G; we are not allowing the dimers to 'hang over' the boundary.

The answer is of course yes, the measure depends very importantly on the precise structure of the boundary. The most well-known example of this phenomenon is the existence of a phase transition in the Ising model. In the Ising model on a large ball in $\mathbb{Z}^2$ and at low temperature, choose spins near the boundary of the ball to all be the same, say $+$. (If we cut a ball out of the Fisher lattice by cutting only bonds not interior to butterflies, any matching corresponds to an Ising configuration where all the spins along the boundary are equal. By convention we can call that spin $+$.) Then the probability that a spin near the origin will be $+$ is greater than $\frac{1}{2}$ by an amount which does not tend to zero as the size of the ball increases. On the other hand at sufficiently high temperature this probability tends to $\frac{1}{2}$. In fact, there is a unique temperature T_c at which this change in behavior occurs. We will see below examples of similar boundary influence in the tiling models as well, even in the case of *equal activities*.

This paper deals with the question of how the choice of boundary influences the Gibbs measure. We reiterate that we are using only 'exact' boundary conditions: we want perfect matchings of our graph. In physics it is more common (and often more relevant) to use 'free' boundary conditions, where the dimers are allowed to overlap the boundary. These two different kinds of boundary conditions give rise to distinct mathematical problems, although they are related by the fact that free boundary conditions can be obtained by summing over all possible 'exact' boundaries. One other set of boundary conditions which is common is 'periodic' boundary conditions, where one takes perfect matchings of a graph on a torus (obtained as a quotient of the infinite periodic graph by a subgroup of $\mathbb{Z}^2$). The main advantage of such boundary conditions is that calculations are usually easier.

1.3. Outline. The following sections each touch on one topic, and are ordered from elementary considerations to more complicated. The first section deals with purely combinatorial questions of the type: which boundary conditions allow the *existence* of a tiling? Section 3 discusses Kasteleyn's method of computing the partition function Z in each of our examples. In Section 4 we discuss how to compute *densities of local patterns*. In Section 5 we discuss how the boundary of a region influences the *local entropies* and local densities in a region. In Section 6 we discuss in the domino model how the boundary, even when it does not have an influence on the local statistics, can still affect the long-range properties of the limiting measure.

2. A Lipschitz Criterion

What regions can be tiled with dominos? There is a surprisingly clean answer to this question [**Th, Fo**]. Figure 5 illustrates three elementary observations. The first region cannot be tiled because it has odd area: only even-area regions can be tiled since each tile has even area. The second region cannot be tiled despite having even area: if you color the unit squares black and white in a checkerboard coloring then the number of black and white squares differ, yet every domino covers exactly one square of each color. So if a region can be tiled then in a checkerboard coloring it has the same number of black squares as white squares. This condition is still not sufficient for existence of a tiling, as is illustrated by the third region in Figure 5. Here the principle is that if we cut the region into two parts then the excess of black squares over white squares in either part cannot exceed the length of the cut. For

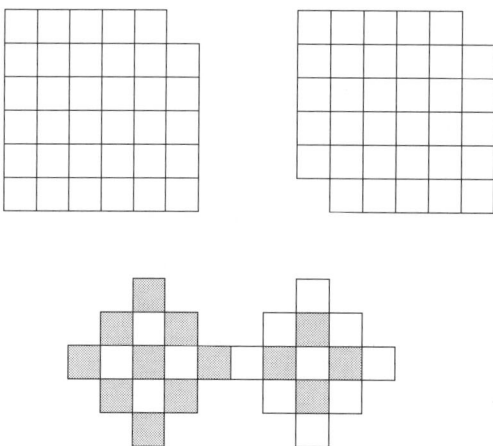

FIGURE 5. Three domino tiling problems

general bipartite matching problems this is the essence of Hall's marriage theorem [**Ha**]. This notion was made precise for dominos by Fournier [**Fo**] in an argument based on the *height function* of Thurston [**Th**].

2.1. Height Function. Given a domino tiling of a simply-connected region, the height function is an integer-valued function on the vertices of the dominos, defined as follows. Fix a checkerboard coloring of the whole plane. Pick some vertex v_0 (say on the boundary for simplicity) and define the height there to be zero. For any other vertex v take a lattice path from v_0 to v which does not pass through the middle of any domino. On each step of the path the height increases by 1 if the square to its left is black and decreases by 1 if the square to its left is white. The height of v is therefore defined to be the number of black squares adjacent to and to the left of the path, minus the number of white squares adjacent to and to the left of the path. The value obtained at a vertex is independent of the path taken (because the height change going around a tile is zero), so the height is well-defined. See Figure 6. The height depends on the tiling and in fact determines the tiling: the tiles cross exactly those edges whose vertices have height difference 3. However, the height on the *boundary* of the region is independent of the tiling; it depends only on the choice of starting point (a different starting point just shifts the height by a constant additive amount). So you don't need to know a tiling to define the boundary height function; in fact the boundary height function gives a *necessary* condition for tilability: if a tiling exists the height function on the boundary must be well-defined, that is, when you define heights along the path which winds around the boundary you must come back to the same height you started with when you get back to the origin. One can show that this will happen exactly when the region contains the same number of 'black' squares as 'white' squares.

Actually the height function gives the following *necessary and sufficient* condition for tilability: a simply-connected region can be tiled if and only if between any two boundary points x, y the difference in the height functions $|h(x) - h(y)|$ is not too large compared to their distance in the graph. More precisely, for x, y vertices in U define $d(x, y)$ to be the shortest length of a path in U from x to y which has

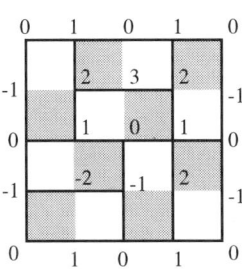 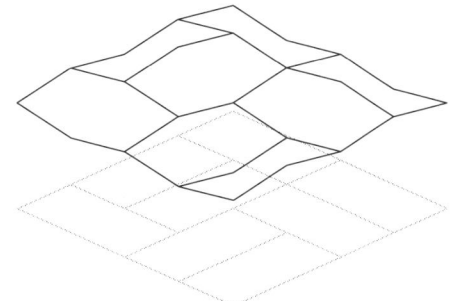

FIGURE 6. Heights and interface of a domino tiling.

only black squares on its left. Then (see [**Fo**]) U is tilable if and only if for all x, $y \in \partial U$ we have $d(x,y) \geq |h(x) - h(y)|$.

2.2. Tiling As Interface. The height function allows one to think of a tiling as a surface in $\mathbb{R}^3$: indeed, the graph of the height function is a surface spanning the graph of the boundary height function. See Figure 6 for an example. Thurston [**Th**] showed that the max of two height functions and the min of two height functions are again height functions, so the set of height functions under the natural partial order is a lattice (in the partially-ordered sense). In particular, there is a unique highest tiling and a unique lowest tiling.

A similar height function exists for both lozenge tilings and diabolo tilings. In the lozenge case use a black-and-white coloring of the underlying equilateral triangulation; the height along a lattice path increases by 1 on an edge with a black triangle to its left and otherwise decreases by 1. In the diabolo case use the black-and-white coloring of the underlying isosceles-right-triangulation. Indeed, for dimer coverings of any *bipartite* planar graph there is a height function with a similar definition and properties [**Cha, Pr1, KPW**]. In all these cases dimer coverings can be represented as interfaces.

2.3. Nonbipartite Case. The nonbipartite case, e.g., for bibones or the Fisher lattice, seems to be genuinely different from the bipartite case. There is no locally-defined notion of long-range combinatorial order similar to the height function. In fact, regarding the existence of a tiling with given boundary, it may be that in all periodic examples any "sufficiently fat" region (i.e., region without narrow isthmus) has a dimer covering. For example, C. Kenyon and E. Remila [**KR**] analyzed completely the bibone tiling problem, concluding that a "simply connected" subgraph of the triangular lattice with an even number of vertices has a perfect matching if it has no vertex with only 1 or 2 neighbors.

3. Kasteleyn's Determinant

Let G be one of the weighted graphs of Figure 7 and $A = (A_{jk})$ be the associated adjacency matrix, that is, the matrix indexed by vertices of G, where A_{jk} is 0 if vertices j and k are not adjacent, and otherwise A_{jk} is the weight on the corresponding edge, multiplied by -1 in case the edge is directed from k to j.

The matrices A are called *Kasteleyn matrices* for the corresponding graphs; Kasteleyn showed that the Pfaffian (square root of the determinant) of a Kasteleyn

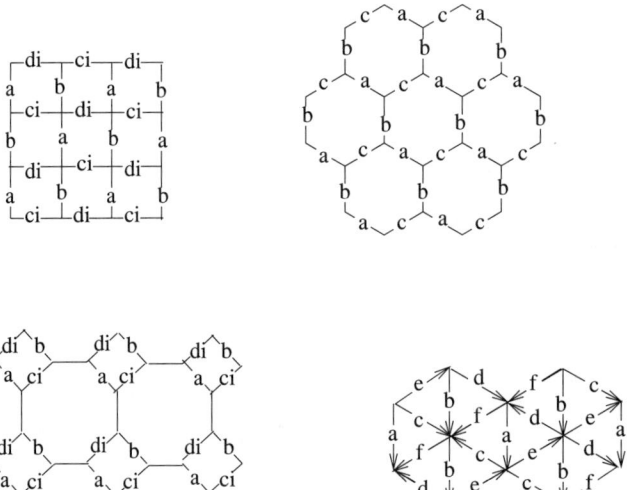

FIGURE 7. 'Natural' edge activities for dominos, lozenges, diabolos and bibones, respectively. The numbers a, b, c, d, e, f are positive reals and $i = \sqrt{-1}$.

matrix is the partition function for the corresponding dimer model whose activities are given by the *absolute values* of the edge weights. See, e.g., [**Tho**].

The reason for the strange choices of signs (and directions in the last case) is to make all the terms in the expansion of the Pfaffians come out with the right sign. In particular, Kasteleyn showed that on any planar graph there is a way to direct the edges so that the associated antisymmetric "adjacency matrix" is a Kasteleyn matrix, i.e., its Pfaffian is the partition function [**K3**]: it suffices to direct the edges so that around any face the number of clockwise-directed edges is odd. There are in fact many ways to direct the edges of a planar graph to make a Kasteleyn matrix [**LL**]. On bipartite graphs instead of directing the edges one can get by with putting unit complex number weights on the edges (see [**Per**]); this has certain advantages over the direction scheme. In either case, a choice of signs or weights on edges of G is called *Kasteleyn flat* if it gives rise to a Kasteleyn matrix. For more information see [**Ku**].

This explains (at least in a vague sense) the signs and directions in Figure 7; what about the activities? Surprisingly enough, in the bipartite cases when one is interested in *uniform* random dimer coverings of finite regions (i.e., even when all edge activities are 1) one is led naturally to consider the above activities. We'll see this below.

It turns out that in the lozenge case any two matchings have the same weight. One way to see this is to notice that, first, with these activities, the two possible matchings of a basic hexagon (i.e, the boundary of a single face) have the same weight (abc), and second, any two matchings can be obtained from one another by local rearrangements around a basic hexagon (as in Figure 8).

This second fact follows from the (partially ordered) lattice property of the height function: there is a lowest tiling and it has the property that it is the unique tiling whose height function has no "local maximum" in the interior. One can move

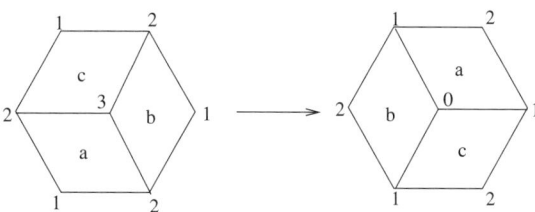

FIGURE 8. This local rearrangement of a lozenge tiling has a number of different names depending on who you talk to. It is 'rotating a hexagon', a 'simpleton flip', 'phason exchange', etc. At the vertices we indicated how the height changes.

from any tiling to the lowest tiling by "rotating" basic hexagons downwards as in Figure 8. See [**Th**].

So in a sense these activities are superfluous for lozenges: if each matching has the same weight, we may as well set all the activities to be 1. The activities are *not* superfluous when the graph is not "simply connected", for example if we have periodic boundary conditions. For then it is not true that all matchings have the same energy. We'll see the effect of this in the next section.

In a similar fashion one can show (see, e.g., [**Th**]) that if we take $ab = cd$ in the domino case, or $ab = cd = 1$ in the diabolo case, then any two matchings of a simply connected subgraph of $\mathbb{Z}^2$, respectively of the square-octagon lattice, will have the same weight.

In the bibone case if $ab = cd = ef$ then all matchings of a simply-connected subgraph of the triangular grid will have the same weight (this is shown in [**KR**]). We may as well set $ab = cd = ef = 1$ by a suitable normalization.

Conditional uniformity is a property of measures on matchings of infinite graphs, which says that when one fixes the matching outside a finite set of vertices S, the conditional measure on matchings of S is the uniform measure. We can extend this definition to finite graphs by saying a measure is conditionally uniform if, when one fixes the matching outside a *simply-connected* set of vertices S, the conditional measure on matchings of S is uniform. If G is simply connected then a conditionally uniform measure is necessarily uniform.

The above-defined Gibbs measures (when defined on not-necessarily simply connected graphs, for example on graphs with periodic boundary conditions) are conditionally uniform in the lozenge case, but only if $ab = cd$ in the domino case, $ab = cd = 1$ in the diabolo case, or $ab = cd = ef$ in the bibone case. In the Ising model the measure is easily seen to be conditionally uniform only in the limit $T \to \infty$.

3.1. Limiting Partition Functions. It is natural to ask oneself what good these determinants are. For an arbitrary planar region we cannot hope to get a closed-form expression for the determinant of one of these big matrices. So it is not a surprise that people early in this subject started out by concentrating on very simple boundary conditions like periodic boundary conditions.

If we take a subgraph modelled on those of Figure 7 having *periodic* boundary conditions, we can compute in each case, using the method of Kasteleyn, the partition function Z_n. This is more complicated than indicated in the previous section since the underlying graph is not planar but toroidal: nonetheless Kasteleyn indicated how to compute the partition function of a graph on the torus as a sum of four Pfaffians of matrices derived from the Kasteleyn matrices [**K1, Tes, RZ**]. The advantage of this approach is that with periodic boundary conditions one can explicitly diagonalize the Kasteleyn matrices and thereby compute their Pfaffians. Another advantage of doing this computation for periodic boundary conditions will become clear in Section 5.2.

Let G_n be the toroidal graph obtained from the infinite graph (depending on the model, $\mathbb{Z}^2$, the honeycomb lattice, square-octagon lattice, triangular grid, Fisher lattice) by taking the quotient by the subgroup $n\mathbb{Z}^2$ of $\mathbb{Z}^2$ (recall that $\mathbb{Z}^2$ acts by translational isomorphisms). Let Z_n be the partition function of dimers on G_n.

We define the *limiting partition function per fundamental domain* to be $Z = \lim_{n\to\infty} Z_n^{1/n^2}$. (In [**CKP**], the authors compute rather the limiting partition function *per dimer*, which is related to this by a constant power.)

Values for $\log Z$ in each case have the impressive-looking formulas:

(1) $\quad \log Z_{\text{dom}} = \dfrac{1}{4\pi^2} \int_0^{2\pi} \int_0^{2\pi} \log\left(\dfrac{(a+be^{i\theta})^2}{e^{i\theta}} + \dfrac{(c+de^{i\phi})^2}{e^{i\phi}}\right) d\theta\, d\phi,$

(2) $\quad \log Z_{\text{loz}} = \dfrac{1}{4\pi^2} \int_0^{2\pi} \int_0^{2\pi} \log(a + be^{i\theta} + ce^{i\phi})\, d\theta\, d\phi,$

(3) $\quad \log Z_{\text{diab}} = \dfrac{1}{4\pi^2} \int_0^{2\pi} \int_0^{2\pi} \log\left(\dfrac{(2a+be^{i\theta})(a+2be^{i\theta})}{e^{i\theta}} + \dfrac{(2c+de^{i\phi})(c+2de^{i\phi})}{e^{i\phi}}\right) d\theta\, d\phi,$

(4) $\quad \log Z_{\text{bib}} = \dfrac{1}{4\pi^2} \int_0^{2\pi} \int_0^{2\pi} \log\bigl(6 - 2\cos(\theta) - 2\cos(\phi) + 2\cos(\theta - \phi)\bigr) d\theta\, d\phi$

(5) $\quad \log Z_{\text{Ising}} = \dfrac{1}{4\pi^2} \int_0^{2\pi} \int_0^{2\pi} \log\bigl((t^2+1)^2 + 2(t - t^3)(\cos(\theta) + \cos(\phi))\bigr) d\theta\, d\phi.$

In the last case $t = e^{-1/(kT)}$. These expressions were obtained by taking a continuum limit of sums of four Pfaffians. It is nontrivial to make these limits rigorous. Indeed, it is not known if the convergence is more than C^1 in the variables a, b, c, d [**CKP**].

Note that in the bibone case Z is independent of the weights! This is a consequence of the fact that even on the torus any two matchings can be obtained from one another using local transformations. The same is true in the Ising model, but there is still a dependence of Z_{Ising} on temperature since the Gibbs measure we chose is not conditionally uniform.

3.2. Phase Transitions. Let us first consider the domino case. If $a > b+c+d$, then the integral in (1) is easy to evaluate by contour integration and $\log Z_{\text{dom}} = \log(a)$. It is *independent* of b, c and d! This implies that the expected number of 'a'-edges per fundamental domain is 1, and the expected number of non-'a'-edges is zero. So the system is "frozen" into a brickwork pattern of all 'a'-type edges. Similarly when any one of a, b, c, d is greater than the sum of the other three, Z is frozen into a brickwork pattern. On the other hand when each of a, b, c, d is less than the sum of the others, Z depends on all four variables and the probability of each edge type is nonzero.

Let us interpret this in more physical terms. Note that if a is the largest of the four activities then as T decreases there will be a point at which $a = e^{-\beta E_a}$ surpasses $b + c + d = e^{-\beta E_b} + e^{-\beta E_c} + e^{-\beta E_d}$. So this temperature is a critical temperature for the system: below this temperature the system is frozen.

In the lozenge case the same phenomenon occurs: there is a phase transition at the temperature where one of the activities equals the sum of the other two. The system is frozen whenever one of the three activities is greater than the sum of the other two. This phenomenon was first noted in Kasteleyn [**K2**]. We'll discuss the diabolo case in Section 5.5.

The Ising model does not 'freeze' at any temperature except $T = 0$. However, there is a phase transition of a different type at the temperature where $t = e^{-1/(kT)} = \sqrt{2} - 1$. This is the phase transition we discussed in the introduction. From (5) one can detect something interesting at this point because Z_{Ising} has discontinuous derivative there. There is a lot of literature on the Ising model so we have chosen not to discuss this transition here: see, e.g., [**MCW**].

4. Local Statistics

Counting tilings is nice, but we'd really like to compute *densities of patterns* to understand the Gibbs measure. In [**Ken1**] we showed how the Kasteleyn matrix (or more precisely, its inverse) can be used to compute these densities in the planar dimer model. The result is as follows.

Let G be a planar graph and K a Kasteleyn matrix for G. Let E be a set of k dimers covering $2k$ distinct vertices $v_1, \ldots, v_{2k}$. The probability of all the dimers of E occurring in a random matching is the square root of

$$a_E^2 \det \begin{pmatrix} K^{-1}(v_1, v_1) & \ldots & K^{-1}(v_1, v_{2k}) \\ \vdots & \ddots & \vdots \\ K^{-1}(v_{2k}, v_1) & \ldots & K^{-1}(v_{2k}, v_{2k}) \end{pmatrix}$$

where a_E is the product of the activities of the dimers E. (Actually, in [**Ken1**] we only dealt with the bipartite case, but the general case follows from the same argument.)

So to compute the density of any local pattern we need "only" to compute the inverse Kasteleyn matrix, called the *coupling function*. Again this can be done explicitly only for simple regions like rectangles (for tori it can also be done using a sum of 4 inverses). Nonetheless with the help of a computer one can experiment on some reasonably-sized regions.

By way of example, in Figure 9 we computed for each $k \in [1, n]$ the probability of the vertical domino $\{(k, n/2), (k, n/2 + 1)\}$ occurring in a domino tiling of an $n \times n$ square.

The probability is close to $\frac{1}{4}$ throughout most of the region. In fact, one can show that as $n \to \infty$ the probability of any given domino e is $\frac{1}{4} + O(1/d^2)$, where d is the combinatorial distance from e to the boundary [**Ken2**]. On the contrary the probability of a boundary edge being covered is asymptotic to $1/\pi$.

Using the coupling function (inverse Kasteleyn matrix) one can prove the following *strong homogeneity property* for uniform domino tilings of the square. If we tile the unit square $[0, 1] \times [0, 1]$ with $\epsilon \times 2\epsilon$ dominos, where $\epsilon = 1/(2n)$, then at every

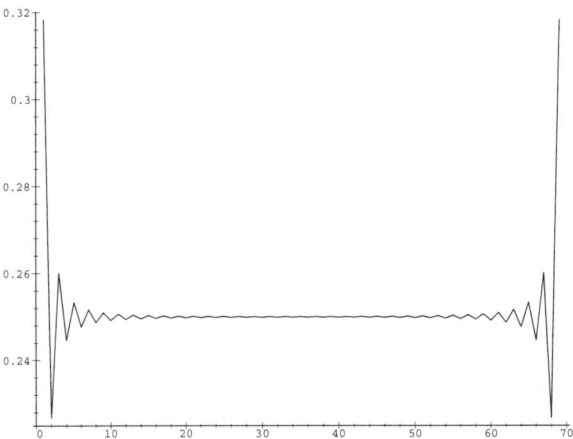

FIGURE 9. Vertical edge probabilities across the middle horizontal line of a square, in a random domino tiling of a $2n \times 2n$ square.

interior point $(x,y) \in (0,1)^2$, the density of any local pattern in a uniform random tiling converges as $n \to \infty$ to a constant independent of (x,y). In particular, for large n if you zoom in to any interior point of the square then the measure that you see (as defined by its densities of patterns) is the same. The limiting measure (on tilings of $\mathbb{Z}^2$) is called the *Burton-Pemantle measure* μ_{BP}: Burton and Pemantle [**BP**] proved that μ_{BP} is the unique translation-invariant measure of maximal entropy on domino tilings of $\mathbb{Z}^2$.

So tilings of the square are very homogeneous. The situation is not at all the same for more general regions. Figure 10 shows a random uniform domino tiling of a tilted square region called an *Aztec diamond*.

Random domino tilings of an Aztec diamond are quite nonhomogeneous, as can be seen from the figure. For example, Figure 11 shows the (approximate) probability of a vertical domino on the horizontal bisector of the region. Not only is the probability distinctly different from $\frac{1}{4}$ except at the center (even when n tends to ∞ these probabilities do not approach $\frac{1}{4}$ except at the origin), but the probabilities at odd coordinates are completely different from those at even coordinates. In [**CEP**] the limiting probabilities of all edges were computed. We rescale the coordinates so that the Aztec diamond is defined by $\{(x,y) : |x| + |y| \leq 1\}$ and we are tiling with $\epsilon \times 2\epsilon$ dominos. Then as $\epsilon \to 0$ the probability of a vertical domino near coordinate $(x, 0)$ converges to either

$$\begin{cases} 0 & \text{if } x \leq -1/\sqrt{2} \\ \frac{1}{2} + \frac{1}{\pi}\arctan([2x-1]/\sqrt{1-2x^2}) & \text{if } -1/\sqrt{2} < x < 1/\sqrt{2} \\ 1 & \text{if } 1/\sqrt{2} \leq x, \end{cases}$$

or

$$\begin{cases} 1 & \text{if } x \leq -1/\sqrt{2} \\ \frac{1}{2} + \frac{1}{\pi}\arctan([2x-1]/\sqrt{1-2x^2}) & \text{if } -1/\sqrt{2} < x < 1/\sqrt{2} \\ 0 & \text{if } 1/\sqrt{2} \leq x, \end{cases}$$

depending on the parity of its distance (in multiples of ϵ) to the rightmost edge. (We used this formula in the graph of Figure 11, which was easier than computing

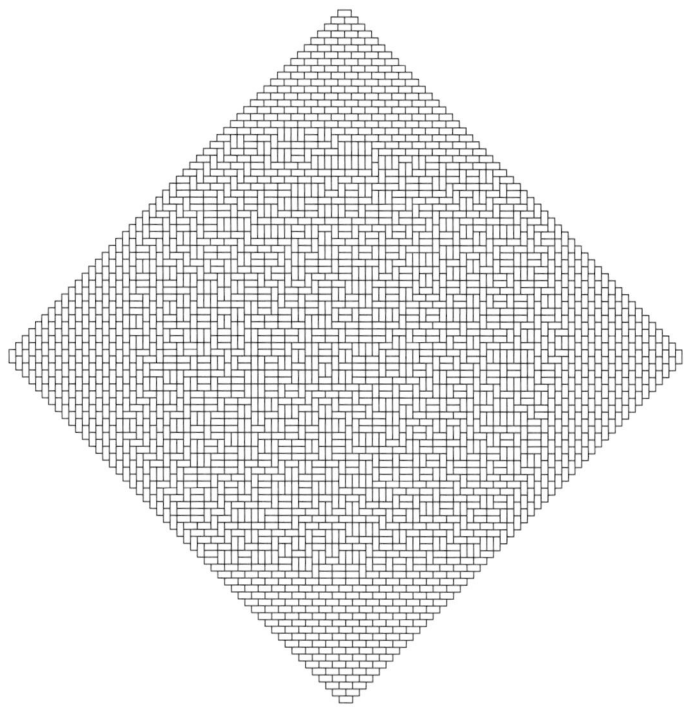

FIGURE 10. Random tiling of an Aztec diamond.

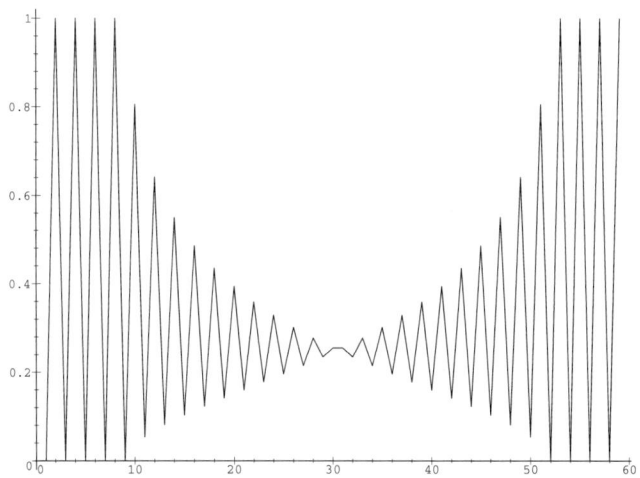

FIGURE 11. Vertical edge probabilities across the middle horizontal line of an Aztec diamond.

the exact probabilities of a size-60 Aztec diamond. The probabilities shown are nonetheless close to their actual values.)

In [**CEP**], Cohn, Elkies and Propp actually compute the limiting probabilities of edges near *any* point (x, y), but were not able to compute the probabilities of other local patterns, only those of individual edges. Recently Helfgott [**Hel**] gave a formula for the inverse Kasteleyn matrix of the Aztec diamond, which would in principle allow one to compute densities of any pattern, but its asymptotic form has not yet been worked out.

Note that when $|x| > 1/\sqrt{2}$ the edge probabilities tend to 0 or 1. In fact, [**JPS**] showed that outside the inscribed circle $x^2 + y^2 = \frac{1}{2}$, with probability tending to 1 the tiling is 'frozen' into one of the four brickwork patterns seen in the figure. One can think of this phenomenon as a kind of "spatial" phase transition.

5. Boundary Effects

It is clear from the previous section that the boundary has a long-range effect on the random dimer covering–at least in the domino tiling case. Quantifying this effect for general regions is a nontrivial problem. In the domino and lozenge case Cohn, Kenyon and Propp [**CKP**] gave the first treatment of this problem, computing the *average height function* and *entropy* of dominos or lozenges on large regions. The following sections 5.1 and 5.2 are essentially a rephrasing of the introduction of [**CKP**]. We concentrate for concreteness on the domino case. The lozenge case is similar.

5.1. The Variational Principle. The starting point is the following result. If we take a large, tilable, fat region (say an approximate square) whose boundary is chosen in such a way that the height function on the boundary approximates a linear function of slope (s,t), then the number of tilings of the region grows like $\exp(A \cdot \text{ent}(s,t))$, where A is the area of the region and $\text{ent}(s,t)$ is a constant depending only on the slope (s,t). In other words the *slope of the boundary height function determines the asymptotic number of tilings*, at least when the boundary height function is flat (approximated by a linear function). The function $\text{ent}(s,t)$ is the entropy of tilings with slope (s,t), see below.

What happens when the boundary height function is not approximately flat? Suppose we have a large tilable region U and we select a random tiling. This is like selecting a random height function with fixed boundary values.

As we just noted, the number of tilings whose height function lies near a given one depends roughly on the average slope of the height function: there are many more ways to vary a tiling near a place where the height function is horizontal (slope $(0,0)$) than near places where the slope is nonzero. The steeper the slope, the fewer tilings there are lying close to that slope (indeed there is a maximal slope beyond which there are no tilings: one must have $|s|+|t| \leq 2$). So the idea is that the height function of the random tiling tries to be as horizontal as possible while maintaining its fixed boundary values. Let us make this more precise.

Since we are discussing limiting behavior, we need a growing sequence of tilable regions. Alternatively we can take a fixed domain and tile it with dominos on finer and finer lattices, as we did in the previous section. So, let $U \subset \mathbb{R}^2$ be a Jordan domain and let $b\colon \partial U \to \mathbb{R}$. For each $\epsilon > 0$ let U_ϵ be a tilable region in $\epsilon\mathbb{Z}^2$ approximating U in the Hausdorff topology on their boundaries[1] such that the scaled boundary height function ϵb_ϵ of U_ϵ approximates b after adding a constant (this presupposes that b has a nice Lipschitz extension to U). Note that we rescale

---

[1]Every point in ∂U_ϵ is close to a point of ∂U

b_ϵ at the same rate (ϵ) as the lattice. We are interested in the behavior of a uniform random tiling of U_ϵ.

The idea that the number of tilings lying close to a given one depends on the slope of the height function is quantified in the following way. The set of possible rescaled height functions of tilings of U_ϵ is converging to the set $L = L(U, b)$ of Lipschitz functions h on U with boundary values b, whose slope (h_x, h_y) satisfies $|h_x| + |h_y| \leq 2$. For each limiting height function $h \in L$ we can approximate the logarithm of the number of tilings of U_ϵ whose rescaled height lies close to h by an integral

$$(6) \qquad \frac{1}{\epsilon^2} \int_U \operatorname{ent}(h_x, h_y) \, dx \, dy.$$

Here $\operatorname{ent}(s, t)$ is the entropy function (defined above). Recall that a Lipschitz function is differentiable almost everywhere so this integral makes sense.

Because L is compact and the "entropy" function $\operatorname{ent}(s, t)$ is strictly concave one can show that there is a unique function $h_{\max} \in L$ maximizing the integral (6). When $\epsilon \to 0$ the number of tilings whose height lies close to $h_{\max}$ overwhelmingly dominates all the other tilings. So we may conclude two things: First, the logarithm of the number of tilings of U_ϵ is given, up to lower order corrections, by the value of the integral (6) evaluated for $h_{\max}$. Secondly, the average height function of a tiling of U_ϵ converges to $h_{\max}$.

The function $\operatorname{ent}(s, t)$ can be explicitly computed (see the next section) and the maximization property of $h_{\max}$ can be turned into an elliptic PDE: the function $h = h_{\max}$ is the unique function with boundary values b satisfying

$$\left(2(1 - R^2) - \sin^2\left(\frac{\pi h_x}{2}\right)\right) h_{xx} + 2\sin\left(\frac{\pi h_x}{2}\right)\sin\left(\frac{\pi h_y}{2}\right) h_{xy}$$
$$+ \left(2(1 - R^2) - \sin^2\left(\frac{\pi h_y}{2}\right)\right) h_{yy} = 0,$$

where $R = \frac{1}{2}(\cos[\pi h_x/2] - \cos[\pi h_y/2])$ and $h_{xx} = \partial^2 h/\partial x^2$, etc. The function h may be only C^1, not C^2, in which case this equation holds only in a distributional sense.

5.2. The Entropy As a Function of Slope. The function $\operatorname{ent}(s, t)$ for domino tilings is by definition the per-unit-area logarithm of the number of tilings of a big tilable region whose boundary height function has approximate slope (s, t).

It can be computed in a roundabout fashion using a graph on a torus. We use the activities of Figure 7, and the partition function for tilings with periodic boundary conditions (1). For this choice of weights, the average slope of a tiling of the torus is $(s, t) = (2(p_a - p_b), 2(p_c - p_d))$ where p_a, p_b, p_c, p_d are the edge probabilities of edges weighted a, b, c, d respectively (by average slope (s, t) we mean simply the total height change as you go horizontally or vertically around the torus respectively, divided by the length of the corresponding curves). The average weight of a tiling is directly related to its average slope, since if a tiling has N_a edges of type a, N_b edges of type b, etc. then $p_a = N_a/N$, $p_b = N_b/N$, etc., where N is the total number of edges.

Furthermore, and this is the key, most of the tilings for this Gibbs measure have average slope close to this constant value (s, t). In fact, a second moment calculation shows that there are so many with this average slope that we can ignore the remaining tilings and make the approximation that all tilings have the same slope, and therefore the same weight. In particular, after dividing by this weight factor, the corresponding measure is the same as the measure on the *unweighted* tilings whose average slope is restricted to be (s,t).

These probabilities p_a, p_b, p_c, p_d can be computed using derivatives of (1), for example $p_a = (a/Z)\partial Z/\partial a$. When none of a, b, c, d is greater than the sum of the others, the probability p_a turns out to be

$$(7) \quad p_a = \frac{1}{\pi} \arcsin\left(\frac{a\sqrt{(a+b+c-d)(a+b-c+d)(a-b+c+d)(-a+b+c+d)}}{(ab+cd)(ac+bd)(ad+bc)}\right),$$

and similar expressions hold for p_b, p_c, p_d.

There is an amusing geometric interpretation of the above formula (7). Take a quadrilateral with edge lengths a, c, b, d in cyclic order, and which is cyclic, that is, inscribed in a circle. There is a unique such quadrilateral up to congruence. Then $p_a = \theta_a/(2\pi)$ where θ_a is angle of arc cut off by the 'a' edge of the quadrilateral. This geometric interpretation is in fact *better* than (7), since it automatically selects the correct branch of the arcsine. There is an even more amusing geometric interpretation of the corresponding entropy $\text{ent}(a,b,c,d)$. In the upper half-space model of 3-dimensional hyperbolic space, take an ideal pyramid with a vertex at ∞ and remaining vertices on the four vertices of the above cyclic quadrilateral on the bounding plane. Then $1/(2\pi)$ times the volume of this pyramid is the entropy $\text{ent}(s,t)$. So far we have no explanation for this fact except that the formulas agree.

5.3. Conditional Uniformity. One may ask, since there is a two-parameter family of slopes (s,t), but a three-parameter family of activities (we are free to choose a normalization factor, so only the ratios $a:b:c:d$ count), which choices of activities give the same slope? This is an important question since it is not clear (and in fact not true) that activities which give the same average slope also give the same entropy. The answer is that for each given slope, the activities which maximize the entropy are those which satisfy $ab = cd$, that is, those whose measures are conditionally uniform. Furthermore for each allowed slope (s,t) there is a unique (up to scale) choice of a, b, c, d whose Gibbs measure satisfies $ab = cd$ and has average slope (s,t).

So we see that, when one considers the problem of understanding *uniform* tilings of general regions, that is, the partition function with *constant* activities, one is naturally led to consider tilings with nonconstant activities, but only those which satisfy conditional uniformity.

One is tempted to think that these conditionally uniform measures $\mu_{(s,t)}$ are the actual measures that occur in a large tiling of the region U, that is, at a point where the average height function has slope (s,t), the 'local' measure there (i.e., densities of configurations) should be given by the unique conditionally uniform Gibbs measure $\mu_{a,b,c,d} = \mu_{(s,t)}$ which we defined for the torus. Unfortunately this is not known to be true. Indeed, in [**CKP**] we conjectured precisely that. That is, we conjectured that a complete description of the (asymptotic) local statistics on a tiling of a general region is obtained from the average height function $h_{\max}$.

5.4. Lozenge Tilings. For lozenge tilings one has results very similar to those of the previous three sections. (This is not too surprising since the lozenge case can be obtained from the domino case with $d = 0$). In this case the entropy as function of a, b, c was first computed by Wannier [**W**] and independently by Houtappel

[**Hout**]. There is a similar PDE describing the asymptotic average height function. The edge probabilities satisfy $p_a = \theta_a/\pi$, where θ_a is the angle opposite the edge of length a in a triangle of sides a, b, c. (When one of a, b, c is greater than the sum of the other two then the edge probabilities are all 1 or 0). Here all Gibbs measures defined using the weights in Figure 7 are conditionally uniform. (The entropy is the volume of an ideal hyperbolic tetrahedron whose vertices are ∞ and the above triangle.)

For diabolo tilings there is an additional surprise.

5.5. Diabolo Tilings. In the diabolo model, with periodic boundary conditions, when the four activities a, b, c, d are nearly equal the edge probabilities are all $\frac{1}{4}$ exactly! These probabilities do not change when a, b, c, d vary while staying close to each other. In fact, this system, unlike the domino or lozenge models, has three phases, which we have denoted solid, liquid, and gaseous. (In contrast the domino or lozenge models have only the 'solid' and 'liquid' phases. Actually the solid phases in dominos and diabolo tilings come in 4 types corresponding to the four brickwork patterns of dominos; there are three flavors of solid phase for lozenges.) See Figure 12. We will restrict ourselves to the conditionally uniform measures: i.e., those which satisfy $ab = cd = 1$. We can then parametrize the phase space using positive reals a, c.

The phase boundary between the 'gaseous' and 'liquid' phase is given by the equation
$$a^2 + a^{-2} + c^2 + c^{-2} = 5,$$
with $a^2 + a^{-2} + c^2 + c^{-2} < 5$ in the gaseous region. The boundary between the 'liquid' and 'solid' phases is given by
$$a^2 + a^{-2} - c^2 - c^{-2} = \pm 5.$$

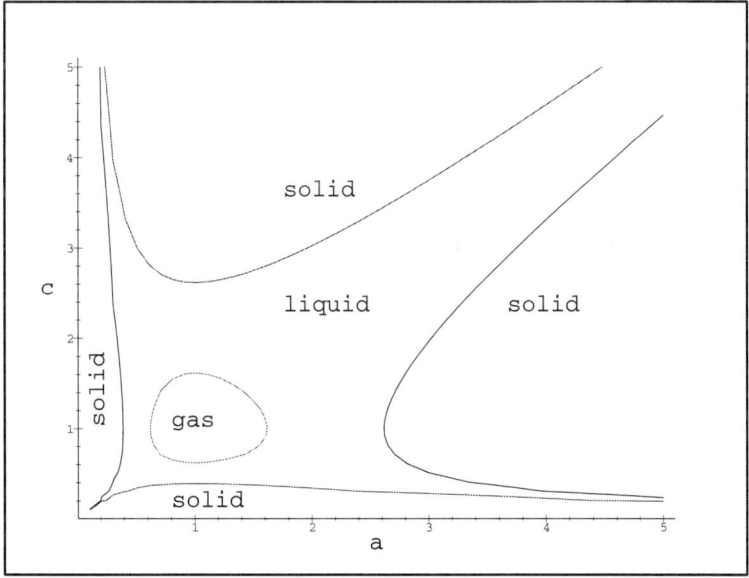

FIGURE 12. Phase boundaries in diabolo tilings with periodic boundary conditions.

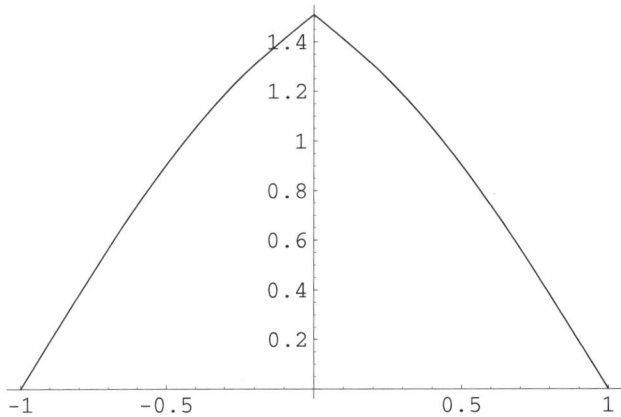

FIGURE 13. Entropy as function of slope $(s,0)$ for diabolo tilings with periodic boundary conditions.

We have not worked out all the details of this model as the computations are cumbersome. However, there are some preliminary results: In the gaseous phase the height function has average slope zero. The measures $\mu_{a,c}$ in this phase are all identical. They appear to have exponential decay of correlations, that is, the probability of local patterns appearing at different points decays exponentially with the distance between the events. In the liquid phase, on the other hand, the measures $\mu_{a,c}$ are all distinct, the height function has nonzero slope and the correlations decay quadratically. In each of the four solid phases the system is frozen, the entropy is zero and the correlations are infinite.

For this system the gaseous phase has the maximum entropy. The entropy as a function of slope $\text{ent}(s,t)$ (see Figure 13 for a numerical plot) has the remarkable feature that it is not differentiable at its maximum.

Furthermore (and perhaps equally seriously) the second derivative tends to zero at this maximum. This implies that the PDE for the entropy-maximizing surface is not elliptic at the point of maximal entropy. In [**Hen**] the assumption is made that (for quasiperiodic random tiling models) the function $\text{ent}(s,t)$ is smooth at its maximum. This example shows that this assumption may not hold in periodic cases. See also [**RHHB**] for another example.

This feature of the entropy function has an interesting consequence for the average height function with fixed boundary conditions. Figure 14 shows a *uniform* random diabolo tiling of a region called a *fortress*. All three phases can be seen to be present: near the corners the system is frozen; near the center the system is in a gaseous phase; in between is a liquid phase. What is remarkable is that the height function is constant in the center region. This is a direct consequence of the nonellipticity of the PDE.

Cohn, Pemantle and Propp [**CPP**] computed the asymptotic boundary curve between the three phases in the fortress; there is a single degree 8 algebraic equation (too long to give here) describing both boundaries simultaneously. At the centers of the sides of the diamond all three phases meet.

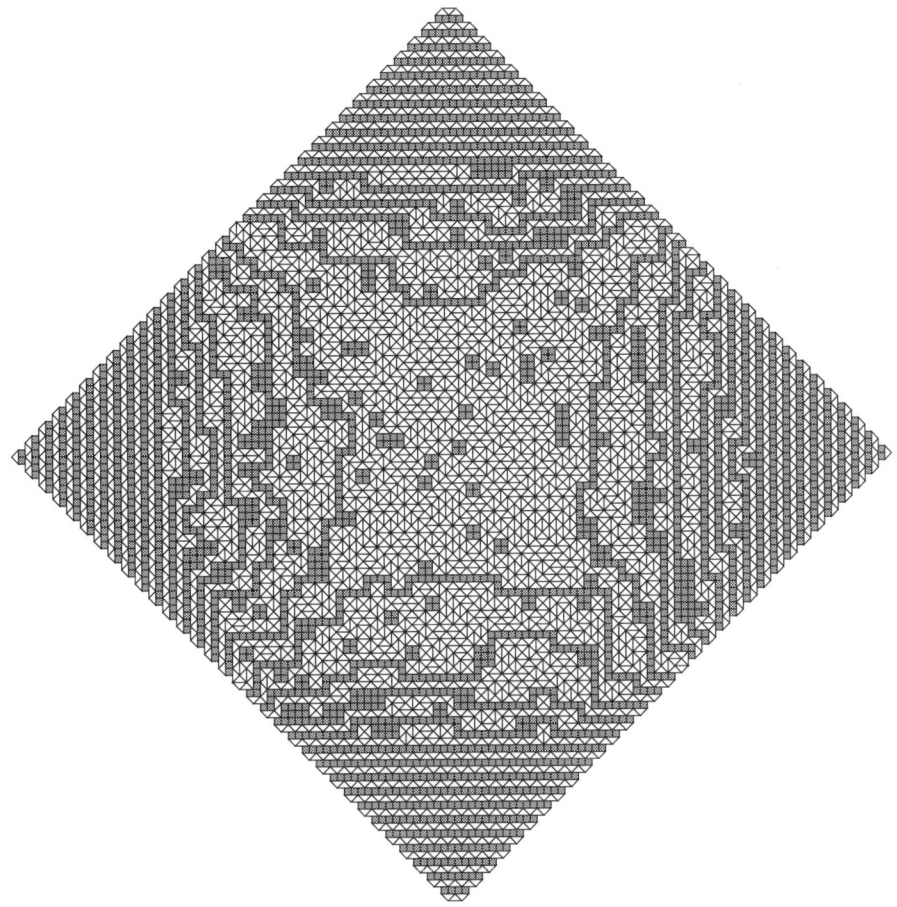

FIGURE 14. A random tiling of a "fortress" with diabolo tiles (courtesy Tilings Research Group, MIT).

6. Temperleyan Boundary Conditions

Finally we would like to return to take a closer look at a domino system with 'bounded' boundary conditions, that is, boundary conditions in which the height function on the boundary stays bounded as the system size grows. Again we concentrate on asymptotic properties, i.e., those which hold in a limiting sense as the lattice spacing ϵ tends to 0.

Unfortunately here again we do not have a complete picture for general boundary conditions. If we restrict ourselves to regions with sufficiently 'nice' horizontal boundary conditions, however, then we can say quite a lot. These special boundary conditions are those which arise from the *uniform spanning tree model* with free boundary conditions via Temperley's bijection (*Temperley's bijection* is a bijection between the dimer model on the square grid and the uniform spanning model on a square grid: see [**T**]). A region with these boundary conditions is therefore called *Temperleyan*. Rather than define precisely here what the bijection is, let us only mention that the set of Temperleyan regions is sufficiently rich to be able to approximate any planar domain.

One beautiful property of Temperleyan regions is that they exhibit a *conformal invariance* property, which for example implies that understanding limiting properties of dominos (with Temperleyan boundary conditions) on a region allows one to understand limiting properties of dominos (with Temperleyan boundary conditions) on any conformally equivalent region [**Ken2**] (two domains are conformally equivalent if there is a conformal bijection from one to the other). For example, understanding domino tilings on a square allows one to understand dominos on any other simply connected Temperleyan domain.

For a Temperleyan region the boundary does not have a long-range effect on the local densities: the local statistics converge to the Burton-Pemantle measure μ_{BP} which we discussed earlier for the square. However, the boundary does still have an effect on certain long-range properties of the measure, among them the height function. The boundary height function on a square is bounded and the average height function in the interior is as well. In fact, the average height function is (in the limit as the mesh size shrinks to 0) the harmonic function whose boundary values are the smoothing of the boundary height function [**Ken2**]. The average height function is not so important for our purposes since the *fluctuations* of the height function for a random tiling are unbounded. Indeed, in [**Ken3**] it is shown that the random variable giving the height function at the center (or any other 'interior' point) of an $n \times n$ square is converging as $n \to \infty$ to a Gaussian with variance $8 \log n/\pi^2 + O(1)$. It is tempting to think of the height as a random interface as we did in Section 5.1, but this would be incorrect. In Section 5.1 we rescaled the height by ϵ, and the interface converged to a fixed surface. Here, however, the (unrescaled) heights at distinct points in the square are essentially independent: the covariance $\mathbb{E}(h(v_1)h(v_2))$ remains bounded as $n \to \infty$. So the fluctuations are too wild to define a continuous (but nonzero) interface, even if one tries to rescale.

It might nonetheless be possible to average the height over small balls so as to make a smooth interface, and then try to understand what happens when the size of the balls shrinks. In fact, one of the current challenges in this theory is to develop a "scaling limit", that is, a continuous object with an intrinsic, conformally invariant description, which has the properties of the limiting random tiling (see a recent attempt in [**ABNW**]). Another challenge is to prove some kind of *universality*, that is, independence of the scaling limit from the local structure of the lattice. For example, it would seem that both dominos and lozenges, when taken with Temperleyan boundary conditions, have the same limiting structure.

7. Afterthoughts

As we have seen, the planar dimer model, while seemingly very special as statistical mechanical models go, displays a wide variety of interesting behavior when one varies the underlying graph and/or the edge activities. We are still very far from a complete description of the possible behaviors of planar dimer models, and it seems reasonable to expect many surprises in the future.

References

[ABNW] M. Aizenman, A. Burchard, C. M. Newman, and D. B. Wilson, *Scaling limits for minimal and random spanning trees in two dimensions*, Random Structures Algorithms **15** (1999), 319–367.

[Bax]   R. J. Baxter, *Exactly solved models in statistical mechanics*, Academic Press, London, 1982.

[BP]   R. Burton and R. Pemantle, *Local characteristics, entropy and limit theorems for spanning trees and domino tilings via transfer-impedances*, Ann. Probab. **21** (1993), 1329–1371.

[Cha]   T. Chaboud, *Pavage par des dominos dans des graphes planaires biréguliers*, C. R. Acad. Sci. Paris Sér. I Math. **318** (1994), 591–594.

[CEP]   H. Cohn, N. Elkies, and J. Propp, *Local statistics for random domino tilings of the Aztec diamond*, Duke Math. J. **85** (1996), 117–166.

[CKP]   H. Cohn, R. Kenyon, and J. Propp, *A variational principle for domino tilings*, J. Amer. Math. Soc. (to appear).

[CPP]   H. Cohn, R. Pemantle, and J. Propp (in preparation).

[DMB]   N. Destainville, R. Mosseri, and F. Bailly, *Configurational entropy of codimension-one tilings and directed membranes*, J. Statist. Phys. **87** (1997), 697–754.

[EKLP]   N. Elkies, G. Kuperberg, M. Larsen, and J. Propp, *Alternating-sign matrices and domino tilings*. I, J. Algebraic Combin. **1** (1992), 111–132.

[Els]   V. Elser, *Solution of the dimer problem on a hexagonal lattice with boundary*, J. Phys. A **17** (1984), 1509–1513.

[FW]   C. Fan and F. Y. Wu, *General lattice model of phase transitions*, Phys. Rev. B **2** (1970), 723–733.

[Fi]   M. E. Fisher, *On the dimer solution of planar Ising models*, J. Mathematical Phys. **7** (1966), 1776–1781.

[Fo]   J.-C. Fournier, *Pavage des figures planes sans trous par des dominos: fondement graphique de l'algorithme de Thurston et parallélisation*, C. R. Acad. Sci. Paris Sér. I Math. **320** (1995), 107–112.

[GCZ]   D. Grensing, I. Carlsen, and H.-C. Zapp, *Some exact results for the dimer problem on plane lattices with nonstandard boundaries*, Philos. Mag. A **41** (1980), 777–781.

[Ha]   P. Hall, *On representatives of subsets*, J. London Math. Soc. **10** (1935), 26–30.

[Hel]   H. Helfgott, *Edge effects on local statistics in lattice dimers: a study of the Aztec diamond (finite case)*, Ph.D. thesis, Brandeis, 1998.

[Hen]   C. L. Henley, *Random tiling models*, Quasicrystals: The State of the Art (D. P. DiVincenzo and P. J. Steinhardt, eds.) Dir. Condensed Matter Phys., vol. 11, World Scientific, Singapore, 1991, pp. 429–524.

[Hout]   R. M. F. Houtappel, *Order-disorder in hexagonal lattices*, Physica **16** (1950), 425–455.

[JPS]   W. Jockusch, J. Propp, and P. Shor, *Random domino tilings and the arctic circle theorem*, Research report, 1995; http://www.math.wisc.edu/~propp/arctic.ps.

[K1]   P. W. Kasteleyn, *The statistics of dimers on a lattice*. I. *The number of dimer arrangements on a quadratic lattice*, Physica **27** (1961), 1209–1225.

[K2]   _____, *Dimer statistics and phase transitions*, J. Mathematical Phys. **4** (1963), 287–293.

[K3]   _____, *Graph theory and crystal physics*, Graph Theory and Theoretical Physics, Academic Press, London, 1967, pp. 43–110.

[Ke]   G. Keller, *Equilibrium states in ergodic theory*, London Math. Soc. Stud. Texts, vol. 42, Cambridge Univ. Press, Cambridge, 1998.

[KR]   C. Kenyon and E. Remila, *Perfect matchings in the triangular lattice*, Discrete Math. **152** (1996), 191–210.

[Ken1]   R. Kenyon, *Local statistics of lattice dimers*, Ann. Inst. H. Poincaré Probab. Statist. **33** (1997), 591–618.

[Ken2]   _____, *Conformal invariance of domino tiling*, Ann. Probab. (to appear).

[Ken3]   _____, *Long-range properties of uniform spanning trees*, 1999 (preprint).

[KPW]   R. Kenyon, J. Propp, and D. Wilson, *Trees and matchings*, Electron. J. Combin. **7** (2000) R25 (electronic).

[Ku]   G. Kuperberg, *An exploration of the permanent-determinant method*, Electron. J. Combin. **5** (1998) R46 (electronic).

[LL]   E. H. Lieb and M. Loss, *Fluxes, Laplacians, and Kasteleyn's theorem*, Duke Math. J. **71** (1993), 337–363.

[MCW]   B. M. McCoy and T. T. Wu, *The two-dimensional Ising model*, Harvard Univ. Press, Cambridge, MA, 1973.

[NYB] J. F. Nagle, C. S. O. Yokoi, and S. M. Bhattacharjee, *Dimer models on anisotropic lattices*, Phase Transitions and Critical Phenomena, vol. 13 (C. Domb and J. L. Lebowitz, eds.), Academic Press, London, 1989, pp. 235–297.

[Per] J. K. Percus, *One more technique for the dimer problem*, J. Mathematical Phys. **10** (1969), 1881–1888.

[Pr1] J. Propp, *Lattice structure for orientations of graphs*, research report, 1994; http://www.math.wisc.edu/~propp.

[Pr2] _____, *Boundary-dependent local behavior for 2-D dimer models*, Internat. J. Modern Phys. B **11** (1997), 183–187.

[RZ] T. Regge and R. Zecchina, *Exact solution of the Ising model on group lattices of genus $g > 1$*, J. Mathematical Phys. **37** (1996), 2796–2814.

[RHHB] C. Richard, M. Höffe, J. Hermisson, and M. Baake, *Random tilings: concepts and examples*, J. Phys. A **31** (1998), 6385–6408; cond-mat/9712267.

[T] H. Temperley, *Enumeration of graphs on a large periodic lattice*, Combinatorics (Aberystwyth, 1973) (T. P. McDonough and V. C. Mavron, eds.), London Math. Soc. Lecture Note Ser., vol. 13, Cambridge Univ. Press, London, 1974, pp. 202–204.

[TF] H. N. V. Temperley and M. E. Fisher, *The dimer problem in statistical mechanics—an exact result*, Philos. Mag. (8) **6** (1961), 1061–1063.

[Tes] G. Tesler, *Matchings in graphs on nonorientable surfaces*, J. Combin. Theory **78** (2000), no. 2, 198–231.

[Tho] C. J. Thompson, *Mathematical statistical mechanics*, Princeton Univ. Press, Princeton, NJ, 1972.

[Th] W. P. Thurston, *Conway's tiling groups*, Amer. Math. Monthly **97** (1990), 757–773.

[W] G. H. Wannier, *Antiferromagnetism. The triangular Ising net*, Phys. Rev. (2) **79** (1950), 357–364; Erratum, Phys. Rev. B **7** (1973) 5017.

LABORATOIRE DE TOPOLOGIE ET DYNAMIQUE, BÂTIMENT 425, UNIVERSITÉ PARIS-SUD, 91405 ORSAY, FRANCE

E-mail address: kenyon@topo.math.u-psud.fr

Digit Tiling of Euclidean Space

Andrew Vince

ABSTRACT. This is an expository paper on digit tiling of Euclidean space, a special kind of self-affine tiling by translates of a single tile. In particular, the following topics are discussed: the construction of digit tiles and the construction of the boundary, the Hausdorff dimension of the boundary, the relation between digit tiles and positional number systems, the self-replicating properties of digit tiling, and lattice and crystallographic digit tiling. In the last sections digit tiling is placed into the broader context of both periodic and nonperiodic self-affine tiling of Euclidean space by a finite set of proto-tiles. In particular, the following topics are discussed: general results on hierarchical tiling, results specific to self-affine and self-similar tiling, the construction of self-affine and self-similar tilings using graph iterated function systems, and some illustrative examples.

1. Introduction

Self-similar tilings of $\mathbb{R}^d$ have attracted the interest of mathematicians in recent years for a variety of reasons that are discussed in this paper. One primary reason, especially relevant in the context of this volume, is that many of these tilings are "quasiperiodic" and serve as models for real quasicrystals. The discovery of quasicrystals in 1984 [**SBGC**] was the impetus for, not just intensified research on tilings, but for much of the recent work on the mathematics of long-range aperiodic order. In this paper there is a shift of emphasis between the first and second parts. Sections 1–7 deal mainly with periodic tilings; Sections 8–10 mainly with nonperiodic tilings. It is worth noting that self-similar tilings are a relatively recent addition to the large body of work on the geometry and symmetry of tilings, a topic surveyed, beginning with the mosaics in the Alhambra at Granada in Spain, in the book [**GS**] by Grünbaum and Shephard.

Two compact sets in $\mathbb{R}^d$ are said to be *nonoverlapping* if their interiors are disjoint. A *tiling* of $\mathbb{R}^d$ is a decomposition of $\mathbb{R}^d$ into nonoverlapping compact sets, each the closure of its interior and each with boundary having Lebesgue measure 0.

This paper is organized as follows. The definitions of self-affine and self-similar tile and, in particular, digit tile are given in Section 2. Digit tiles possess a self-similar property like that of the "Gosper flowsnake" shown in Figure 1. The union of the seven tiles is similar to each small tile. The construction of digit tiles in

---

2000 *Mathematics Subject Classification*. Primary 52C20, 28A80; Secondary 52C23, 37B10.

©2000 Andrew Vince

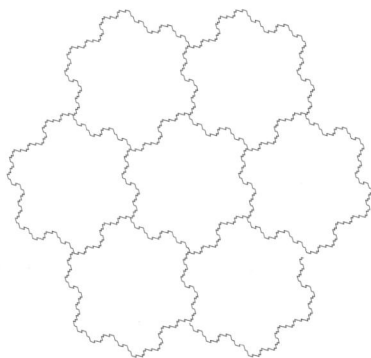

FIGURE 1. Gosper flowsnake tiling.

Section 2 is by way of iterated function systems, a standard method for constructing fractals. The boundary of a digit tile is usually fractal.

The construction of a digit tile $T := T(A, D)$ depends only on an expanding matrix A and a finite set D of lattice points in $\mathbb{R}^d$. The terminology "digit tile" comes from this data, which is analogous to the usual base and digits used to represent the integers. This connection to positional number systems (radix systems) is discussed in Section 3. In particular, the tiling in Figure 1 is related to a certain radix system with applications to image processing.

Sections 2 and 3 concern the digit tiles themselves; Section 4 concerns tilings by translates of a single digit tile. Every digit tile admits a tiling of $\mathbb{R}^d$ with a strong global property called self-replication. When this self-replicating tiling is a lattice tiling there are applications to the construction of wavelet bases. The main theorem of Section 4 gives ten conditions, all equivalent to the self-replicating lattice tiling property. One of the conditions is measure theoretic; some concern the behavior of the boundary; some concern unique radix representation of lattice points; and some are algorithmic, allowing for efficient testing procedures. One such condition involves the A-adic numbers, a generalization of the p-adics, p prime.

The boundary of a digit tile is the subject of Sections 5 and 6. In the book *Classics on fractals* [**E**], Edgar asked what the Hausdorff dimension of the boundary of the Lévy dragon might be. In general Edgar asked what could be said about the dimension of the boundary of a self-similar tile. In Section 5 an easily computable formula is provided for the Hausdorff dimension of the boundary of a self-similar digit tile.

Section 6 concerns the construction of the boundary of a digit tile. The recurrent set method for constructing a fractal curve, due to Dekking [**De1, De2**], is related to L-systems, the "L" for Lindenmayer who used the method to model biological growth [**Lin**]. Given an alphabet and a rewriting rule, the idea is to iterate the rule to produce progressively longer strings of symbols. Each symbol is then interpreted geometrically, producing a figure in the plane. The main theorem in Section 6 gives a bijection between the parameters used to construct a digit tile in $\mathbb{R}^2$ by an iterated function system and the parameters used to construct a curve by the recurrent set method. The bijection is such that the curve constructed by the recurrent set method is the boundary of the tile constructed by the iterated function system. Figure 1 in this paper was constructed by the recurrent set method.

It is an open question as to whether every tile T that admits a tiling of $\mathbb{R}^d$ by translates of T also admits a periodic tiling by translates of T. Related to this question is the *Lattice Tiling Question* of Gröchenig and Haas [**GH**]: every digit tile admits a tiling of $\mathbb{R}^d$ by translates; does every digit tile admit a (not necessarily self-replicating) lattice tiling? This question and its solution by Lagarias and Wang [**LW4**] are discussed in Section 7. It is also open whether every tile T that admits a tiling of $\mathbb{R}^2$ (not necessarily by translates) also admits a periodic tiling by copies of T. In other words, does there exist an aperiodic proto-tile? Gummelt's solution of the analogous problem for coverings of $\mathbb{R}^2$ [**Gu**], where tiles are allowed to overlap, has received considerable attention recently because of its implications for the structure of quasicrystals. This is also discussed briefly in Section 7.

Sections 1–7 are restricted to digit tiling. The remaining three sections concern generalizations. The intent is to place digit tiling into a broader context. The following topics are briefly discussed: (Section 8) crystallographic digit tiling; (Section 9) hierarchical tiling; and (Section 10) self-affine and self-similar tiling by copies of tiles taken from a finite set of proto-tiles.

Crystallographic digit tiling, due to Gelbrich [**Ge1**], is a generalization from tilings by the image of single tile under the action of a lattice group to tilings by a single tile under the action of a crystallographic group.

Sections 9 and 10 extend the subject from tiling by copies of a single tile to tilings by copies of tiles taken from a finite set of proto-tiles. These include the nonperiodic Penrose tilings [**P1**] and the Pisot tilings of Thurston [**Th**]. We introduce the notion of a hierarchy. Associated with a given hierarchy $\mathcal{P}$ are hierarchical tilings, called $\mathcal{P}$-tilings. General properties of hierarchies and their tilings are discussed in Section 9. Included are results about codes of $\mathcal{P}$-tilings, number of tilings, nonperiodicity, and quasiperiodicity. Two special types of hierarchies are self-affine and self-similar hierarchies. Miscellaneous properties of their tilings are surveyed in Section 10. A constructive approach to self-affine and self-similar tiling, by way of graph iterated function systems, is also discussed.

Many of the tilings in Section 10 are quasiperiodic and serve as models for real quasicrystals. Except for brief comments in Section 7 and on X-ray diffraction in Section 10.2, the physics of quasicrystals is not discussed. For an introductory account of quasicrystals we refer the reader to M. Senechal's book [**Sen**] and M. Baake's paper [**Ba**]. For current research trends refer to the papers in [**M**] and in this volume.

This paper is basically expository. Proofs of theorems that are readily found elsewhere are omitted. The subject of self-similar sets is vast. The paper is not intended to be comprehensive, and we apologize for any favorite topics or results that are omitted.

2. Digit Tiles

The systematic study of self-similarity properties goes back at least to 1964. Golomb [**Go**] defined a set T in the plane to be *rep-N* if T can be tiled by N congruent similar sets. Three rep-4 figures are shown in Figure 2. These examples are somewhat misleading because the boundary of a rep-N figure is often fractal.

Fractal tiles were constructed early on, for example, by Mandelbrot [**Ma**] and in the replicating superfigures of Giles [**Gil**]. But perhaps the best known method of constructing fractals at this time is by iterated function systems [**Hu**]. Many

 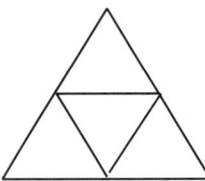 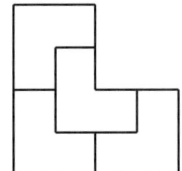

FIGURE 2. Rep-4 figures.

of the illustrations of fractals in the popular literature use this method; see, for example, the nice expositions by Barnsley [**Bar**] and Falconer [**F1**]. An *iterated function system* (IFS) is a finite set $\{f_1, f_2, \ldots, f_N\}$ of contractions:

$$f_i \colon \mathbb{R}^d \to \mathbb{R}^d.$$

A function $f \colon \mathbb{R}^d \to \mathbb{R}^d$ is a *contraction* if there is a number c with $0 < c < 1$ such that $|f(x) - f(y)| \leq c|x - y|$ for all $x, y \in \mathbb{R}^d$. Let $\mathcal{C}(\mathbb{R}^d)$ denote the collection of all nonempty compact subsets of $\mathbb{R}^d$. The Hausdorff metric h on $\mathcal{C}(\mathbb{R}^d)$ is defined as follows:

$$h(A, B) = \inf\{\epsilon \mid A \subset B_\epsilon \text{ and } B \subset A_\epsilon\},$$

where $A_\epsilon = \{x \in \mathbb{R}^d : |x - y| \leq \epsilon \text{ for some } y \in A\}$. With respect to this metric the function

$$F \colon \mathcal{C}(\mathbb{R}^d) \to \mathcal{C}(\mathbb{R}^d)$$

$$F(X) = \bigcup_{i=1}^{N} f_i(X)$$

is a contraction on the complete metric space $\mathcal{C}(\mathbb{R}^d)$ and thus, by the contraction mapping theorem, has a unique fixed point or *attractor* T that satisfies

$$(2.1) \qquad T = \bigcup_{i=1}^{N} f_i(T).$$

There is an alternative representation for the attractor given by

$$(2.2) \qquad T = \lim_{n \to \infty} F^{(n)}(T_0),$$

where $F^{(n)}$ denotes the nth iterate of F and T_0 is an arbitrary compact subset of $\mathbb{R}^d$. The limit is with respect to the Hausdorff metric. The set

$$(2.3) \qquad T_n = F^{(n)}(T_0)$$

is an nth approximation to T and is easy to express in algorithmic form. It is usually such an algorithm (or a randomized version) that is used to produce the fractal graphics that appear in many books and papers on the subject.

Consider the special case of an IFS where the contractions f_i are affine with the same linear part A^{-1} and with translational parts $D = \{d_1, d_2, \ldots, d_N\}$:

$$(2.4) \qquad f_i(x) = A^{-1}(x + d_i).$$

Let A be an expanding matrix, where *expanding* means that the modulus of each eigenvalue is greater than 1. With respect to an appropriate metric related to the Euclidean metric [**Li**], A^{-1} is a contraction. The inverse is used merely as a

FIGURE 3. Approximations to the twin dragon.

convenience for stating certain results. The functional equation (2.1), for example, is equivalent to

$$A(T) = \bigcup_{i=1}^{N} (T + d_i). \tag{2.5}$$

Let m denote d-dimensional Lebesgue measure and ∂ the boundary. If

(1) the attractor T is the closure of its interior and $m(\partial T) = 0$ and
(2) the union in Eq. (2.5) is nonoverlapping,

then T is called a *self-affine tile*. If, in addition, A is a similarity, then T is called a *self-similar tile*. A linear map is a *similarity* with expansion factor c if $\|Ax\| = c\|x\|$ for some $c > 1$ and for all $x \in \mathbb{R}^d$. The term self-affine refers to the geometric interpretation of (2.5): the large tile $A(T)$ is the nonoverlapping union of translates of the small tile T. The Gosper flowsnake in Figure 1 is a self-similar tile. Figure 3 shows the first 12 approximations to the self-similar "twin dragon," where the IFS is given in Example 2.1 and T_0 in (2.3) is a unit square.

EXAMPLE 2.1 (Twin dragon).

$$f_1 \begin{pmatrix} x \\ y \end{pmatrix} = \begin{pmatrix} 1 & -1 \\ 1 & 1 \end{pmatrix}^{-1} \left(\begin{pmatrix} x \\ y \end{pmatrix} + \begin{pmatrix} 0 \\ 0 \end{pmatrix} \right)$$

$$f_2 \begin{pmatrix} x \\ y \end{pmatrix} = \begin{pmatrix} 1 & -1 \\ 1 & 1 \end{pmatrix}^{-1} \left(\begin{pmatrix} x \\ y \end{pmatrix} + \begin{pmatrix} 1 \\ 0 \end{pmatrix} \right)$$

The attractor of the IFS (2.4) usually does not satisfy conditions (1) and (2) in the definition of self-affine tile. One case for which it does is a digit tile. Such tiles have been the subject of research by, among others, Bandt [**B2**], Dekking [**De1, De2**], Gelbrich [**BGe, Ge2**], Gröchenig and Haas [**GH**], Gröchenig and Madych [**GM**], Kenyon [**Ke2**], Lagarias and Wang [**LW2, LW3, LW4**], Solomyak

Figure 4. Gasket, rocket and shooter.

[**So1, So2**], Strichartz [**Str**] and Vince [**V1, V2, V3, V4**]. By a *lattice* in $\mathbb{R}^d$ is meant the set of all integer linear combinations of d linearly independent vectors. If A is a linear map and L is a lattice, we say that L is *A-invariant* if $A(L) \subset L$. If, for some expanding matrix A, there exists a lattice L, invariant under A, then a set D of coset representatives of the quotient $L/A(L)$ is called a *digit set*. It is assumed that $0 \in D$. By standard results in algebra, for D to be a digit set it is necessary that

$$|D| = |\det A|.$$

If A is expanding and $D = \{d_1, \ldots, d_N\}$ is a digit set, then the attractor of the affine IFS in (2.4) is called a *digit tile*. Note that a digit tile is completely determined by the pair (A, D) and will be denoted $T(A, D)$. Theorem 2.5 below states that a digit tile is indeed a self-affine tile. Figures 1 and 3 are self-similar digit tiles, based on the hexagonal and integer lattices, respectively. Both of these tiles are homeomorphic to a disk. Topologically more complicated self-similar digit tiles (Examples 2.2–2.4) appear in Figure 4. The last example in this figure shows the large tile as the nonoverlapping union of the nine small tiles.

EXAMPLE 2.2 (Gasket).
$$A = \begin{pmatrix} 2 & 0 \\ 0 & 2 \end{pmatrix}$$
$$D = \{(0,0), (1,0), (0,1), (-1,-1)\}$$

EXAMPLE 2.3 (Rocket).
$$A = \begin{pmatrix} 3 & 0 \\ 0 & 3 \end{pmatrix}$$
$$D = \{(0,0), (1,1), (2,2), (-1,0), (-2,0), (-1,1), (0,-1), (0,-2), (1,-1)\}$$

EXAMPLE 2.4 (Shooter).
$$A = \begin{pmatrix} 3 & 0 \\ 0 & 3 \end{pmatrix}$$
$$D = \{(0,0), (1,0), (2,0), (0,1), (0,2), (2,2), (4,4), (2,1), (1,2)\}$$

What we call a digit set D, Lagarias and Wang [**LW3**] call a *standard* digit set. They call D *nonstandard* if $|D| = |\det A|$ but D is not a set of coset representatives of $L/A(L)$ for any lattice L. For example, in 1-dimension $D = \{0, 1, 8, 9\}$ is nonstandard for matrix $A = (4)$. The attractor of the corresponding IFS is $[0,1] \cup [2,3]$.

FIGURE 5. Nonprimitive digit tile.

For most nonstandard digit sets D, however, the attractor T has Lebesgue measure 0; in particular, the interior of T is empty. For example, if $|\det A|$ is prime, then this is always the case [**LW3**]. In general, it seems a nontrivial problem to determine whether a nonstandard digit tile has positive Lebesgue measure. For (standard) digit tiles this is not an issue; a proof of the follow result can be found in [**GH, LW2, V1**].

THEOREM 2.5. *A digit tile T is a self-affine tile. Namely T is compact; T is the closure of its interior; $m(\partial T) = 0$; and the union in (2.5) is nonoverlapping.*

For ease of exposition and with essentially no loss of generality, we will often make the following three assumptions concerning digit tiles. A pair (A, D), consisting of an expanding matrix A and a digit set D, will be called *basic* if the following three statements hold. In this case the tile $T(A, D)$ is also called *basic*.

(1) A is an integer matrix and D is a set of coset representatives of $\mathbb{Z}^d/A(\mathbb{Z}^d)$.
(2) (A, D) is pure.
(3) (A, D) is primitive.

By assumption (1) the invariant lattice is the integer lattice. By *pure* is meant that 0 is contained in the interior of $T(A, D)$. By *primitive* is meant that D is contained in no proper A-invariant sublattice of $\mathbb{Z}^d$. Example 2.6 is a pair (A, D) in $\mathbb{R}^2$ that is not primitive, and Figure 5 shows the first three approximations to the corresponding digit tile $T(A, D)$, which is a square. Note that the sublattice of $\mathbb{Z}^2$ consisting of all lattice points with even coordinate sum is a proper A-invariant sublattice.

EXAMPLE 2.6 (Nonprimitive digit tile).
$$A = \begin{pmatrix} 3 & 0 \\ 0 & 3 \end{pmatrix}$$
$$D = \{(0,0), (2,0), (1,-1), (-1,1), (1,1), (3,1), (0,2), (2,2), (1,3)\}$$

That little loss of generality is incurred by restricting to basic digit tiles is the statement of the following result [**LW2, V1**].

THEOREM 2.7. *Let $A\colon \mathbb{R}^d \to \mathbb{R}^d$ be a linear expanding map, L an A-invariant lattice, and D a digit set. There exists a basic pair (A', D') such that A' is similar to some power of A and $T(A', D') = \phi(T(A, D))$, where ϕ is an invertible affine map. Moreover, if $L = \mathbb{Z}^d$, then "similar to" can be replaced by "equal to", and if (A, D) is already primitive, then ϕ is just a translation.*

It is usually the case that a digit tile is not homeomorphic to a ball. Bandt and Gelbrich call two digit tiles T and T' *isomorphic* if there is an affine bijection ϕ from

T to T' that preserves pieces at all levels. For a precise definition of "preserving the pieces" see [**BGe**].

THEOREM 2.8 (Bandt & Gelbrich [**BGe**]; Gelbrich [**Ge2**]).

(1) *For any $N \geq 2$ there are finitely many isomorphism classes of digit tiles in $\mathbb{R}^2$ with N pieces and homeomorphic to a disk.*
(2) *There are finitely many isomorphism classes of digit tiles in $\mathbb{R}^d$ with 2 pieces and homeomorphic to a d-ball.*

The authors have determined, for example, that there are three types of disklike digit tiles in $\mathbb{R}^2$ with two pieces and seven types with three pieces. In [**GH**] Gröchenig and Haas give, in the -dimensional case, a sufficient condition on the pair (A, D) for $T(A, D)$ to be connected. In any dimension, Hacon, Salanha and Veerman [**HSV**] prove that any digit tile with two pieces is connected.

3. Radix Representation

To justify the terminology "digit" tile, consider an expanding matrix A as a base for an A-invariant lattice L (so $A(L) \subset L$), and a digit set $D \subset L$ as a set of digits for L. Use the Minkowski sum notation $X + Y = \{x + y \mid x \in X, y \in Y\}$ and $A(D) = \{A(d) \mid d \in D\}$, and let

$$(3.1) \qquad D_n = \sum_{i=0}^{n-1} A^i(D) \quad \text{and} \quad D_\infty = \bigcup_{i=1}^\infty D_n.$$

Then D_n is the subset of the lattice that can be expressed using at most n digits, and D_∞ is the set of lattice points that can be expressed using any finite sequence of digits. Let the initial approximation to the tile $T = T(A, D)$ be a single point: $T_0 = \{0\}$. In this case Eq. 2.2 becomes

$$(3.2) \qquad T := T(A, D) = \lim_{n \to \infty} \sum_{i=1}^n A^{-i}(D),$$

where the limit is with respect to the Hausdorff metric. Then $T(A, D)$, according to (3.2), is the set of points in $\mathbb{R}^d$ that can be expressed using digits only to the "right" of the decimal point. In particular, consider the 1-dimensional case where $A = (10)$, $L = \mathbb{Z}$, and $D = \{0, 1, \ldots, 9\}$. Then D_n is the set of integers that can be represented in the ordinary base 10 system using at most n digits; D_∞ is the set of nonnegative integers; and T is the closed interval $[0, 1]$. For the above reasons we refer to the pair (A, D) as a *radix system* (positional number system) for lattice L.

The representation of numbers using positional number systems has an extensive literature prior to the advent of fractals. Knuth's classic [**Kn**] contains early references dating back to Cauchy, who noted that negative digits make it unnecessary for a person to memorize the multiplication table past 5×5. Gilbert [**Gi1, Gi2, Gi3**] considered radix representation for the Gaussian integers $\mathbb{Z}[i] = \{a + bi \mid a, b \in \mathbb{Z}\}$ and for integers in other algebraic number fields. For example, every Gaussian integer has a unique base $\beta = -1 + i$ representation of the form $\sum_{i=0}^n d_i \beta^i$, where $d_i \in D = \{0, 1\}$. This is analogous to a binary system for the Gaussian integers. Of course, this is just the radix system (A, D) for the lattice $\mathbb{Z}^2 = \mathbb{Z}[i]$, where A is the linear map given by $Ax = \beta x$.

Two obvious questions concerning radix representation are as follows.

QUESTION 3.1. Given an expanding integer matrix A and digit set D, when is it the case that every lattice point x has a unique representation of the form

$$(3.3) \qquad x = \sum_{i=0}^{n-1} A^i(e_i), \; e_i \in D.$$

QUESTION 3.2. Given an expanding integer matrix A, does there exist a digit set D such that every lattice point x can be uniquely represented in the radix form (3.3).

Question 3.1 will be addressed as part of Theorem 4.3 in Section 4. The answer to Question 3.2 is "no, but almost." We mention two particular results. Here I is the identity matrix.

THEOREM 3.3 (Vince [**V1**]). *If* $\det(I - A) = \pm 1$, *then there is no digit set* D *such that every lattice point* x *has a unique representation of the form* (3.3).

Examples of such matrices include

$$\begin{pmatrix} 1 & -1 \\ 1 & 1 \end{pmatrix}, \quad \begin{pmatrix} 2 & a \\ 0 & 2 \end{pmatrix}, \quad \begin{pmatrix} 0 & -a \\ 1 & a \end{pmatrix}.$$

Recall that for any matrix A there exist orthogonal matrices U and V such that $U^T A V$ is diagonal [**GL**]. The diagonal entries are called the *singular values of* A. Let C denote the canonical fundamental domain of the the origin with respect to the cubic lattice (the closure of C is a unit cube centered at the origin).

THEOREM 3.4 (Vince [**V1**]). *Let* A *be a d-dimensional matrix and* L *an* A*-invariant lattice. If the singular values of* A *are greater than* $3\sqrt{d}$ *and* $D = A(C) \cap L$, *then every lattice point* x *has a unique representation of the form* (3.3). *In the 1- and 2-dimensional cases, the bound* $3\sqrt{d}$ *can be improved to* 2.

The following previously known result follows directly from the two theorems above.

COROLLARY 3.5. *For any Gaussian integer* $\beta \in \mathbb{Z}[i]$, *except* 0, ± 1, $\pm i$, 2 *and* $1 \pm i$, *there is a digit set* D *such that every Gaussian integer has a unique radix representation of the form* $\sum_{i=0}^{n-1} e_i \beta^i$, $e_i \in D$. *No such digit set exists for* $\beta = 2$ *and* $\beta = 1 \pm i$.

It follows from (3.2) that

$$T = \lim_{n \to \infty} A^{-n}(D_n);$$

so it should not be surprising that properties of the set D_∞ on a large scale are directly related to properties of the digit tile T on a small scale. The following theorem is an example. A set $S \in \mathbb{R}^d$ is *uniformly discrete* if there is a bound $r > 0$ such that distinct $x, y \in S$ satisfy $|x - y| \geq r$. Note that, in this theorem, it is not assumed that D is a digit set.

THEOREM 3.6 (Lagarias and Wang [**LW2**]). *Assume that* A *is a real expanding matrix with* $|\det A| = m \in \mathbb{Z}$ *and* D *a subset of* $\mathbb{R}^d$ *with* $|D| = m$ *and* $0 \in D$. *Then the following statements are equivalent.*

(1) $T := T(A, D)$ *is the closure of its interior and* $m(\partial T) = 0$.
(2) *All* m^n *elements of* D_n *are distinct and* D_∞ *is a uniformly discrete set.*

We now consider in more detail a radix system, called the generalized balanced ternary, which has applications to image processing. Consider a monic polynomial $f(x) = x^d + a_{d-1} x^{d-1} + \cdots + a_0 \in \mathbb{Z}[x]$. In the quotient ring $\Lambda_f = \mathbb{Z}[x]/(f)$ let

$\alpha = x+(f)$. Then Λ_f has the structure of a $\mathbb{Z}$-module with basis $(1, \alpha, \alpha^2, \ldots, \alpha^{d-1})$. In other words Λ_f is a lattice which can be realized (in many ways) in $\mathbb{R}^d$ by embedding the d basis elements as d linearly independent vectors in $\mathbb{R}^d$.

If $f(x)$ is irreducible over $\mathbb{Z}$ then, as rings, $\Lambda_f = \mathbb{Z}[x]/(f) \cong \mathbb{Z}[\alpha]$ where α is any root of $f(x)$ in an appropriate extension field of the rationals. For example, if $f(x) = x^2 + 1$ then the lattice Λ_f is the ring of Gaussian integers $\mathbb{Z}[i]$ with basis $(1, i)$ and can be realized as the square lattice in the complex plane.

Consider the special case $f(x) = 1 + x + x^2 + \cdots + x^d$. Let $\omega = x + (f)$. In the ring $\mathbf{\Lambda}_d = \mathbb{Z}[x]/(f)$ we have $1 + \omega + \cdots + \omega^d = 0$ and $\omega^{d+1} = 1$. For the sake of symmetry we take as a generating set for the lattice $\mathbf{\Lambda}_d$ the set $(1, \omega, \omega^2, \ldots, \omega^d)$ although it is linearly dependent. Embed the lattice $\mathbf{\Lambda}_d$ in d-dimensional Euclidean space by defining an inner product on pairs of basis elements $(1, \omega, \omega^2, \ldots, \omega^d)$ by

$$(\omega^i, \omega^j) = \begin{cases} 1 & \text{if } i = j \\ -1/d & \text{if } i \neq j. \end{cases}$$

In dimension $d = 1$ this is the integer lattice; for $d = 2$ it is the hexagonal lattice; and for $d = 3$ it is the lattice that consists of the centers of the tiling of space by truncated octahedra. In general it is the dual of the classical d-dimensional root lattice $\mathbf{A}_d$; so the weight lattice $\mathbf{A}_d^* = \mathbf{\Lambda}_d$ [CS]. Now let $\beta = 2 - \omega$ and define a linear expanding map

$$A_\beta \colon \mathbf{\Lambda}_d \to \mathbf{\Lambda}_d$$

by

$$A_\beta(\mathbf{x}) = \beta \mathbf{x}.$$

Although not well-defined, a matrix for A_β with respect to the generating set $(1, \omega, \omega^2, \ldots, \omega^d)$ is

$$A_\beta = \begin{pmatrix} 2 & 0 & 0 & \ldots & 0 & -1 \\ -1 & 2 & 0 & \ldots & 0 & 0 \\ 0 & -1 & 2 & \ldots & 0 & 0 \\ \vdots & \vdots & \vdots & \ddots & \vdots & \vdots \\ 0 & 0 & 0 & \ldots & 2 & 0 \\ 0 & 0 & 0 & \ldots & -1 & 2. \end{pmatrix}$$

Let

$$D_\beta = \{\epsilon_0 + \epsilon_1 \omega + \epsilon_2 \omega^2 + \cdots + \epsilon_d \omega^d : \epsilon_i \in \{0, 1\}\},$$

where not all $\epsilon_i = 0$. Note that $|D_\beta| = 2^{d+1} - 1$. It can be shown that D_β is a digit set for A_β with respect to the lattice $\mathbf{\Lambda}_d$. Hence (A_β, D_β) is a radix system for lattice $\mathbf{\Lambda}_d$, called the *generalized balanced ternary* (GBT). Moreover, if follows from a variant of Theorem 3.4 that every lattice point in $\mathbf{\Lambda}_d$ has a unique representation in the GBT.

In dimension 1

$$A_\beta = (3), \quad D = \{-1, 0, 1\}.$$

This is a base 3 system classically called the *balanced ternary*. Every integer (positive or negative) can be uniquely expressed base three using the three digits (trits). For $d = 2$ the digit set D_β is the subset of the hexagonal lattice consisting of the origin and all 6th roots of unity. The corresponding tile $T(A_\beta, D_\beta)$ is shown in Figure 1. The GBT radix system has been suggested for spatial addressing of images as a viable alternative to a rectangular grid—for both geometric reasons (the round shape of the pixels) and algebraic reasons (the efficient algorithmic properties of the radix system [**GL, KVW, Roe**]).

4. Self-Replicating Tiling

Let A be an expanding integer matrix, D a digit set, and assume that the pair (A, D) is basic. Section 2 concerns the self-affine tile $T(A, D)$. The term "tile" was used, rather than "set", because, given any self-affine tile T, there always exists a tiling of $\mathbb{R}^d$ by translates of T. To see this, iterate functional equation (2.5) to obtain
$$A^n(T) = \bigcup_{d \in D_n} (T + d).$$
Since (A, D) is pure, 0 lies in the interior of $A^n(T)$. Since A is an expansion, any ball centered at the origin lies in $A^n(T)$ for some n. In the notation of (3.1), the sets $D_1 \subset D_2 \subset \ldots$ are nested because $0 \in D$ and hence

(4.1) $$\mathcal{T}_\infty := \{T + d \mid d \in D_\infty\}$$

is a tiling of $\mathbb{R}^d$.

In fact, $\mathcal{T}_\infty$ is a special type of tiling by digit tiles, called a self-replicating tiling. A tiling $\mathcal{T}$ of $\mathbb{R}^d$ by copies of a single tile is called *self-replicating* if, for some linear expansion A, the expanded tile $A(T)$ is, for each $T \in \mathcal{T}$, tiled by elements of $\mathcal{T}$. Note that the self-replicating property is a global property of the tiling, not a property of the tile. This self-replicating property was investigated by Thurston [**Th**] for more general tilings to be discussed in Section 10. The tiling by twin dragons in Figure 6 is self-replicating; the image of each dragon under the mapping A is the union of two horizontally adjacent dragons. The following proposition is an easy consequence of (2.5).

PROPOSITION 4.1. *Given any basic digit tile T, the corresponding tiling $\mathcal{T}_\infty$ is self-replicating.*

According to Proposition 4.1, for any basic digit tile T there is a self-replicating tiling by translates of T. The proposition does not imply, however, that this tiling is by translation by the integer lattice, as is the case in Figure 6. A tiling $\mathcal{T}$ is a *lattice*

FIGURE 6. Lattice tiling by twin dragons.

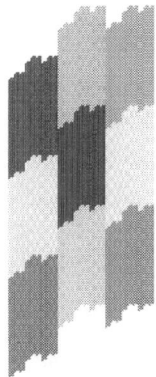

FIGURE 7. Not a lattice tiling.

tiling of $\mathbb{R}^d$ if $\mathcal{T}$ is a tiling by translation by a lattice, i.e., $\mathcal{T} = \{T + x \mid x \in L\}$ for some lattice L. Consider Example 4.2; the corresponding tiling $\mathcal{T}_\infty$ shown in Figure 7 is not a lattice tiling.

EXAMPLE 4.2.
$$A = \begin{pmatrix} 3 & 0 \\ 0 & 3 \end{pmatrix}$$
$$D = \{(-1,-3),(-1,-1),(-1,1),(0,-2),(0,0),(0,2),(1,-2),(1,0),(1,2)\}$$

The following theorem [**V4**] is central and gives ten equivalent conditions for the existence of a self-replicating lattice tiling. All terms not yet defined will be discussed after the statement of the theorem.

THEOREM 4.3. *Let $T = T(A, D)$ be a basic digit tile. Let $T_n = F^{(n)}(T_0)$ be the approximating tiles, where T_0 is the unit d-cube centered at the origin with edges parallel to the axes. The following statements are equivalent. Limits are with respect to the Hausdorff metric.*

(1) $\mathcal{T} := \{T + x \mid x \in \mathbb{Z}^d\}$ *is a tiling of $\mathbb{R}^d$.*
(2) $\mathcal{T} := \{T + x \mid x \in \mathbb{Z}^d\}$ *is a self-replicating tiling of $\mathbb{R}^d$.*
(3) $m(T) = 1$.
(4) *The characteristic function $\chi_T(x)$ is a scaling function of a multiresolution analysis.*
(5) $\lim_{n \to \infty} \partial T_n = \partial T$.
(6) $\lim_{n \to \infty} \partial T_n$ *is not space filling.*
(7) $D_\infty = \mathbb{Z}^d$.
(8) *Every lattice point has a unique finite address.*
(9) *Every lattice point in the ball $\mathcal{B}(A, D)$ has a finite address.*
(10) $\lambda(A, D) < |\det A|$.

Condition (3) states that the Lebesgue measure of T is 1. This is clearly necessary if statement (1) is to hold. The converse appears in [**GH**]. It is known that $m(T)$ is always an integer [**LW3**], but is not always 1. For the tile of Example 4.2 the measure is 2. Figure 8, showing the first few approximations to this tile, may provide insight into why the tile is "stretched."

The equivalence of conditions (1) and (4) is due to Gröchenig and Madych [**GM**]. An important application of digit tiling is to wavelets, the construction of

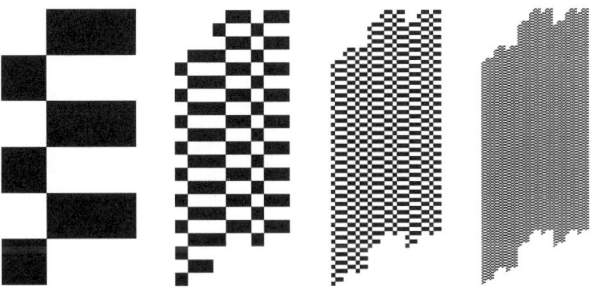

Figure 8. Digit tile with Lebesgue measure 2.

orthonormal wavelet bases in $\mathbb{R}^d$. The multiresolution analysis machinery produces an orthonormal wavelet bases of $L^2(\mathbb{R}^d)$. We refer the reader to [**GM**, **Str**] and any number of introductory texts, for example [**Ch**], rather than elaborating on wavelets in this paper.

Conditions (5) and (6) concern the boundary of the approximating tiles; proof of their equivalence to the other conditions appears in [**V4**]. Condition (5) states that the boundaries of the approximating tiles approach the boundary of the limit tile in the Hausdorff topology. It is easy to see that this is not the case for the tile in Figure 8. Condition (6) states that, if the conditions of Theorem 4.3 fail, then the behavior of the boundary is indeed pathological; the limit of the boundaries of the approximates is space filling—contains some open set. In the case of Example 4.2, the limit is the whole tile T.

Conditions (7) and (8) relate to Question 3.1 in Section 3. They state that every lattice point x has a unique base A representation with digits D. In other words $x = \sum_{i=0}^{n-1} A^i(e_i)$, $e_i \in D$. The proof of the equivalence of conditions (7), (8) and (9) to the other conditions in [**V1**] relies on the concept of A-adic integer, analogous to the classical p-adic integer, p a prime (see [**Se**] for background on the p-adic integers). The set of *A-adic integers* is the completion of $\mathbb{Z}^d$ with respect to the metric induced by the norm

$$|x| = \frac{1}{|\det A|^\nu},$$

where ν is the greatest integer such that $x \in A^\nu(\mathbb{Z}^d)$. Analogous to the p-adic case, there is a canonical representation of each A-adic number in the form

$$x = \sum_{i=0}^{\infty} A^i(e_i), \quad e_i \in D.$$

Define the *address* of such an A-adic as

$$\cdots e_3 e_2 e_1 e_0.$$

It can be shown that, given a digit set D for A, each point in $\mathbb{Z}^d$ has an address that eventually repeats, in the same sense as an ordinary repeating decimal. A lattice point is said to have a *finite address* if $e_n = 0$ for all n sufficiently large. In fact, there is an easy algorithm to obtain the address of any lattice point x.

ALGORITHM ($x_0 = x$).
$$e_n \equiv x_n \bmod A(\mathbb{Z}^d)$$
$$x_{n+1} = A^{-1}(x_n - e_n).$$

Moreover there is a computable bound on the number of iterations of this algorithm sufficient to determine whether or not the lattice point has a finite address.

EXAMPLE 4.4. In 1-dimension the 3-adic address, i.e., $A = (3)$, of the integer 2 with respect to digit set $\{-1, 0, 4\}$ is $(-1)(4)(-1)$:

$$\begin{aligned} x_0 &= 2 & e_0 &= -1 \\ x_1 &= 1 & e_1 &= 4 \\ x_2 &= -1 & e_2 &= -1 \\ x_3 &= 0 & e_3 &= 0. \end{aligned}$$

Note that condition (8) together with formula (3.2) imply that every point in $\mathbb{R}^d$, except those on the overlap of two tiles in $\mathcal{T}$, can be uniquely expressed in the form

$$e_n \cdots e_1 e_0 . e_{-1} e_{-2} \cdots := \sum_{i=-\infty}^{n} A^i(e_i), \quad e_i \in D.$$

The representation of points on the overlap of two tiles is not unique; for example for $A = (10)$ and $D = \{0, 1, \ldots, 9\}$, we have $.999\cdots = 1$.

Conditions (9) and (10) are algorithmic. They provide efficient methods to check that all conditions in Theorem 4.3 hold. The number $\lambda(A, D)$ in condition (10) is the largest eigenvalue of a certain easily computable matrix. A definition and discussion of this matrix appears in Section 5. Condition (9) states that there is a ball $\mathcal{B}(A, D)$ centered at the origin, whose radius $r(A, D)$ depends only on (A, D), such that, if every lattice point in $\mathcal{B}(A, D)$ has a finite address, then all lattice points do. In the case that A is a similarity with expansion factor c, an explicit value of the radius is easy to express:

$$r(A, D) = \frac{\max\{|d| : d \in D\}}{c - 1}.$$

Applying the formula for $r(A, D)$ to Example 4.5, the only lattice point in $\mathcal{B}(A, D)$ is the origin, which obviously has a finite address. By condition (9) in Theorem 4.3, the corresponding lattice tiling is a self-replicating lattice tiling; it is shown in Figure 9.

EXAMPLE 4.5.
$$A = \begin{pmatrix} 2 & 1 \\ -1 & 2 \end{pmatrix}$$
$$D = \{(0,0), (1,0), (0,1), (-1,0), (0,-1)\}$$
$$r(A, D) = \frac{1}{\sqrt{5} - 1} = .8090\ldots$$

It should be remarked that either condition (7) or (8) automatically implies that (A, D) is basic [**V1**]. In Example 4.6 there are 21 points in $\mathcal{B}(A, D)$ to check using condition (9), including the point $(-1, 0)$. The algorithm gives the repeating address $(1, 0), (0, 0), (1, 0), (0, 0), \ldots$ for the point $(-1, 0)$, not a finite address. The problem in this case is that (A, D) is not basic; it is not pure. As pointed out in

FIGURE 9. Self-replicating tiling.

Section 2, there is a related basic pair (A', D') such that $T(A, D)$ and $T(A', D')$ are the same up to translation. Then $T(A', D')$ does satisfy the conditions of Theorem 4.3. The corresponding tiling is the twin dragon tiling in Figure 6, which is indeed a self-replicating lattice tiling.

EXAMPLE 4.6.
$$A = \begin{pmatrix} 1 & -1 \\ 1 & 1 \end{pmatrix}$$
$$D = \{(0,0), (1,0)\}$$
$$r(A, D) = \frac{1}{\sqrt{2}-1} = 1 + \sqrt{2} = 2.4142\ldots$$

5. Dimension of the Boundary

For some well studied tiles, like the twin dragon in Figure 3, the Hausdorff dimension of the boundary is known and has been computed by various means. More recently Duvall and Keesling [**DK**] determined the Hausdorff dimension of the boundary of a particular tile, the Lévy dragon. In [**Kees**] Keesling showed that the Hausdorff dimension of the boundary of any self-similar tile in $\mathbb{R}^d$ is less than d, but that this dimension could be arbitrarily close to d. This section outlines a method due to Duvall, Keesling and Vince [**DKV**] for determining the Hausdorff dimension of the boundary of any self-similar digit tile. After our results were obtained we came across unpublished preprints by Veerman [**Ve**] and by Strichartz and Wang [**SW**] which contain similar results obtained by different methods. The only condition that is needed on the digit tile T for our formula in Theorem 5.1 is that one of the equivalent conditions given in Theorem 4.3 holds for T. This is not unexpected in light of conditions (5) and (6) of that theorem. The method given below either determines precisely the Hausdorff dimension of the boundary of T or it determines that condition (10) of Theorem 4.3 fails. The problem of determining an exact formula for a self-affine (not necessarily self-similar) digit tile remains open.

Recall the definition of Hausdorff dimension; an introductory treatment can be found, for example, in [**F1**]. An ϵ-cover of a set $X \subset \mathbb{R}^d$ is a collection of sets of diameter at most ϵ such that X is contained in their union. Let $|U|$ denote the diameter of the set U, and let s be a nonnegative number. For any $\epsilon > 0$ define

$$H_\epsilon^s(X) = \inf\left\{\sum_{i=1}^\infty |U_i|^s : \{U_i\} \text{ is an } \epsilon\text{-cover of } X\right\}.$$

As ϵ decreases, the collection of possible covers is reduced; hence $H_\epsilon^s(X)$ decreases. Define the *s-dimensional Hausdorff measure* of X by

$$H^s(X) = \lim_{\epsilon \to 0} H_\epsilon^s(X).$$

It is easy to show that there is a critical value of s at which this limit jumps from ∞ to 0. Define the *Hausdorff dimension* by

$$\dim_H(X) = \inf\{s : H^s(X) = 0\} = \sup\{s : H^s(X) = \infty\}.$$

To state the main result, the contact matrix, first defined by Gröchenig and Haas [**GH**], is introduced. Given an expanding integer matrix A and digit set D for the integer lattice in $\mathbb{R}^d$, a set $N = N(A, D)$ of integer lattice points, called the *neighborhood* for (A, D), is used to index the rows and columns of the contact matrix.

The neighborhood $N(A, D)$ is defined as follows. Let $\{e_1, \ldots, e_d\}$ denote the canonical basis of $\mathbb{R}^d$ and let $N_0 = \{0\} \cup \{\pm e_1, \ldots, \pm e_d\}$. Then $N(A, D)$ is the unique smallest finite set $N \subset \mathbb{Z}^d$ such that $N_0 \subseteq N$ and $D + N \subseteq A(N) + D$. The neighborhood can easily be computed using the following algorithm, and it is easy to show that the algorithm terminates after a finite number of steps. Because D is a set of coset representatives of $\mathbb{Z}^d/A(\mathbb{Z}^d)$, for any lattice point y the equation $Ax + d = y$ has a unique solution pair (x, d), where $x \in \mathbb{Z}^d$ and $d \in D$.

ALGORITHM.
- $N = N_0$.
- Repeat until the two sets are equal:

$$N \leftarrow N \cup \{x \in \mathbb{Z}^d \mid Ax + d = y \text{ for some } d \in D \text{ and } y \in D + N\}.$$

For each $x \in N$ and $d \in D$, let x_d denote the unique lattice point such that $d + x \in Ax_d + D$. By the definition of N we have $x_d \in N$. Let C' be the $k \times k$ matrix whose rows and columns are indexed by the elements in N and whose entries are as follows. For $x, y \in N$

$$c_{xy} = |\{d \in D \mid x_d = y\}|.$$

By convention let the first index of C' correspond to the element $0 \in N$. Note that $c_{00} = |D|$ and $c_{0y} = 0$ for $y \neq 0$. Thus the first row of C' consists of all zeros except for one entry. Let C denote the $(k-1) \times (k-1)$ matrix obtained from C' by removing the first row and column. Call C the *contact matrix* for the pair (A, D). (In [**GH**] it is actually C' that is referred to as the contact matrix.)

According to the Perron-Frobenius Theorem for nonnegative matrices, C has a real eigenvalue λ such that, for any other eigenvalue μ, we have $\lambda \geq |\mu|$. In other words, the spectral radius of C is an eigenvalue.

THEOREM 5.1 (Duval, Keesling and Vince [**DKV**]). *Let $T = T(A, D)$ be a self-similar digit tile where A has expansion factor c and the contact matrix has largest eigenvalue $\lambda := \lambda(A, D)$. Under any of the conditions in Theorem 4.3 we have*

$$\dim_H(\partial T) = \frac{\log \lambda}{\log c}.$$

Examples.

Twin dragon. The dimension of the boundary of the Twin Dragon (Example 2.1 and Figure 3) has been calculated by various means. Using our method the neighborhood is the following set of lattice points:

$$N = \{(0,0), (0,1), (1,0), (1,-1), (0,-1), (-1,0), (-1,1)\}$$

Ordering the elements of $N \setminus \{0\}$ as above (clockwise around a hexagon) the contact matrix C, computed using the definition, is the following integer matrix with cyclical structure.

$$C = \begin{pmatrix} 1 & 1 & 0 & 0 & 0 & 0 \\ 0 & 0 & 1 & 0 & 0 & 0 \\ 0 & 0 & 0 & 2 & 0 & 0 \\ 0 & 0 & 0 & 1 & 1 & 0 \\ 0 & 0 & 0 & 0 & 0 & 1 \\ 2 & 0 & 0 & 0 & 0 & 0 \end{pmatrix}.$$

The characteristic polynomial is easy to compute because of the near diagonal structure of the matrix:

$$\det(C - \lambda I) = \lambda^4(1-\lambda)^2 - 4 = (\lambda+1)(\lambda^2 - 2\lambda + 2)(\lambda^3 - \lambda^2 - 2).$$

So the largest eigenvalue of C is the real root of $\lambda^3 - \lambda^2 - 2$. Hence the Hausdorff dimension of the twin dragon is

$$\dim_H \partial T = \frac{\log \lambda}{\log \sqrt{2}} \simeq 1.523627$$

Gasket. For the gasket (Example 2.2 and Figure 4), the neighborhood N is again in a hexagonal pattern:

$$N = \{(0,0), (1,0), (1,1), (0,1), (-1,0), (-1,-1), (0,-1)\}$$

The contact matrix is a cyclic matrix with three ones in each row:

$$C = \begin{pmatrix} 1 & 1 & 0 & 0 & 0 & 1 \\ 1 & 1 & 1 & 0 & 0 & 0 \\ 0 & 1 & 1 & 1 & 0 & 0 \\ 0 & 0 & 1 & 1 & 1 & 0 \\ 0 & 0 & 0 & 1 & 1 & 1 \\ 1 & 0 & 0 & 0 & 1 & 1 \end{pmatrix}.$$

Hence the Perron-Frobenius eigenvector, the unique eigenvector with positive entries, is the all ones vector. The corresponding eigenvalue is $\lambda = 3$.

$$\dim_H \partial K = \frac{\log 3}{\log 2} = 1.5849625\ldots$$

Lander. The lander is the digit tile $T(A, D)$ where

$$A = \begin{pmatrix} 3 & 0 \\ 0 & 3 \end{pmatrix}$$

$$D = \{(0,0), (1,1), (1,-1), (2,0), (-1,-2), (3,-2), (-1,2), (3,2), (1,3)\}.$$

The dimension of the boundary of the lander in Figure 10 is somewhat greater than for the other examples,

$$\dim_H \partial T \simeq 1.913624.$$

FIGURE 10. Lander.

SKETCH OF THE PROOF OF THEOREM 5.1. Let T_0 be the unit cube centered at the origin with edges parallel to the axes and let $T_n = F^{(n)}(T_0)$ denote the nth approximation to the tile $T := T(A, D)$ as given in (2.3). Then T_n is the nonoverlapping union of copies of cubes of edge length $1/c^n$. For each lattice point, consider the unit cube centered at that point. Hence the neighborhood $N := N(A, D)$ can also be regarded as the nonoverlapping union of cubes. Let N_n denote the neighborhood N contracted by a factor of $1/c^n$. Then it can be shown by induction that the sum of the elements in the nth power C^n of the contact matrix C is approximately equal to the number α_n of small cubes q in T_n such that the neighborhood, centered at q, lies both inside and outside of T_n. In other words, α_n counts the number of small cubes in T_n "close" to ∂T_n. When we use the term "approximately" here we mean that there are upper and lower bounds of one quantity by a constant multiple of the other quantity, where the constants do not depend on n.

What simplifies the calculation of the Hausdorff dimension of ∂T is that, for the boundary of a self-similar digit tile, the Hausdorff dimension coincides with the box-counting dimension. This is a consequence of a result of Falconer [**F2**] on sub-self-similar sets. Consider the collection of cubes in the ϵ-coordinate mesh of $\mathbb{R}^d$. For a given set $X \in \mathbb{R}^d$ let $\beta_\epsilon(X)$ denote the number of such cubes that intersect X. The *box-counting dimension* is defined by

$$\dim_B(X) = \lim_{\epsilon \to 0} \frac{\log \beta_\epsilon(X)}{-\log \epsilon}.$$

Letting $\epsilon = 1/c^n$, it can be proved, in the case of our digit tile T, that the number of small cubes that intersect ∂T is approximately equal to the number of small cubes α_n in T_n that are "close" to ∂T_n. Therefore

$$\dim_H(\partial T) = \dim_B(\partial T) = \lim_{n \to \infty} \frac{\log \alpha_n}{n \log c} = \lim_{n \to \infty} \frac{\log |C^n|}{n \log c},$$

where $|C|$ denotes the sum of all the entries in a matrix C. What completes the theorem is the fact that the largest eigenvalue of any nonnegative matrix C is given by the formula $\lambda(A, D) = \lim_{n \to \infty} |C^n|^{1/n}$. □

6. Construction of the Boundary

The main result of this section is an explicit correspondence between two known methods for constructing digit tiles in the plane. The IFS method produces the tile itself; the recurrent set method, due to Dekking [**De1, De2**], produces the boundary of the tile.

The proof of the theorem in this section appears in [**V4**]. Another connection between the IFS and recurrent set method appears in Bedford [**Be1, Be2**] in the context of constructing Markov partition boundaries for hyperbolic toral endomorphisms. Kenyon [**Ke3**] uses the recurrent set method in a setting discussed in Section 10.

The IFS "data" from which a digit tile $T = T(A, D)$ is constructed is simply the expanding matrix A and the digit set D. The pair (A, D) will be referred to as *tile data* if

(1) A is an expanding 2×2 integer matrix and
(2) D is a set of coset representatives of $\mathbb{Z}^2/A(\mathbb{Z}^2)$.

We use an integer matrix to keep the exposition simple. As explained in Section 2, all results are easily extended to the case of a tile based on a general lattice.

The "data" for the recurrent set method is a free group endomorphism ([**Lo**] is an introductory text on combinatorial group theory). Let $G := G\langle a, b\rangle$ be the free group on two generators a and b. Thus G consists of all words in the letters $\{a, b, a^{-1}, b^{-1}\}$, including the empty word e. The operation is concatenation, and the only relations are $aa^{-1} = e = a^{-1}a$ and $bb^{-1} = e = b^{-1}b$. Consider an endomorphism $\sigma: G \to G$. Note that σ is determined by its action on a and b. Define a matrix

$$A_\sigma = \begin{pmatrix} m_{aa} & m_{ab} \\ m_{ba} & m_{bb} \end{pmatrix},$$

where $m_{\alpha\beta}$ is the number of occurrences of α in $\sigma(\beta)$, counting α^{-1} as occurring -1 time. Here α and β are each either a or b. This process is called Abelianization.

EXAMPLE 6.1 (Twin dragon).

$$\sigma(a) = ab$$
$$\sigma(b) = a^{-1}b$$
$$A_\sigma = \begin{pmatrix} 1 & -1 \\ 1 & 1 \end{pmatrix}$$

EXAMPLE 6.2 (Gasket).

$$\sigma(a) = a^{-1}b^{-1}abaa$$
$$\sigma(b) = ba^{-1}ba$$
$$A_\sigma = \begin{pmatrix} 2 & 0 \\ 0 & 2 \end{pmatrix}$$

Denote by $f: G \to \mathbb{R}^2$ the homomorphism determined by $f(a) = (1, 0)$ and $f(b) = (0, 1)$. Let $w = \alpha_1\alpha_2 \ldots \alpha_n$ be any word in which each α_i is an element of $\{a, b, a^{-1}, b^{-1}\}$, and consider the sequence of points $x_i \in \mathbb{R}^2$, $i = 0, 1, \ldots, n$, given by $x_0 = (0, 0)$ and $x_i = f(\alpha_1\alpha_2 \ldots \alpha_i) = f(\alpha_1) + f(\alpha_2) + \cdots + f(\alpha_i)$, $i \geq 1$. Join the points $x_0, x_1, \ldots, x_n$ sequentially by line segments to obtain a polygonal path

FIGURE 11. Approximations to the boundary of the twin dragon.

$p(w)$ and let

(6.1) $$K_n := K_n(\sigma) = A_\sigma^{-n} p\big(\sigma^n(aba^{-1}b^{-1})\big).$$

Basically the path is obtained by traveling one unit left or right for an occurrence of a or a^{-1}, resp., in the string and one unit up or down for an occurrence of b or b^{-1}, resp.; then the path is contracted by A^{-n}. It is known [**De1**] that, if A_σ is expanding, then the sequence $\{K_n\}$ converges with respect to the Hausdorff metric to a closed curve

$$K := K(\sigma) = \lim_{n \to \infty} K_n.$$

Some line segments may be traversed by K_n more than one time. We impose the convention that each traversal of a line segment in one direction cancels a traversal of that line segment in the opposite direction. Thus K_n can consist of several closed curves, and hence K_n, and also K, may be disconnected. It can happen that the winding number of K_n about a point is more than 1. In this case there is no well defined region enclosed by K_n. The following result makes this situation easy to detect [**V4**].

LEMMA 6.3. *If the winding number of K_1 about every point of $\mathbb{R}^2 \setminus K_1$ is either 0 or 1, then the same is true of K_n, $n > 1$.*

The endomorphism $\sigma: G \to G$ will be referred to as *boundary data* if

(1) A_σ is expanding, and
(2) the winding number of K_1 about every point of $\mathbb{R}^2 \setminus K_1$ is either 0 or 1.

From (6.1) the path $A\big(K_1(\sigma)\big)$ has sides that are parallel to the axes and joins integer lattice points. Let D_σ be the set of lattice points that are the lower left corners of unit squares that lie inside $A\big(K_1(\sigma)\big)$.

THEOREM 6.4 (Vince [**V4**]). *The mapping $\Theta: \sigma \mapsto (A_\sigma, D_\sigma)$ induces a bijection from the collection of all boundary data to the collection of all tile data such that*

$$\partial T_n(A_\sigma, D_\sigma) = K_n(\sigma).$$

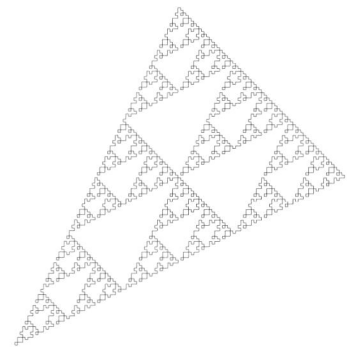

FIGURE 12. Boundary of the gasket.

Moreover, if any of the conditions in Theorem 4.3 hold, then

$$\partial T(A_\sigma, D_\sigma) = K(\sigma).$$

The bijection is algorithmic and was used to draw Figure 11, which gives the first approximations to the boundary of the twin dragon corresponding to the approximations in Figure 3 drawn by the IFS method. The endomorphism is that of Example 6.1. Figure 12 shows the boundary of the topologically more complicated gasket originally pictured in Figure 4. The endomorphism is that of Example 6.2.

7. Lattice Tiling Problem and Aperiodic Proto-tile Problem

One part of Hilbert's 18th problem asks whether there exists a polyhedron, copies of which tile space, but which is not the fundamental region of a group of isometries. In other words, the symmetry group of the tiling is not transitive on tiles. Examples were discovered early on, a polyhedron in 3 dimensions by Reinhardt [**Re**] in 1928 and a convex pentagon in 2 dimensions by Kershner [**Ker**] in 1968. All known examples, however, are periodic. A tiling of $\mathbb{R}^d$ is *periodic* if its symmetry group contains translations in d linearly independent directions.

A strong version of Hilbert's question is whether there exists a single tile which admits only nonperiodic tilings. A *nonperiodic* tiling is one that admits no translations. The Penrose tiles comprise a set of two tiles, copies of which tile the plane in uncountably many ways, but no such tiling is periodic. A set of *proto-tiles*, copies of which tile $\mathbb{R}^d$ but only nonperiodically, is called *aperiodic*. The Schmitt-Conway-Danzer (SCD) tile [**Da, Sch**], for example, is a single, convex, aperiodic tile in $\mathbb{R}^3$ (under the restriction that mirror image copies of the proto-tile are not allowed and screw symmetry does not count as a periodic symmetry). The SCD tile provides a solution to the above question, but the following questions remain open.

QUESTION 7.1. Does there exists a single aperiodic proto-tile in $\mathbb{R}^2$?

QUESTION 7.2. Does there exist an aperiodic proto-tile that tiles R^d by translation?

The answer to Question 7.2 in dimension 1 is no [**LW1**]. Venkov [**Ven**] answered Question 7.2 in 1954 in any dimension for the case of a convex proto-tile, a result independently rediscovered by McMullen [**McM**]. Their result: if a convex T tiles $\mathbb{R}^d$ by translation, then there is a lattice tiling of $\mathbb{R}^d$ by copies of T. The same result

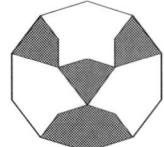 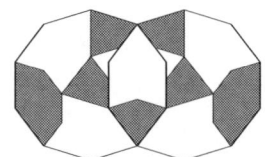 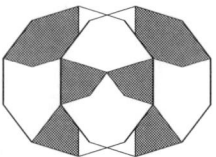

FIGURE 13. Overlapping marked decagons.

is true in dimension 2 for polyominoes[1] [**BN, KVi, WvL**]. However, the Venkov-McMullen result is not true for nonconvex tiles in general. The 1-dimensional tile $[0,1] \cup [2,3]$ allows a a tiling, but no lattice tiling of $\mathbb{R}$. Szabó [**Sz**] constructs a 3-dimensional, centrally symmetric, star polyhedron whose translates tile $\mathbb{R}^3$, but admits no lattice tiling of $\mathbb{R}^3$. A lattice tiling is periodic, but a periodic tiling is not necessarily a lattice tiling. So Question 7.2 remains unresolved in the nonconvex, nonpolyomino case.

A natural place to seek an example that might affirmatively answer Question 7.2 is among the digit tiles. Any digit tile T admits a tiling by translation as given by (4.1) in Section 4. However this tiling is sometimes not periodic, as in Example 4.2 and Figure 7. The tile in Figure 7, however, does admit a lattice tiling—by translation by the lattice generated by vectors $(1,0)$ and $(0,2)$. Gröchenig and Haas [**GH**] conjectured that every digit tile admits a lattice tiling. What makes the conjecture difficult is the existence of tiles, as in Example 4.2, that do not satisfy the conditions of Theorem 4.3. The lattice tiling conjecture was recently verified by Lagarias and Wang; so it is not possible to find an aperiodic digit tile. Note that the tiling guaranteed by their theorem is not necessarily self-replicating in the sense of Section 4.

THEOREM 7.3 (Lagarias and Wang [**LW4**]). *Every digit tile T admits a lattice tiling of $\mathbb{R}^d$ for some lattice $L \subseteq \mathbb{Z}^d$.*

For remarks on Question 7.1 see Penrose's paper [**P2**]. Although there is no known single aperiodic proto-tile in $\mathbb{R}^2$, the analogous problem for coverings of $\mathbb{R}^2$ is solved. Moreover, the result has received considerable attention recently because of its implications for the structure of real quasicrystals. Consider the marked regular decagon on the left in Figure 13. This proto-tile is used to cover the plane with overlap allowed, but only according to the following overlap rule: two decagons may overlap only if shaded regions overlap and the overlap area is greater than or equal to the area of the overlap hexagon in the center illustration in Figure 13. The figure shows the two possible sizes of the overlap. Gummelt [**Gu**] proved that every covering by marked decagons that satisfies the overlap rule is nonperiodic. Moreover, by dissecting each decagon into Penrose acute and obtuse triangles (Figure 14), such decagon coverings can be put into correspondence with the Penrose tilings.

Jeong and Steinhardt [**JS**] subsequently proved that both the Penrose matching rules for Penrose tilings and the overlap rule for decagon coverings can be replaced by a condition on the density of certain clusters. More precisely, the Penrose tilings are the tilings by Penrose rhombs for which the density of certain clusters of tiles (clusters whose union is essentially the Gummelt decagon) is maximum. This

---

[1] A *polyomino* is a rookwise connected tile formed by joining unit squares at their edges.

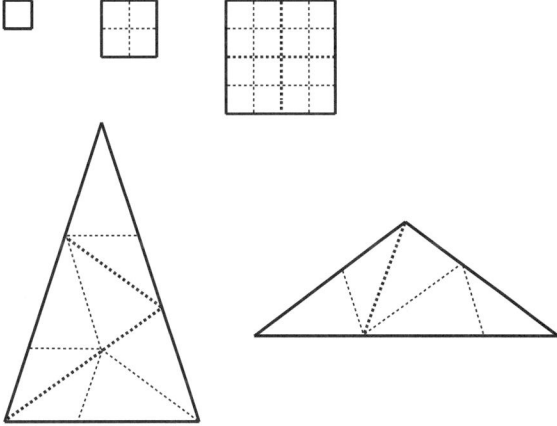

FIGURE 14. Square and Penrose hierarchies.

result led Jeong and Steinhardt to hypothesize that quasicrystals are formed from a single type of atomic cluster that can share atoms with neighboring clusters and that quasicrystals maximize cluster density. Evidence for such a model recently came from electron microscopy [**SJSTAT**]. Electron micrographs of $Al_{72}Ni_{20}Co_8$ show striking similarities to the decagon coverings in Gummelt's paper.

8. Crystallographic Digit Tiling

A *crystallographic group* Γ is a discrete, cocompact group of isometries of Euclidean space. *Discrete* means that any ball contains at most finitely many points in the Γ-orbit of any point. *Cocompact* means that the quotient space $\mathbb{R}^d/\Gamma$ is compact. A lattice group, the group of translations by the points of a lattice, is a special case of a crystallographic group. A fundamental theorem of Bieberbach states that it Γ is a d-dimensional crystallographic group, then Γ contains a *translation subgroup*, a subgroup generated by translations in d independent directions.

Under any of the conditions of Theorem 4.3 a self-replicating digit tiling is a lattice tiling. This means that

$$\mathcal{T} = \{\gamma(T) \mid \gamma \in L\},$$

where L is a lattice group. But a lattice group L is only one of 17 crystallographic groups in the plane and only one of 230 crystallographic groups in 3-space. This section briefly describes a generalization, due to Gelbrich [**Ge1**], from lattice tiling to crystallographic tiling. A *crystallographic tiling* is of the form

$$\mathcal{T} = \{\gamma(T) \mid \gamma \in \Gamma\},$$

where Γ is a crystallographic group.[2]

The basic construction of digit tiles given in Section 2 is based on a lattice L. The linear expansion A maps L into itself; so ALA^{-1} is the subgroup of translations by points of the sublattice $A(L)$. A set of coset representatives of L/ALA^{-1} consists of translations by a digit set D. To generalize, let Γ be any crystallographic

---

[2] The term "crystallographic" is often used interchangeably with the term "periodic." A crystallographic tiling is periodic by Bieberbach's theorem, but a periodic tiling is not necessarily crystallographic. The symmetry group of a periodic tiling may not act transitively on the tiles.

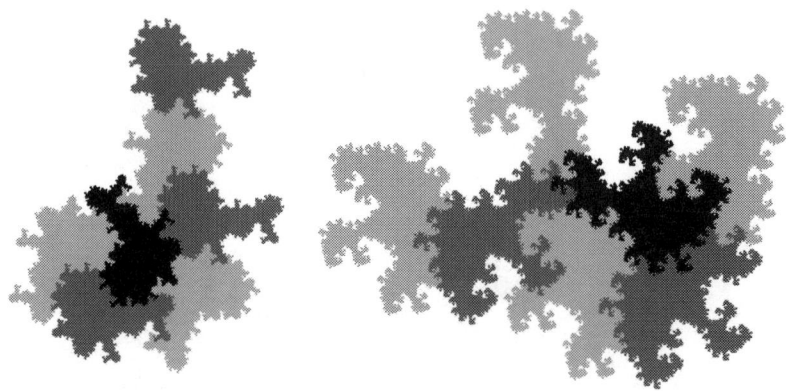

Figure 15. Crystallographic digit tilings: sea horse and coral reef.

group; let $A: \mathbb{R}^d \to \mathbb{R}^d$ be a linear expanding map such that $A\Gamma A^{-1} \subset \Gamma$ and let $D = \{d_1, \ldots, d_N\}$ be a set of right coset representatives of $\Gamma/A\Gamma A^{-1}$. Then the contractions
$$f_i(x) = A^{-1} \circ d_i(x)$$
provide an iterated function system with a unique attractor, say $T := T(\Gamma, A, D)$. The analogue of Theorem 2.5 holds: $T(\Gamma, A, D)$ is a compact set that is the closure of its interior. Call $T(\Gamma, A, D)$ a *crystallographic digit tile*.

Using the same reasoning as for ordinary digit tiles, every crystallographic digit tile in $\mathbb{R}^d$ admits a tiling of $\mathbb{R}^d$ that is self-replicating in the sense of Section 4. Some crystallographic tilings, courtesy of Gelbrich and Giesche [**GG**], are shown in Figure 15 and are reminiscent of fractalized Escher prints. Analogous to (4.1) it can be shown that every self-replicating crystallographic tiling is of the form
$$\mathcal{T} = \{\gamma(T) \mid \gamma \in \Gamma_0\},$$
where Γ_0 is a subset (not necessarily a subgroup) of Γ.

The analogous result to Theorem 7.3, that every crystallographic digit tile admits a crystallographic tiling, seems likely, but is open as far as we know. The issue is, given a crystallographic tile T, whether there exists a tiling $\{\gamma(T) \mid \gamma \in \Gamma_0\}$ where Γ_0 is a crystallographic group. Generalizing the results of Sections 4–6 to crystallographic tiles would also be of interest.

9. Hierarchical Tiling

All tilings in Sections 1–8 are by copies of a single tile. We now turn to tilings by copies of tiles taken from a finite set of *proto-tiles*. Many of the concepts that occur in the remainder of this paper are valid in a general context; so we introduce the notions of hierarchy and hierarchical tiling and frame the theory in this setting. Hierarchy is the basic notion; the tilings will be produced automatically from the hierarchy.

9.1. Hierarchy. Let $\mathcal{P} = (P_0, P_1, P_2, \ldots)$ be a sequence of finite proto-tile sets. Define inradius(P_n) to be the largest r such that each proto-tile in P_n contains a ball of radius r. Call $\mathcal{P}$ a *hierarchy* if the following three conditions are satisfied.

(1) $\lim_{n \to \infty}$ inradius$(P_n) = \infty$.

(2) Each tile in P_{n+1} has a unique subdivision into the nonoverlapping union of isometric copies of tiles in P_n.

The *subdivision rule* in condition (2) must be unique in the sense that each tile in P_{n+1} can be subdivided into the nonoverlapping union of isometric copies of tiles in P_n in a unique way. (If there is ambiguity, for example if a proto-tile has nontrivial symmetry, then it is common to color some points in the tiles so that colors must match. In the IFS approach discussed in Section 10 this coloring is unnecessary.) Let $\mathcal{S}$ be an nonoverlapping set of tiles in $\mathbb{R}^d$ taken from P_n. Using the subdivision rule there is a unique set $\mathcal{S}_{(1)}$ of tiles from P_{n-1} obtained by subdividing each tile in $\mathcal{S}$ according to the subdivision rule. Repeat to obtain from $\mathcal{S}$ the kth *subdivision* $\mathcal{S}_{(k)}$, $k \leq n$, by tiles in P_{n-k}.

(3) For any given m, each tile in P_m appears in the $(n-m)$th subdivision of each tile in P_n for all n sufficiently large.

The square hierarchy example in Figure 14 shows the first three proto-tile sets and the first and second subdivisions. (Each proto-tile set consists of a single tile.) The second hierarchy in Figure 13 is by acute and obtuse Penrose triangles. Each proto-tile set consists of two tiles. The second subdivision is shown. (It can also be considered as the fourth subdivisions in the finer hierarchy shown in [**GS**, p. 540].) To insure uniqueness of the subdivision rule, the vertices of the triangles should be appropriately colored, as is usually done for the Penrose tiles. In both of these examples the proto-tile sets $P_0, P_1, \ldots$ have the same cardinality. Moreover, corresponding tiles in P_n and P_{n+1} are similar, the ratio being 2 in the case of the squares and the golden ratio τ in the case of the Penrose tiles. (These are examples of what are commonly called *local inflation rules*). In general, this does not have to be the case for a hierarchy.

9.2. Hierarchical Tiling. A tiling by copies of tiles taken from a proto-tile set P will be called a P-*tiling*. A *patch* of a tiling is a subset of tiles whose union is a topological ball. The definition of hierarchy concerns the proto-tile sets, not tilings by these proto-tiles. Now define a tiling $\mathcal{T}$ to be *hierarchical* if there exists a hierarchy $P_0, P_1, P_2, \ldots$ and a sequence $\mathcal{T}_0, \mathcal{T}_1, \mathcal{T}_2, \ldots$ of tilings with $\mathcal{T}_0 = \mathcal{T}$ such that

(1) $\mathcal{T}_n$ is a P_n-tiling for all n.
(2) $\mathcal{T}_n$ is the subdivision of $\mathcal{T}_{n+1}$ for each n.
(3) Each patch in $\mathcal{T}$ appears in the nth subdivision of some tile in P_n, for n sufficiently large, n depending only on the size of the patch.

The last condition is to eliminate from consideration tilings such as the following. Combine the square tiling of the left half-plane and the square tiling of the right half-plane offset slightly along a vertical "fault" where the two half-planes meet.

If $\mathcal{T}$ is a tiling with hierarchy $\mathcal{P}$, then $\mathcal{T}$ will be referred to as a $\mathcal{P}$-tiling. We also use the terminology $\mathcal{P}$ *admits* the tiling $\mathcal{T}$. Note that if $\mathcal{P} = (P_0, P_1, P_2, \ldots)$ is a hierarchy then so is any infinite subsequence $\mathcal{P}' = (P_{i0}, P_{i1}, P_{i2} \ldots)$ with the obvious subdivision rule coming from the subdivision rule for $\mathcal{P}$. Moreover, if $P_0 = P_{i0}$, then a tiling $\mathcal{T}$ is a $\mathcal{P}$-tiling if and only if $\mathcal{T}$ is a $\mathcal{P}'$-tiling. Such hierarchies $\mathcal{P}$ and $\mathcal{P}'$ will be considered *equivalent*.

If, for every $\mathcal{P}$-tiling $\mathcal{T}$, the sequence $\mathcal{T}_0, \mathcal{T}_1, \mathcal{T}_2, \ldots$ is uniquely determined, then we say that $\mathcal{P}$ *forces uniqueness*. The hierarchy of squares in Figure 14 does not force uniqueness; for the tiling $\mathcal{T}$ of the plane by squares, there are infinitely many ways to choose the sequence of tilings $\mathcal{T}_0, \mathcal{T}_1, \mathcal{T}_2, \ldots$. The Penrose hierarchy does force uniqueness on any Penrose tiling of the plane by thick and thin triangles. In other words, the subdivision rule for the Penrose hierarchy is locally invertible; the subdivision rule for the square hierarchy is not. If the hierarchy for a tiling forces uniqueness, then the tiling is commonly said to satisfy the *unique composition property* or the *local inflation/deflation property*, compare [**So2, Ba**].

A tiling $\mathcal{T}$ is of *finite type* if, for any positive number r, there are at most finitely many patches, up to congruence, within a ball of radius r. A tiling $\mathcal{T}$ has the *local isomorphism property* if, for any patch $\mathcal{Q}$ of $\mathcal{T}$, there is a number R such that any ball of radius R contains, up to congruence, a copy of $\mathcal{Q}$. Two tilings $\mathcal{T}_1$ and $\mathcal{T}_2$ are said to be *locally isomorphic* if every patch of $\mathcal{T}_1$ can be found in $\mathcal{T}_2$ and vice versa. Local isomorphism is an equivalence relation.[3] Nonperiodic tilings that are both of finite type and satisfy the local isomorphism property have been referred to as quasiperiodic. Since the term quasiperiodic has multiple definitions in the literature we will not use it.[4]

THEOREM 9.1. *Let $\mathcal{P}$ be a hierarchy.*
(1) *Every $\mathcal{P}$-tiling is of finite type.*
(2) *Every $\mathcal{P}$-tiling has the local isomorphism property, and any two $\mathcal{P}$-tilings are locally isomorphic.*
(3) *If $\mathcal{P}$ forces uniqueness, then every $\mathcal{P}$-tiling is nonperiodic.*

PROOF. Let $\mathcal{P} = (P_0, P_1, P_2, \ldots)$ be the hierarchy and $(\mathcal{T}_0, \mathcal{T}_1, \mathcal{T}_2, \ldots)$ the corresponding sequence of tilings with $\mathcal{T}_0 = \mathcal{T}$. By condition (3) in the definition of hierarchical tiling, any ball of radius r in a $\mathcal{P}$-tiling $\mathcal{T}$ is contained in the kth subdivision $p_{(k)}$ of some tile $p \in P_k$, where k depends only on r. Since there are at most finitely many configurations within the kth subdivision of the tiles of P_k, the finite type property is verified.

To verify the local isomorphism property, let $\mathcal{Q}$ be a patch in $\mathcal{T}$. Again, $\mathcal{Q}$ appears in the kth subdivision $p_{(k)}$ of some proto-tile $p \in P_k$. But by condition (3) in the definition of hierarchy, the proto-tile p, in turn, appears in the subdivision of each tile in P_n for n sufficiently large. Finally, since the tiles are compact, there is a number R, depending only on n, such that any ball of radius R contains some tile in $\mathcal{T}_n$. Therefore any ball of radius R contains $\mathcal{Q}$. The same reasoning shows that any two $\mathcal{P}$-tilings are locally isomorphic.

Assume that $\mathcal{P}$ forces uniqueness, and assume, by way of contradiction, that $\mathcal{T}$ admits a translational symmetry. This induces a translational symmetry of $\mathcal{T}_1$; otherwise uniqueness of $\mathcal{T}_1$ is violated. Repeating this argument implies that, for each n, there is a translational symmetry of $\mathcal{T}_n$. But this is impossible because inradius $(P_n) \to \infty$ by condition (1) in the definition of hierarchy. □

Note that, for the set P of Penrose tiles, the standard matching rules guarantee that every P-tiling is a $\mathcal{P}$-tiling. Since the Penrose hierarchy forces uniqueness, it

---

[3]Concerning terminology in the literature, two tiling in the same local isomorphism class are sometimes called *locally indistinguishable*, and a tiling with the local isomorphism property is sometimes called *repetitive*. Another equivalence relation among tilings, *mutual local derivability*, will not come into play in this paper. We use the terms "finite type" and "local isomorphism" with respect to congruence. Analogous versions with respect to translations are also often used.

[4]This paper does not discuss the well known projection method for constructing "quasiperiodic" point sets. It is interesting to note, however, that there exists such sets for which the window system can be interpreted as a self-similar tile with fractal boundary; see [**LGJJ**].

follows from Theorem 9.1 that no P-tiling is periodic. In this case we call P an *aperiodic set*; no tiling by copies of tiles in P is periodic.

So far it has not been assumed that a given hierarchy $\mathcal{P} = (P_0, P_1, \dots)$ admits even a single tiling. The existence of $\mathcal{P}$-tilings is now addressed. If $p_n \in P_n$ and $p_{n+1} \in P_{n+1}$ then, in accordance with the subdivision rule, p_n can possibly appear several times in p_{n+1} (or not at all). Let $S(p_n, p_{n+1})$ be a set of symbols denoting the positions of p_n in the subdivision of p_{n+1}. If p_n does not appear in p_{n+1}, then $S(p_n, p_{n+1})$ is empty. Consider any sequence $C = (c_0, c_1, \dots)$ where each $c_n \in S(p_n, p_{n+1})$ for some $p_n \in P_n$, $p_{n+1} \in P_{n+1}$ and, if $c_{n-1} \in S(q_{n-1}, q_n)$ and $c_n \in S(p_n, p_{n+1})$ then $q_n = p_n$. Construct a tiling from C as follows. Start with $\mathcal{Q}_0 := p_0$; $\mathcal{Q}_0$ is embedded in the subdivision $\mathcal{Q}_1$ of tile p_1 in position c_0; p_1, in turn, is embedded in the subdivision $\mathcal{Q}_2$ of p_2 in position c_1. Continue in this way to obtain a nested sequence $\mathcal{Q}_0 \hookrightarrow \mathcal{Q}_1 \hookrightarrow \mathcal{Q}_2 \hookrightarrow \dots$ of patches. The union $\bigcup_n \mathcal{Q}_n$ is a *partial* tiling. We use the term "partial" because the union may not be all $\mathbb{R}^d$. Call two such sequences C and C' *equivalent* if there is an integer k such that the sequences C and C' agree after the first k terms. Because of the uniqueness of subdivision, equivalent sequences yield the same partial tiling up to congruence. Call an equivalence class of sequences a *code* for the tiling it produces. So there is a well-defined mapping from the set of codes onto the set of partial $\mathcal{P}$-tilings. (The mapping may not be one-to-one; the square tiling of the plane, for instance, has infinitely many codes.)

If, in condition (3) in the definition of hierarchy, it is required that each tile in P_m appears in the interior (not intersecting the boundary) of each tile in P_n, then we call the hierarchy *interior*. The following result is surely known; in particular it has long been known for the Penrose hierarchy [**GS**].

THEOREM 9.2.

(1) *If a hierarchy $\mathcal{P}$ is interior, then $\mathcal{P}$ admits (full) tilings.*
(2) *If $\mathcal{P}$ forces uniqueness, then there is a bijection between the set of codes and the set of partial tilings (up to isometry). In particular $\mathcal{P}$ admits uncountably many partial tilings (uncountably many full tilings if $\mathcal{P}$ is interior).*

PROOF. Concerning (1), the property of being interior insures that for some code the union $\bigcup_n \mathcal{Q}_n$ described above covers all $\mathbb{R}^d$, hence producing a full tiling.

Concerning (2), given a $\mathcal{P}$-tiling $\mathcal{T}$, any code $C(\mathcal{T}) = (c_0, c_1, \dots)$ for $\mathcal{T}$ is obtained as follows. Choose an arbitrary tile $T_0 \in \mathcal{T}$, where T_0 has proto-tile type p_0, Then T_0 is contained at position c_0 in a unique tile T_1 of proto-tile type p_1 at the next level. In general T_n of type p_n is contained at position c_n in a unique tile of type p_{n+1}. Moreover if C and C' are both codes for $\mathcal{T}$ then they must be equivalent because, any two initial tiles in $\mathcal{T}$ are contained in the same single tile at a sufficiently high level.

Because of condition (1) in the definition of hierarchy, there are at least two choices for the next embedding at infinitely many stages. So there are uncountably many codes, hence uncountably many tilings. $\square$

The code for the Penrose tiling by acute and obtuse triangles can be denoted by binary digits 0 or 1 in such a way that each partial tiling is given by a unique binary sequence which contains no subsequence 11. (This code is with respect to the finer hierarchy mentioned in reference to Figure 13.) Every such binary sequence, except $(000\ldots)$, $(10001000\ldots)$ and $(00100010001\ldots)$ yields a tiling of $\mathbb{R}^2$. The exceptions yield partial tilings which can easily be extended to full tilings. Hence by Theorem 9.2 there is a bijection between the set of codes and set of Penrose tilings. The Penrose tiling with code $(000\ldots)$, called the *cartwheel*, has been singled out in the literature. For example it is shown in [**GS**] that, except for seven exceptions, every tile in the cartwheel tiling lies in a patch of tiles whose symmetry group is the dihedral group D_5. A special case of Theorem 10.1 in the next section implies the surprising property that the cartwheel is the unique Penrose tiling $\mathcal{T}$ for which an expansion by the golden ratio sends each tile in $\mathcal{T}$ to the union of tiles in $\mathcal{T}$.

10. Self-affine and Self-similar tiling

The basic concept in Section 9 is a hierarchy $\mathcal{P}$. From a given hierarchy, tilings are produced according to Theorem 9.2, infinitely many in the case that $\mathcal{P}$ forces uniqueness. This section concerns two special types of hierarchies, self-affine and self-similar, and their associated tilings. After defining self-affine and self-similar hierarchy (Section 10.1), a few important results concerning the associated tilings are presented (Section 10.2). An alternative approach based on graph iterated function systems is given in Section 10.3. Examples appear in Section 10.4.

10.1. Definitions. Let $A\colon \mathbb{R}^d \to \mathbb{R}^d$ be a linear expanding map. Let $P = \{T_1, T_2, \ldots, T_N\}$ be a finite set of proto-tiles, and let

(10.1) $$P_n = \{A^n(p) \mid p \in P\}.$$

The subdivision rule for the first level of a hierarchy $\mathcal{P} = (P_0, P_1, \ldots)$ is given explicitly as follows for each $i = 1, 2, \ldots, N$:

(10.2) $$A(T_i) = \bigcup g_{ij}^k(T_j),$$

where the union is nonoverlapping with indices $j = 1, 2, \ldots, N$ and "multiplicities" $k = 1, 2, \ldots, k(i,j)$, and each g_{ij}^k is an isometry. The functional equation (10.2) states that each large tile $A(T_i)$ is the nonoverlapping union of copies of the small tiles $T_1, \ldots, T_N$. In this union, each tile of type T_j can appear one or more times ($k(i,j) \geq 1$) or not at all ($k(i,j) = 0$).

To define the subdivision rule on P_n for $n > 1$, make the following assumption:

(10.3) $$A \circ g_{ij}^k \circ A^{-1} \text{ is an isometry for all } i,\, j,\, k.$$

Assumption (10.3) allows (10.2) to be iterated to obtain a subdivision rule at every level. The matrix $M = \bigl(k(i,j)\bigr)$ of multiplicities from (10.2) is called the *substitution matrix* for the subdivision. Thus $k(i,j)$ is the number of times T_j appears in T_i. Condition (3) in the definition of hierarchy in Section 9.1 is equivalent to some power of M being strictly positive, i.e., M is what is called a *primitive matrix*. If this is the case $\mathcal{P}$ satisfies all three conditions in the definition of hierarchy.

Assumption (10.3) holds if either

(1) g_{ij}^k is a translation for each $i,\,j,\,k$, or
(2) A is a similarity.

In case (1) the hierarchy $\mathcal{P}$ will be called *self-affine* and in case (2) *self-similar*. If both (1) and (2) hold we call the hierarchy *translationally self-similar*. Let P_n be as in (10.1) and let $P'_n = \{A'^n(p) \mid p \in P\}$, where $A' = \phi \circ A$ for some isometry ϕ. Note that $\mathcal{P} = (P_0, P_1, \ldots)$ and $\mathcal{P}' = (P'_0, P'_1, \ldots)$ are the same hierarchy. In particular, in the self-similar case it can be assumed that $A(x) = cx$ where $c > 1$.

In either case, the remarks in Section 9.2 imply that replacing A by $\phi \circ A^s$, where ϕ is an isometry and s any positive integer results in an equivalent hierarchy as defined in Section 9.2.

A $\mathcal{P}$-tiling will be called *self-affine* if $\mathcal{P}$ is a self-affine hierarchy and *self-similar* if $\mathcal{P}$ is a self-similar hierarchy. It is unfortunate that the term "self-similar" has slightly different definitions in various publications on the subject. The definition of self-similar in [**So2**], for example, assumes that both conditions (1) and (2) hold, translationally self-similar in our terminology. A self-similar tiling in [**Ke3, So1, Th**] has an additional property we will call special. A self-affine or self-similar tiling $\mathcal{T}$ is *special* if the image $A'(T)$ is, for any $T \in \mathcal{T}$, the union of tiles in $\mathcal{T}$. Here A' can be any linear map of the form $\phi \circ A^s$, which, as discussed in the paragraph above, results in a hierarchy equivalent to the original hierarchy. This definition of special is a direct generalization from Section 4 of the term self-replicating; in that case $s = 1$.

10.2. Some Results. In this section several miscellaneous results on self-affine and self-similar tilings are presented. Let $\mathcal{P}$ be either a self-affine or a self-similar hierarchy and denote by $\Omega_\mathcal{P}$ the set of all $\mathcal{P}$-tilings. The *subdivision operator* $\sigma \colon \Omega_\mathcal{P} \to \Omega_\mathcal{P}$ is defined as follows. Using the notation $A(\mathcal{T}) = \{A(T) \mid T \in \mathcal{T}\}$ define

$$\sigma(\mathcal{T}) = A(\mathcal{T})_{(1)},$$

the first subdivision of the inflated tiling $A(\mathcal{T})$. According to the next result, the special self-affine and self-similar tilings are the ones with a repeating code.

THEOREM 10.1. *The following statements are equivalent for a self-affine or self-similar tiling $\mathcal{T}$.*

(1) *$\mathcal{T}$ is a fixed point of the subdivision operator σ^s for some positive integer s.*
(2) *There is a repeating code for $\mathcal{T}$ of the form*

$$C(\mathcal{T}) = (c_1, c_2, \ldots, c_s, c_1, c_2, \ldots, c_s, \ldots).$$

(3) *The tiling $\mathcal{T}$ is special.*

PROOF. (1) $\Leftrightarrow$ (2). First, $\sigma^s(\mathcal{T}) = \mathcal{T}$ if and only if the two tilings have the same code (up to equivalence), say $(c_1, c_2, \ldots)$. But if $C(\mathcal{T}) = (c_1, c_2, \ldots)$, then, by the definition of the subdivision operator, $C(\sigma^s(\mathcal{T})) = (c'_1, c'_2, \ldots, c'_s, c_1, c_2, \ldots)$ for some symbols $c'_1, c'_2, \ldots, c'_s$. Hence, by the definition of equivalent codes, $\sigma^s(\mathcal{T}) = \mathcal{T}$ if and only if $c_{k+s} = c_k$ for k sufficiently large. This is the case if and only if $c(T)$ repeats with period s.

(1) $\Leftrightarrow$ (3). The tiling $\mathcal{T}$ is a fixed point of the subdivision operator σ^s if and only if $A^s(\mathcal{T})_{(s)} = \phi(\mathcal{T})$ for some isometry ϕ. This is the case if and only if, for each tile $T \in \mathcal{T}$, we have $A^s(T) = \bigcup_{i=1}^{K} \phi(T_i)$ for some tiles $T_i \in \mathcal{T}$. This equation is equivalent to $(\phi^{-1} \circ A^s)(T) = \bigcup_{i=1}^{K} T_i$; in other words, $\mathcal{T}$ is special. $\square$

COROLLARY 10.2. *Every self-affine or self-similar hierarchy admits a special tiling.*

PROOF. Property (3) in the definition of the hierarchy, i.e., that the substitution matrix is primitive, implies that the hierarchy admits a tiling whose code repeats. The result then follows from Theorem 10.1. $\square$

The following result concerns the unique composition property defined in Section 9.2. The third part of Theorem 9.1 states, in particular, that a self-affine tiling with the unique composition property (local inflation/deflation) must be nonperiodic. A proof of the converse in the 1-dimensional case appeared in [**Mo**]. The converse is true in general.

THEOREM 10.3 (Solomyak [**So2**]). *If a self-affine tiling is nonperiodic then it has the unique composition property.*

The next result concerns tile frequencies. Recall, for example, that the frequencies of the two Penrose tiles in any Penrose tiling exist and the ratio of the two frequencies is the golden ratio [**GS**]. The existence of uniform frequencies of patches in cubes was established by Lunnon and Pleasants for substitution tilings by tiles that are polytopes [**LuP**]. In general, let $\mathcal{Q}$ be a patch in a tiling $\mathcal{T}$. Let $L_{\mathcal{Q}}(X)$ denote the number of translates of $\mathcal{Q}$ in a region $X \subset \mathbb{R}^d$. The frequency freq($\mathcal{Q}$) of the patch is defined as the following limit, if it exists,

$$\lim_{n\to\infty} \frac{L_{\mathcal{Q}}(X_n)}{\text{Vol}(X_n)},$$

where X_n is a region with d-dimension measure $\text{Vol}(X_n)$ that tends to infinity in such a way that the boundary of X_n does not wriggle too much. A precise definition and the following statement appear in [**So1**].

THEOREM 10.4 (Solomyak). *If $\mathcal{T}$ is a self-affine tiling, then the frequencies of patches exist.*

For a nonempty patch $\mathcal{Q}$ in a translationally self-similar tiling $\mathcal{T}$, define the *locator set*

$$L_{\mathcal{Q}}(\mathcal{T}) = \{x \in \mathbb{R}^d \mid \text{there exists } \mathcal{Q}' \subset \mathcal{T} \text{ with } \mathcal{Q} = \mathcal{Q}' - x\}.$$

Voronoï tilings based on these locator sets can be constructed. Priebe [**Pri**] proves an interesting finiteness property concerning the number of these derived Voronoï tilings of $\mathcal{T}$.

There is a growing body of work on the dynamical systems arising from the action by translation on a certain space of tilings. Solomyak [**So1**] gives a comprehensive survey of results on the dynamics of self-affine tilings, including a proof of unique ergodicity. We refer the interested reader to the cited paper and the references therein.

Perhaps the best known property of translationally self-similar tilings concerns possible expansion constants. For a self-similar tiling of the plane $\mathbb{R}^2 \cong \mathbb{C}$ the map A can be represented as multiplication by an *expansion constant* $\lambda \in \mathbb{C}$. The next theorem was announced by Thurston with a proof of necessity. Kenyon gave a constructive proof of sufficiency and a generalization to self-affine tilings in $\mathbb{R}^d$ [**Ke1**].

THEOREM 10.5 (Thurston [**Th**], Kenyon [**Ke3**]). *A translationally self-similar tiling of the plane with expansion constant λ exists if and only if λ is a complex Perron number, that is, an algebraic integer whose Galois conjugates, except $\bar\lambda$, are less than $|\lambda|$ in modulus.*

Concerning Theorem 10.5, it is not hard to show that, for a translationally self-similar tiling, $|\lambda|^2 = \lambda\bar\lambda$ is a real Perron number. In fact, this is essentially what is done in the proof of Proposition 10.7 later in this paper. The proof that $|\lambda|^2$ is a Perron number is based on the fact that the area of each proto-tile increases by a factor of $|\lambda|^2$ under the inflation by λ and this inflated area is an integer linear combination of the areas of the original proto-tiles. To show the stronger result that λ itself is a Perron number, Thurston considers certain distinguished points (capitals or control points) for each proto-tile, and a certain finite set of differences between control points in the tiling. Then λ inflates this set of differences so that the inflated differences are an integer linear combination of the original differences.

We conclude this section with a very brief comment on the diffraction spectrum of a self-similar tiling. One of the common definitions of quasicrystal is that of an atomic structure whose X-ray diffraction shows Bragg peaks—sharp spots in the diffraction pattern. For a discrete set Y of points in $\mathbb{R}^d$ (an atomic arrangement say), consider the distribution $f(x) = \sum_{y \in Y} \delta_y$, where δ_x is the Dirac delta. The X-ray diffraction of Y can be described using the Fourier transform $\widehat{\gamma}$ of a related distribution γ, called the autocorrelation. See [**Ba**] or [**Sen**], for example, for definitions and background. Under mild conditions $\widehat{\gamma}$ can be decomposed into a discrete part (Bragg spectrum) and continuous part (diffuse spectrum). Concerning tilings, by choosing a distinguished point for each type of tile, the spectrum of a $\mathcal{P}$-tiling can be discussed. In several examples of self-similar tilings it was noticed that, for the existence of nontrivial Bragg spectrum, it is necessary that the Perron-Frobenius eigenvalue (the largest eigenvalue) of the substitution matrix be a Pisot number [**BT**]. A *Pisot number* is an algebraic integer $\beta > 1$ such that all its other Galois conjugates lie inside the unit circle. In the generality below, the result is due to Gähler and Klitzing [**GK**].

THEOREM 10.6 (Gähler and Klitzing). *If $c > 1$ is the expansion factor of a self-similar tiling with nontrivial Bragg spectrum, then c must be a Pisot number.*

That c is a Pisot number is equivalent to the Perron-Frobenius eigenvalue of the substitution matrix being a Pisot number. Gähler and Klitzing go on to give a nice description of the Bragg spectrum of a self-similar tiling, which leads to distinguishing three types of such tilings: quasiperiodic, limit-periodic and limit-quasiperiodic.

10.3. Graph Iterated Function Systems. This section concerns a constructive approach to self-affine and self-similar tilings based on graph iterated function systems. Whereas the attractor to an IFS is a single compact set, the attractor of a graph IFS is a finite collection of compact sets. This generalization can be found in [**MW**] as well as in the literature on image compression. Bandt [**B1, B3**] applies the method to tilings.

Using the same notation as in Section 2 let $\mathcal{C} := \mathcal{C}(\mathbb{R}^d)$ denote the space of nonempty compact subsets of $\mathbb{R}^d$, complete with respect to the Hausdorff metric, and let $\mathcal{C}^N$ be the N-fold Cartesian product of copies of $\mathcal{C}$. A *graph iterated function system* (GIFS) is a directed graph G, possibly with loops and multiple edges in which the vertices of G are labeled by $\{1, 2, \ldots, N\}$ and each edge e is labeled with a contraction $f_e \colon \mathbb{R}^d \to \mathbb{R}^d$. It is also assumed that G is *strongly connected*, i.e., that there is a directed path from any vertex to any other. Let E_{ij} denote the set of edges from vertex i to vertex j. Define the function

$$F \colon \mathcal{C}^N \to \mathcal{C}^N$$

as follows. If $\boldsymbol{X} = (X_1, X_2, \ldots, X_N) \in \mathcal{C}^N$, then

$$F(\boldsymbol{X}) = (F_1(\boldsymbol{X}), F_2(\boldsymbol{X}), \ldots, F_N(\boldsymbol{X})),$$

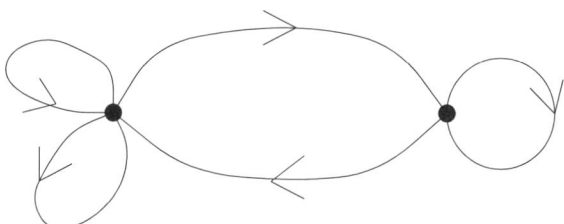

FIGURE 16. Graph iterated function system for the Penrose tiles.

where
$$F_i(\boldsymbol{X}) = \bigcup_{j=1}^{N} \bigcup_{e \in E_{ij}} f_e(X_j).$$
It can be shown that F is a contraction on $\mathcal{C}^N$, and consequently has a unique fixed point $\boldsymbol{T} = (T_1, T_2, \ldots, T_N)$.

Now consider the special case where each contraction is of the form
$$f_e(x) = A^{-1} \circ g_e,$$
where A is an expanding linear map and g_e is an isometry. The definition of fixed point implies

(10.4) $$A(T_i) = \bigcup_{j=1}^{N} \bigcup_{e \in E_{ij}} g_e(T_j), \quad i = 1, 2, \ldots, N,$$

which is precisely (10.2). So, if each g_e is a translation or A is a similarity, then call the GIFS *self-affine* or *self-similar*, respectively. In this case the sequence

(10.5) $$\mathcal{P} = \{\boldsymbol{T}, A(\boldsymbol{T}), A^2(\boldsymbol{T}), \ldots\}$$

is a self-affine or self-similar hierarchy whose substitution rules are determined by (10.4) provided

(1) T_i is the closure of its interior for each i, and
(2) the unions in (10.4) are nonoverlapping.

From [**Kees**] it is known that $m(\partial T_i) = 0$. In the definition of GIFS, the condition that G be strongly connected is equivalent to condition (3) in the definition of hierarchy in Section 9.1. So, assuming conditions (1) and (2) given just above, the $\mathcal{P}$-tilings are self-affine or self-similar tilings and, conversely, every self-affine or self-similar tiling can be obtained by such a GIFS construction.

Figure 16 shows the self similar GIFS whose fixed point is the pair of Penrose tiles shown in Figure 14. The two loops directed from the left node correspond to the two similarities taking the acute Penrose triangle to two smaller similar copies in its first subdivision. The edge directed from the first to the second node corresponds to the similarity taking the obtuse Penrose triangle to a smaller similar copy in the first subdivision of the acute Penrose triangle. Likewise, the two edges directed from the right node correspond to similarities taking each of the two Penrose triangles to smaller similar copies in the first subdivision of the obtuse triangle.

Assuming condition (1) holds, it is not difficult to give a necessary and sufficient condition for condition (2). Note that, in the GIFS terminology, the $N \times N$ matrix $M = (|E_{ij}|)$ is the substitution matrix as defined in Section 10.1.

PROPOSITION 10.7. *Assume that condition* (1) *holds for a self-affine or self-similar* GIFS. *Then condition* (2) *holds if and only if* $|\det A|$ *is the Perron-Frobenius eigenvalue (the largest real eigenvalue) of the substitution matrix* M.

PROOF. Let x_i denote the Lebesgue measure of tile T_i. The unions in (10.4) are nonoverlapping if and only if

$$|\det A|x_i = \sum_{j=1}^{N} |E_{ij}|x_j, \quad i = 1, 2, \ldots, N.$$

This means that $|\det A|$ is an eigenvalue of M. But for a nonnegative matrix, the only eigenvalue with a positive eigenvector is the Perron-Frobenius eigenvalue. □

Given a directed edge path $p = e_1 e_2 \cdots e_n$ and a contraction $f \colon \mathbb{R}^d \to \mathbb{R}^d$ we introduce the notation

$$f_p = f_{e_1} \circ f_{e_2} \circ \cdots \circ f_{e_n}.$$

The following proposition follows directly from (2.3) and allows for an algorithm to produce approximations of each of the proto-tiles $T_1, T_2, \ldots, T_N$ in the self-affine or self-similar hierarchy.

PROPOSITION 10.8. *Let* $\boldsymbol{T} = (T_1, T_2, \ldots, T_N)$ *be the fixed point of a GIFS* G, *and let* $E_i^{(n)}$ *denote the set of all finite, directed edge paths of length* n *in the graph* G *with initial vertex* i. *Then* T_i *is the limit with respect to the Hausdorff metric of the sets* $\{f_p(0) \mid p \in E_i^{(n)}\}$ *as* $n \to \infty$.

According to the proposition above, the graph G can be regarded as a finite state machine. If the initial state is vertex i, then the tile T_i is the language accepted by the machine. (In fact, this is the point of view taken by Thurston [**Th**] in the Pisot tiling example in Section 10.4.) Recall that a *finite state machine* M over the alphabet F is a finite set S (the states of the machine), a map $t : F \times S \to S$ (the state transition map), together with a distinguished element $I \in S$ (the initial state), and a distinguished set $OK \subset S$ (the accepting states). A finite state machine can be represented as a directed graph in which each state is represented by a node and each transition $(f, s) \mapsto s'$ is represented by an arc from s to s' labeled f. A word w in the alphabet F is *accepted* by M if, when you start at I and go along the direction given by w, you end up in OK. An infinite word is accepted if each finite prefix is accepted. The GIFS graph G is made into finite state machine by declaring the vertices of G accepting states and adding "fail states" so that the transition map is defined for on all $F \times S$.

In a code $(c_0, c_1, \ldots)$ for a self-affine or self-similar tiling, the position c_n of a tile $A^n(T_j)$ in tile $A^{n+1}(T_i)$ is completely determined by f_e where e is the appropriate edge from vertex i to vertex j in the graph G. Therefore, a code for such a tiling corresponds to (the equivalence class of) an infinite directed path in G with a given terminal vertex. (Two edge paths with the same terminal vertex are equivalent if they coincide except possibly for the last finite number m of edges.) If the hierarchy forces uniqueness, then there is a bijection between such equivalences classes of directed paths and the (partial) tilings. In fact, the tilings can be given explicitly. In the self-affine case each contraction can be written in the form

(10.6) $$f_e(x) = A^{-1}x + d_e,$$

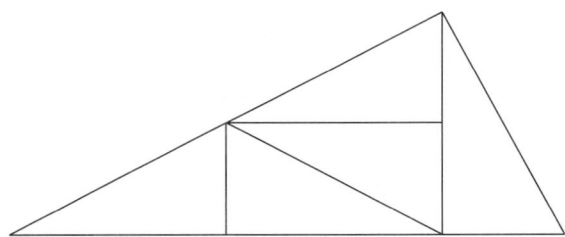

FIGURE 17. Subdivision rule for the pinwheel tiling.

where $d_e \in \mathbb{R}^d$. In the self-similar case each contraction is of the form

(10.7) $$f_e(x) = c\, g_e(x),$$

where $c < 1$ and g_e is an isometry. If p is an infinite, directed edge path in the graph G with fixed terminal vertex, let $p(n)$ denote the finite, directed edge path with the same terminal vertex consisting of the last n edges in p. By carefully applying the definitions we obtain the following tilings.

PROPOSITION 10.9. *Let $\boldsymbol{T} = (T_1, T_2, \ldots, T_N)$ be the proto-tiles of a self-affine or self-similar hierarchy corresponding to the graph iterated function system G.*

(1) *If $\mathcal{T}$ is a self-affine tiling (contractions of the form (10.6)) with code given by path $p = \cdots e_2 e_1 e_0$, then*

$$\mathcal{T} = \bigcup \left\{ \sum_{j=0}^{n} (A^j (d_{e_j'} - d_{e_j}) + T_i) \right\},$$

where the union is over all n and all edge paths $q = e_n' \cdots e_1' e_0'$ with the same initial vertex as $p(n)$, and i is the terminal vertex of q.

(2) *If $\mathcal{T}$ is a self-similar tiling (contractions of the form (10.7)) with code given by path p, then*

$$\mathcal{T} = \bigcup \{(g_{p(n)}^{-1} \circ g_q)(T_i)\},$$

where the union is over all n and all edge paths q that have the same length and initial vertex as $p(n)$, and i is the terminal vertex of q.

10.4. Examples. Four types of examples of self-similar hierarchies are mentioned in this section. Recall that a self-similar hierarchy is completely determined by the first subdivision rule, that is, by (10.4) of the GIFS graph G.

Polygonal Hierarchies. Numerous sporadic self-similar hierarchies using a single polygonal tile have been constructed [**GS**]. A simple example is the **L**-shaped triomino hierarchy with subdivision rule as given by the third diagram in Figure 2. This particular hierarchy forces uniqueness; so by the results of Section 9 there are uncountably many **L**-shaped triomino tilings, all nonperiodic, of finite type and locally isomorphic. This is called the *chair tiling*, and it has obvious analogues in higher dimensions.

The best known polygonal self-similar hierarchy is the Penrose hierarchy in Figure 13—already discussed in Section 9. Another important hierarchy is the the pinwheel hierarchy [**R1**] based on 1, 2, $\sqrt{5}$ right triangles, the subdivision rule shown in Figure 17. This hierarchy has the property that, up to congruence, there is one proto-tile, but in any of the uncountably many, nonperiodic pinwheel tilings, the tile appears in (countably) infinitely many orientations.

Hierarchies Using a Free Group Endomorphism. For certain special cases, Kenyon [**Ke3**] has extended the recurrent set method of Section 6.[5] Using essentially the same notation as in Section 6, let $G := G\langle a_1, \ldots, a_N \rangle$ denote the free group on N generators; let $\sigma: G \to G$ be an endomorphism. Using the notation $[ab] = aba^{-1}b^{-1}$ for the commutator, assume that each $\sigma([a_i a_j])$ is the product of conjugates of various $[a_{i_r} a_{j_r}]$. Kenyon finds a family of endomorphisms that satisfy this assumption. Take, for example, the case $N = 3$ and let

$$\sigma(a) = b$$
$$\sigma(b) = c$$
$$\sigma(c) = c^q a^{-s} b^{-r}$$

where $q, r \geq 0$, $s \geq 1$. Then there is a complex root λ of $x^3 - qx^2 + rx + s = 0$ such that, if $f : G \to \mathbb{C}$ is the homomorphism determined by $f(a) = 1$, $f(b) = \lambda$, $f(c) = \lambda^2$ and p denotes the corresponding polygonal path, then

$$A_n = \lambda^{-n} p(\sigma^{(n)}([ab]))$$
$$B_n = \lambda^{-n} p(\sigma^{(n)}([bc]))$$
$$C_n = \lambda^{-n} p(\sigma^{(n)}([ac]))$$

converge in the Hausdorff metric to closed curves A, B, C, respectively. Let T_a, T_b, T_c denote the enclosed compact tiles. Then λT_a is T_b; λT_b is the nonoverlapping union of s translates of T_c and r translates of T_b; and λT_c is the nonoverlapping union of q translates of T_b and s translates of T_a. This gives a subdivision rule for a translationally self-similar hierarchy. Some associated tilings are illustrated in [**Ke3, So1**]. Figure 18 is an example with six types of tiles, courtesy of R. Kenyon, whose expansion is a complex root of $x^4 + x + 1$. It is also a Pisot tiling as defined in the next paragraph.

Pisot tilings. Thurston [**Th**] considers radix representation of a real number on the line or complex number in the plane in the form $z = \sum_{i=i_0}^{n} a_i \beta^{-i}$, where β is a fixed real (complex) number and the a_i are chosen from a finite set D of algebraic integers in $\mathbb{Q}(\beta)$, and D contains 0. In general, D is not a digit set in the sense of Section 2. It is not difficult to choose D so that every number z has such a radix representation, but the representation is usually not unique.

The first step in constructing a self-similar hierarchy is to choose an ordering of D: $d_1 < d_2 < \cdots < d_N$. A *proper* representation of a number z is the one which is greatest in the corresponding lexicographic order. A representation of z is *weakly proper* if every finite initial segment of z can be extended to a proper representation. As a one dimensional example consider base $\tau = \frac{1}{2}(1 + \sqrt{5})$ and $D = \{0, 1\}$ with $0 < 1$. Then .101010... is weakly proper, but not proper because $1 = .101010\ldots$. In this example the weakly proper representations are exactly those that contain no two consecutive 1's.

Thurston shows that if β is a complex (or real) Pisot number, an algebraic integer such that all its Galois conjugates except β and $\overline{\beta}$ lie inside the unit circle, then there exists a finite state machine $M(\beta, D)$, as defined in Section 10.3, which

---

[5]Also related is the work of Garcia-Escudero and Kramer [**G-EK**] concerning an interpretation of certain 2-dimensional tilings using automorphisms of free groups.

FIGURE 18. Pisot tiling.

will recognize whether a sequence of elements from D gives a weakly proper representation for some number z. (In the one dimensional case, the finite state machine can be explicitly constructed from the *carry sequence*, which is the sequence of digits in the weakly proper representation of 1. If β is a Pisot number then the carry sequence is eventually periodic. In the example above .101010... is the carry sequence.)

A self-similar hierarchy can be constructed from the finite state machine. Turn the finite state machine into a graph iterated function system as follows. Given a pair (β, D), where β is a Pisot number, first remove all the FAIL states (the states that are not OK) from the associated finite state machine $M(\beta, D)$. Then relabel the edges as follows. On each edge e replace its label d_e by the contraction $f_e(z) = \beta^{-1}(z + d_e)$. This graph G, with say N nodes, determines a GIFS. The attractor of this GIFS is $(T_1, T_2, \ldots, T_N)$, where T_j can be described as follows. According to Proposition 10.2, the tile T_j consists of all points $z = \sum_{i=1}^{\infty} a_i \beta^{-i}$, where $a_i \in D$ for all i, and where the word $a_1 a_2 a_3 \ldots$ is accepted by the finite state machine $M(\beta, D)$ with vertex j as the initial state. In other words, T_j consists of all real (complex) numbers with decimal expansion only to the right of the decimal point and with weakly proper representation corresponding to a directed path in G starting at vertex j. Let E_{ji} denote the set of edges from vertex j to vertex i. Since multiplication by β is just a right shift of the decimal point we have the subdivision rule for each j:

$$\beta(T_j) = \bigcup_{i=1}^{N} \bigcup_{e \in E_{ji}} T_i + d_e.$$

The union is nonoverlapping because of the uniqueness of weakly proper representation.

The Pisot tiling of R. Kenyon in Figure 18 has six types of tiles and uses radix β where β is a complex root of $x^4 + x + 1$ with modulus greater than 1.

Dual hierarchies. Given a self-affine or self-similar hierarchy $\mathcal{P}$ in terms of a GIFS graph G, the construction of a dual hierarchy $\mathcal{P}^*$ is outlined by Thurston [**Th**] and expanded on and generalized by Gelbrich [**Ge3**] and by Praggastis [**Pra**]. It also appears in a paper on the construction of sofic partitions of hyperbolic toral automorphisms by Kenyon and Vershik [**KVe**]. We sketch the basic idea of the construction given in [**Ge3**].

Given a GIFS graph G define a dual graph G^* as follows. If G has vertex set $\{1, 2, \ldots, N\}$, let G^* have vertex set $\{1^*, 2^*, \ldots, N^*\}$. Each edge in G labeled with a contraction $f : \mathbb{R}^d \to \mathbb{R}^d$ is replaced by an oppositely directed edge in G^* labeled by its dual f^*, which is defined in the next paragraph.

A toral automorphism $\tilde{A} : \mathbb{R}^M \to \mathbb{R}^M$ is a linear map leaving some lattice L invariant and such that $|\det \tilde{A}| = 1$. If each eigenvalue of $\tilde{A}$ has modulus $\neq 1$, then $\mathbb{R}^M = E_s \oplus E_u$ such that $\tilde{A}_s = \tilde{A}|_{E_s}$ is a contraction and $\tilde{A}_u = \tilde{A}|_{E_u}$ is an expansion. It is known that, for a map $A : \mathbb{R}^d \to \mathbb{R}^d$ that is the expansion for certain self-similar or self-affine hierarchies, there exists a toral automorphism $\tilde{A} : \mathbb{R}^M \to \mathbb{R}^M$ that is a lifting of A. This means that there is an embedding

$$i : \mathbb{R}^d \hookrightarrow \mathbb{R}^M$$

such that $i(\mathbb{R}^d) = E_u$ and $\tilde{A} \circ i = i \circ A$. Let $A^* = \tilde{A}^{-1}|_{E_s}$ be the inverse of the lifting restricted to the complementary space.

More generally, for such a self-affine hierarchy (and sometimes for a self-similar hierarchy) an affine contraction $f : \mathbb{R}^d \to \mathbb{R}^d$ with linear part A can also be lifted to an affine map $\tilde{f} : \mathbb{R}^M \to \mathbb{R}^M$ given by $\tilde{f}(x) = \tilde{A}(x - b)$ where $\tilde{A}$ leaves E_u and E_s invariant and maps L bijectively onto itself. Let $f^* : E_s \to E_s$ be defined by

$$f^*(x) = A^*(x) + \mathrm{proj}_{E_s} b.$$

Now the dual graph G^*, and thus the dual hierarchy $\mathcal{P}^*$, is defined.

Some examples of this dual construction appear in [**So1, Th**]. Figure 19, courtesy of R. Kenyon, shows the 2-dimensional dual of a 1-dimensional Pisot tiling that uses the real root of $x^3 - x^2 - 1$ as base and $\{0, 1\}$ as digit set. The subdivision rule for the three types of tiles is of the form: $T_1 = f_1(T_2)$; $T_2 = f_2(T_3)$; $T_3 = f_1(T_1 \cup T_3)$.

Gelbrich [**Ge3**] computes the dual of the Penrose hierarchy and gives illustrations of some associated tilings. These tilings appeared previously in [**BGu**] and have the following appealing property. For the Penrose tiles (kite and dart, thick and thin rhombs, or acute and obtuse triangles), somewhat artificial matching rules guarantee that the tilings are self-similar and, consequently, that the proto-tile set is aperiodic. For the dual proto-tile set, the matching rules are a direct consequence of the fractal shape of the boundaries of the two proto-tiles. Every tiling by copies of the dual proto-tiles must be a self-similar tiling.

FIGURE 19. Dual of a Pisot tiling.

It is not always the case that (1) the dual tiles have nonempty interior and (2) the union in (10.4) for the dual is nonoverlapping. But these two conditions turn out to be equivalent [**Ge4**].

11. Concluding Remark

A main question at this point is how, in general, to construct the self-affine and self-similar hierarchies—and hence tilings. Any such hierarchy is the attractor of a GIFS. So from the GIFS point of view the issue is how to choose the parameters (the linear map A and translations d_e in (10.6) or the expansion factor c and isometries g_e in (10.7)) so that the tiles in the attractor of the GIFS have nonempty interior (condition (1) in Section 10.3). In the case of a single proto-tile this was done in Section 2 by choosing the set of translations d_e as a digit set D. In the absence of periodicity, however, there is no obvious analogue of the quotient $D = L/A(L)$ of a lattice by the sublattice. There are known sufficient conditions to insure nonempty interior, including the "open set condition" [**F1**] and an equivalent algebraic condition due to Bandt and Graf [**BGr**], but these are usually not readily applicable in practice. A reasonable approach to the problem appears open at this time.

References

[Ba]   M. Baake, *A guide to mathematical quasicrystals*, Quasicrystals (J.-B. Suck, M. Schreiber, and P. Häussler, eds.), Springer, Berlin, 2001 (to appear); math-ph/9904014.

[B1]   C. Bandt, *Self-similar sets. III. Constructions with sofic systems*, Monatsh. Math. **108** (1989), 89–102.

[B2]   _____, *Self-similar sets. V. Integer matrices and fractal tilings of $\mathbb{R}^n$*, Proc. Amer. Math. Soc. **112** (1991), 549–562.

[B3]   _____, *Self-similar tilings and patterns described by mappings*, The Mathematics of Long-Range Aperiodic Order (Waterloo, ON, 1995) (R. V. Moody, ed.), NATO ASI Ser. C: Math. Phys. Sci., vol. 489, Kluwer, Dordrecht, 1997, pp. 45–83.

[BGe]   C. Bandt and G. Gelbrich, *Classification of self-affine lattice tilings*, J. London Math. Soc. (2) **50** (1994), 581–593.

[BGr]   C. Bandt and S. Graf, *Self-similar sets. VII A characterization of self-similar sets with positive Hausdorff measure*, Proc. Amer. Math. Soc. **114** (1992), 995–1001.

[BGu]   C. Bandt and P. Gummelt, *Fractal Penrose tilings I. Construction and matching rules*, Aequationes Math. **53** (1997), 295–307.

[Bar]   M. Barnsley, *Fractals everywhere*, Academic Press, Boston, MA, 1988.

[BN]   D. Beauquier and M. Nivat, *On translating one polyomino to tile the plane*, Discrete Comput. Geom. **6** (1991), 575–592.

[Be1]   T. Bedford, *Dimension and dynamics for fractal recurrent sets*, J. London Math. Soc. (2) **33** (1986), 89–100.

[Be2]   _____, *Generating special Markov partitions for hyperbolic toral automorphisms using fractals*, Ergodic Theory & Dynam. Systems **6** (1986), 225–233.

[BT]   E. Bombieri and J. Taylor, *Quasicrystals, tiling, and algebraic number theory: some preliminary connections*, The Legacy of Sonya Kovalevskaya (Cambridge, MA, and Amherst, MA, 1985) (L. Keen, ed.) Contemp. Math., vol. 64, Amer. Math. Soc., Providence, RI, 1987, pp. 241–264.

[Br]   N. G. de Bruijn, *On bases for the set of integers*, Publ. Math. Debrecen **1** (1950), 232–242.

[Bu]   S. E. Burkov, *Structure model of the AL-Cu-Co decagonal quasicrystal*, Phys. Rev. Lett. **67** (1991), 614–618.

[Ch]   C. Chui, *An introduction to wavelets*, Wavelet Anal. Appl., vol. 1, Academic Press, Boston, MA, 1992.

[CS]   J. H. Conway and N. J. A. Sloane, *Sphere packings, lattices and groups*, 3rd ed., Grundlehren Math. Wiss., vol. 290, Springer, New York, 1999.

[CuSa]   K. Culik and A. Salomaa, *Ambiguity and decision problems concerning number systems*, Inform. and Control **56** (1983), 139–153.

[Da]   L. Danzer, *A family of 3D-spacefillers not permitting any periodic or quasiperiodic tilings*, Aperiodic '94 (Les Diablerets, 1994) (G. Chapuis and W. Paciorek, eds.), World Scientific, Singapore, 1995, pp. 11–17.

[De1]   F. M. Dekking, *Recurrent sets*, Adv. in Math. **44** (1982), 78–104.

[De2]   _____, *Replicating superfigures and endomorphisms of free groups*, J. Combin. Theory Ser. A **32** (1982), 315–320.

[DK]   P. Duvall and J. Keesling, *The Hausdorff dimension of the Lévy dragon*, Geometry and Topology in Dynamics (Winston-Salem, NC, 1998/San Antonio, TX, 1999), Contemp. Math., vol. 246, Amer. Math. Soc., Providence, RI, 1999, pp. 87–97.

[DKV]   P. Duvall, J. Keesling, and A. Vince, *The Hausdorff dimension of the boundary of a self-similar tile*, J. London Math. Soc. (to appear).

[E]   G. Edgar, *Classics on fractals*, Addison-Wesley, Reading, MA, 1993.

[F1]   K. Falconer, *Fractal geometry. Mathematical foundations and applications*, John Wiley & Sons, Chichester, 1990.

[F2]   _____, *Sub-self-similar sets*, Trans. Amer. Math. Soc. **347** (1995), 3121–3129.

[GK]   F. Gähler and R. Klitzing, *The diffraction pattern of self-similar tilings*, The Mathematics of Long-Range Aperiodic Order (Waterloo, ON, 1995) (R. V. Moody, ed.), NATO ASI Ser. C: Math. Phys. Sci., vol. 489, Kluwer, Dordrecht, 1997, pp. 141–174.

[G-EK]   J. Garcia-Escudero and P. Kramer, *An interpretation of 2D quasiperiodic patterns in terms of automorphisms of free groups*, J. Phys. A **26** (1993), L1029–L1035.

[Ge1]   G. Gelbrich, *Crystallographic reptiles*, Geom. Dedicata **51** (1994), 235–256.

[Ge2]   _____, *Self-affine lattice reptiles with two pieces in $\mathbb{R}^n$*, Math. Nachr. **178** (1996), 129–134.

[Ge3] _____, *Fractal Penrose tiles. II. Tiles with fractal boundary as duals of Penrose triangles*, Aequationes Math. **54** (1997), 108–116.

[Ge4] G. Gelbrich, *Galois duals of self-similar tilings*, Quasicrystals. Proceedings of the 5th International Conference (Avignon, 1995) (C. Janot and R. Mosseri, eds.), World Scientific, Singapore, 1995, pp. 76–79.

[GG] G. Gelbrich and K. Giesche, *Fractal Escher and salamanders and other animals*, Math. Intelligencer **20** (1998), 31–35.

[GL] L. Gibson and D. Lucas, *Spatial data processing using generalized balanced ternary*, IEEE Conference on Pattern Recognition and Image Processing (Las Vegas, 1982), IEEE Computer Society, 1982, pp. 566–571.

[Gi1] W. J. Gilbert, *Fractal geometry derived from complex bases*, Math. Intelligencer **4** (1982), 78–86.

[Gi2] _____, *Geometry of radix representations*, The Geometric Vein, Springer, New York, 1981, pp. 129–139.

[Gi3] _____, *Radix representations of quadratic fields*, J. Math. Anal. Appl. **83** (1981), 264–274.

[Gil] J. Giles, Jr., *Construction of replicating superfigures*, J. Combin. Theory Ser. A **26** (1979), 328–334.

[Go] S. W. Golomb, *Replicating figures in the plane*, Math. Gaz. **48** (1964), 403–412.

[GL] G. H. Golub and C. F. van Loan, *Matrix computations*, 3rd ed., Johns Hopkins Stud. Math. Sci., Johns Hopkins Univ. Press, Baltimore, MD, 1996.

[GH] K. Gröchenig and A. Haas, *Self-similar lattice tilings*, J. Fourier Anal. Appl. **1** (1994), 131–170.

[GM] K. Gröchenig and W. R. Madych, *Multiresolution analysis, Haar bases and self-similar tilings of* $\mathbb{R}^n$, IEEE Trans. Inform. Theory **38** (1992), 556–568.

[GS] B. Grünbaum and G. C. Shephard, *Tilings and patterns*, Freeman, New York, 1987.

[Gu] P. Gummelt, *Penrose tilings as coverings of congruent decagons*, Geom. Dedicata **62** (1996), 1–17.

[HSV] D. Hacon, N. C. Saldanha, and J. J. P. Veerman, *Remarks on self-affine tilings*, Experiment. Math. **4** (1994), 317–327.

[Ho] J. Honkala, *Bases and ambiguity of number systems*, Theoret. Comput. Sci. **31** (1984), 61–71.

[Hu] J. E. Hutchinson, *Fractals and self-similarity*, Indiana Univ. Math. J. **30** (1981), 713–747.

[JS] H.-C. Jeong and P. J. Steinhardt, *Constructing Penrose-like tilings from a single prototile and the implicatings for quasicrystals*, Phys. Rev. B **55** (1997), 3520–3532.

[KK] I. Kátai and B. Kovács, *Canonical number systems in imaginary quadratic field*, Acta Math. Acad. Sci. Hungar. **37** (1981), 159–164.

[KS] I. Kátai and J. Szabó, *Canonical number systems for complex integers*, Acta Sci. Math. (Szeged) **37** (1975), 255–260.

[KVi] K. Keating and A. Vince, *Isohedral polyomino tiling of the plane*, Discrete Comput. Geom. **21** (1999), 615–630..

[Kees] J. Keesling, *The boundaries of self-similar tiles in* $\mathbb{R}^n$, Top. Appl., to appear.

[Ke1] R. Kenyon, *Self-similar tilings*. Ph.D. thesis, Princeton University, 1990.

[Ke2] _____, *Self-replicating tilings*, Symbolic Dynamics and Its Applications (New Haven, CT, 1991) (P. Walters, ed.), Contemp. Math., vol. 135, Birkhäuser, Boston, MA, 1992, pp. 239–264.

[Ke3] _____, *The construction of self-similar tilings*, Geom. Funct. Anal. **6** (1996), 471–488.

[KLSW] R. Kenyon, Jie Li, R. S. Strichartz, and Y. Wang, *Geometry of self-affine tiles. II*, 1998 (preprint).

[KVe] R. Kenyon and A. Vershik, *Arithmetic construction of sofic partitions of hyperbolic toral automorphisms*, Ergodic Theory & Dynam. Systems **18** (1998), 357–372.

[Ker] R. B. Kershner, *On paving the plane*, Amer. Math. Monthly **75** (1968), 839–844.

[KVW] W. Z. Kitto, A. Vince, and D. C. Wilson, *An isomorphism between the p-adic integers and a ring associated with a tiling of n-space by permutohedra*, Discrete Appl. Math. **52** (1994), 39–51.

[Kn] D. E. Knuth, *The art of computer programming*. Vol. 2, *Seminumerical algorithms*, 3rd ed., Addison-Wesley, Reading, MA, 1997.

[LW1]   J. C. Lagarias and Y. Wang, *Tiling the line with translates of one tile*, Invent. Math. **124** (1996), 341–365.
[LW2]   _____, *Self-affine tiles in $\mathbb{R}^n$*, Adv. Math. **121** (1996), 21–49.
[LW3]   _____, *Integral self-affine tiles in $\mathbb{R}^n$. I. Standard and nonstandard digit sets*, J. London Math. Soc. (2) **54** (1996), 161–179.
[LW4]   _____, *Integral self-affine tiles in $\mathbb{R}^n$. II. Lattice tilings*, J. Fourier Anal. Appl. **3** (1997), 83–102.
[Li]   D. A. Lind, *Dynamical properties of quasihyperbolic toral automorphisms*, Ergodic Theory & Dynam. Systems **2** (1982), 49–68.
[Lin]   A. Lindenmayer, *Mathematical models for cellular interaction in development*, Parts I and II, J. Theoret. Biol. **18** (1968), 280–315.
[Lo]   M. Lothaire (ed.), *Combinatorics on words*, 2nd ed., Cambridge Math. Lib., Cambridge Univ. Press, Cambridge, 1999.
[LGJJ] J. M. Luck, C. Godrèche, A. Janner, and T. Janssen, *The nature of the atomic surfaces of quasiperiodic self-similar structures*, J. Phys. A **26** (1993), 1951–1999.
[LuP]   W. F. Lunnon and P. A. B. Pleasants, *Quasicrystallographic tilings*, J. Math. Pures Appl. (9) **66** (1987), 217–263.
[Ma]   B. B. Mandelbrot, *The fractal geometry of nature*, Freeman, San Francisco, CA, 1982.
[MW]   R. D. Mauldin and S. Williams, *Hausdorf dimension in graph directed constructions*, Trans. Amer. Math. Soc. **309** (1988), 811–823.
[Mat]   D. W. Matula, *Basic digit sets for radix representations*, J. Assoc. Comput. Mach. **4** (1982), 1131–1143.
[McM]   P. McMullen, *Convex bodies which tile space by translation*, Mathematika **27** (1980), 113–121. Acknowledgment of priority, **28** (1981), 191.
[M]   R. V. Moody (ed.), *The mathematics of long-range aperiodic order* (Waterloo, ON, 1995), NATO ASI Ser. C: Math. Phys. Sci., vol. 489, Kluwer, Dordrecht, 1997.
[Mo]   B. Mossé, *Puissances de mots et reconnaissabilité des points fixes d'une substitution*, Theoret. Comput. Sci. **99** (1992), 327–334.
[Od]   A. M. Odlyzko, *Nonnegative digit sets in positional number systems*, Proc. London Math. Soc. (3) **37** (1978), 213–229.
[P1]   R. Penrose, *Pentaplexity*, Math. Intelligencer **12** (1965), 247–248.
[P2]   _____, *Remarks on tiling: details of a $(1 + \epsilon + \epsilon^2)$-aperiodic set*, The Mathematics of Long-Range Aperiodic Order (Waterloo, ON, 1995) (R. V. Moody, ed.), NATO ASI Ser. C: Math. Phys. Sci., vol. 489, Kluwer, Dordrecht, 1997, pp. 467–497.
[Pra]   B. L. Praggastis, *Markov partitions for hyperbolic toral automorphisms*, Ph.D. thesis, University of Washington, Seattle, WA, 1992.
[Pri]   N. M. Priebe, *Detecting hierarchy in tiling dynamical systems via derived Voronoï tessellalations*, Ph.D. thesis, University of North Carolina, Chapel Hill, NC, 1997.
[R1]   C. Radin, *The pinwheel tiling of the plane*, Ann. of Math. (2) **139** (1994), 661–701.
[R2]   _____, *Space tilings and substitutions*, Geom. Dedicata **55** (1995), 257–264.
[Re]   K. Reinhardt, *Zur Zerlegung der euklidischen Räume in kongruente Polytope*, S.-Ber. Preuß. Akad. Berlin (1928), 150–155.
[Roe]   J. W. van Roessel, *Conversion of Cartesian coordinates from and to generalized balanced ternary addresses*, Photogrammetric Engineering & Remote Sensing **54** (1988), 1565–1570.
[Sch]   P. Schmitt, *An aperiodic prototile in space*, 1998 (preprint).
[Sen]   M. Senechal, *Quasicrystals and geometry*, Cambridge Univ. Press, Cambridge, 1995; corr. reprint, 1996.
[Se]   J.-P. Serre, *A course in arithmetic*, Grad. Texts in Math., vol. 7, Springer, New York, 1973.
[SBGC] D. Shechtman, I. Blech, D. Gratias, and J. Cahn, *Metallic phase with long-range orientational order and no translational symmetry*, Phys. Rev. Lett. **53** (1984), 1951–1954.
[So1]   B. Solomyak, *Dynamics of self-similar tilings*, Ergodic Theory & Dynam. Systems **17** (1997), 695–738.
[So2]   _____, *Nonperiodicity implies unique composition for self-similar translationally finite tilings*, Discrete Comput. Geom. **20** (1998), 265–279.

[SJSTAT] P. J. Steinhardt, H.-C. Jeong, K. Saitoh, M. Tanaka, E. Abe, and A. P. Tsai, *Experimental verification of the quasi-unit-cell model of quasicrystal structure*, Nature **55** (1998), 55–57.
[Str] R. S. Strichartz, *Wavelets and self-affine tilings*, Constr. Approx. **9** (1993), 327–346.
[SW] R. S. Strichartz and Y. Wang, *Geometry of self-affine tiles. I*, Indiana Univ. Math. J. **48** (1999), 1–23.
[Sz] S. Szabó, *A star polyhedron that tiles but not as a fundamental domain*, Intuitive Geometry (Siófok, 1985) Colloq. Math. Soc. János Bolyai, vol 48, North-Holland, Amsterdam, 1987, pp. 531–544.
[Th] W. P. Thurston, *Groups, tilings, and finite state automata* (Boulder, CO, 1989) Amer. Math. Soc. Colloq. Lectures.
[Ve] J. J. P. Veerman, *Hausdorff dimension of boundaries of self-affine tiles in $\mathbb{R}^n$*, Bol. Soc. Mat. Mexicana (3) **4** (1998), 159–182.
[Ven] B. A. Venkov, *On a class of Euclidean polyhedra*, Vestnik Leningrad. Univ. Ser. Mat. Fiz. Him. **9** (1954), 11–31 (Russian).
[V1] A. Vince, *Replicating tessellations*, SIAM J. Discrete Math. **6** (1993), 501–521.
[V2] _____, *Radix representation and rep-tiling*, Congr. Numer. **98** (1993), 199–212.
[V3] _____, *Rep-tiling Euclidean space*, Aequationes Math. **50** (1995), 191–215.
[V4] _____, *Self-replicating tiles and their boundary*, Discrete Comput. Geom. **21** (1999), 463–476.
[WvL] H. A. G. Wijshoff and J. van Leeuwen, *Arbitrary versus periodic storage schemes and tessellations of the plane using one type of polyomino*, Inform. and Control **62** (1984), 1–25.

DEPARTMENT OF MATHEMATICS, UNIVERSITY OF FLORIDA, GAINESVILLE, FL 32611, USA
E-mail address: vince@math.ufl.edu

Centre de Recherches Mathématiques
CRM Monograph Series
Volume **13**, 2000

A Guide to Quasicrystal Literature

Michael Baake and Uwe Grimm

ABSTRACT. This collection is meant as a help to find the mathematical contributions in the area of aperiodic order. We focus on books, proceedings and review volumes, and indicate where further reference lists can be found.

Some Introductory Remarks

The literature on aperiodic order and quasicrystals is scattered over a vast number of journals, and due to the history of the field, a large number of articles is actually published in physics journals. This is also true of many mathematically oriented contributions in the eighties, but review volumes such as [**SO**] or some of the later lists of references should help to locate them. Also, many of those contributions can be found via either *Mathematical Reviews* or *Zentralblatt für Mathematik*, since (fortunately) several journals that are relevant in this context are actually covered by at least one of these review journals.

Scientific journals with regular contributions to the field of aperiodic order and mathematical quasicrystals include (in alphabetical order)

- Canadian Journal of Mathematics
- Communications in Mathematical Physics
- Discrete and Computational Geometry
- Ergodic Theory and Dynamical Systems
- Geometriae Dedicata
- Journal of Mathematical Physics
- Journal of Physics A: Mathematical and General
- Journal of Statistical Physics
- Letters in Mathematical Physics

Finally, let us mention that a decent fraction of the articles, both old and new, are available through the standard electronic archives, both in Los Alamos (http://arXiv.org/, with contributions in both the physics and the mathematics part of it) and in Austin (http://www.ma.utexas.edu/mp_arc/). We also plan to keep a little collection of links to sites on quasicrystal people and research (reachable from http://www.math.ualberta.ca/People/Facultypages/Moody.R.html, or

---

2000 *Mathematics Subject Classification.* 52C23.

©2000 Michael Baake and Uwe Grimm

from one of our own homepages[1]—wherever they may be by the time you look for them).

References

[ADG] F. Axel, F. Denoyer, and J.-P. Gazeau (eds.), *From quasicrystals to more complex systems*, Springer, Berlin, and EDP Sciences, Les Ulis, 2000.
Proceedings of a summer school at Les Houches, France; a follow-up of the following entry.

[AGr] F. Axel and D. Gratias (eds.), *Beyond quasicrystals* (Les Houches, 1994), Springer, Berlin, and EDP Sciences, Les Ulis, 1995.
Proceedings of a summer school at Les Houches, France, with review articles mainly on theoretical and mathematical issues. A good source also for open problems.

[Ba] M. Baake (ed.), *Selected topics in the theory of quasicrystals*, Internat. J. Modern Phys. B **7** (1993), no. 6-7 (special issue).
Collection of theoretical research articles dedicated to Peter Kramer on the occasion of his 60th birthday. Available through the editor.

[BM] M. Baake and R. V. Moody (eds.), *Directions in mathematical quasicrystals*, CRM Monograph Series, Amer. Math. Soc., Providence, RI, 2000, this volume.
Another example of self-reference... This book is meant as a continuation of the volumes **[Mo]** and **[Pa]**.

[Be] R. Berger, *The undecidability of the domino problem*, Mem. Amer. Math. Soc., vol. 66, Amer. Math. Soc., Providence, RI, 1966.
The origin of the connection between fundamental questions of logic and undecidability with the theory of tilings, see **[GS]** for further hints.

[BVC] M. de Boissieu, J.-L. Verger-Gaugry, and R. Currat (eds.), *Aperiodic '97* (Alpe d'Huez, 1997), World Scientific, Singapore (1998).
Proceedings of the International Conference on Aperiodic Crystals, Alpe d'Huez, France.

[CP] G. Chapuis and W. Paciorek (eds.), *Aperiodic '94* (Les Diablerets, 1994), World Scientific, Singapore, 1995.
Proceedings of the International Conference on Aperiodic Crystals, Les Diablerets, Switzerland. This conference series has a mathematical section on quasicrystals and aperiodic order.

[DVS] D. P. DiVincenzo and P. J. Steinhardt (eds.), *Quasicrystals: the state of the art*, Directions in Condensed Matter Physics, vol. 11, World Scientific, Singapore, 1991.
Collection of original review articles; updated version with additional articles (1999).

[DSS] H.-D. Doebner, W. Scherer, and C. Schulte (eds.), *GROUP 21—Physical applications and mathematical aspects of geometry, groups, and algebras*, vol. 2 (Goslar, 1996), World Scientific, Singapore, 1997.
Proceedings of the XXI International Colloquium on Group Theoretical Methods in Physics, Goslar, Germany (1996). This volume contains a section on 'Crystallography and Aperiodical Order' (pp. 937–1002).

[GS] B. Grünbaum and G. C. Shephard, *Tilings and patterns*, Freeman, New York, 1987.
The tiling bible.

[HG] F. Hippert and D. Gratias (eds.), *Lectures on quasicrystals* (Aussois, 1994), EDP Sciences, Les Ulis, 1994.
Proceedings of a winter school at Aussois, France (1994), with articles on experimental and theoretical aspects of quasicrystals.

[Jan] C. Janot, *Quasicrystals. A primer*, 2nd ed., Monographs on the Physics and Chemistry of Materials, Clarendon Press, Oxford, 1994.
The physical primer.

[JM] C. Janot and R. Mosseri (eds.), *Quasicrystals* (Avignon, 1995), World Scientific, Singapore, 1995.
Proceedings of the 5th International Conference on Quasicrystals, Avignon, France (1995).

[Ja1] M. V. Jarić (ed.), *Introduction to quasicrystals*, Aperiodicity and Order, vol. 1, Academic Press, San Diego, CA, 1988.
First volume of a book series with invited review articles on quasicrystals.

---

[1] At present, the addresses are http://solid13.tphys.physik.uni-tuebingen.de/baake/ and http://www.tu-chemnitz.de/~ugr/, respectively.

[Ja2] _____, *Introduction to the mathematics of quasicrystals*, Aperiodicity and Order, vol. 2, Academic Press, San Diego, CA, 1989.

[JG] M. V. Jarić and D. Gratias (eds.), *Extended icosahedral structures*, Aperiodicity and Order, vol. 3, Academic Press, San Diego, CA, 1989.

[LR] L. Loreto and M. Ronchetti (eds.), *Topics on contemporary crystallography and quasicrystals*, Per. Mineral. **LIX** (1990), Nos. 1–3 (special issue).
Contains an extensive bibliography on quasicrystals until 1991 (pp. 219–257).

[LMW] J.-M. Luck, P. Moussa, and M. Waldschmidt (eds.), *Number theory and physics*, Springer Proceedings in Physics, vol. 47, Springer, Berlin, 1990.
Part II (pp. 77–137) deals with quasicrystals and related geometric structures.

[Me] Y. Meyer, *Algebraic numbers and harmonic analysis*, North-Holland Math. Library, vol. 2, North-Holland, Amsterdam, 1972.
Many results on the structure of quasiperiodic point sets are already contained in this book because they are special cases of Meyer's harmonious sets. A classic.

[Mi] J. Miękisz, *Quasicrystals: microscopic models of nonperiodic structures*, Lecture Notes in Mathematical and Theoretical Physics, vol. A5, Leuven University Press, Leuven, 1993.
These lecture notes deal with the connection between quasicrystals, sphere packings and lattice gas models, with focus on non-periodic ground states and frustration.

[Mo] R. V. Moody (ed.), *The mathematics of long-range aperiodic order* (Waterloo, ON, 1995), NATO ASI Ser. C: Math. Phys. Sci., vol. 489, Kluwer, Dordrecht, 1997.
Proceedings of a NATO Advanced Study Institute at the Fields Institute, Waterloo, Canada (1995). The present mathematical state of the art, part I.

[Pa] J. Patera (ed.), *Quasicrystals and discrete geometry* (Toronto, ON, 1995), Fields Inst. Monogr., vol. 10, Amer. Math. Soc., Providence, RI, 1998.
Invited research articles from a long-term research program at the Fields Institute, Toronto, Canada (1995). The present mathematical state of the art, part II.

[R] C. Radin, *Miles of tiles*, Amer. Math. Soc., Providence, RI, 1999.
Introductory text with special emphasis on tilings that are of finite type with respect to Euclidean motions, but not with respect to translations.

[Se] M. Senechal, *Quasicrystals and geometry*, Cambridge Univ. Press, Cambridge, 1995; corrected reprint, 1996.
The geometric primer.

[SO] P. J. Steinhardt and S. Ostlund (eds.), *The physics of quasicrystals*, World Scientific, Singapore, 1987.
Reprint volume which contains many of the early articles until 1987, and also a bibliography.

[SSH] J.-B. Suck, M. Schreiber, and P. Häussler (eds.), *Quasicrystals—An introduction to structure, physical properties and applications of quasicrystalline alloys* (Chemnitz, 1997), Springer, Berlin, 2001 (to appear).
Lecture notes of a WE-Heraeus summer school in Chemnitz, Germany (1997). Contains further bibliographic notes.

[TF] S. Takeuchi and T. Fujiwara (eds.), *Quasicrystals* (Tokyo, 1997), World Scientific, Singapore, 1998.
Proceedings of the 6th International Conference on Quasicrystals, Tokyo, Japan (1997).

[YT] M. J. Yacamán and M. Torres (eds.), *Crystal-quasicrystal transitions*, North-Holland, Amsterdam, 1993.
Collection of invited research and review articles, in particular on the theory of periodic approximants.

INSTITUT FÜR THEORETISCHE PHYSIK, UNIVERSITÄT TÜBINGEN, AUF DER MORGENSTELLE 14, 72076 TÜBINGEN, GERMANY
E-mail address: michael.baake@uni-tuebingen.de, baake@miles.math.ualberta.ca

INSTITUT FÜR PHYSIK, TECHNISCHE UNIVERSITÄT CHEMNITZ, 09107 CHEMNITZ, GERMANY
E-mail address: u.grimm@physik.tu-chemnitz.de

Index

A-adic integer, 341
AF-algebra, 198
C^*-algebra
  of a groupoid, 241
  of a tiling
    continuous case, 184
    discrete case, 186
  of observables, 189, 217
K_0, 190
K_1, 193
L^1, 12
M-topology, 237
3-adic numbers, 31

Abelian group
  compactly generated, 145
  locally compact, 9, 144
acceptance domain, 105, 235
address of a lattice point, 341
admissible Lipschitz mappings, 5
admissible set, 236
affine map, 9
algebraic number field, 96–98, 125, 128, 130, 138
almost periodic function
  Besicovitch class B^2, 77, 87
  uniform, 74, 84
almost sure absence of eigenvalues, 292
alphabet, 280
Ammann
  bars, 96, 98, 102, 118, 129
  cells, 134
  grid, 102
  hyperplanes, 97, 123, 124, 134, 137
Ammann-Beenker pattern, 31
Ammann-Beenker set, 117
aperiodic proto-tile set, 349
approximate unit, 12
archives, 371
attractor, 4
autocorrelation, 55, 102, 146
automorphism, 9
averaging operator, 16
Aztec diamond, 318

Besicovitch B^2-distance, 87
bibones, 308

Bohr almost periodic spectrum, 85
books, 372
Boolean combination, 113, 114, 121
boundary conditions, 311
Bragg peak, 359
Brillouin zone, 210
  noncommutative, 218

canonical transversal, 225
cartwheel tiling, 356
Cayley-Hamilton theorem, 286
chair tiling, 362
character, 10
coefficient ring, 129, 130, 132
commensurate lattices, 110, 114, 119, 128
compact IFS, 5
compact sets of contractions, 3
complexity function, 280
conditional uniformity, 315
conformal invariance, 326
Connes' Thom isomorphism, 195
consistent phase property, 73
contact matrix, 344
continuously representable, 28
contraction, 4
control points, 26
convergence
  compact, 11
  uniform, 11
convolution, 144
convolution algebra, 12
coupling constant, 283
covering the plane by a single proto-tile, 350
crystallographic
  group, 351
  restriction, 98, 138
  tiling, 351
cut and project, 95–98, 105, 106, 108, 110, 113, 138
  method, 235
  plain, 95, 96, 105, 106, 109, 132
  scheme, 23
CW-complex, 182

deflation, 96, 103
Delone measures, 222
Delone set, 23, 61, 98, 212

Besicovitch almost periodic, 77
Bohr almost periodic set, 75
cut and project set, 62
finite type, 62, 212
Meyer set, 62
model set, 63, 71
perfectly diffractive, 66
self-replicating, 63, 71
densities, 28
density, 24
density deviation, 263, 264, 269
diabolo tilings, 323
diabolos, 308
diameter, 7
difference set, 99
diffraction, 96, 97, 102, 138
    measure, 154, 217
    pattern, 55
    spectrum, 102, 154
digit set, 334
digit tiling, 339
dimer, 307
dimer model, 307
Diophantine approximation, 97, 111
distribution
    B^2 almost periodic, 77
    $\mathcal{B}$-almost periodic, 74
    positive type, 83
    Schwartz, 82
    tempered, 82
    uniformly almost periodic, 74
dodecahedron, 273
dominos, 307
dynamical spectrum, 154
dynamical system
    continuous eigenfunction, 70, 78
    homogeneous, 71, 78
    measurable eigenfunction, 70
    metric, 70
    minimal, 70
    pure point spectrum, 70
    spectrum of, 70
    topological, 70
    uniquely ergodic, 70
dynamical system of an FLC set, 151

Ehrhart polynomial, 264
embedding dimension, 130
ergodic family of Schrödinger operators, 281
ergodic measure, 279
evaluation, 4
expanding affine map, 63

facet, 262
factor, 280
family of model sets, 2
Fibonacci chain
    deformed, 58
Fibonacci operator, 284

finite local complexity, 62, 99, 101, 135, 137, 148, 179
finite pattern condition, 179
finite state machine, 361
finitely generated, 103
flation, 96, 97, 103, 119, 128
FLC set, 148
forcing the border, 182
Fourier transform, 102
Fourier-Bohr coefficient, 46
frequency, 280
    uniform, 99, 101–103, 108, 119
    upper uniform, 99, 102
fringe, 3

gap labeling, 193
    group, 194
    theorem, 247
Gaussian integers, 336
generalized balanced ternary (GBT), 338
generic, 157
generic absence of eigenvalues, 292
genus, 173
Gibbs measure, 214, 310
Gibbs-Curie energy, 262
graph iterated function system (GIFS), 359
groupoid, 226

Haar measure, 9
half-open, 96, 101, 106, 107, 114, 116, 117, 129
Hausdorff
    dimension, 344
    measure, 297
    metric, 3
height function, 312
hierarchical tiling, 353
hierarchy, 352
    dual, 365
    equivalence, 353
    forces uniqueness, 353
    interior, 355
    Penrose, 353
    self-affine, 357
    self-similar, 357
    translationally self-similar, 356
homogeneous operator, 228
hull, 214, 226, 228, 234
    continuous, 179
    discrete, 187
    strong operator, 189
Hutchinson metric, 6
hyperbolic dynamics, 181

icosian ring, 162
ideal crystal, 62
infinite word, 280
inflation, 30, 96, 103, 106, 117, 129, 130, 132, 136, 267

local, 96, 98, 103, 121, 122, 129, 132, 134, 137
refining, 96, 104, 119, 120, 122, 129, 132, 134, 137, 138
local, 97
symmetry, 37
integrated density of states (IDOS), 244
internal space, 96, 105, 114, 122, 127–129
invariant set, 279
Ising model, 309
iterated function system (IFS), 4, 5, 331
just touching, 29
nonoverlapping, 29

journals, 371

Kohmoto model, 244, 252
Kotani theory, 281

Lévy's continuity theorem, 40
lattice, 23, 95–97, 101, 105, 108, 110, 112, 114, 121, 125, 128, 131, 132, 210, 262
tiling, 340
tiling problem, 350
lattice-like, 97, 102
LCAG, 9, 144
Lebesgue decomposition theorem, 82
length, 280
letter, 280
LI-class, 96, 98–100, 103–105, 120, 139, 151
linear operator, 97, 106, 119, 121, 124, 128, 129, 132, 133
Lipschitz constant, 4
Lipschitz function, 4
local derivability, 97, 100, 104, 114, 115, 121, 123, 130, 133, 138
mutual, 104
local isomorphism, 354
local rules, 96, 103
perfect, 97, 104, 123, 125
weak, 96, 97, 104, 123, 124, 129, 138
locally indistinguishable, 99, 151
locally isomorphic, 99, 104, 105, 117
lozenge tilings, 322
lozenges, 307
Lyapunov exponent, 282

Markov process, 136
matching rules, 96, 104
measurability
Lebesgue, 95
Riemann, 95, 106, 111, 119
measure, 6, 81
absolute value, 81
Besicovitch almost periodic, 89
Borel, 82
bounded, 81
diffraction, 66, 154, 157
positive, 81

positive definite, 83
product, 16
Radon, 82
regular Borel, 82
self-similar, 11
tempered, 83
transformable, 83
translation-bounded, 66, 83
translationally bounded, 144
unbounded, 81
uniformly almost periodic, 86
vague topology, 81
Meyer set, 62, 99, 102, 103, 132, 134, 212
minimal, 151, 279
minimal tiling, 180
minimality, 101, 104
model set, 23, 63, 95–98, 105, 106, 116–118, 120, 123–125, 127, 129–133, 138, 156, 235, 265
deformed, 46
full submodel, 108, 110, 117, 118, 122, 124, 127, 134
generic, 35
multi-component, 2
nonsingular, 108, 117, 119
plain, 106, 109, 113, 114, 128, 129, 138
quotient model, 96, 109, 110, 118, 134
regular, 35, 46, 156
singular, 96, 103, 108, 114, 120
submodel, 96, 97, 108, 110, 117, 127, 129
module, 97, 98, 125, 127–131, 138
modulus, 9
Morita equivalent, 188

natural topology, 70
near-crystallographic, 102
nonperiodic, 102, 117, 349
norm topology, 12

octagonal quasicrystal, 253
optimal imbedding, 170
orbit, 279
order, 129, 163
order on K_0, 191
orientation conditions, 96, 97, 109, 110, 113, 127
generic, 97
nongeneric, 110

packings, 259
parametric density, 262
Parseval property, 77, 84
partition, 291
patch, 353
patch frequencies, 358
Patterson set, 66
Penrose
set, 99, 100, 102–104, 117, 131
tiles, 349, 350, 353, 356, 358

tiling, 100, 123, 124, 131, 265, 269
perfect matching, 307
periodic state, 38
periodic tiling, 349
Perron number, 358
phase information, 71, 80
physical space, 96, 105, 114
pinwheel tiling, 363
Pisot
   number, 98, 104, 120, 121, 129, 132, 359
   tiling, 363
   unit, 121, 131, 132
point set
   discrete, 98, 100, 101, 138
   relatively dense, 98
   uniformly discrete, 98
Poisson summation formula, 67
polyomino, 350
polytope, 262
primitive, 284
primitive nonnegative matrix, 63
proto-tile set
   aperiodic, 349
pseudo-torus, 237
pyrite, 264

quasicrystal, 95–98, 100, 106, 109, 113, 117, 119, 123, 124, 130, 132, 138, 351, 359
   multicomponent, 106
   summation formula, 69
quasicrystallographic restriction, 98, 138
quasilattice, 64
quaternion algebra, 161

radix system, 336
RAGE theorem, 297
recognizable substitution, 181
recurrent set, 347
references, 372
refinement operator, 13
regular dodecahedron, 273
relatively dense, 62, 98, 103, 104
repetitive, 151, 180
repetitivity, 96, 101, 104–106, 132, 133
   linear, 101
rhombic triacontahedron, 273
Riemann hypothesis, 81
Riemann measurable, 24
Riesz-Fischer property, 77
Riesz-Markov representation theorem, 82
root system, 166
rotation number, 283

Salem number, 120
Schwartz space, 65
self-affine tiling, 357
self-replicating tiling, 339
self-similar tiling, 357
shelling, 167

Shubin's formula, 245
silver mean, 32
   chain, 31
Smale space, 181
spectrum
   diffraction, 154, 156
   dynamical, 154, 156, 157
sphere packings, 259
star map, 23, 154
   regular, 154
stationary measure, 279
Steiner formula, 264
stricly ergodic, 279
strip, 106, 108, 113, 114, 116, 119
strongly aperiodic, 180
Sturmian potential, 284
Sturmian sequence, 284
subdivision operator, 357
subdivision rule, 353
subshift, 280
subspace
   affine, 102, 106
   lattice, 108
substitution, 284
   binary non-Pisot, 284
   dynamical system, 285
   Fibonacci, 284
   matrix, 63
   period doubling, 284
   Rudin-Shapiro, 284
   sequence, 284
   Thue-Morse, 284
   tilings, 180
subword, 280
symbol, 280
symmetry, 96–98, 101, 103, 107, 119, 123, 124, 128, 129, 131, 132, 134, 137, 138
   overt, 139

Temperleyan boundary conditions, 325
three-block method, 287
tight binding approximation, 189
tight binding representation, 240
tile, 179
   digit, 334
   self-affine, 333
   self-similar, 333
tiling, 26, 64, 179, 307
   chair, 71, 362
   code, 355
   crystallographic, 351
   digit, 339
   finite type, 354
   hierarchical, 353
   inflation, 30
   lattice, 340
   periodic, 349
   pinwheel, 363

Pisot, 363
self-affine, 357
self-replicating, 339
self-similar, 357
special, 357
sphinx, 71
topological dynamical system, 279
toral automorphism, 365
torus parametrization, 155
trace, 191
trace map, 287
transfer matrix, 282, 285
translation, 9
transversal, 187
triacontahedron, 273
two point correlation function, 66
two-block method, 286

uniform
  absence of eigenvalues, 292
  frequency, 113
  structure, 147
uniformity, 95–97, 101, 106, 134, 136, 137, 147
uniformly discrete, 61, 98, 104
union maps, 4
unique composition property, 354
unique ergodicity, 152, 279

vague topology, 6, 144
van Hove sequence, 145

wavelet, 341
web sites, 371
Weyl's Theorem, 24, 25
Williams solenoid, 183
window, 95, 96, 105–108, 114, 116, 118, 120, 124, 129, 130, 132
word, 280
Wulff-shape, 263, 267